# Laser Cleaning

Sijie Wang

# Laser Cleaning

## Key Technology of Substrate Surface Morphology and Performance Control

Springer

Sijie Wang
National Engineering Research Center for Remanufacturing of Mechanical Products
Army Armored Forces Academy
Beijing, China

ISBN 978-981-95-2934-6 ISBN 978-981-95-2935-3 (eBook)
https://doi.org/10.1007/978-981-95-2935-3

Jointly published with Harbin Institute of Technology Press

The print edition is not for sale in China (Mainland). Customers from China (Mainland) please order the print book from: Harbin Institute of Technology Press.



This Springer imprint is published by the registered company Springer Nature Singapore Pte Ltd.
The registered company address is: 152 Beach Road, #21-01/04 Gateway East, Singapore 189721, Singapore

# Introduction

Laser cleaning technology, widely acknowledged as a preeminent green cleaning solution for sustainable manufacturing in the twenty-first century, has gained substantial traction in both academic research and industrial applications owing to its non-contact operation, micron-scale precision, and exceptional environmental compatibility. This technology has demonstrated transformative potential across mission-critical sectors including aerospace systems, next-generation rail transport, cultural heritage preservation, and offshore marine engineering, where its implementation has become indispensable. Nevertheless, several formidable challenges persist in practical implementation, particularly concerning the optimization of cleaning kinetics, mitigation of substrate thermomechanical damage, and development of robust process parameterization frameworks. Addressing these technological bottlenecks requires concerted interdisciplinary innovation, with particular emphasis on establishing synergistic paradigms that simultaneously optimize cleaning efficacy while preserving substrate integrity—a research priority now dominating both academic and industrial discourse.

Drawing upon a decade of original research by the author's team and synthesizing cutting-edge global advancements, this monograph establishes an integrated theoretical-practical framework for laser cleaning technology, bridging fundamental principles with industrial implementation. Comprising 12 thematically organized chapters, this work presents a systematic progression from fundamental principles (including surface characterization and mechanistic studies) through computational modeling and process parameterization to real-time monitoring techniques and field applications, thereby forming a closed-loop knowledge architecture. The study provides exhaustive coverage of laser-based surface engineering across diverse material systems including metallic alloys, carbon fiber-reinforced resin substrates, glass fiber-reinforced resin substrates, and other materials, with particular focus on nanoscale morphology modulation and in situ quality assessment methodologies. This treatise offers a state-of-the-art critical review of contemporary advances in laser-assisted surface pretreatment technologies, incorporating multidisciplinary insights from materials science, photonics, and industrial engineering.

This book comprises 12 thematically focused chapters, each addressing distinct yet interconnected scientific and technological challenges in laser surface engineering, thereby establishing a coherent knowledge network with modular yet synergistic architecture. Chapter 1 presents a comprehensive historiographical analysis of laser cleaning technological evolution, while elucidating fundamental surface modification mechanisms across ferrous alloys, aluminum/titanium systems, and advanced composite matrices. Chapter 2 establishes a unified theoretical framework for laser–matter interactions, quantitatively characterizing the synergistic interplay of vibrational, ablative, and plasma effects through multiscale experimental validation of their dynamic coupling phenomena. Chapter 3 deciphers the laser–carbon fiber reinforced plastics (CFRP) interaction physics, uncovering the thermal-mechanical genesis of HAZ formation and pioneering a laser-induced breakdown spectroscopy (LIBS)-guided adaptive control paradigm for thermal damage mitigation. Chapter 4 develops a materials-specific pretreatment methodology, establishing optimized laser parameter matrices for diverse alloy systems through mechanistic analysis of surface activation phenomena. Chapter 5 pioneers a coupled thermomechanical predictive model synergizing LIBS spectroscopy with photoacoustic tomography for real-time damage prognosis. Chapter 6 implements multimodal process tomography combining vibrational spectroscopy with infrared thermography, enabling holographic process characterization. Chapter 7 devises a defect-distribution-adaptive algorithm via computational modeling of heterogeneous coating thermal dynamics. Chapter 8 introduces a breakthrough dual-stage cleaning protocol employing low-fluence secondary processing to simultaneously maximize cleaning efficiency and substrate preservation. Chapter 9 elucidates the laser-induced oxide autogenesis mechanism in marine steels and establishes a novel corrosion-abrasion resistant surface engineering approach. Chapter 10 conducts comparative analysis of ns/ps laser-induced interface reconstruction and develops a wettability-engineered recoating strategy for aerospace applications. Chapter 11 redefines the damage threshold paradigm for CFRP through resin-layer-preservation strategies, enabling nondestructive performance-enhanced refurbishment. Chapter 12 demonstrates a novel continuous-pulse hybrid regime enabling concurrent paint stripping and nanostructured oxide fabrication on Mg alloys, with enhanced corrosion resistance via grain boundary engineering.

This book synthesizes cutting-edge multidisciplinary research achievements and pioneering developments from the authors' laboratory, delivering advanced theoretical frameworks for fundamental mechanism exploration alongside empirically validated process optimization datasets for industrial implementation. The work establishes a comprehensive pedagogical architecture for academic capacity building in advanced surface engineering while serving as both a technological accelerator for next-generation laser surface engineering innovations and a transformative driver for China's advanced manufacturing ecosystem. Particularly through its contributions to precision surface modification technologies, this treatise promises to significantly foster sustainable industrial upgrading and technological advancement.

This scholarly work was authored by Sijie Wang, with significant intellectual contributions from Zhe Li, Wanqing Zheng, Shiyu Wang, Liangpeng Hou, Yunpeng

Song, Teng Shan, Fengshi Yin, Yaokun Pan, and Pengzhan Wang, who collectively provided comprehensive experimental datasets and analytical frameworks essential for this research compendium. The research team extends profound appreciation to all collaborators for their meticulous efforts in multidimensional data acquisition, computational modeling, experimental methodology development, and rigorous case study validation. The authors gratefully acknowledge sustained research funding from the National Natural Science Foundation of China (Grant 52275226) and the Beijing Natural Science Foundation (Grant 3222022), which were instrumental in facilitating this scientific endeavor.

Given the exponential pace of advancement in laser surface engineering and the inherent limitations of a single research perspective, certain aspects of this dynamic field may not be fully encompassed within the present volume. The scholarly community is cordially invited to provide constructive feedback and scholarly discourse to enhance future iterations of this work.

# Preface

Laser cleaning technology is widely recognized as an outstanding green cleaning solution for sustainable manufacturing in the 21st century. Due to its non-contact operation, micron-scale precision, and excellent environmental compatibility, it has gained significant traction in both academic research and industrial applications.

Beijing, China　　　　Sijie Wang

# Contents

**1 Exploring the Progress of Laser Cleaning Technology from the Perspective of Chinese Patents** . . . . 1
1.1 Introduction . . . . 1
1.2 Overall Overview of Laser Cleaning Patents . . . . 2
1.2.1 Basic Trend of Laser Cleaning Patents . . . . 3
1.2.2 Classification and Proportion of Laser Cleaning Patents . . . . 5
1.2.3 Distribution of Mechanism Attributes of Laser Cleaning Patents . . . . 6
1.2.4 Distribution of the Patent Application Areas . . . . 6
1.3 Patent Analysis of Laser Cleaning for Different Matrix Materials . . . . 7
1.3.1 Patent Analysis of Laser Cleaning Iron Matrix . . . . 8
1.3.2 Patent Analysis of Laser Cleaning Aluminum Alloy Substrate . . . . 10
1.3.3 Patent Analysis of Laser Cleaning Titanium Alloy Substrate . . . . 12
1.3.4 Patent Analysis of the Laser-Cleaning Composite Matrix . . . . 13
1.4 Patent Analysis of Laser Cleaning Different Coatings . . . . 15
1.4.1 Patent Analysis of the Laser Cleaning Rust Layer . . . . 15
1.4.2 Patent Analysis of the Laser Cleaning Paint Layer . . . . 17
1.4.3 Patent Analysis of Laser Cleaning Composite Coating . . . . 19
1.4.4 Patent Analysis of Laser Cleaning Other Coatings . . . . 22
1.5 Patent Analysis of Laser Cleaning Equipment and Monitoring Device . . . . 26
1.5.1 Laser Cleaning Equipment . . . . 27
1.5.2 Terminal Device . . . . 29
1.5.3 Monitoring Device . . . . 30
1.6 Patent Analysis of the Laser Cleaning Process . . . . 32

1.6.1 Surface Integrity Control of Laser Cleaning . . . . . . . . . . . . 33
1.6.2 Surface Functionalization of Laser Cleaning. . . . . . . . . . . . 35
1.6.3 Composite Laser Cleaning. . . . . . . . . . . . . . . . . . . . . . . . . . 36
1.7 Conclusion . . . . . . . . . . . . . . . . . . . . . . . . . . . . . . . . . . . . . . . . 38
References. . . . . . . . . . . . . . . . . . . . . . . . . . . . . . . . . . . . . . . . . . 39

**2 Typical Applications of Laser Cleaning and Its Impact on Substrate Surface Integrity** . . . . . . . . . . . . . . . . . . . . . . . . . . . . . . . . . 45
2.1 Introduction . . . . . . . . . . . . . . . . . . . . . . . . . . . . . . . . . . . . . . . 45
2.2 Classification of Laser Cleaning . . . . . . . . . . . . . . . . . . . . . . . . 46
2.3 Mechanism of Laser Cleaning . . . . . . . . . . . . . . . . . . . . . . . . . . 47
2.3.1 Vibration Effect . . . . . . . . . . . . . . . . . . . . . . . . . . . . . . . . 48
2.3.2 Ablation Effect. . . . . . . . . . . . . . . . . . . . . . . . . . . . . . . . . 48
2.3.3 Plasma Shock Effect . . . . . . . . . . . . . . . . . . . . . . . . . . . . 48
2.4 Typical Application of Laser Cleaning Technology. . . . . . . . . . . . 49
2.4.1 Laser Paint Removal . . . . . . . . . . . . . . . . . . . . . . . . . . . . 49
2.4.2 Laser Derusting . . . . . . . . . . . . . . . . . . . . . . . . . . . . . . . . 50
2.4.3 Cultural Relics Restoration . . . . . . . . . . . . . . . . . . . . . . . 52
2.4.4 Laser Degreasing . . . . . . . . . . . . . . . . . . . . . . . . . . . . . . . 54
2.4.5 Laser Removing Adhesion of Micron and Submicron Particles . . . . . . . . . . . . . . . . . . . . . . . . . . . . . . . . . . . . . . 56
2.4.6 Typical Application of Laser Cleaning in Its Field . . . . . . . 57
2.5 Influence of Laser Cleaning on Surface Integrity . . . . . . . . . . . . . 59
2.5.1 Effect of Laser Cleaning on Surface Structure and Microstructure of Substrate . . . . . . . . . . . . . . . . . . . . . . . . 60
2.5.2 Effect of Laser Cleaning on Surface Properties of Substrate . . . . . . . . . . . . . . . . . . . . . . . . . . . . . . . . . . 62
2.6 Conclusion . . . . . . . . . . . . . . . . . . . . . . . . . . . . . . . . . . . . . . . . 64
References. . . . . . . . . . . . . . . . . . . . . . . . . . . . . . . . . . . . . . . . . . 65

**3 Effects of Laser Cleaning on the Surface Morphology of Carbon Fiber Reinforced Polymer** . . . . . . . . . . . . . . . . . . . . . . . . . . . 69
3.1 Introduction . . . . . . . . . . . . . . . . . . . . . . . . . . . . . . . . . . . . . . . 69
3.2 Laser Cleaning Mechanism . . . . . . . . . . . . . . . . . . . . . . . . . . . . 71
3.3 Effect of Laser Cleaning on Mechanical Properties of CFRP . . . . . . . . . . . . . . . . . . . . . . . . . . . . . . . . . . . . . . . . . . . 75
3.3.1 Effect of Laser Cleaning on the Shear Strength of CFRP . . . . . . . . . . . . . . . . . . . . . . . . . . . . . . . . . . . . . . . 75
3.3.2 Effect of Laser Cleaning on CFRP Tensile Properties. . . . . 83
3.3.3 Effect of Laser Cleaning on Other Mechanical Properties of CFRP . . . . . . . . . . . . . . . . . . . . . . . . . . . . . 85
3.4 Effect of Laser Cleaning on Bonding Properties of CFRP . . . . . . . 87
3.4.1 Effect of Specific Surface Microstructures on Bonding Properties. . . . . . . . . . . . . . . . . . . . . . . . . . . . . . . 87
3.4.2 Effect of Heterogeneous Joint Connections on Bonding Performance . . . . . . . . . . . . . . . . . . . . . . . . . . . . . 88
3.5 Online Monitoring of Laser Cleaned CFRP . . . . . . . . . . . . . . . . . 91

3.5.1 Online Monitoring of Laser-Induced Breakdown Spectroscopy . . . 91
3.5.2 Temperature Monitoring . . . 92
3.6 Numerical Simulation of Laser Cleaning of CFRP . . . 96
3.6.1 Numerical Simulation of the Temperature Field . . . 96
3.6.2 Numerical Simulation of the Stress Field . . . 99
3.7 Conclusions and Outlook . . . 101
References . . . 102

**4 Laser Cleaning Technology for Surface Pretreatment** . . . 109
4.1 Introduction . . . 109
4.2 Laser Cleaning Surface Pretreatment for Welding . . . 110
4.2.1 Pre-treatment Before Laser Cleaning Welding . . . 110
4.2.2 Laser Cleaning Welded Joints . . . 113
4.2.3 Laser Cleaning Weld Performance . . . 114
4.3 Laser Cleaning Surface Pretreatment for Coating . . . 117
4.3.1 Surface Cleaning Treatment for Spraying . . . 117
4.3.2 Surface Cleaning Treatment Before Painting . . . 119
4.4 Laser Cleaning Surface Pretreatment for Anticorrosion . . . 124
4.4.1 Laser Surface Cleaning of Aluminum Alloy . . . 124
4.4.2 Laser Surface Cleaning of Ferroalloy . . . 126
4.4.3 Laser Surface Cleaning of Copper Alloy . . . 128
4.4.4 Laser Surface Cleaning of Titanium Alloy . . . 129
4.5 Laser Cleaning Surface Pretreatment for Plating . . . 131
4.5.1 Electroplating Surface Cleaning Treatment . . . 131
4.5.2 Coating Surface Cleaning Treatment . . . 133
4.6 Laser Cleaning Surface Pretreatment for Adhesive Bonding . . . 135
4.6.1 Surface Cleaning Treatment of Carbon Fiber Reinforced Composites . . . 135
4.6.2 Surface Cleaning Treatment of Metal Bonding Joints . . . 137
4.7 Conclusion and Prospect . . . 139
References . . . 141

**5 Thermal Effects on the Surface-Interface During Laser Cleaning and Substrate Damage Control** . . . 151
5.1 Introduction . . . 151
5.2 Thermal Effects of Laser Cleaning . . . 152
5.2.1 Thermal Effects in Contaminants . . . 153
5.2.2 Thermal Effects in Substrates . . . 156
5.3 Influence of Laser Cleaning Parameters on Thermal Effects . . . 163
5.3.1 Influence of Pulse Width on Thermal Effects . . . 164
5.3.2 Effect of Cleaning Speed on Thermal Effects . . . 167
5.3.3 Influence of Laser Energy Density on Thermal Effects . . . 171
5.4 Numerical Simulation of Thermal Effects . . . 175
5.5 Online Monitoring of Laser Cleaned Surface . . . 178
5.5.1 Laser-Induced Breakdown Spectroscopy . . . 179

5.5.2 Photoacoustic Monitoring Techniques . . . 181
5.6 Conclusion and Prospects . . . 184
References . . . 185

**6 Acoustic-Optical-Thermal Dynamic Monitoring of Surface-Interface in Laser Cleaning** . . . 191
6.1 Introduction . . . 191
6.2 Wave Vibration Monitoring Technology . . . 193
6.3 Spectral Monitoring Technology . . . 200
6.4 Visual-Image Monitoring Technology . . . 209
6.5 Thermal Monitoring Technology . . . 216
6.6 Composite Monitoring Technology . . . 219
6.7 Conclusion . . . 226
References . . . 227

**7 Thermomechanical Coupling Modeling and Simulation in Laser Cleaning Surface-Interface** . . . 231
7.1 Introduction . . . 231
7.2 Laser Cleaning Simulation Process . . . 234
7.2.1 Model Establishment and Material Properties . . . 234
7.2.2 Governing Equations . . . 236
7.2.3 Calculation Method . . . 242
7.3 Results and Discussion . . . 242
7.3.1 The Influence of Particles in Laser Cleaning . . . 243
7.3.2 The Influence of Pits in Laser Cleaning . . . 251
7.3.3 The Influence of Crack in Laser Cleaning . . . 254
7.4 Conclusion . . . 259
References . . . 260

**8 Substrate Surface Morphology Control After Laser Paint Removal on Aluminum Alloy** . . . 261
8.1 Introduction . . . 261
8.2 Experimental Materials and Methods . . . 262
8.2.1 Laser Cleaning Equipment . . . 262
8.2.2 Sample Preparation . . . 263
8.2.3 Parameter Selection . . . 264
8.2.4 Characterization Methods . . . 265
8.3 Results and Discussion . . . 266
8.3.1 Influence of Laser Cleaning on Surface Microstructure . . . 266
8.3.2 Characterization After Surface Cleaning . . . 274
8.4 Research Findings and Technical Highlights . . . 282
References . . . 283

**9 Control of Substrate Surface Morphology During Laser Paint Removal from EH36 Steel Surface** . . . 285
9.1 Introduction . . . 285

9.2 Influence of Different Laser Cleaning Parameters on the Surface Morphology of EH36 Steel . . . . . 286
9.2.1 Macroscopic Morphology . . . . . 286
9.2.2 Microscopic Morphology . . . . . 287
9.2.3 Cleanliness Characterization . . . . . 290
9.3 Cross-Sectional Analysis . . . . . 291
9.4 Corrosion Resistance Analysis . . . . . 293
9.5 Wear Resistance Analysis . . . . . 294
9.6 Dynamic Cleaning Behavior Under Different Laser Parameters . . . 296
9.7 Key Findings and Application Potential . . . . . 296
References . . . . . 297

**10 Control of Substrate Surface Morphology After Laser Cleaning of CFRP Coating** . . . . . 299
10.1 Introduction . . . . . 299
10.2 Experimental Materials and Testing Methods . . . . . 300
10.2.1 Experimental Materials . . . . . 300
10.2.2 Experimental Equipment and Parameters . . . . . 300
10.2.3 Characterization Methods . . . . . 301
10.3 Characterization of CFRP Surface Cleanliness After Nanosecond Laser Cleaning . . . . . 303
10.3.1 Surface Micromorphology . . . . . 303
10.3.2 Three-Dimensional Morphology and Surface Roughness Analysis . . . . . 304
10.3.3 FT-IR Spectroscopic Characterization . . . . . 305
10.4 Evaluation of CFRP Surface Properties After Nanosecond Laser Cleaning . . . . . 305
10.4.1 Changes in Surface Wettability . . . . . 305
10.4.2 Adhesion Strength of Repainted Samples . . . . . 306
10.5 Characterization of CFRP Surface Cleanliness After Picosecond Laser Cleaning . . . . . 308
10.5.1 Surface Micromorphology . . . . . 308
10.5.2 Three-Dimensional Morphology and Surface Roughness Analysis . . . . . 308
10.5.3 FT-IR Spectroscopic Characterization . . . . . 310
10.6 Evaluation of CFRP Surface Properties After Picosecond Laser Cleaning . . . . . 311
10.6.1 Changes in Surface Wettability . . . . . 311
10.6.2 Adhesion Strength of Repainted Samples . . . . . 312
10.7 Online Monitoring of CFRP During Nanosecond Laser Cleaning . . . . . 313
10.7.1 Temperature Monitoring . . . . . 314
10.7.2 Acoustic Monitoring . . . . . 315
10.7.3 High-Speed Imaging Monitoring . . . . . 316
10.8 Experimental Results and Technical Contributions . . . . . 316
References . . . . . 318

**11 Surface Morphology Control of the Matrix After Laser Cleaning of GFRP Coatings** . . . . 319
11.1 Characterization of Surface Cleanliness of GFRP After Nanosecond Laser Cleaning . . . . 320
11.1.1 Microscopic Surface Morphology . . . . 320
11.1.2 Three-Dimensional Morphology and Surface Roughness Analysis . . . . 321
11.1.3 FT-IR Characterization . . . . 323
11.2 Evaluation of GFRP Surface Properties After Nanosecond Laser Cleaning . . . . 324
11.2.1 Variation in Surface Wettability . . . . 324
11.2.2 Bonding Strength of Recoated Samples . . . . 324
11.3 Online Monitoring of Nanosecond Laser Cleaning of GFRP . . . . 326
11.3.1 Temperature Monitoring . . . . 326
11.3.2 Acoustic Monitoring . . . . 327
11.3.3 High-Speed Imaging Monitoring . . . . 328
11.4 Application Results and Significance of Nanosecond Laser Cleaning of GFRP . . . . 330
References . . . . 331

**12 Controllable Texturing and Performance of Magnesium Alloy Substrate Surface After Laser Cleaning** . . . . 333
12.1 Materials and Methods . . . . 335
12.1.1 Experimental Materials . . . . 335
12.1.2 Hybrid Laser Cleaning Equipment . . . . 335
12.1.3 Path Design for Hybrid Laser Cleaning . . . . 337
12.1.4 Structural Characterization and Performance Testing of Samples . . . . 337
12.2 Results . . . . 339
12.2.1 Surface Texture Morphology Analysis After Hybrid Laser Cleaning . . . . 339
12.2.2 Effects of Hybrid Laser Cleaning on Substrate Microstructure . . . . 343
12.2.3 Effects of Hybrid Laser Cleaning on Substrate Surface Properties . . . . 347
12.3 Discussion . . . . 350
12.4 Conclusion . . . . 354
References . . . . 354

# Chapter 1
# Exploring the Progress of Laser Cleaning Technology from the Perspective of Chinese Patents

## 1.1 Introduction

Traditional cleaning methods, such as mechanical polishing, are easy to cause irreversible damage to the workpiece, and chemical cleaning is seriously polluted. Ultrasonic cleaning, high-pressure water jet cleaning, sandblasting cleaning and other methods are time-consuming, consumable, cumbersome and environmentally unfriendly, and it is difficult to meet the current requirements of high-quality, efficient and green industrial cleaning. Laser cleaning technology attracts more and more researchers' attention because of its advantages of no damage to the substrate, no pollution, safety and intelligence, high cleaning efficiency and high quality. Its principle is to use laser to irradiate the surface of pollutants, so that pollutants can be stripped from the surface of the workpiece under the combined action of vibration effect [1], ablation effect and photochemical effect, and the laser power can be further improved, which can induce plasma shock waves to penetrate the air medium and act on the surface of the workpiece to remove pollutants [2]. Laser cleaning technology can also strengthen the substrate surface [3], improve corrosion resistance [4] and improve hydrophobicity [5] while cleaning. In recent years, laser cleaning technology has been widely used in cultural relics protection [6], rubber molds [7], road transport [8], aerospace [9], ocean-going operations [10], semiconductor workpieces [11] and other fields.

Although laser cleaning technology has obvious advantages, it still faces some challenges in practical application. For example, the improvement of laser cleaning efficiency, the in-depth understanding of cleaning mechanism, the development of online monitoring technology and the research and development of intelligent equipment all need further research and innovation. Zhu et al. [8] employed dry and wet nanosecond laser cleaning techniques to remove road markings in various colors. They investigated the best cleaning protocols by adjusting the laser energy density and the overlap ratio of the laser spots. However, in the actual road cleaning, although the cost of laser cleaning technology is low, the efficiency is still an



S. Wang, *Laser Cleaning*, https://doi.org/10.1007/978-981-95-2935-3_1

important factor limiting its replacement of traditional road cleaning technology. Zhang et al. [12] discussed the removal mechanism of blue and red polyurethane paint on aluminum alloy surface by nanosecond pulse laser cleaning, which provided parameter information and theoretical basis for selecting appropriate laser cleaning technology for different colors and types of paint. Li [13] and Zhu et al. [14] summarized the research progress of monitoring technology of laser cleaning technology, indicating that the monitoring technology is vulnerable to environmental interference in practical application, and there are some problems such as relying on light source illumination and poor real-time performance. Ouyang et al. [15] optimized the energy consumption and performance of laser cleaning to remove coating. Hou et al. [16] reviewed the thermal effect and substrate damage control in laser cleaning. It shows that the diversification of cleaning methods, the selection of laser parameters and the technology of real-time monitoring are the keys to achieve efficient cleaning and maintain the integrity of the substrate. According to different cleaning objects, how to select appropriate laser parameters, optimize cleaning process, and how to realize real-time monitoring and intelligent control of cleaning process are the hot and difficult points in current research, and also the key points that must be overcome in knowledge capitalization and technology transfer and commercialization.

All valuable technologies usually seek a strong authority for exclusive protection, so the patents cover almost all of the most advanced technology research and development. Moreover, the patent often reflects whether the applicant and the inventor expect the technology fully, and reflects the development direction of the application in related fields of the technology. The National Intellectual Property Administration of the People's Republic of China is an institution directly under the State Council, which is responsible for collecting and managing patents. The National Intellectual Property Administration of the People's Republic of China is an institution directly under the State Council, which is responsible for collecting and managing patents. Through the "patent retrieval" function provided by the China National Intellectual Property Administration, this paper analyzes the relevant patents in the past 6 years (January 1, 2018–December 31, 2023), so as to explore the research status and development trend in the field of laser cleaning in China.

## 1.2 Overall Overview of Laser Cleaning Patents

On January 24, 2024, China's utility model and invention patents related to laser cleaning were inquired. As of December 31, 2023, there were 3507 pieces of data. It was added to China National Intellectual Property Administration web analytics Library, and the overall situation of China's laser cleaning patents was obtained. Due to the lag between patent application and publication, only partial incomplete data were retrieved in 2023, but it did not affect the analysis of the overall development trend of laser cleaning. The following focuses on the development trend of

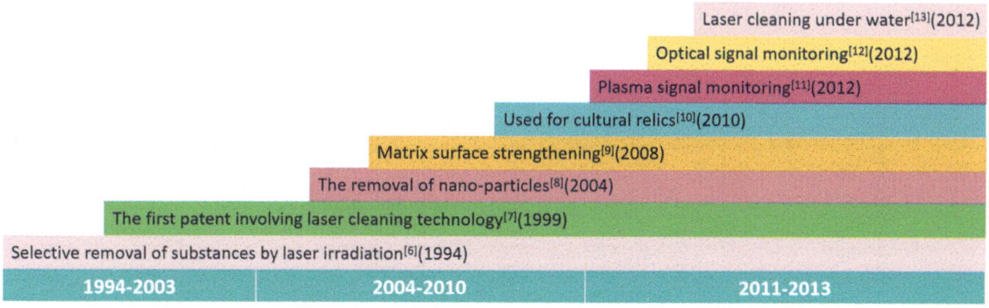


**Fig. 1.1** The development process of laser cleaning before 2013

laser cleaning related patents, the proportion of classification, the distribution of institutional attributes and the number of patent applications in each province.

### *1.2.1 Basic Trend of Laser Cleaning Patents*

Figure 1.1 shows the development process of laser cleaning before 2013, in 1994, Cauldron Co., Ltd. [17] proposed to selectively remove substances by irradiation, which can remove impurities such as grease, paint and oxide on the surface of the substrate without affecting the physical characteristics of the substrate. This is the first patent in China to remove pollutants by laser irradiation. Due to the beginning of laser cleaning technology, the attention was low in China, as shown in Fig. 1.2, it was not until 1999 that the first patent involving laser cleaning technology appeared [18]. In 2004, The Institute of Engineering Thermophysics of the Chinese Academy of Sciences [19] has developed a device for laser cleaning of micro and nanoparticles, which is the first patent in China to achieve nanoscale particle removal and named after laser cleaning. In 2008, Tsinghua University [20] discloses an all-solid ultraviolet laser washer and laser cleaning method, in the cleaned AlN plate silver, compared with the plate without laser cleaning, the silver layer adhesion increased by 120%, is the first patent of laser cleaning for matrix surface reinforcement in China.

In 2010, Laser Fusion Research Center of China Academy of Engineering Physics [21] proposed a portable laser cleaning machine for stone cultural relics, which is the first laser equipment dedicated to cultural relics cleaning in China. In 2012, Jiangsu University [22, 23] proposed the method of automatically judging the laser cleaning effect and testing the laser cleaning threshold, monitored the cleaning situation through optical signals and plasma signals to prevent the occurrence of incomplete cleaning and excessive damage to the matrix. In the same year, Jiangsu University [24] also invented an underwater laser cleaning method. Under the action of high-energy laser, a large number of cavitation bubbles were generated in water, and the contaminated layer on the surface of underwater workpiece was cleaned by the shock wave generated by cavitation effect.

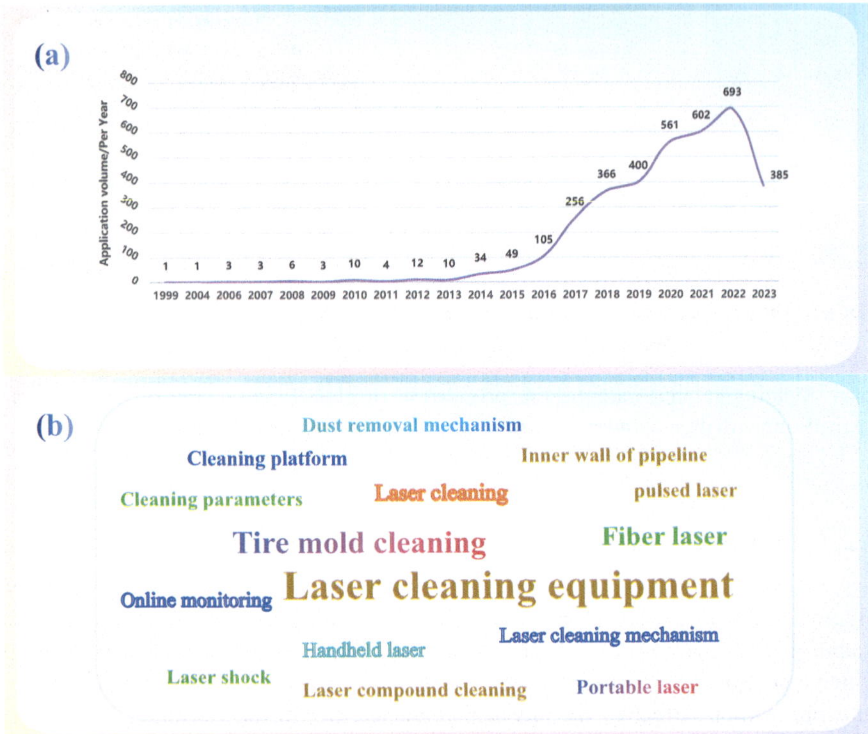


**Fig. 1.2** Overview of laser cleaning patent applications. (**a**) Trend of laser cleaning patent applications. (**b**) Ranking chart of the number of keywords

With the gradual maturity of laser cleaning technology, its application fields are more and more extensive. Through the development trend of laser cleaning patents in China (Fig. 1.2), it is not difficult to find that since 2013, the number of patent applications for laser cleaning has increased by spurt, and now laser cleaning technology is in a high-speed development stage. The ranking of keyword quantity in 2018 and mid-2023 was analyzed. As shown in Fig. 1.2b, the more prominent the keyword, the more significant advantage it indicates in the field of patents. The focus of patents related to laser cleaning mainly focuses on the research and development of laser cleaning equipment suitable for different types of workpieces, among which there are relatively many patents applied to tire mold cleaning. Under the influence of the huge market of automobile industry, the technology related to laser cleaning tire mold will have broad development prospects.

In general, laser cleaning can not only ensure the surface integrity of the substrate, but also conform to the global environmental protection strategy to achieve green and efficient surface cleaning, which has a long-term development prospect.

**Table 1.1** Main classification of patents related to laser cleaning

| IPC class number | Description of class number | Application volume | Percentage |
|---|---|---|---|
| B08 | Cleaning | 2548 | 52% |
| B23 | Machine tools; metalworking not included in other categories | 578 | 12% |
| H01 | Basic electrical components | 256 | 5% |
| G01 | Measurement; test | 131 | 3% |
| B29 | Processing of plastics; processing of substances in plastic state | 107 | 2% |
| G02 | Optics | 101 | 2% |
| | Other | 1198 | 24% |

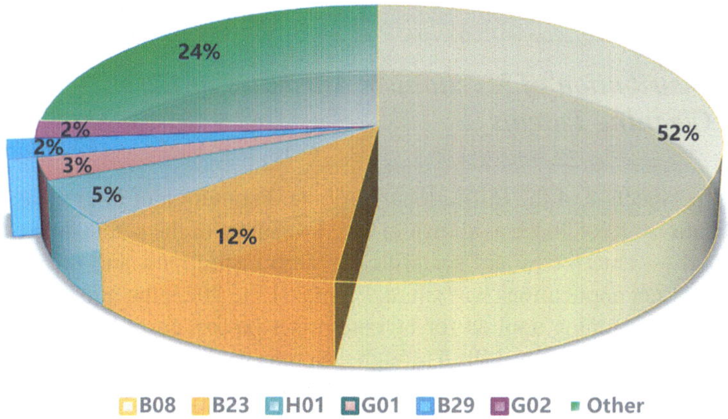


**Fig. 1.3** The classification proportion of laser cleaning patents

## 1.2.2 *Classification and Proportion of Laser Cleaning Patents*

According to the International Patent Classification System (IPC), the laser cleaning patents are inquired, among which there are more than 80 categories related to the laser cleaning patents, and the main classification information is shown in Table 1.1. Because a patent will be divided into multiple classification numbers, the number searched is more than the actual number. This paper only analyzes the classification proportion of laser cleaning patents. As shown in Fig. 1.3, it can be seen that B08 (cleaning) accounts for about 52%, which indicates that the current focus of laser cleaning patent is how to effectively remove pollutants. The disruptive improvement of laser cleaning efficiency depends on the research and development of high-power lasers, but the relevant hardware patents only account for 12%, as shown in B23 (machine tools; metalworking not included in other categories), mainly due to the difficulty of equipment development, long cycle and high cost, which hinder the engineering application of laser cleaning.

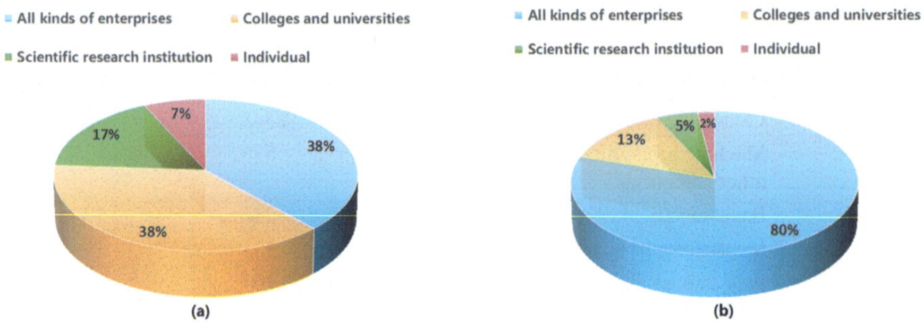


**Fig. 1.4** Overview of laser cleaning patent applications. (**a**) Distribution of mechanism attributes of laser cleaning patents up to 2013. (**b**) Distribution of institutional properties for laser cleaning patents from 2013 to 2023

### *1.2.3 Distribution of Mechanism Attributes of Laser Cleaning Patents*

From the perspective of the institutional attribute distribution of laser cleaning patent applications, as of 2013, as shown in Fig. 1.4a, most of the patents were applied by universities, enterprises and scientific research institutions, and the number of individual patent applications was small. From 2013 to 2023, the number of patents related to laser cleaning applied for by enterprises has accounted for the majority, about 80%, as shown in Fig. 1.4b. Through the comparison of (a) and (b) in Fig. 1.4, it can be seen that with the increase of market demand for this technology, more enterprises have paid attention to it, and people's in-depth understanding of its potential and application value has promoted the sustainable development of this technology. As can be seen from Fig. 1.5, there is only one university among the top ten core applicants in the number of applications, which indicates that enterprises have gradually begun to become the main force of patent applications in the field of laser cleaning, and laser cleaning technology has shown a good momentum of development in the commercialization.

### *1.2.4 Distribution of the Patent Application Areas*

From the distribution of patent applications by province (Fig. 1.4), the main subjects of patent applications in the field of laser cleaning in China are mainly concentrated in the southern provinces, among which Jiangsu and Guangdong provinces occupy the main position, guiding the investment direction of Chinese enterprises and institutions of higher education in laser cleaning research and development, and provides a good development driver for laser cleaning technology.

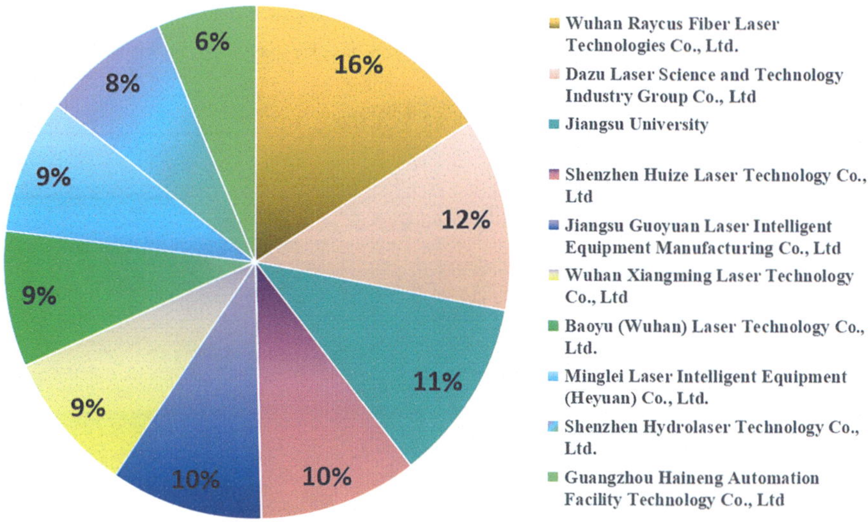


**Fig. 1.5** The top ten applications of core applicants from 2013 to 2023

In developed cities (regions), the support of local governments and the close cooperation between enterprises and scientific research institutions have promoted the innovation and progress of laser cleaning technology.

For example, Jiangsu University in Jiangsu Province, Wuhan Ruike Fiber Laser Technology Co., Ltd. in Hubei Province and Shenzhen Hydrolaser Technology Co., Ltd. in Guangdong Province, have many excellent laser cleaning research teams, the research results in the field of laser cleaning not only promote the development of related enterprise technology, but also bring new breakthrough points for the whole industry. Through the in-depth analysis of the laser cleaning patent application data in Chinese provinces (Fig. 1.6), we can see that the performance of southern China is particularly outstanding in this field, which contains huge development potential behind it. In the future, with the participation of more excellent enterprises and scientific research forces, China's laser cleaning technology will continue to grow and occupy a place in the global scope.

## 1.3 Patent Analysis of Laser Cleaning for Different Matrix Materials

Due to the different shape of the workpiece, the surface film layer and the processing environment, the laser cleaning process methods and devices for dealing with the surface pollutants of various matrix materials will also change, which brings great challenges to the laser cleaning technology. In order to cope with these challenges, many enterprises and universities have carried out targeted research.

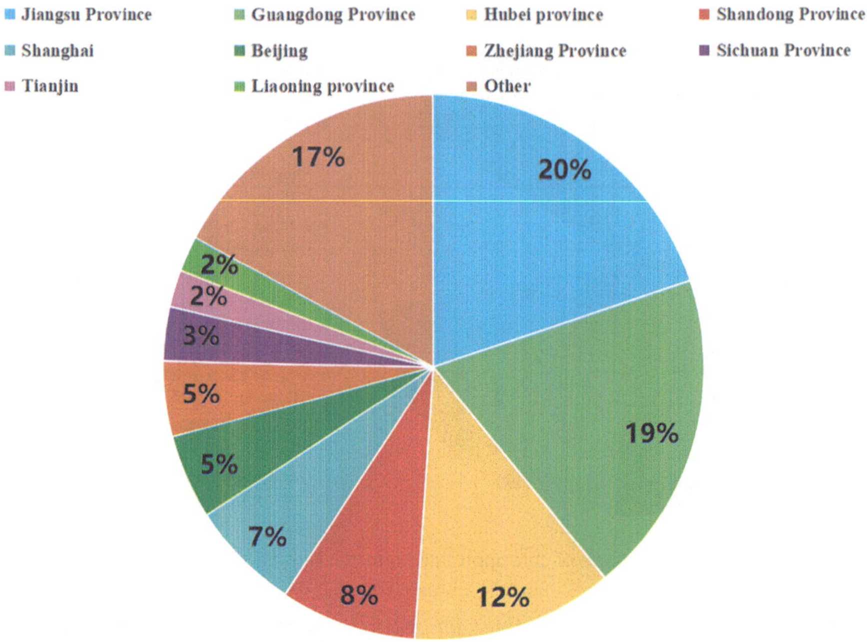


**Fig. 1.6** Distribution of patent applications for laser cleaning in different provinces

## *1.3.1 Patent Analysis of Laser Cleaning Iron Matrix*

Iron-based materials have the characteristics of high strength and high hardness, which are widely used in ships, railway traffic, automobile manufacturing and other fields. It is one of the important basic materials of China's advanced manufacturing technology.

The corrosion of parts in railway affects their performance and service life in the use process, and also involves railway traffic safety, so regular inspection and cleaning maintenance are required. At present, shot blasting equipment is often used to remove rust from brake beam and other parts of railway vehicles. There are some problems in the process of removing rust, such as loud noise, low efficiency and serious environmental pollution. Laser cleaning can very well avoid the above problems.

Henan Sancheng Mechanical and Electrical Equipment Co., Ltd. [25] has developed a laser derusting device for railway tracks, which makes full use of the characteristics of high efficiency of laser cleaning, and makes the laser cleaning head move between two railway tracks through a horizontal moving mechanism, thus realizing the function of cleaning two railway tracks at the same time by a single laser derusting device. Harbin Wick Rail Transit Technology Development Co., Ltd. [26] has developed a laser rust removal device for brake beams of railway vehicles.

During the cleaning process of double brake beams, a high-pressure dust removal system was installed to purify the dust generated during the cleaning process and ensure the working environment. Hebei university of technology [27] designed a three-axis support and two axis scanning vibration laser cleaning track machine, body carrying laser and equipped with intelligent measurement and control system, accurate fixed-point operation surface, through the camera real-time monitoring cleaning surface, achieve accurate and efficient all-round cleaning, effectively solve the problem of rail surface rust and pollution, improve the efficiency of railway transport, to ensure the safety of transportation.

The contamination on the tire mold will reduce the quality of the tire and shorten the service life of the tire. Therefore, the tire mold needs to be cleaned regularly to ensure the qualified rate of the tire manufacturing. However, the tire mold cleaning needs to stop and dismantle, time-consuming and laborious, reduce the production efficiency. Thus, Shenyang University of Technology [28] discloses a laser equipment that can clean the tire mold online. It uses a six-axis robot to realize the multi-angle cleaning of the mold on the production line, which can be operated continuously and greatly improves the production efficiency. Hangzhou Innovation Research Institute of Beihang University [29] developed a laser cleaning device and method based on 3D point cloud, combined with the point cloud data of the inner wall of the tire mold and the space 3D coordinate system, establishes a three-dimensional column model and determines the cleaning path, which greatly improves the cleaning quality and reduces the wear of the tire mold in the cleaning process.

Ductile cast iron pipe is often used in urban water supply and gas supply, but it is prone to corrosion and erosion caused by external factors such as sediment and water. Zinc spraying can solve the above problems and improve the beauty and gloss of the pipeline. Laser cleaning the outer wall of the pipeline before spraying zinc can effectively improve the adhesion of zinc layer. However, after cleaning, the surface temperature of the substrate is high, so it is necessary to let it stand and cool, and then spray zinc, which will easily lead to secondary oxidation. In order to solve this problem, Hubei Zhengxin Pipe Fitting Co., Ltd. [30] added a cooling system to the laser cleaning equipment to realize the rapid cooling of the substrate surface after laser cleaning, and zinc can be sprayed directly without cooling. Xinxing Cast Pipe Co., Ltd. [31] has developed a pretreatment device for inner wall spraying, which can clean and roughen the inner wall of ductile cast iron pipe at the same time, thus improving the adhesion of the paint after cleaning.

Compared with the traditional cleaning method, laser cleaning technology has the characteristics of high efficiency, precision and environmental protection, which can improve the cleaning efficiency and cleaning quality. Especially for iron-based alloys, which are sensitive to laser irradiation, laser cleaning technology has great potential in cleaning iron-based materials, which is worth digging deeply.

### *1.3.2 Patent Analysis of Laser Cleaning Aluminum Alloy Substrate*

Anodizing and coating paint can improve the wear resistance and corrosion resistance of aluminum alloy surface, but with the increase of service time, the oxide film or paint on aluminum alloy surface will be damaged and fall off, which needs to be cleaned and replaced in time, so there is a high requirement for process control of aluminum alloy surface cleaning. Sichuan University [32] proposed a selection method of laser parameters. According to the thermal stress and ablation removal threshold of paint layer, the appropriate laser process parameters are selected for cleaning, which provides a process choice for nondestructive cleaning on the surface of aluminum alloy matrix. Wuhan WISCO Huagong Laser Large Equipment Co., Ltd. [33] has developed a laser cleaning device for the composite layer of aluminum alloy anodic oxide film and paint film, using nanosecond pulse laser to clean the damaged area paint layer and anodic oxide film layer to make it ablation and gasification, and finally peel off the surface of the substrate.

The above two patents only clarify the process and parameter selection method of paint removal on aluminum alloy anodic oxide film, including the traditional mechanical grinding, chemical cleaning, etc., and do not involve the problem of how to retain the anodized oxide film on aluminum alloy surface after paint removal. For example, for the removal of the coating of aviation equipment skin, the complete retention of oxide film is the basic index requirement for its cleaning. Armored Force Academy of the Chinese People's Liberation Army [34] proposed a method to protect the anodic oxide film on the surface of aluminum alloy substrate during laser cleaning. First, a fiber laser with high cleaning efficiency was used to remove the thick paint layer, and then a $CO_2$ gas laser was used to quantitatively remove the remaining paint layer by layer. The patent protects the anodic oxide film on the surface of aluminum alloy to the greatest extent while completely removing the paint layer, and opens up a new direction of laser cleaning with intact oxide film.

In order to ensure the cleaning efficiency of aluminum alloy, dry laser cleaning is generally used. However, the cleaning process will produce a large amount of residue and dust, deteriorating the working environment. To address this problem, Dongguan Ruiyong Hardware Products Co., Ltd. [35] has developed a laser dry cleaning equipment for aluminum alloy anodic oxide film, which is equipped with dust removal device and residue collection device to reduce the labor burden of users. Hunan Vocational College of Railway Technology [36] has developed a laser cleaning process and device for cleaning aluminum alloy car bodies. It is equipped with smoke removal equipment, and the smoke is introduced into the purification device through a fan for treatment, so as to prevent the air pollution caused by the untreated spilled smoke, and ensure the health of the staff.

After welding aluminum alloy workpieces, the nondestructive flaw detection liquid is usually used for penetrant inspection. After inspection, the nondestructive flaw detection liquid will remain on the surface of the workpiece, which will affect the subsequent processes if it is not cleaned in time.

CRRC Nanjing Puzhen Co., Ltd. [37] proposed a device and method for dry laser cleaning nondestructive flaw detection liquid on aluminum alloy surface. According to different thickness flaw detection liquid layers, the laser cleaning parameters were adjusted, so that the flaw detection liquid could be evaporated instantly without damaging the substrate to achieve the cleaning purpose.

Cooperating with other processing (maintenance) technologies can give full play to the advantages of laser cleaning technology. As shown in Table 1.2, many patents combine laser cleaning technology with laser welding technology to realize real-time cleaning of oxide film on aluminum alloy surface before welding, which effectively avoids the problem of reoxidation during processing and provides convenience for subsequent material processing.

From this point of view, laser cleaning technology can efficiently and quickly remove aluminum alloy surface paint layer, oxide film and other pollutants. Through patent analysis, it can be found that the key to distinguish laser cleaning technology from other traditional cleaning methods is that it can ensure the integrity of anodic oxidation layer on aluminum alloy surface while removing paint layer, and it shows important application value in welding pretreatment and purification of working environment.

**Table 1.2** Patent of laser cleaning assisted laser welding aluminum alloy matrix

| Title of invention | Invention content | Beneficial effect | | Applicant |
|---|---|---|---|---|
| Welding pretreatment method of aluminum alloy spiral welded pipe based on laser cleaning technology | Laser cleaning pretreatment of aluminum alloy before welding | Reduce the welding pore rate and improve the welding quality | Laser cleaning technology to clean and preheat the workpiece | Laser Institute of Shandong Academy of Sciences [38] |
| Equipment for laser pretreatment of oxidation film before longitudinal seam welding of aluminum alloy material | | | Laser cleaning to remove the aluminum alloy surface oxide film | Wuhan Xiangming Laser Technology Co., Ltd. [39] |
| Laser grinding equipment for aluminum alloy car body profile before and after welding | | | Removing oxide film is synchronous with welding | Wuhan Xiangming Laser Technology Co., Ltd. [40] |
| Welding method and device for aluminum alloy shell of high-voltage switch based on laser cleaning | | | | Laser Institute of Shandong Academy of Sciences [41] |
| Welding device and method of 6-series aluminum alloy profile based on laser cleaning | | | Add dust removal system | CRRC Nanjing Puzhen Co., Ltd. [42] |

### 1.3.3 Patent Analysis of Laser Cleaning Titanium Alloy Substrate

Titanium alloy is a new structural material with high strength and lightweight, and its alloy powder is commonly used in the field of additive manufacturing (3D printing). However, titanium alloy powder is easy to be oxidized by moisture, and forms an oxide film on the surface when it comes into contact with air. The oxide film of titanium alloy will prevent the powder from contacting with the next layer of material, affect the bonding between layers of material, and lead to the decrease of the compactness of printed workpiece, thus seriously affecting the quality of 3D printing. China Aviation Manufacturing Technology Research Institute [43] has developed a laser cleaning device for titanium alloy powder. The oxide film on the surface of titanium alloy powder is removed by laser beam under protective atmosphere. After the powder is cleaned, the workpiece printed by it has low porosity and high compactness.

Titanium alloy is mostly used in workpieces with complex structures. However, a dense oxide film will be formed on the surface of titanium alloy workpiece during high temperature thermoforming, which will affect the bonding strength of workpiece coating. Shenyang Heshitai General Titanium Industry Co., Ltd. [44] has developed a laser cleaning device, which can efficiently clean the oxide layer on the surface of titanium alloy bar after heat treatment.

China Aviation Manufacturing Technology Research Institute [45] has developed a laser cleaning device and method for titanium alloy welding wire. Different from the traditional method of pickling titanium alloy before welding, this method can realize real-time cleaning of welding, and it is not necessary to equip wire winding equipment to rewind the wire like pickling method. Wuhan Xiangming Laser Technology Co., Ltd. [46] proposed a laser cleaning method to remove the oxide layer on the surface of titanium alloy. The thickness of the oxide layer on the surface of titanium alloy was divided into three different thickness ranges: 1–20, 20–30 and 30 μm, and suitable laser parameter ranges were selected for different thickness ranges to achieve better cleaning effect. Chengdu Mairuijie Laser Technology Co., Ltd. [47] has developed an automatic laser cleaning device for the weld of titanium alloy tube sheet, which uses laser to effectively clean the oxide layer of the weld of titanium alloy tube sheet. Wuhan Tianyu Intelligent Manufacturing Co., Ltd. [48] has developed a device to realize titanium alloy welding under the double guarantee of inert atmosphere and laser cleaning, which ensures no oxidation welding in each pass and improves welding quality. Beijing Xinghang Electromechanical Equipment Co., Ltd. [49] proposed a method of electron beam welding titanium alloy based on laser cleaning treatment. Laser cleaning is used to treat the welding surface, which can not only remove the oxide layer, but also clean the surface greasy dirt and non-metallic impurities, so that the welding quality is significantly improved.

The production cost of titanium alloy is relatively high, and a lot of money needs to be invested in its processing, production and other links. Therefore, the use of laser cleaning technology to clean titanium alloy matrix materials and realize the recycling of workpieces can reduce the production cost of enterprises and improve production efficiency to a certain extent. The advantages of laser cleaning technology applied to titanium alloy materials cleaning are attracting more and more attention from all walks of life.

### 1.3.4 Patent Analysis of the Laser-Cleaning Composite Matrix

Composite materials have excellent properties such as light weight, high strength, good toughness, corrosion resistance and wear resistance, and have been widely used in industrial fields. The traditional cleaning methods of composite surface coating, such as chemical cleaning and mechanical polishing, are not only easy to damage the matrix, but also have low cleaning efficiency.

In order to solve this problem, Hunan Xinao Photoelectric Technology Co., Ltd. [50] proposed a method of laser cleaning to remove the paint layer on the surface of composite materials, and adjusted the laser cleaning parameters according to the difference of the thickness of the paint layer on the surface of composite materials to achieve the effect of effectively removing the paint layer on the surface. Minglei Laser Intelligent Equipment (Heyuan) Co., Ltd. [51] disclosed a system for cleaning the surface coating of composite materials and an ultraviolet laser cleaning machine. The surface coating of glass fiber composite materials and carbon fiber composite materials was cleaned by ultraviolet laser with a scanning speed of 1800 mm/s, a frequency of 20 kHz and a pulse width of 1 ns, which not only did not damage the substrate, but also achieved high efficiency in cleaning the coating. The patent also compared the mechanical properties of the composite materials before and after laser cleaning, and found that the mechanical properties of the substrate materials did not change after cleaning. Nanchang Hangkong University [52] puts forward a surface modification method of carbon fiber surface functionalization, which uses dopamine as surface modifier to increase the specific surface area and roughness of carbon fiber, so as to improve the compatibility between carbon fiber and resin matrix. Experiments show that this method obviously improves the mechanical properties of carbon fiber resin matrix composites.

The treatment of carbon fiber composites by the above technologies is mainly a single cleaning or surface performance improvement. Jiangsu University [53] has developed a device and method for composite processing of laser cleaning and surface modification of carbon fiber composite materials, which can not only ensure that the surface pollutants can be cleaned, but also make the surface of carbon fiber form concave-convex morphology, increase the surface roughness and improve the surface adhesion. From this point of view, laser cleaning can effectively clean and modify the surface at the same time, simplify the workflow and save costs. This patent provides a new idea for laser cleaning and modified composite processing.

Spraying metal powder on the surface of aircraft skin to form a conductive layer can prevent lightning strike. In order to cope with the harsh service environment, the method of spraying paint on the conductive layer of aircraft skin is generally used to improve the impact resistance and corrosion resistance of the skin. However, after long-term service, the paint layer on the surface of aircraft skin will be damaged and ineffective, so it is necessary to remove the paint regularly and spray it again. As shown in Fig. 1.7, it is easy to damage the conductive layer when cleaning the paint layer. To solve this problem, Nanjing University of Aeronautics and Astronautics [54] proposed a method for removing paint from the surface of carbon fiber reinforced resin-based composites containing conductive layers, which adopted a multilayered removal method, namely first using high-power, low-frequency and wide-pulse laser parameters to remove part of the paint layer on the surface of the materials, and then using low-power, high-frequency and narrow-pulse laser parameters to remove the remaining paint layer on the surface of the materials, so as to remove the paint layer without damaging the conductive layer, and also to strengthen the conductive layer and further improve its conductivity. This method effectively promotes the transformation of laser cleaning technology to modern aviation material cleaning application. Jiangsu University [55] has disclosed a device and method for laser cleaning aircraft composite radar topcoat. The device is equipped with a double guide rail system, which can realize multi-directional and comprehensive cleaning of the workpiece. The cleaning position and temperature can be monitored in real time through CCD imaging system and infrared temperature sensor, and the nondestructive cleaning of composite radar mask can be realized.

Aiming at the problem that it is difficult to predict the temperature field change of materials in the existing laser cleaning process, Shenyang Institute of Automation Chinese Academy of Sciences [56] proposed a simulation method of temperature field of release agent on the surface of carbon fiber material based on laser cleaning. By using COMSOL simulation software, the temperature changes of release agent and carbon fiber material during cleaning were monitored in real time, and the effects of different laser power, energy density and scanning speed on cleaning depth were explored, which provided a theoretical basis for laser cleaning process parameters of carbon fiber composite materials.

Laser cleaning can efficiently and quickly remove pollutants and damaged paint layers on the surface of composite materials, it can also be modified in the cleaning

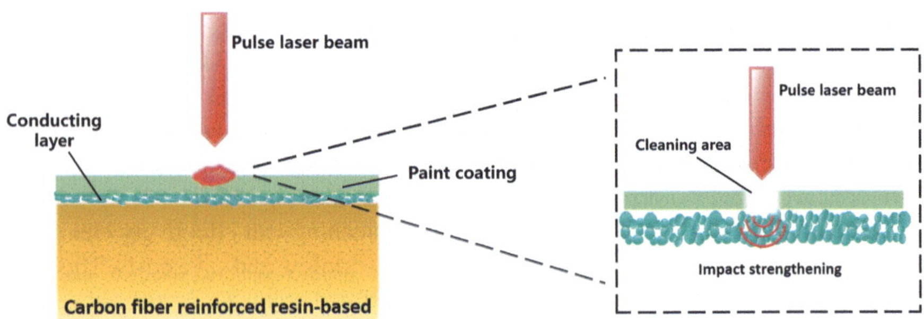


**Fig. 1.7** Structural diagram of carbon fiber reinforced resin-based composites containing conductive layers

process to improve the material density and surface strength. Therefore, laser cleaning is expected to become an indispensable technology of composite material cleaning in the modern manufacturing industry.

## 1.4 Patent Analysis of Laser Cleaning Different Coatings

In the process of manufacturing, processing and using the workpiece, the surface is often covered with a variety of different coatings, such as rust, paint, oxide film, etc. Due to the different properties of the cladding, it is difficult to control the laser cleaning process parameters, which is easy to cause damage to the surface of the workpiece, so it is necessary to comprehensively consider the properties of the covering and the requirements of the workpiece, in order to select the appropriate laser cleaning parameters.

### *1.4.1 Patent Analysis of the Laser Cleaning Rust Layer*

Compared with traditional rust removal methods, laser cleaning technology has the characteristics of high precision, high efficiency and not easy to damage the matrix. Using laser to clean the rust layer of metal parts is one of the important directions of engineering application. For different materials of the workpiece, choose the appropriate process parameters for laser cleaning, not only can completely remove rust, but also can improve the hardness of the matrix surface, corrosion resistance and wear resistance and other mechanical properties, to maximize the potential of laser cleaning technology [57–60].

The patent of laser cleaning applied to rust removal in China was first applied to the narrow pulse width laser decontamination machine developed by Nankai University [61] in 2007. The equipment uses narrow pulse width laser to clean various pollutants such as rust layer and paint layer at the same time, the use of instruments and devices to monitor and analyze the ultrasonic wave formed by vibration during the cleaning process to ensure the integrity of the medium surface and realize the automation of the cleaning process, which shows the advantages of high efficiency and automation of laser cleaning technology in removing rust. After 2017, the number of patent applications for laser cleaning technology for rust removal began to increase.

At present, laser cleaning mostly adopts line scanning, and the effect is difficult to be satisfactory when cleaning the seriously corroded surface. Jiangsu University [62] coated an energetic coating composed of acrylic epoxy resin, hydrogen peroxide and nano carbon powder particles on the corrosion layer (Fig. 1.8a). During laser cleaning, the energetic coating will explode under the instantaneous heat of the laser beam and produce a shock wave, which will destroy the rust layer and drive it off (Fig. 1.8b). With the help of the special function of the auxiliary coating, this

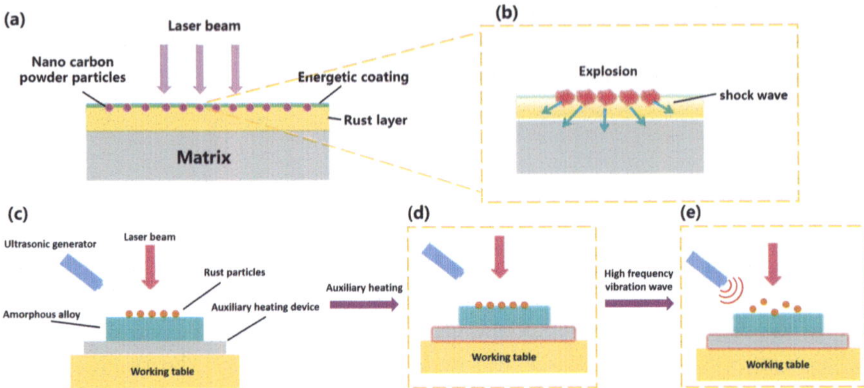


**Fig. 1.8** Schematic diagram of auxiliary laser cleaning in the field. (**a**) Schematic diagram of laser cleaning energetic coating to remove rust layer. (**b**) The shock wave drives the rust layer to fall off. (**c**) Structural diagram of integrated laser cleaning equipment with heating and ultrasound. (**d**) Auxiliary heating reduces the strength and hardness of amorphous alloy. (**e**) Ultrasonic wave loosens corroded particles

technology can efficiently and quickly remove the thick rust layer, paint layer and other coatings on the surface, which can greatly improve the efficiency of laser cleaning and provide a new idea and direction for the development of laser cleaning field.

In the process of long-term use, the oil pipeline will be influenced by environmental factors and the aging of the pipeline itself, which will lead to the corrosion of the pipeline. In order to ensure the safe and stable operation of the oil pipeline, the surface of the pipeline is generally coated with a protective layer to prevent corrosion. Baoyu (Wuhan) Laser Technology Co., Ltd. [63] has developed a laser cleaning equipment for derusting oil pipelines, which not only has good cleaning effect, but also can produce uniform anchor lines during derusting, providing good adhesion for repainting and fully ensuring the close combination of subsequent anti-corrosion coatings and pipeline surfaces.

In atmospheric environment, aircraft skin and some key parts will be corroded, and how to effectively remove them without damage is a difficult problem. In order to solve this problem, Qingdao Campus of Naval Aeronautical University [64] put forward a cleaning method of nanosecond pulse laser to clean the surface corrosion of 7B04 aluminum alloy. Through in-depth analysis of the surface corrosion mechanism of laser cleaning aluminum alloy, combined with process test, an optimized process parameter was successfully designed, which eliminate the uncontrollable factors of mechanical grinding to remove the rust layer and significantly improved the cleaning efficiency. This achievement has important application value in the maintenance of military equipment.

In order to protect the surface of alloy substrate from damage during cleaning, the common method is to use low-power laser to clean, but reducing power means giving up high efficiency. For this reason, Dongguan Eontec Co., Ltd. [65] added an auxiliary heating device and an ultrasonic generator to the low-power laser cleaning equipment (Fig. 1.8c), firstly heated the amorphous alloy sample to reduce its strength and hardness (Fig. 1.8d), and then used the ultrasonic generator to emit high-frequency vibration waves to loosen the corrosion particles embedded in the surface (Fig. 1.8e), thus reducing the cleaning difficulty. On the premise of ensuring high cleaning efficiency, it also avoids matrix damage. Shenhua Group Zhungeer Energy Sources Co., Ltd. [66] couples the laser beam into the deionized water jet, so that the laser cleaning processing area is in the deionized water environment. Even under the high-power laser cleaning, thermal damage can be avoided, and the water jet can be relied on to wash away the corroded particles generated in the cleaning process in time to prevent secondary pollution.

### *1.4.2 Patent Analysis of the Laser Cleaning Paint Layer*

Paint layer is mostly used to protect the workpiece from the external environment, but it will be damaged after long-term service, so it is necessary to remove and spray a new paint layer regularly [9]. The composition of paint layer is complex, and the parameters such as laser power and pulse width are difficult to control during cleaning, and the substrate is easy to be damaged [67]. At present, the most common methods to remove paint layer are chemical cleaning and sand blasting, which are easy to cause stress damage to the substrate and pollute the environment. Laser cleaning is to remove the paint layer on the surface of the object to be cleaned by laser high-energy beam irradiation, without contacting with the surface of the substrate, thus avoiding the secondary pollution of chemical reagents and mechanical polishing damage. Minglei Laser Intelligent Equipment (Heyuan) Co., Ltd. [68] discloses a method and a system for laser paint removal, which can generate a laser beam according to the thickness of the paint layer on the surface of the workpiece to be cleaned, effectively remove the paint layer on the surface, and realize the rapid removal of the paint layer and the intact and accurate cleaning of the substrate.

The application of laser cleaning technology in marine field is more and more common. If necessary, laser cleaning operation needs to be carried out in water. The cleaning process should consider not only the physical properties of paint layer and substrate, but also the acoustic impedance of water layer. Sichuan University [69] discloses a method and device for removing paint in water. According to the impact strength of the pulsed laser in the water on the paint layer, the reflection and transmission coefficients of the interface between water and paint layer and between paint layer and substrate are determined, so as to find out the optimal energy density range required by laser pulse, ensure that the original substrate structure will not be destroyed, and efficiently clean the paint layer, and expand the new direction of laser underwater cleaning.

Aiming at materials such as stainless steel, which are highly reflective of light and sensitive to heat input, Shanghai Ruirong Laser Welding Technology Co., Ltd. [70] have developed a laser cleaning device for the paint layer of Nero-engine stainless steel parts. This device uses the thermal effect of laser and the difference of thermal expansion coefficient between the paint layer and the substrate to crack the paint layer, and then peels the paint layer by laser plasma impact, so as to completely remove the paint layer without damaging the surface of the substrate. Durr Coating System Engineering (Shanghai) Co., Ltd. [71] invented a laser paint removal device for automobile manufacturing. By irradiating the high-energy pulsed laser emitted by the laser on the surface of the workpiece to be cleaned, the paint layer quickly gasified and separated from the surface of the automobile body, ensuring the stability and sustainability of the automobile production process.

Aiming at the problems of high cost, toxicity and environmental pollution in the process of using paint remover to remove the paint layer of aircraft skin, China Hi-Tech Co., Ltd. [72] has developed an aircraft skin cleaning system, which uses high-energy laser pulse to vaporize the paint layer and impurities on the skin surface, and high-intensity shock wave to make the paint layer fall off. The cleaning cost is low, and it can clean complex-shaped workpieces. However, a large amount of solid powder will be produced during laser cleaning. If it is not treated in time, it will easily lead to inhalation by workers and affect their health. Sichuan Minghang Lander Technology Co., Ltd. [73] is equipped with special dust removal equipment in the laser cleaning aircraft skin system. Through the matching of dust cover and vacuum cleaner, the dust generated in the cleaning process can be sucked away in time, which highlights the environmental friendliness of laser cleaning.

Aircraft skin has aerodynamic characteristics, which can be divided into single curvature skin, double curvature skin and complex shape skin. In view of the complexity of the skin shape and the difference of the head and tail curvature, it is challenging to maintain the consistency of the laser and aircraft skin spacing during cleaning, and it is also difficult to plan the cleaning path and regulate the process parameters.

Hebei University of Technology [74] combines the laser cleaning device with the cross-frame moving device, so that the cleaning process is close to the surface of the aircraft, with high safety factor, and the paint layer can be removed in layers, so as not to damage the substrate. Nanchang Hangkong University [75] plans the running path of the robot arm by establishing the aircraft skin model, and can use laser to clean the paint layer of aircraft skin with different curvatures and irregular large areas. It is also found that the best cleaning effect of the surface paint layer is achieved when the power is 50 W, the frequency is 200 kHz, the scanning speed is 2000 mm/s, the robot arm speed is 2 mm/s, and the cleaning time is six times. Hebei Hanguang Heavy Industry Co., Ltd. [76] has developed a laser cleaning device for cleaning curved objects. The laser ranging module is used to measure the distance between the cleaning head and the cleaning target, which can automatically focus and fit the curved surface. The operation is simple and the surface cleanliness after cleaning is high. The proposal and application of these technologies can effectively prevent the uneven cleaning effect caused by the excessive change of laser energy density, solve the cleaning problem of large, curved and irregular surfaces, provide a good technical foundation for the large-scale application of laser cleaning technology to aircraft skin cleaning. It is expected to replace the traditional cleaning method to realize low-cost and high-efficiency green cleaning.

As shown in Table 1.3, a variety of patents of laser cleaning devices and methods applied to paint removal of specific workpieces or objects are listed. It can be seen that compared with traditional cleaning methods, laser cleaning technology can realize efficient cleaning of the surface of workpieces, and will not produce chemical substances, which is environmentally friendly. With the miniaturization, portability and light weight of laser cleaning device, laser cleaning technology will be closer to people's daily life and work, and it has a wider application prospect.

### *1.4.3 Patent Analysis of Laser Cleaning Composite Coating*

Composite coating refers to a coating composed of two or more materials with comprehensive properties, in which the composite coating made of metal or ceramic powder has the characteristics of high performance, high durability and easy

**Table 1.3** Patent of laser cleaning paint layers of different workpieces

| Title of invention | Purpose of invention | Beneficial effect | Applicant |
|---|---|---|---|
| Laser cleaning device for brake regulator of railway vehicle | Cleaning the paint layer on the surface of railway vehicle brake regulator | No contact, no damage and no pollution | Harbin Wick Rail Transit Technology Development Co., Ltd. [77] |
| Laser cleaning device for small advertisements | City clean | Portable | Yingtan Zhihui Internet of Things Application Research Institute Co., Ltd. [78] |
| Self-propelled wire laser cleaning device and method | Wire cleaning | 360° laser cleaning | Shandong Zhongshi Yitong Group Co., Ltd. [79] |
| Device for laser cleaning inner wall of pipeline | Clean the inner wall of the pipeline | Realize the efficient laser cleaning of the pipe inner wall | Jiangsu University [80] |
| Automatic device for removing paint coating of flat wire of motor by laser | Laser paint removal for flat wire of motor | Automatically finish the paint removal of flat wire | Suzhou Jinghai Laser Intelligent Technology Co., Ltd. [81] |
| Automatic laser cleaning paint equipment | Laser cleaning paint | Suitable for laser cleaning of pipes with different sizes | Hunan Xinao Photoelectric Technology Co., Ltd. [82] |
| Laser cleaning device for inner wall paint of flat tail big shaft | Laser cleaning paint layer on the inner wall of flat tail big shaft | Real-time observation of cleaning effect and laser light-emitting condition | State-owned Wuhu machinery factory [83] |

maintenance. Laser Institute of Shandong Academy of Sciences [84] discloses a device and a method for laser cleaning composite coatings. According to the specific requirements of aircraft skin paint stripping, single-layer coatings in composite coatings on aircraft surfaces can be removed specifically, which solves the problem of poor uniformity and accuracy of traditional sandblasting and chemical paint stripping at present.

In the process of aircraft assembly, it is necessary to carry out anti-electric shock and anti-interference lap joint on the designated part, and it is essential to clean the non-conductive coating on this part, which is very important to ensure the good conductivity of the lap joint. At present, manual sandpaper grinding or grinder grinding are widely used in cleaning methods, which cannot accurately control the grinding shape and depth. Wuhan Jindun Laser Technology Co., Ltd. [85] proposed an in-situ laser cleaning method for aircraft non-conductive composite coating. The laser process parameters were adjusted according to the material type and processing requirements of the area to be cleaned, so that the energy density of the laser beam was between the cleaning threshold and the damage threshold of the substrate, and protective tape was covered around the area to be cleaned in advance, which avoided the additional damage of the internal structure by laser. The composite coating can be cleaned efficiently, accurately and non-destructively.

Micro-arc oxidation coating is a composite ceramic coating with high temperature resistance and corrosion resistance, which is produced by instantaneous high temperature and high pressure on the surface of valve metal and its alloy through micro-arc discharge under the conditions of electrolyte and corresponding electrical parameters. When the micro-arc oxidation coating is damaged locally, it is necessary to remove the damaged coating and generate a new coating in situ. Zhejiang University of Technology [86] has disclosed a device and method for local repair of micro-arc oxidation coating on titanium alloy (Fig. 1.9a). The residual thickness of the damaged film layer is measured by eddy current thickness gauge. Under appropriate laser cleaning parameters, it is not necessary to completely immerse the workpiece in electrolyte. In addition, the residue after cleaning can be washed away by flowing electrolyte.

Thermal barrier coating has excellent oxidation resistance and high temperature resistance. The coating will have the problem of crack propagation due to excessive internal pores in the manufacturing process, so the unqualified coating needs to be removed for re-preparation. As shown in Fig. 1.9b, Nanchang Hangkong University [87] proposed a method for cleaning thermal barrier coating of zirconia (YSZ), which used ultrasonic vibration to weaken the particle sputtering in the laser cleaning process. Under the synergistic effect of ultrasonic cleaning and laser cleaning, the molten pool generated by overheating of the surface interface was reduced, which blocked the splashing and enrichment of molten pool particles under the action of plasma shock wave and ensured the cleanliness of the substrate surface. With the service temperature of a new generation of engine turbine blades greatly increased, YSZ thermal barrier coatings will have problems such as phase transformation and sintering when the temperature is higher than 1200 °C. Deposits attached to engine turbine blades (such as CaO, MgO, $Al_2O_3$ and $SiO_2$, abbreviated as CMAS)

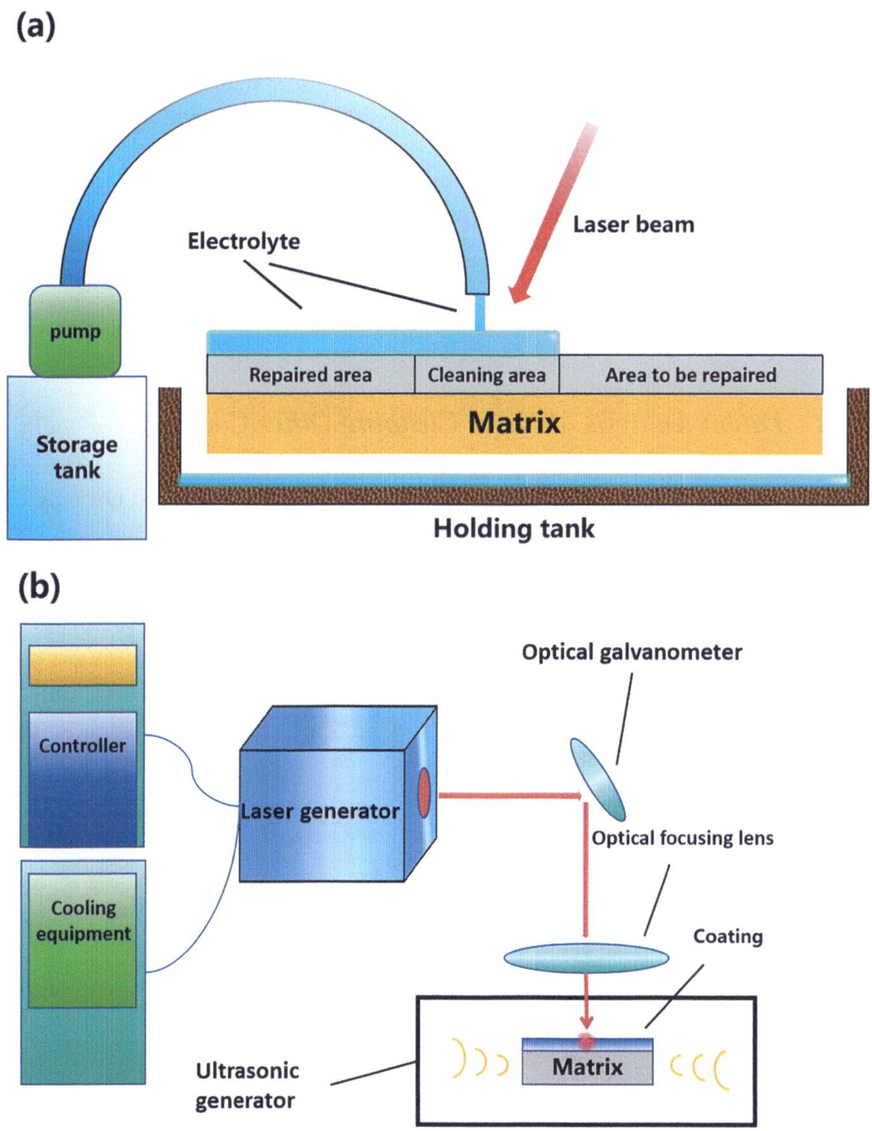


**Fig. 1.9** (**a**) Device and method for locally repairing titanium alloy microarc oxidation coating. (**b**) Schematic diagram of laser cleaning YSZ thermal barrier coating

will transform into molten state and penetrate into the coating at high temperature, which will lead to the degradation of coating performance. Therefore, Tianjin University [88] proposed a method of laser cleaning to remove the deposits on the surface of thermal barrier coating of Nero-engine. Using pulsed laser with power of 50–200 W, the CMAS layer was removed several times in layers, with the thickness of 1/2–3/4 of the total thickness removed for the first time, and the subsequent single removal thickness was less than 5 μm. After the CMAS layer is completely removed, the thermal barrier coating of 5–30 μm will be removed continuously to ensure that the thermal barrier coating thickness is more than 2/3 of the original thickness, so that the engine blade can still play its original role and meet the daily service requirements.

It can be seen that laser cleaning technology can accurately remove the single coating in the composite coating by adjusting the energy density, which provides a new process for micro-arc oxidation coating micro-area repair, provides a good guarantee for the maintenance and use of thermal barrier coating, and has important practical application value.

### *1.4.4 Patent Analysis of Laser Cleaning Other Coatings*

Laser cleaning is an efficient cleaning technology, which can be used not only to clean the rusted layer, paint layer and composite coating, but also to remove carbon deposits in engines, tiny particles of optical devices, greasy dirt and metal oxides and other coatings (Fig. 1.10). This is because laser cleaning technology can accurately locate the surface of the workpiece to be cleaned, the surface of pollutants is

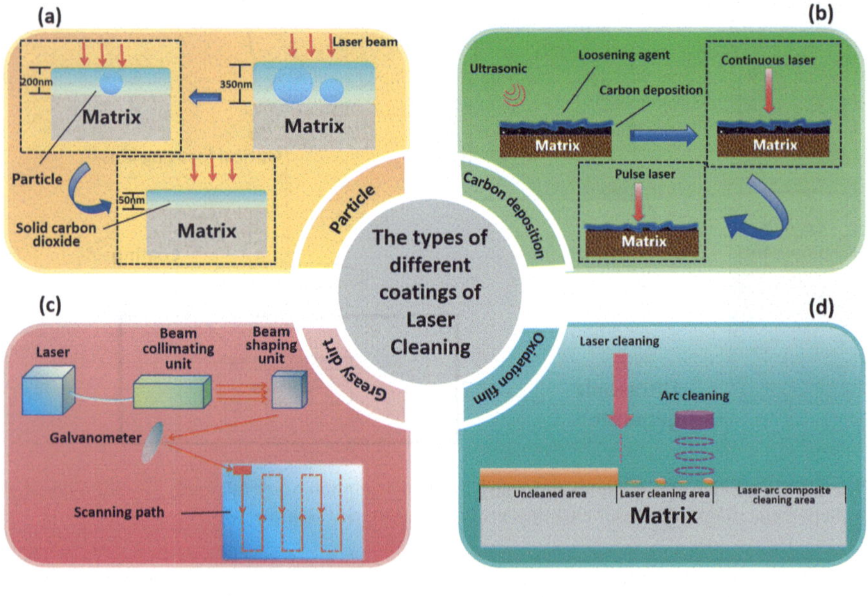


**Fig. 1.10** Schematic diagram of laser cleaning carbon deposit, particles, greasy dirt and oxidation film. (**a**) Method for laser cleaning semiconductor surface particles by using solid carbon dioxide as buffer medium [73]. (**b**) Method for laser cleaning surface carbon deposit of titanium alloy by using loosening agent [76]. (**c**) Method and system for laser cleaning oil stain on metal plate surface by rectangular spot [78]. (**d**) Laser-arc compound cleaning of oxide film on aluminum alloy surface [80]

instantly heated by using a laser beam with high energy density, and the pollutants on the matrix surface are removed under the synergistic effect of ablation, vibration, plasma impact and other effects.

#### 1.4.4.1 Particles

Ductile iron has the characteristics of high strength, high toughness and wear resistance, and can maintain stable performance in harsh environments such as high temperature and high pressure, but the embedded spherical graphite particles will affect the welding performance. At present, pickling, alkali washing and other methods cannot effectively remove graphite particles. Laser Fusion Research Center of China Academy of Engineering Physics [89] can effectively remove graphite particles on the surface of nodular cast iron by adjusting laser pulse frequency, pulse width, average power and spot size. This technology has the characteristics of selectively removing graphite particles, which can effectively avoid the damage of nodular cast iron matrix while removing graphite particles.

When using 3D printer to print layer by layer, impurities and metal particles usually appear on the surface of semi-finished metal prints. The accumulation of impurity particles on the surface will lead to gaps between layers of finished printed parts, and the binding force will decrease, which will affect the quality of printed parts. ShanDong CharmRay Laser Technology Co., Ltd. [90] has developed a laser cleaning device for metal 3D printing layer by layer, which uses the characteristics of non-contact, accuracy and high efficiency of laser cleaning to quickly clean the impurity particles on the surface of metal printed matter layer by layer.

In the process of processing optical devices and semiconductors, it is inevitable to introduce dust and other pollutant particles on the surface of components, which will lead to the failure of semiconductors and affect the reflection and transmission of laser beams of optical devices. SOL Electronic Technology Co., Ltd. [91] proposed a laser cleaning method to remove nano-particles from semiconductors. Solid carbon dioxide was used as a buffer medium to clean particles of different sizes on the surface of silicon wafers. The solid film is thinned from thick to thin, and the thickness of the film is reduced by 150 nm at a time to 50 nm. Particles of different sizes are removed by solid films with different thicknesses.

It is worth reminding that the laser cleaning process will produce thermal ablation, plasma radiation and other phenomena, so it is necessary to detect the optical components after cleaning. Dongguan Feichuang Laser Intelligent Equipment Co., Ltd. [92] put forward a method to detect the effect of laser cleaning on the surface impurities of optical components. The changes of substrate elements before and after cleaning were analyzed by EDS energy spectrum, observe the laser cleaning effect and the damage condition, and constantly adjust the cleaning parameters, and finally realize the non-damage cleaning.

#### 1.4.4.2 Carbon Deposit

After long-term use, the Nero-engine will generate a layer of carbon deposit on the inner wall, which will continue to accumulate and block the nozzle, air inlet, piston ring or other components, resulting in the decline of engine performance. Although carbon deposit is a product of high temperature, it will still burn at high temperature, and even explode when it burns violently, causing serious accidents. Xi'an Lanxiang New Materials Technology Co., Ltd. [93] has developed a laser cleaning device to clean the carbon deposits on Nero-engine blades. According to the carbon deposits status on the surface of Nero-blades, the corresponding cleaning parameters are preset. The carbon deposits are burned and vaporized by high-energy laser beams. Armored Force Academy of the Chinese People's Liberation Army [94] uses ultrasonic wave to disperse and coat the loosening agent on the carbon deposit layer of titanium alloy, which can well weaken the force between carbon deposit molecules, then uses pulsed laser cleaning to reduce the thickness of carbon deposit layer, and then uses low energy density continuous laser to clean the remaining carbon deposit. The cleaning temperature is monitored in real time during the cleaning process to avoid damage to the titanium alloy matrix. This patent shows the advantages of the combined use of chemical and laser cleaning methods, which greatly improves the cleaning efficiency.

#### 1.4.4.3 Greasy Dirt

Lubricating oil will be added to the workpiece in the process of processing and production to reduce friction, so it is difficult to avoid contamination of the workpiece with greasy dirt. Aiming at the problem of greasy dirt, Shengtong Intelligent Machinery Equipment (Shanghai) Co., Ltd. [95] specially developed a laser cleaning equipment for greasy dirt cleaning on metal surfaces. However the traditional circular spot laser has a Gaussian distribution with high center energy and low edge energy, which makes the laser cleaning efficiency low. Beijing University of Technology [96] shapes the circular laser beam into a rectangular beam, and uses the rectangular beam to clean the greasy dirt on the metal plate surface, which effectively increases the cleaning efficiency and cleaning area. In addition to the above two patents for laser cleaning greasy dirt, there are many similar patents. Considering the innovation of the patents, I will not elaborate too much here.

#### 1.4.4.4 Oxidation Film

In the process of 3D printing aluminum alloy parts, oxides often appear with each additional layer, which leads to the decrease of the bonding force between layers. Wuhan Xiangming Laser Technology Co., Ltd. [97] discloses a device and method for laser cleaning assisted arc additive manufacturing, which uses laser cleaning technology to remove surface oxides generated during machining. The traditional additive manufacturing process generally uses a single laser cleaning or arc cathode cleaning to remove the oxide film on the surface of the substrate. However, the single cleaning method is prone to the problems of incomplete cleaning and secondary oxidation. Beijing University of Technology [98] puts forward a laser-arc compound cleaning method for aluminum alloy oxide film, which removes the oxide film with a fixed depth by using the characteristic that the laser removes the oxide film at the same depth, provide preconditions for fixed-point cleaning of oxide film with arc cathode. Under the protective atmosphere, the cathode spots with extremely high current density in the arc will automatically find the residual oxide film and evaporate it to clean it, realizing high-quality cleaning, effectively preventing the oxide film from regenerating, and obviously improving the working efficiency. It can be seen that the combination of laser cleaning and arc cathode cleaning makes up for the deficiency of single cleaning method, and also provides ideas for the combination of multiple cleaning methods in the future.

Mirrors made by magnetorheological polishing are widely used in aviation infrared optical systems, but black oxide film will be formed on the mirror surface during magnetorheological polishing, which will affect the mirror performance. National University of Defense Technology [99] discloses a method and a device for ultrafast laser cleaning of aluminum alloy mirror surface. The photo-thermal absorption experiment of aluminum alloy mirror and its surface oxide film was carried out. Based on the difference of photo-thermal absorption coefficient between the two, select the wavelength range where the photothermal absorption rate of aluminum alloy oxide film is greater than that of aluminum alloy matrix. Then femtosecond laser was used to clean the aluminum alloy surface oxide film, which removed the surface impurities and pollutants without damaging the substrate, greatly improving the reflectivity of aluminum mirror. This achievement can be applied to the production and maintenance of optical components, improve the accuracy and performance of optical components, and meet the needs of modern optical mirrors.

As an efficient and environment-friendly cleaning technology, laser cleaning can play a good auxiliary role in metal processing such as additive manufacturing and welding. Konfoong Materials International Co., Ltd. [100] put forward a welding method of aluminum target assembly, which promoted the diffusion between atoms and enhanced the bonding effect by making a brazing metal infiltrating layer on the surface of the material. However, aluminum is likely to oxidized during the production of the infiltration layer. In order to avoid this situation, laser cleaning is the most convenient and effective way before welding. Aifaco Electronic Materials (Suzhou) Co., Ltd. [101] put forward a new welding method, which can quickly remove oxides and other pollutants on the aluminum surface by laser cleaning parameters with power of 100–400 W, wavelength of 1064 nm, spot diameter of 60 μm and scanning speed of 1.2–3 m/min. Shanghai Chip Electronic Technology Co., Ltd. [102] has developed a laser cleaning mechanism for aluminum alloy oxide film before welding, which relies on the device table to clean the inner wall of aluminum alloy parts, and is equipped with a dust suction head to suck away the debris generated after cleaning, so as to ensure the cleaning of aluminum alloy oxide film before welding and improve the working environment.

**Table 1.4** Patents of laser cleaning oxide

<table>
<tr><th>Title of invention</th><th>Invention content</th><th colspan="2">Beneficial effect</th><th>Applicant</th></tr>
<tr><td>Welding gun integrating laser cleaning and welding</td><td rowspan="3">Laser cleaning removes impurities and pollutants such as oxide film on the surface before and after welding</td><td rowspan="2">Avoid welding defects</td><td rowspan="5">Improve welding quality</td><td>Borunte Robot Co., Ltd. [103]</td></tr>
<tr><td>Welding method for automatically cleaning welding seams before and after welding</td><td>Sichuan Aerospace Changzheng Equipment Manufacturing Co., Ltd. [104]</td></tr>
<tr><td>Continuous narrow gap gas shielded welding method with on-line cleaning</td><td>Realize no oxidation before welding</td><td>Dongfang Electric Group Dongfang Steam Turbine Co., Ltd. [105]</td></tr>
<tr><td>Underwater welding device and method</td><td>Removal of water film by thermal effect of laser cleaning</td><td>Realize oxygen-free and hydrogen-free before welding</td><td>Huazhong University of Science and Technology [106]</td></tr>
<tr><td>Integrated equipment and method for laser cleaning and welding combined processing</td><td>Device and method capable of performing laser welding immediately after laser cleaning</td><td>Realize no oxidation before welding</td><td>Shenyang University of Technology [107]</td></tr>
</table>

Laser cleaning technology complements other laser technologies, especially as a surface pretreatment before laser welding, which is a very good cleaning process. As shown in Table 1.4, many patents combine laser cleaning technology with laser welding technology, and integrate the laser cleaning head into the robotic arm for automatic welding, which can make the whole cleaning process accurate, efficient and highly automated. It has high practical value and industrial application prospect.

## 1.5 Patent Analysis of Laser Cleaning Equipment and Monitoring Device

Laser cleaning equipment and supporting monitoring devices are the basic guarantee to obtain high-quality, efficient, green and intelligent cleaning effect. In recent years, the research and development of laser cleaning related hardware has been accelerated, and the number of related patent applications has increased exponentially, which has effectively promoted the leap-forward development of laser cleaning technology.

## 1.5.1 *Laser Cleaning Equipment*

### 1.5.1.1 Functionalization of Laser Equipment

Generally speaking, a high-power laser needs a high-power vibrating mirror, and the power of the vibrating mirror is proportional to the volume, and the corresponding high-power laser cleaning head is also more bulky, which brings great burden to the hand-held operation. Shenzhen Hydrolaser Technology Co., Ltd. [108] combines low-power continuous laser and pulsed laser, and only needs a vibrating mirror and a bracket with extremely small volume and weight, which reduces the weight of the laser cleaning head and is convenient to hold. Strictly speaking, hand-held laser cleaning is still dangerous. Baoyu (Wuhan) Laser Technology Co., Ltd. [109] discloses a portable hand-held intelligent laser cleaning device, which uses gyroscope to monitor the hand-held state solves the problem that unstable hand-held can easily cause uneven cleaning or fall off from the hand and hurt the staff, which fully guarantees the safety of the workers.

The above invention optimizes the volume and weight of the laser cleaning head, solves the problem of handheld instability, and provides convenience for manual operation, but it does not have the function of blowing dust. However, it does not have the function of blowing dust. After cleaning, the dust particles will attach to the surface of the laser lens, absorbing part of the laser energy and affecting the cleaning effect. Jiangsu Guoyuan Laser Intelligent Equipment Manufacturing Co., Ltd. [110] has developed a handheld composite laser cleaning gun equipped with a ventilation system that can blow gas from air channels on both sides of the gun, not only blowing away dust particles, but also for air cooling. Nanjing University of Aeronautics and Astronautics [111] discloses a pollution monitoring method and device for a protective mirror of a laser cleaning head which are internally provided with detecting light. The pollution degree is judged by the change of light refraction and reflection by pollutants. When the pollution degree reaches a preset value, it automatically stops working, which avoids the damage of protective glasses caused by the absorption of energy by pollutants. Baoyu (Wuhan) Laser Technology Co., Ltd. [112] has developed an annular laser cleaning head, which is different from the existing annular laser head in the pipeline on the market. It has high transmission precision, and can rotate 360° to realize cleaning without dead angle in the pipe.

### 1.5.1.2 Continuous Laser Equipment

Continuous laser has high cleaning efficiency because of its durable and high energy characteristics. However, due to its concentrated and continuous energy, it is easy to cause thermal damage to the material matrix, especially in the cleaning process of precision devices. Therefore, continuous laser cleaning technology is usually suitable for those application scenarios that are not very strict with surface quality. Wu Xiang county Hong Chen Wan Ju Environmental Protection Technology Co., Ltd.

[113] uses continuous laser to treat the surface of concrete imitation brick or slab, which can make the solid surface expand by heating and realize the crushing and peeling of cement attached to the top surface of surface sand. China Energy Longyuan Environmental Protection Co., Ltd. [114] uses the characteristics of continuous laser, such as high energy density, controllable direction and strong convergence ability, to remove organic substance from waste salt. It should be noted that the laser energy should not be too high, otherwise the waste salt particles will melt and agglomerate. South China University of Technology [115] put forward a laser remelting technology, which mentioned the use of continuous laser technology to clean and repair the surface of damaged waste steel. This method avoids using pulsed laser for initial cleaning and then switching to continuous laser for remelting repair, thus simplifying the processing flow and improving the processing efficiency. The integrated treatment of continuous laser reduces the process steps, avoids the processing interruption and inconsistency that may be introduced due to frequent switching of laser modes, and ensures the consistency of the processing process and the consistency of steel repair quality.

#### 1.5.1.3 Pulsed Laser Equipment

In cleaning tasks that require high precision and surface integrity, it is necessary to consider using more gentle technologies such as pulsed laser. Shanghai Yingfeng Electronic Technology Co., Ltd. [116] uses laser cleaning instead of traditional methods to process the evaporated metal film of thin film capacitors, because the laser action time is below nanosecond level, and nanosecond pulse laser cleaning is a cold working technology, which will not affect the film substrate. Similarly, Kunming Sachuang Photoelectricity Technology Co., Ltd. [117] chose to use pulsed laser with high peak power of 200 kW and ultra-short pulse width of 4 ns to clean the mask plate, which can control the thermal damage to the target in a small range and avoid deformation caused by thermal influence.

Although pulse laser has lower thermal effect than continuous laser in cleaning process, it will still cause different degrees of damage when facing the surface with high precision. China Yangtze Power Co., Ltd. [118] controls the laser to form a sine wave shape in the X–Y direction through the galvanometer system, and uses the Fourier transform technology to form any complex sine wave shape in a specific plane size. Through innovative scanning mode and focal length compensation mechanism, the problems of uneven cleaning and surface damage in traditional methods are effectively avoided.

Judging from the current development trend, laser cleaning equipment is improving in the direction of miniaturization, portability and intelligence. These improvements not only improve the efficiency and convenience of the equipment, but also make the laser cleaning technology easier to popularize and more universal. With the characteristics of continuous high energy input, continuous wave laser cleaning technology is mainly used in engineering fields with relatively low surface accuracy requirements and high efficiency requirements. However, the pulsed laser cleaning technology is more suitable for application scenarios with high requirements on surface cleanliness and accuracy because of its high peak energy and short pulse characteristics.

### 1.5.2 Terminal Device

In addition to improving the performance of laser cleaning equipment, the intelligent upgrade of terminal equipment is also an important way to improve the cleaning efficiency [14]. With the gradual improvement of the intelligent level of terminal equipment, its influence effect has been widely recognized by researchers in various fields.

China Construction Third Division First Construction Engineering Co., Ltd. [119] provides a control method of laser cleaning robot based on machine vision. By collecting image information such as workpiece position and surface impurity state, the corresponding model of laser cleaning robot is selected. There is no need for workers to go to the site to operate, which not only improves the work efficiency, but also ensures the safety of workers. Fanachi Laser Technology (Shandong) Co., Ltd. [120] discloses an efficient and intelligent laser cleaning machine, which reasonably plans the placement position of the equipment and provides a special protective box for the laser cleaning gun, thus improving the convenience and safety of the equipment. China Aviation Xi'an Aircraft Industry Group Co., Ltd. [121] has developed an intelligent laser cleaning system and method based on robot, which uses the intelligent module of laser cleaning head instead of manual operation to realize adaptive and intelligent tracking cleaning, and solves the problem that desktop equipment is cumbersome and difficult to be used for large-scale surface treatment at present.

Similarly, Suzhou Xiangming Laser Technology Co., Ltd. [122] designed a laser cleaning equipment for aluminum alloy surface paint, which used longitudinal transfer module and positioning bracket to fix the position of the workpiece, so as to keep the workpiece stable in the cleaning process. It is equipped with components such as transverse transfer module and cleaning bracket to realize multi-angle and all-round cleaning. Shandong Zotye Construction Technology Co., Ltd. [123] invented a laser cleaning device for aluminum alloy template processing. Through the combination of the cleaning mechanism and the clamping component, both sides of the aluminum alloy template can be cleaned without flipping. On this basis, Guangdong Longyue Construction Engineering Co., Ltd. [124] is equipped with a sweeping device on the laser cleaning equipment to sweep the surface debris residue after cleaning. The environmental pollution problem caused by laser cleaning is completely solved, and the green characteristics of laser cleaning technology are more obvious.

While laser cleaning technology focuses on improving cleaning efficiency, it is also necessary to consider the safety of operators. The intelligence of terminal equipment improves the production efficiency of enterprises, which is the development direction that must be paid attention to in the modernization construction. At the same time, we should also do a good job in protective facilities to ensure the safety and health of workers [125].

### *1.5.3 Monitoring Device*

The traditional monitoring technology is to monitor the cleaning quality and effect off-line after laser cleaning, which usually lacks timeliness and accuracy and cannot improve the efficiency and quality of laser cleaning. Because of the complex components of pollutants, it is difficult to control the cleaning parameters. It is necessary to monitor the cleaning state in real time during the cleaning process, so as to timely control the laser power, scanning path and other parameters, achieve the requirements of complete cleaning without damaging the substrate [126].

In the process of laser cleaning, a large number of sound and light signals will be generated. Researchers will summarize these signals and express them in the form of sound waves, images and spectra for monitoring. Hebei University of Science and Technology [127] discloses a real-time monitoring method and system for laser cleaning, which realizes real-time monitoring by collecting acoustic signals, optical signals and video stream signals during laser cleaning. Dongguan Feichuang Laser Intelligent Equipment Co., Ltd. [128] can identify the cleaning state of aircraft skin in real time through the difference of acoustic signals generated by laser cleaning top coat, primer and substrate. Nanjing University of Aeronautics and Astronautics [129] discloses a real-time monitoring system and method for laser cleaning based on visual and tactile perception, which uses the electric signal transmitted by the friction between the sensor block and the workpiece surface to realize online monitoring of the cleaning process. CRRC Nanjing Puzhen Co., Ltd. [130] CRRC Nanjing Puzhen Co., Ltd. [113] discloses a method and a device for visual inspection feedback control in the laser cleaning process of train exterior wall paint. By using the visual inspection device, different paint layers can be intelligently identified, and then the best cleaning parameters can be selected.

Laser-Induced Breakdown Spectroscopy (LIBS), that is, high-energy laser beam is irradiated on the surface of the sample to excite plasma, and then the change of characteristic peaks of each element in the cleaning process is detected by laser plasma spectroscopy, which can accurately and real-timely monitor the cleaning situation and is favored by researchers [131]. Civil Aviation Flight University of China [132] discloses a laser cleaning control method and control system based on online monitoring of LIBS technology, which uses laser-induced plasma spectrum to quickly detect and analyze sample components and accurately control cleaning quality. Dongguan Feichuang Laser Intelligent Equipment Co., Ltd. [133] discloses a real-time monitoring device and method for laser cleaning aircraft skin based on spectral monitoring method. The cleaning process is divided into four stages: single-pulse laser to remove top coat, multi-pulse laser to remove primer, substrate to be exposed and paint layer to be completely removed. According to the EDS test results

(Fig. 1.11i): Ba is selected as the main characteristic element of top coat. Si and Ti are selected as the main characteristic elements of primer; Cr is selected as the main characteristic element of chemical conversion coating; Al is selected as the main characteristic element of the matrix. As shown in Fig. 1.11ii, from the 16th pulse, the characteristic peak intensity of Ti began to weaken, while that of Al gradually increased. The results showed that the paint layer was almost cleaned at the 16th pulse, but the substrate was not damaged. After the 16th pulse, the primer was completely removed, and the damage degree of the substrate gradually increased with the increase of pulse number. The element calibration of the corresponding characteristic peaks of different stages in the LIBS spectrogram can monitor the removal process of different paint layers under different pulse numbers in real time, that is, when a specific signal appears in the plasma spectrogram, it is proved that the cleaning task of the previous stage has been completed, making the monitoring process is real-time and accurate.

The existing LIBS technology is mostly dual laser beam mode, a high frequency laser beam used for cleaning and a low frequency laser beam used for stimulating plasma, but the dual beam method has high energy, which increases the risk of matrix damage. The strong continuous background radiation caused by the laser paint removal process masks the characteristic spectra of atoms and ions, so it is difficult to collect the plasma spectrum, so it is difficult to use only a single laser to give consideration to cleaning and plasma excitation. Civil Aviation Flight University of China [134] discloses a system and method for detecting the effect of high-frequency laser paint removal. By adjusting the process parameters of high-frequency laser paint removal in the industrial personal computer, the laser cleaning threshold and the peak power density excited by plasma are balanced, which solves the problem of mutual interference between thermal radiation and plasma during cleaning. It promotes the application and development of high-frequency laser in the field of LIBS paint removal detection.

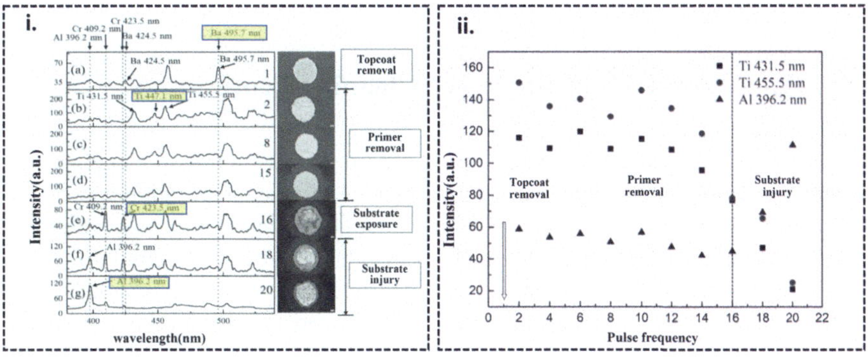


**Fig. 1.11** Plasma spectra at different pulse numbers [133]. (**i**) Plasma spectra under different pulse numbers (a) 1 pulse. (b) 2 pulse. (c) 8 pulse. (d) 15 pulse. (e) 16 pulse. (f) 18 pulse. (g) 20 pulse. (**ii**) The relationship between the characteristic peak intensity and the number of pulses

Beijing University of Technology [135] invented a method and a system for monitoring the quality of laser cleaning of thermoplastic resin based on the plume angle, which can monitor and analyze the plume angle of spatter in the cleaning process, suck the spatter pollutants away in time, and effectively ensure the sustainability of cleaning. Shanghai Hangyi High-Tech Development Research Institute Co., Ltd. [136] puts forward an online monitoring method of laser cleaning effect based on visual inspection. By graying the RGB image on the surface of the cleaned workpiece, the gray image is divided into multiple sub-images and compared with the preset gray standard value, so as to find out the untreated areas and make overall plans for re-cleaning.

The diversification of monitoring means enables researchers to control the state of laser cleaning more accurately, and the combination of multiple monitoring methods can improve the accuracy doubt caused by a single monitoring method to some extent, which is indispensable for the intelligent judgment of laser cleaning and the independent adjustment of process parameters, and will be a major trend in the development of laser cleaning monitoring technology in the future [137].

To sum up, the laser cleaning equipment is constantly updated and upgraded, developing towards a more convenient and lightweight direction. The subsequent development may not be limited to enterprises and factories, but will be used in all aspects of daily life. The terminal equipment and the cleaning equipment are linked and controlle, through the robot to achieve intelligent automatic cleaning, which is helpful to reduce the cost of manpower and time, improve the production efficiency and quality, and reduce the production cost and risk of enterprises. At the same time, the laser cleaning equipment is equipped with advanced sensors and data processing systems. The real-time monitoring and data analysis of the cleaning process can be realized through acoustic and optical sensors, which will significantly improve the controllability of the cleaning process. Through the analysis of China patents, it is found that laser cleaning equipment not only improves the cleaning efficiency and quality, but also makes great progress in cleaning speed, stability, energy saving and environmental protection.

## 1.6 Patent Analysis of the Laser Cleaning Process

The control of laser cleaning process is the basic way to improve cleaning efficiency, avoid substrate damage and obtain the best surface quality. In recent years, the protection of related intellectual property rights has attracted the attention of technical researchers. Through the analysis of related patents, it is found that the precise regulation of laser cleaning process has great potential for functional modification of the cleaned substrate surface, which is worth exploring. Of course, the process requirements for cleaning efficiency, matrix surface integrity and functional collaborative control are very demanding. Among them, the laser cleaning method using laser composite, liquid assistance and surface microtexture preparation has a significant effect (Fig. 1.12), which is well confirmed by the rapid increase of relevant patent applications.

### 1.6.1 *Surface Integrity Control of Laser Cleaning*

The common laser beam intensity is Gaussian distribution, the center intensity of the spot is high and the periphery is weak. There are some problems such as irregular edge of the spot and uneven energy distribution, so it is impossible to remove the coating evenly while ensuring the surface integrity of the substrate. China Aviation Manufacturing Technology Research Institute [138] has disclosed a device and method for accurately and uniformly cleaning aircraft composite coating by laser. The problem of poor roundness of light spot is solved by laser spinning. At the same time, the common spinning and scanning mode are combined, which increases the cleaning range equivalently and improves the accuracy and uniformity of laser cleaning.

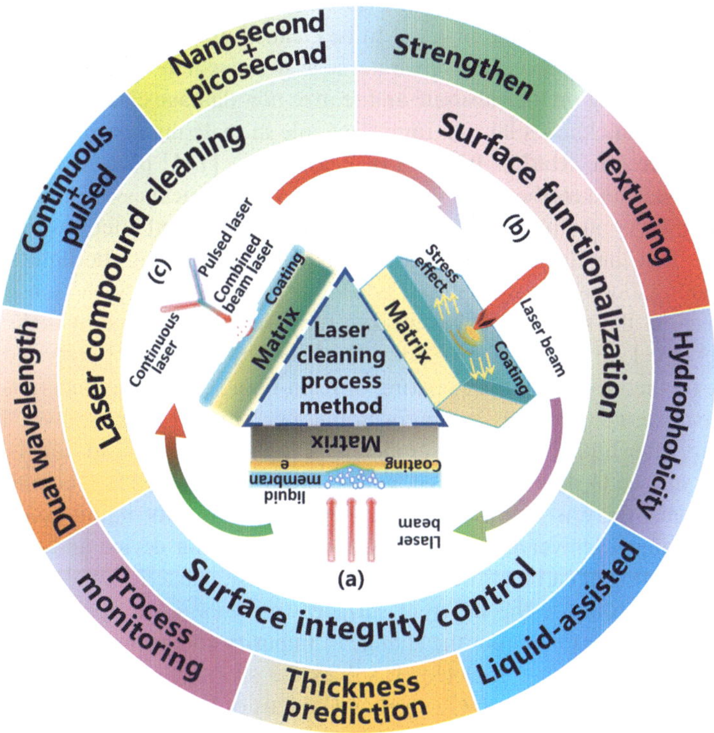


**Fig. 1.12** Schematic diagram of the laser cleaning process. (**a**) Liquid-assisted laser cleaning. (**b**) Laser cleaning for surface modification. (**c**) Composite laser cleaning

The quality of repainting after laser cleaning is related to the thickness of molten layer produced on the cleaned surface. In order to solve this problem, Armored Force Academy of the Chinese People's Liberation Army [139] proposed a control method that can effectively control the thickness of the molten layer of the paint layer on the surface of the substrate after cleaning. Firstly, the thickness of paint layer is predicted, and then the laser parameters are adjusted accordingly. Finally, the thickness of surface melting layer after cleaning is effectively controlled without affecting the mechanical properties of matrix materials. Qilu University of Technology [140] established a PSO-SVR prediction model for laser paint removal thickness to predict the relationship between paint removal thickness and cleaning parameters, so as to accurately control the paint removal thickness and protect the substrate from damage.

The surface of carbon fiber resin matrix composites is susceptible to be aged by light, so it needs to be coated with paint to improve its life. Dry laser cleaning method facing the cleaning object, often the substrate ablation damage threshold is higher than the ablation damage threshold of the surface paint, while the resin-based composite is just the opposite, so it is easy to damage itself when using dry laser to clean the surface paint of the resin-based composite. Therefore, liquid-assisted laser cleaning can be considered by taking advantage of the good corrosion resistance of resin-based composites. As shown in Fig. 1.12a, the existing liquid-assisted laser cleaning method is to put the material to be cleaned in a suitable liquid layer, and use the liquid layer to protect the substrate, so as to prevent the laser focusing energy from directly acting on the substrate and realize non-damage cleaning. The control of liquid layer thickness plays an important role in realizing non-destructive laser cleaning, Therefore, Harbin Institute of Technology [141] proposed a wet laser cleaning device and method. In this method, the laying thickness of liquid film can be adjusted by controlling valve, and the liquid film will boil after absorbing laser energy, so that the cleaned pollutants will be stripped. During the liquid-assisted laser cleaning, liquid film will take away a lot of laser heat energy, the substrate will not damage at the same time also weaken the laser impact force, it is difficult to effectively peel off the surface coating, Nankai University [142] proposed a laser cleaning method for the surface coating of resin-based composite materials. Based on the characteristic that the decomposition temperature of resin is lower than that of coating, after liquid-assisted laser cleaning, high-pressure liquid flow is used to spray the surface to separate the decomposed resin with the surface coating. In the above patents, there is still a lack of process monitoring for liquid-assisted laser cleaning. Jiangsu University [143] provides a monitoring device and method for laser wet cleaning of carbon fiber resin-based composite materials. Select the appropriate lens according to the different surface structures to be cleaned, and add a water quality sensor to monitor the change of impurity content in the water to judge whether the cleaning is completed. It solves the shortage of process monitoring research of wet laser cleaning. Changzhou Campus of Hohai University [144] proposed a laser cleaning device and method for aluminum alloy, which used liquid nitrogen for pre-cooling to avoid the change of the original internal structure caused by excessive heat accumulation in the base material. This cleaning device can reduce the heat accumulation on the surface of aluminum alloy by laser cleaning on the premise of ensuring the cleaning effect.

For composite materials, it is sensitive to laser, and the possibility of ablation damage is great. Therefore, laser cleaning technology should choose appropriate cleaning process according to different situations. While improving the cleaning efficiency and cleaning quality, the excessive thermal impact and impact stress caused by laser irradiation on the material surface are accurately controlled to achieve the integrity and stability of the material surface after cleaning.

### *1.6.2 Surface Functionalization of Laser Cleaning*

Some researchers will laser cleaning and laser shock reinforcement combination, the use of pulse laser irradiation on the surface of the matrix, so that the matrix surface pollutants ablation, gasification, phase explosion and other physical processes, to achieve the purpose of cleaning at the same time, the phase explosion induced plasma shock wave to modify the surface, improve its comprehensive performance (Fig. 1.12b). China Aviation Manufacturing Technology Research Institute [145] has developed a compound processing method for laser cleaning and strengthening of aluminum alloy materials. By adopting laser cleaning parameters with short pulse width of 10–20 ns and power density of 20–100 $GW/cm^2$, the metal surface dual effects of removing corrosion layer and improving fatigue resistance are realized. Shenzhen Institute of Information Technology [146] discloses a compound processing equipment and method for laser cleaning and laser shock strengthening. First, lasers with wavelengths of 1064 and 532 nm are used for cleaning, and then 355 nm ultraviolet laser with high single photon energy is used to shock strengthen the surface of the workpiece, so that its properties such as hardness, fatigue resistance, wear and stress corrosion are strengthened. Shenyang Institute of Automation Chinese Academy of Sciences [147] put forward a compound method of laser cleaning and strengthening for plate heat exchanger, which used two laser beams with different wavelengths and pulse widths to clean the sample surface first and then strengthen it, thus realizing the integration of cleaning and strengthening, and solving the technical problem of single function of traditional laser.

The hydrophobic paint layer can't be attached effectively because the metal surface is too smooth after cleaning. Harbin Institute of Technology [148] puts forward a laser cleaning-texturing compound machining method, which puts forward two schemes: the first one is to use a gun head to divide a laser beam into two beams, which are used for laser cleaning and laser texturing respectively; the second is to directly use two lasers to generate two lasers, which are cleaned in front and roughened in the back at the same time, thus achieving the purpose of simultaneous implementation of cleaning and strengthening. Compared with painting on metal surface to achieve hydrophobicity, Suzhou Xingbo Laser Technology Co., Ltd. [149] proposed a one-step preparation method for femtosecond laser cleaning-superhydrophobic micro-nano surface of aluminum alloy. The patent through the femtosecond laser low power cleaning, and then increase the power etching way to prepare the micro-nano texture with ultra-hydrophobicity small scale, then the hydrophobic effect is achieved on the metal surface; by controlling the laser power, cleaning and etching are combined, and the whole process is connected with the information speed, which can well avoid the re-oxidation after cleaning.

Armored Force Academy of the Chinese People's Liberation Army [150] proposed a method of cleaning stainless steel surface with pulsed laser to form micro-nano structure layer. In the process of cleaning stainless steel surface paint layer with pulsed laser, the micro-nano structure layer was formed on stainless steel surface by using the impact effect and thermal effect of laser, which improved the deformation resistance of stainless steel surface and could be expanded as a new functional material. Taiyuan University of Science and Technology [151, 152] invented a magnetic pulse welding method for lap joints of Mg/Ti alloy plates and a method for preparing magnesium/titanium layered wave interface composite materials based on a rolling method. In both patents, laser is used to clean the welded joint of Mg alloy plate and Ti alloy plate before welding, and micro-texture is prepared while removing the oxide film on the surface. In the patent example, the interface between Mg alloy plate and Ti alloy plate will form a wave-like mechanical interlocking bonding mode after welding, which greatly improves the bonding strength of the welding interface.

By analyzing the above patents, we can find that laser cleaning technology can not only effectively clean, but also improve the properties of metal materials through impact strengthening. Preparing the corresponding interface micro-texture at the same time of laser cleaning can realize the functionalization of the substrate surface after cleaning, and better meet the requirements of industrial production for multi-functional special surfaces.

### *1.6.3 Composite Laser Cleaning*

The existing laser cleaning equipment generally uses a single wavelength continuous laser or pulsed laser, which uses the difference of different laser absorption rates between the substrate and the coating to decompose or evaporate the pollutant layer from the surface of the substrate [153]. When faced with pollutants with complex components, the cleaning effect of a single type of laser is limited. Therefore, in order to improve the cleaning efficiency some researchers began to consider the combination of multiple lasers and cooperative cleaning (Fig. 1.12c) [154].

Experience shows that the nanosecond pulse laser mainly relies on the thermal expansion vibration effect of the contaminant particles and the substrate to achieve cleaning, so it is suitable for cleaning large particle pollutants; on the other hand, picosecond pulse laser has high instantaneous energy, and the energy absorbed by pollutant particles is ablated and gasified before it is conducted to the substrate by heat conduction, so the removal effect of small particles is the best. Xiamen University of Technology [155] invented an intelligent dual-beam laser cleaning device and its use method. The light sources of nanosecond laser and picosecond laser are overlapped, and the advantages of two lasers with different pulse widths are utilized to realize complementary advantages and disadvantages, which can be applied to cleaning pollutant particles with different sizes and improve cleaning efficiency and quality.

According to the different working conditions of laser, it can be divided into continuous laser and pulsed laser. Under the continuous injection of continuous laser, it is easy to cause matrix damage due to heat accumulation. Pulsed laser has high instantaneous power but low overall heat input, so it is usually cleaned by multi-point lap, and the cleaning efficiency is worse than that of continuous laser cleaning. Jiangsu University [156] adopts the method of laser cleaning to remove paint, which includes fast scanning cleaning with continuous laser and finishing with pulsed laser, thus avoiding the damage of continuous laser to the substrate and overcoming the problem of slow cleaning efficiency with pulsed laser. Wuhan Raycus Fiber Laser Technologies Co., Ltd. [157] invented a new type of laser cleaning device, which coupled the continuous laser beam with the pulsed laser beam through the beam combiner, Not only significantly improved the cleaning efficiency, but also had good rust and paint removal effects. China Aviation Manufacturing Technology Research Institute [158] disclosed a dual-light source composite laser cleaning method and device, which used the characteristics of pulsed laser single point peak power to destroy the coating at multiple points. And then through continuous laser continuous output energy to achieve rapid cleaning; in addition, a multi-mode fiber laser for laser cleaning has been developed [159], which can switch between continuous laser and pulsed laser through seed light module. According to the specific cleaning requirements, the light mode can be adjusted to achieve a variety of continuous laser, full pulse laser and continuous pulse alternating cleaning modes, which can meet the needs of different substrate cleaning thresholds and adapt to a wide range.

In order to efficiently remove pollutants without damaging the radar cover substrate, Beijing University of Technology [160] discloses a laser cleaning system and method which can synchronously output dual-wavelength lasers, and uses the difference of absorption rate of different paint layers to remove the laser with different wavelengths in layers. Suzhou Usiland Optronics Co., Ltd. [161] discloses a dual-wavelength laser cleaning machine for cultural relics, which is compatible with dual-wavelength output of 1064 and 532 nm. Compared with single-wavelength laser cleaning equipment, the cleaning color range is wider. Wenzhou Polytechnic [162] disclosed a desktop laser precise cleaning device with dual-wavelength composite energy distribution. This device divided a Gaussian laser beam into a Bessel ring laser beam and a short-wavelength laser beam, and then coupled the Bessel ring laser beam and the short-wavelength Gaussian laser beam to form a dual-wavelength composite beam, which improved the cleaning accuracy and uniformity. The above patents combine different wavelengths of laser without the need to cooperate between the two kinds of lasers, which makes the laser cleaning process more convenient and efficient.

It can be seen that efficient and accurate cleaning can be achieved by using lasers with different wavelengths to clean different components in composite materials, or combining pulsed laser with continuous laser to complement each other, which provides a new idea for the subsequent combination of different lasers.

To sum up, the current laser cleaning and strengthening composite technology makes full use of the characteristics of different laser sources, but the complex structure of optical system and high processing cost restrict the subsequent development. When using lasers with different wavelengths and different pulse widths, the laser power density is quite different, which is difficult to control accurately and easy to cause matrix damage. Therefore, if the liquid-assisted laser cleaning, monitoring device and surface modification are combined under real-time monitoring, the corrosion resistance and wear resistance of the material surface can be improved by laser shock strengthening technology while avoiding the thermal ablation damage of the substrate, which will greatly expand the application field and scope of the equipment, which is worthy of further study.

## 1.7 Conclusion

From the perspective of patent application, this paper makes a systematic analysis of the patents in the field of laser cleaning in China. At present, the number of laser cleaning patents shows a straight upward trend, and the industrial layout has been basically formed. However, there are still some problems in transforming the intellectual property of laser cleaning into engineering application, such as imperfect mechanism and poor universality, which is an urgent problem to be solved in the field of laser cleaning in China. Therefore, strengthening the cooperation between enterprises and universities, intellectual property innovation incentive and special policy support are effective measures to promote the high-quality development of laser cleaning production, studying, research and application. Laser cleaning technology has been highly concerned by the government, universities (Scientific Research Institutes) and enterprises, showing a momentum of vigorous development.

1. Laser cleaning can meet the cleaning requirements of different substrates such as iron, aluminum, titanium and composite materials, and has the ability to effectively clean corrosion layers, paint layers and carbon deposits. Laser cleaning equipment is developing in a more convenient and lightweight direction, the compatibility between terminal equipment and cleaning equipment has been greatly improved, online monitoring means are more abundant and intelligent cleaning has gradually become the mainstream. The precise control of laser cleaning process and laser compound cleaning method have great potential in maintaining the surface integrity of substrate and realizing surface functionalization.
2. There are still many areas to be improved in the field of laser cleaning. For example, off-line monitoring methods are generally poor in timeliness, and it is

difficult to adjust the cleaning scheme according to the cleaning effect in time, and it is difficult to control the cleaning threshold of pollutants and the damage threshold of matrix, so it is urgent to develop real-time monitoring equipment and means. Laser cleaning with high energy density is easy to cause damage to the substrate due to heat accumulation. Liquid-assisted laser cleaning and liquid nitrogen (precooled gas) assisted laser cleaning should be adopted to avoid secondary damage to the substrate. We should focus on the research and development of pipe fittings, precision instruments and large components to expand the application field of laser cleaning.

3. Reducing cost, increasing efficiency and improving quality of laser cleaning are the direction of continuous efforts in the future. Laser compound cleaning, intelligent cleaning and "washing-repairing" integration are expected to become the research focus in the field of laser cleaning. It can be predicted that the related patent application will become a hot spot in the industrial field, thus promoting the laser cleaning technology to enter a new stage of faster development. In addition, as a green cleaning technology, laser cleaning will play an increasingly important role in China's "carbon peaking and carbon neutrality" strategy, but also contribute to the global sustainable development.

## References


1. Guodong Z, Zhenhai X, Yang J, Chen X, Yang L, Xu J, Shan D, Chen Y, Guo B. Mechanism and application of laser cleaning: A review [J]. Optics and Lasers in Engineering, 2022, 157.
2. Nie J, Zhang H, Zhang D, Xu J, Zhang J, Shan D, Guo Bin. Removal mechanism of laser cleaning for inorganic thermal control coatings on aluminum alloys [J]. Applied Surface Science, 2023, 633: 157578.
3. Shi T, Wang C, Mi G, Yan F. A study of microstructure and mechanical properties of aluminum alloy using laser cleaning [J]. Journal of Manufacturing Processes, 2019, 42: 60-66.
4. Zhang F D, Liu H, Suebka C, Liu Y X, Liu Z, Guo W, Cheng Y M, Zhang S L, Li L. Corrosion behaviour of laser-cleaned AA7024 aluminium alloy [J]. Applied Surface Science, 2018, 435: 452-461.
5. Wang W, Li X, Liu W, Xing F, Wang J, Zhang K. Experimental study on hydrophobic properties and corrosivity of laser cleaned 7075 aluminum alloy anodized film surface [J]. Optics & Laser Technology, 2023, 166: 109615.
6. Maingi E M, Alonso M P, de la Fuente G F, Dubernet S, Lefrais Y, Chapoulie R, Vally E, Angurel LA. UV femtosecond laser cleaning of encrusted historical stained-glasses [J]. Journal of Cultural Heritage, 2023, 61: 100–108.
7. Ye Y, Jia B, Chen J, Jiang L, Tang H, Wang H, Luan X, Liao W, Zhang C, Yao C. Laser cleaning of the contaminations on the surface of tire mould [J]. International Journal of Modern Physics B, 2017, 31(16–19): 1744100.
8. Zhu G, Zhang D, Zhang L, Xu J, Shan D, Guo B. Environmentally friendly photothermal reconstruction of cement/asphalt road: Can road markings also be cleaned in such unexpected ways? [J]. Journal of Cleaner Production, 2024, 434: 140347.
9. Zhu G, Wang S, Cheng W, Ren Y, Wen D. Corrosion and wear performance of aircraft skin after laser cleaning [J]. Optics & Laser Technology, 2020, 132: 106475.

10. Liu Y, Liu W, Zhang D, Tian Z, Sun X, Wei Z. Experimental investigations into cleaning mechanism of ship shell plant surface involved in dry laser cleaning by controlling laser power [J]. Applied Physics A, 2020, 126: 1–17.
11. Zhou Z, Sun W, Wu J, Chen H, Zhang F, Wang S. The Fundamental Mechanisms of Laser Cleaning Technology and Its Typical Applications in Industry [J]. Processes, 2023, 11(5).
12. Zhang D, Xu J, Li Z, Jin Y, Su X, Shan D, Guo B. Removal mechanisms of nanosecond pulsed laser cleaning of blue and red polyurethane paint [J]. Applied Physics A, 2022, 128(2): 170.
13. Li Y, Wu Z, Chu D, Yang H, Deng G, Zhou S. Research progress of laser cleaning and monitoring technology (Invited) [J]. Infrared and Laser Engineering, 2023, 52 (2): 20220784-1–20220784-19.
14. Zhu G, Zhang D, Li Z, Su X, Jin Y, Xu J, Shan D, Guo B. Research Progress and Challenges of Laser Cleaning Technology(Invited) [J]. Chinese Journal of Lasers, 2024, 51(4):97–120.
15. Ouyang J, Mativenga P, Goffin N, Liu W, Liu Z, Mirhosseini N, Jones L, Woolley E, Li L. Energy consumption and performance optimisation of laser cleaning for coating removal [J]. CIRP Journal of Manufacturing Science and Technology, 2022, 37: 245–257.
16. Hou L, Yin F, Wang S, Sun J, Yin H. A review of thermal effects and substrate damage control in laser cleaning [J]. Optics & Laser Technology, 2024, 174: 110613.
17. Cauldron Co., Ltd. Chinese patent, CN1125917A; HK1013270A, 1994.
18. University of Missouri. Chinese patent, CN1317155A, 1999.
19. The Institute of Engineering Thermophysics of the Chinese Academy of Sciences. Chinese patent, CN2783534Y, 2004.
20. Tsinghua University. Chinese patent, CN101219430A, 2008.
21. Laser Fusion Research Center of China Academy of Engineering Physics. Chinese patent, CN102327885A, 2010.
22. Jiangsu University. Chinese patent, CN103063167A, 2012.
23. Jiangsu University. Chinese patent, CN103090969A, 2012.
24. Jiangsu University. Chinese patent, CN103100537A, 2012.
25. Henan Sancheng Mechanical and Electrical Equipment Co., Ltd.; China Railway Design Group; Feitai Communications Technology Co., Ltd. Chinese patent, CN216338780U, 2021.
26. Harbin Wick Rail Transit Technology Development Co., Ltd. Chinese patent, CN217775034U, 2022.
27. Hebei University of Technology. Chinese patent, CN110899253A, 2019.
28. Shenyang University of Technology. Chinese patent, CN113910507A, 2021.
29. Hangzhou Innovation Research Institute of Beihang University. Chinese patent, CN116689408A, 2023.
30. Hubei Zhengxin Pipe Fitting Co., Ltd. Chinese patent, CN218110180U, 2022.
31. Xinxing Cast Pipe Co., Ltd. Chinese patent, CN207272437U, 2017.
32. Sichuan University. Chinese patent, CN106001928A, 2016.
33. Wuhan WISCO Huagong Laser Large Equipment Co., Ltd. Chinese patent, CN210676213U, 2019.
34. Armored Force Academy of the Chinese People's Liberation Army. Chinese patent, CN110508563A, 2019.
35. Dongguan Ruiyong Hardware Products Co., Ltd. Chinese patent, CN210936254U, 2019.
36. Hunan Vocational College of Railway Technology. Chinese patent, CN116727367A, 2023.
37. CRRC Nanjing Puzhen Co., Ltd., Chinese patent, CN110434124A, 2019.
38. Laser Institute of Shandong Academy of Sciences. Chinese patent, CN112548477A, 2020.
39. Wuhan Xiangming Laser Technology Co., Ltd. Chinese patent, CN211305212U, 2019.
40. Wuhan Xiangming Laser Technology Co., Ltd. Chinese patent, CN110976438A, 2019.
41. Laser Institute of Shandong Academy of Sciences. Chinese patent, CN111496430A, 2020.
42. CRRC Nanjing Puzhen Co., Ltd., Chinese patent, CN110434502A, 2019.
43. China Aviation Manufacturing Technology Research Institute. Chinese patent, CN115055447A, 2022.

44. Shenyang Heshitai Tongyong Titanium Industry Co., Ltd. Chinese patent, CN214348352U, 2020.
45. China Aviation Manufacturing Technology Research Institute. Chinese patent, CN115178883A, 2022.
46. Chengdu Aircraft Industry (Group) Co., Ltd.; Wuhan Xiangming Laser Technology Co., Ltd. Chinese patent, CN112371653A, 2020.
47. Chengdu Mairuijie Laser Technology Co., Ltd. Chinese patent, CN214441505U, 2021.
48. Wuhan Tianyu Intelligent Manufacturing Co., Ltd. Chinese patent, CN211028501U, 2019.
49. Beijing Xinghang Electromechanical Equipment Co., Ltd. Chinese patent, CN110385517A, 2019.
50. Hunan Xinao Photoelectric Technology Co., Ltd. Chinese patent, CN108787636A, 2018.
51. Minglei Laser Intelligent Equipment (Heyuan) Co., Ltd. Chinese patent, CN216225902U, 2021.
52. Nanchang Hangkong University. Chinese patent, CN110016807A, 2019.
53. Jiangsu University. Chinese patent, CN110586579A, 2019.
54. Nanjing University of Aeronautics and Astronautics. Chinese patent, CN114985939A, 2022.
55. Jiangsu University. Chinese patent, CN112090866A, 2020.
56. Shenyang Institute of Automation Chinese Academy of Sciences. Chinese patent, CN112380677A, 2020.
57. Song Y, Zhang T, Fan W, Zhang Y, Yang W, Wang A. Effect of the overlap ratio on surface properties of 7B04 aluminum alloy for aviation during laser derusting [J]. Journal of Materials Research and Technology, 2022, 20.
58. Wang Z, Zeng X, Huang W. Parameters and surface performance of laser removal of rust layer on A3 steel [J]. Surface and Coatings Technology, 2003, 166.
59. He Y, Cai X, Ye J. Research on pulse laser cleaning and rust removal technology in power systems. AIP Advances 1 September 2023; 13 (9): 095101.
60. Sun Q, Zhou J, Meng X, Li P. Mechanical properties and microstructure characteristics of 2024-T351 aluminum alloy specimen subjected to paint removal by laser cleaning [J]. Vacuum, 2023, 211.
61. Nankai University. Chinese patent, CN101143364A, 2007.
62. Jiangsu University. Chinese patent, CN109226102A, 2018.
63. Baoyu (Wuhan) Laser Technology Co., Ltd. Chinese patent, CN116651863A, 2023.
64. Qingdao Campus of Naval Aeronautical University. Chinese patent, CN115837380A, 2022.
65. Dongguan Eontec Co., Ltd.; Dongguan Yihao Metal Material Technology Co., Ltd.; Zhuzhou Yi Anxin Material R&D Co., Ltd. Chinese patent, CN111185441A, 2021.
66. Shenhua Group Zhungeer Energy Sources Co., Ltd. Chinese patent, CN117000695A, 2023.
67. Sun Q, Zhou J, Meng X, Yang J, Guo Z, Zhu M, Guo S. Mechanism and threshold fluence of nanosecond pulsed laser paint removal [J]. Rare Metals, 2022, 41(03): 1022–1031.
68. Minglei Laser Intelligent Equipment (Heyuan) Co., Ltd. Chinese patent, CN112827947A, 2021.
69. Sichuan University. Chinese patent, CN113070287A, 2021.
70. Shanghai Ruirong Laser Welding Technology Co., Ltd.; China hi-tech technology Co., Ltd. Chinese patent, CN212121073U, 2020.
71. Durr Coating System Engineering (Shanghai) Co., Ltd. Chinese patent, CN212494343U, 2020.
72. China Hi-Tech Technology Co., Ltd. Chinese patent, CN213384777U, 2020.
73. Sichuan Minghang Lander Technology Co., Ltd. Chinese patent, CN217571320U, 2022.
74. Hebei University of Technology. Chinese patent, CN114713570A, 2022.
75. Nanchang Hangkong University. Chinese patent, CN116060384A, 2023.
76. Hebei Hanguang Heavy Industry Co., Ltd. Chinese patent, CN113695315A, 2021.
77. Harbin Wick Rail Transit Technology Development Co., Ltd. Chinese patent, CN218610797U, 2022.
78. Yingtan Zhihui Internet of Things Application Research Institute Co., Ltd. Chinese patent, CN112404043A, 2020.

79. Shandong Zhongshi Yitong Group Co., Ltd. Chinese patent, CN113042463A, 2021.
80. Jiangsu University. Chinese patent, CN114054272A, 2021.
81. Suzhou Jinghai Laser Intelligent Technology Co., Ltd. Chinese patent, CN217513073U, 2022.
82. Hunan Xinao Photoelectric Technology Co., Ltd. Chinese patent, CN217748471U, 2022.
83. State-Owned Wuhu Machinery Factory. Chinese patent, CN113042463A, 2023.
84. Laser Institute of Shandong Academy of Sciences. Chinese patent, CN111167804A, 2020.
85. Wuhan Jindun Laser Technology Co., Ltd. Chinese patent, CN111940423A, 2020.
86. Zhejiang University of Technology. Chinese patent, CN112410847A, 2020.
87. Nanchang Hangkong University. Chinese patent, CN115945465A, 2023.
88. Tianjin University. Chinese patent, CN116984314A, 2023.
89. Laser Fusion Research Center of China Academy of Engineering Physics. Chinese patent, CN109821820A, 2019.
90. ShanDong CharmRay Laser Technology Co., Ltd. Chinese patent, CN115742312A, 2022.
91. SOL Electronic Technology Co., Ltd. Chinese patent, CN115430665A, 2022.
92. Dongguan Feichuang Laser Intelligent Equipment Co., Ltd. Chinese patent, CN115950907A, 2023.
93. Xi'an Lanxiang New Materials Technology Co., Ltd. Chinese patent, CN213194883U, 2020.
94. Armored Force Academy of the Chinese People's Liberation Army. Chinese patent, CN116899981A, 2023.
95. Shengtong Intelligent Machinery Equipment (Shanghai) Co., Ltd. Chinese patent, CN115283365A, 2022.
96. Beijing University of Technology. Chinese patent, CN115889340A, 2022.
97. Wuhan Xiangming Laser Technology Co., Ltd. Chinese patent, CN116551120A, 2023.
98. Beijing University of Technology; Laser Institute of Shandong Academy of Sciences; Shandong chan Yan Qiang Yuan Laser Technology Co., Ltd. Chinese patent, CN116727862A, 2023.
99. National University of Defense Technology. Chinese patent, CN110523715A, 2019.
100. Konfoong Materials International Co., Ltd. Chinese patent, CN104588810A, 2013.
101. Aifaco Electronic Materials (Suzhou) Co., Ltd. Chinese patent, CN112475505A, 2020.
102. Shanghai Chip Electronic Technology Co., Ltd. Chinese patent, CN219683471U, 2023.
103. Borunte Robot Co., Ltd. Chinese patent, CN116586771A, 2023.
104. Sichuan Aerospace Changzheng Equipment Manufacturing Co., Ltd. Chinese patent, CN112453710A, 2020.
105. Dongfang Electric Group Dongfang Steam Turbine Co., Ltd. Chinese patent, CN114289832A, 2021.
106. Huazhong University of Science and Technology. Chinese patent, CN113070572A, 2021.
107. Shenyang University of Technology. Chinese patent, CN116673272A, 2023.
108. Shenzhen Hydrolaser Technology Limited. Chinese patent, CN215543383U, 2021.
109. Baoyu (Wuhan) Laser Technology Co., Ltd. Chinese patent, CN218873124U, 2022.
110. Jiangsu Guoyuan Laser Intelligent Equipment Manufacturing Co., Ltd. Chinese patent, CN115889343A, 2022.
111. Nanjing University of Aeronautics and Astronautics. Chinese patent, CN116651856A, 2023.
112. Baoyu (Wuhan) Laser Technology Co., Ltd. Chinese patent, CN219758591U, 2023.
113. Wuxiang county Hong Chen Wan Ju Environmental Protection Technology Co., Ltd. Chinese patent, CN118003443A, 2024.
114. China Energy Longyuan Environmental Protection Co., Ltd. Chinese patent. CN117505407A, 2024.
115. South China University of Technology. Chinese patent, CN117888101A, 2023.
116. Shanghai Yingfeng Electronic Technology Co., Ltd. Chinese patent, CN117583741A, 2023.
117. Kunming Sachuang Photoelectricity Technology Co., Ltd. Chinese patent, CN116689405A, 2023.
118. China Yangtze Power Co., Ltd. Chinese patent, CN116765613A, 2023.

119. China Construction Third Division First Construction Engineering Co., Ltd. Chinese patent, CN116441244A, 2023.
120. Fanachi Laser Technology (Shandong) Co., Ltd. Chinese patent, CN219169124U, 2023.
121. China Aviation Xi'an Aircraft Industry Group Co., Ltd. Chinese patent, CN115889341A, 2022.
122. Suzhou Xiangming Laser Technology Co., Ltd. Chinese patent, CN219899450U, 2023.
123. Shandong Zotye Construction Technology Co., Ltd. Chinese patent, CN216679378U, 2022.
124. Guangdong Longyue Construction Engineering Co., Ltd. Chinese patent, CN218574494U, 2022.
125. Liu J. research on intelligent laser derusting system based on machine vision[D]. Suzhou University, 2020. DOI:https://doi.org/10.27351/d.cnki.gszhu.2020.000751.
126. Wen J, Sun L, Lu Y, Qi L, Wang W, Qiao H. Laser induced breakdown spectroscopy online monitoring of laser cleaning quality on carbon fiber reinforced plastic [J]. Optics and Laser Technology, 2022, 145.
127. Hebei University of Science and Technology; Institute of Remanufacturing Industry Technology, Jing-Jin-Ji; Shanghai Xinfumei Gearbox Technology Service Co., Ltd. Chinese patent, CN115979588A, 2023.
128. Dongguan Feichuang Laser Intelligent Equipment Co., Ltd. Chinese patent, CN115945462A, 2022.
129. Nanjing University of Aeronautics and Astronautics. Chinese patent, CN116441247A, 2023.
130. CRRC Nanjing Puzhen Co., Ltd., Chinese patent, CN111951257A, 2020.
131. Li X, Guan Y. Real-time monitoring of laser cleaning for hot-rolled stainless steel by laser-induced breakdown spectroscopy [J]. Metals, 2021, 11(5): 790.
132. Civil Aviation Flight University of China. Chinese patent, CN114367498A, 2021.
133. Dongguan Feichuang Laser Intelligent Equipment Co., Ltd. Chinese patent, CN116183583A, 2022.
134. Civil Aviation Flight University of China. Chinese patent, CN115839944A, 2023.
135. Beijing University of Technology; Institute of Semiconductors Chinese Academy of Sciences. Chinese patent, CN116371825A, 2023.
136. Shanghai Hangyi High-Tech Development Research Institute Co., Ltd. Chinese patent, CN114119535A, 2021.
137. Li Y, Wu Z, Chu D, Yang H, Deng G, Zhou S. Research progress of laser cleaning and monitoring technology [J]. Infrared and laser engineering, 2023, 52(2): 20220784-1–20220784-19.
138. China Aviation Manufacturing Technology Research Institute. Chinese patent, CN115532737A, 2022.
139. Armored Force Academy of the Chinese People's Liberation Army. Chinese patent, CN112044872A, 2020.
140. Qilu University of Technology (Shandong Academy of Sciences). Chinese patent, CN116644525A, 2023.
141. Harbin Institute of Technology. Chinese patent, CN110586580A, 2019.
142. Nankai University. Chinese patent, CN112605067A, 2020.
143. Jiangsu University. Chinese patent, CN112077053A, 2020.
144. Changzhou Campus of Hohai University. Chinese patent, CN111672832A, 2020.
145. China Aviation Manufacturing Technology Research Institute. Chinese patent, CN116162937A, 2023.
146. Shenzhen Institute of Information Technology. Chinese patent, CN112317458A, 2020.
147. Shenyang Institute of Automation Chinese Academy of Sciences. Chinese patent, CN114505297A, 2020.
148. Harbin Institute of Technology. Chinese patent, CN109967881A, 2019.
149. Suzhou Xingbo Laser Technology Co., Ltd. Chinese patent, CN113059324A, 2021.
150. Armored Force Academy of the Chinese People's Liberation Army. Chinese patent, CN111979557A, 2020.
151. Taiyuan University of Science and Technology. Chinese patent, CN115026402A, 2022.

152. Taiyuan University of Science and Technology. Chinese patent, CN115026129A, 2022.
153. Shan T, Wang S, Yin F, Qiao Y, Liu P. Typical applications of laser cleaning and effects on matrix surface integrity [J]. Material guide, 2021, 35(11):11164–11173.
154. Chu W, Zeng B, Li Z, Yao J, Xie H, Li G, Wang Z, Cheng Y. Range extension in laser-induced breakdown spectroscopy using femtosecond–nanosecond dual-beam laser system [J]. Applied Physics B, 2017, 123(6).
155. Xiamen University of Technology. Chinese patent, CN116511167A, 2023.
156. Jiangsu University. Chinese patent, CN114653686A, 2022.
157. Wuhan Raycus Fiber Laser Technologies Co., Ltd. Chinese patent, CN209680715U, 2019.
158. China Aviation Manufacturing Technology Research Institute. Chinese patent, CN115958020A, 2023.
159. China Aviation Manufacturing Technology Research Institute. Chinese patent, CN115395358A, 2022.
160. Beijing University of Technology. Chinese patent, CN116571512A, 2023.
161. Henan Museum; Suzhou Usiland Optronics Co., Ltd. Chinese patent, CN213367026U, 2020.
162. Wenzhou Polytechnic. Chinese patent, CN107297365A, 2017.

# Chapter 2
# Typical Applications of Laser Cleaning and Its Impact on Substrate Surface Integrity

## 2.1 Introduction

With the rapid development of industrial society, national science and technology, culture, military, aerospace and ocean exploration have also developed rapidly, which puts forward higher requirements for the surface quality of aircraft, ships, military weapons, various precision instruments and other equipment. The surfaces of these equipment need to be cleaned and maintained regularly. The removal of the old paint layer on the surface of aircraft, the cleaning of marine microorganisms on the surface of ships, the cleaning of carbon deposits inside engines, the cleaning of weapons and the removal of micron-sized particles on the surface of precision instruments all have extremely high requirements on the surface quality after cleaning, so the demand for high-quality cleaning technology is increasing [1].

Traditional cleaning technologies mainly include mechanical cleaning, chemical cleaning, ultrasonic cleaning and high-pressure water jet cleaning [2]. Mechanical cleaning mainly adopts manual polishing, scraping and wiping, which is not only time-consuming, labor-intensive and inefficient, but also has a low surface accuracy after cleaning [3]. The biggest problem faced by chemical cleaning is environmental protection, which will pollute the environment. At the same time, chemical cleaning agents with different components will inevitably be mixed for different pollutants, which greatly limits the application of chemical cleaning [4]. Ultrasonic cleaning can effectively avoid the problems of low cleaning efficiency, environmental pollution and poor accuracy, but this cleaning method is basically powerless to micron and submicron particles, and cannot meet the required accuracy requirements for cleaning some precision instruments [5]. The most common raw material for high-pressure water jet cleaning is water. This cleaning method is low in cost and will not pollute the environment, but it depends on the pressure control of the water pump. If the pressure is too small, the pollutants cannot be completely removed. If the pressure is too high, it will damage the substrate to be cleaned, and some substrates that are not waterproof cannot be cleaned at all [6]. It can be seen that the traditional



S. Wang, *Laser Cleaning*, https://doi.org/10.1007/978-981-95-2935-3_2

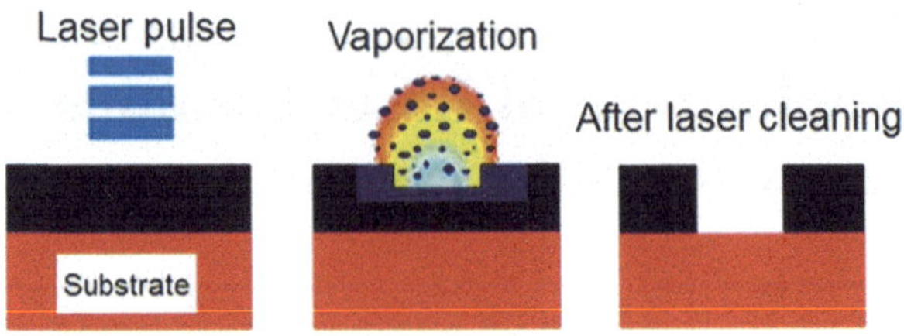


**Fig. 2.1** Schematic diagram of laser cleaning process [9]

cleaning methods cannot effectively guarantee the cleaning quality and surface integrity due to technical limitations. In this case, a new cleaning technology is urgently needed to solve this series of problems, and the development of laser cleaning technology has made scholars all over the world see the dawn.

Laser cleaning is to irradiate a laser beam with high energy density on the surface of the object to be cleaned, and make the pollutants on the surface of the object fall off through the effects of vibration, decomposition and gasification, so as to achieve the cleaning effect. Laser cleaning will produce cleaning effect only when it reaches the cleaning threshold, and it can't exceed the damage threshold, otherwise it will damage the substrate, destroy the surface integrity of the substrate, and the surface performance will also decline and fail to meet the use requirements [7, 8]. A typical schematic diagram of laser cleaning process is shown in Fig. 2.1 [9].

Laser cleaning is a new surface cleaning technology, which has the advantages of environmental protection, non-contact, wide application range and high cleaning accuracy [9, 10]. It has good cleaning effect in paint removal [11], rust removal and compound [12], micron and submicron particles removal [13] and cultural relics protection [14], and is widely used in aerospace, ships, automobiles and precision instruments. At present, foreign laser cleaning technology is very mature and can be industrialized in many fields, while domestic laser cleaning technology started late and developed rapidly, and has made remarkable application progress. However, there is little research on the surface state of the substrate after laser cleaning, which limits the application transformation of this technology to the field of high-end equipment manufacturing and remanufacturing [15, 16].

## 2.2 Classification of Laser Cleaning

Laser methods can be divided into: dry laser cleaning [17], wet laser cleaning [18] and laser plasma shock wave cleaning [19]. Dry laser cleaning mainly uses laser to irradiate pollutants on the surface of the substrate, so that the absorbed energy of pollutant particles expands and leaves the surface of the substrate or the surface of the substrate expands after absorbing laser energy, overcoming the binding force between pollutant particles and the surface and achieving the purpose of removing pollutants [20]; wet laser cleaning is to coat a layer of liquid film (ethanol, acetone,

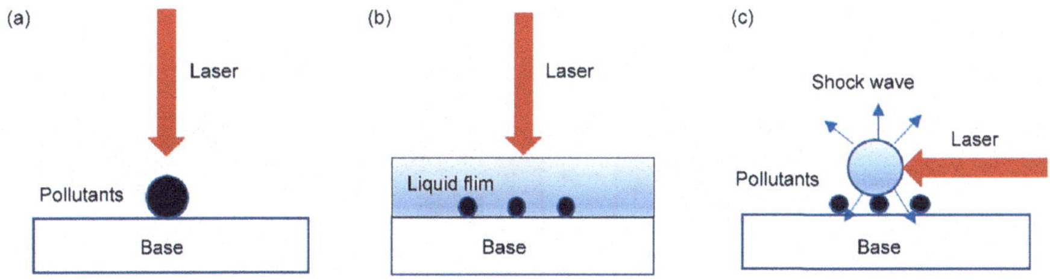


**Fig. 2.2** Schematic diagram of three laser cleaning methods. (**a**) Dry laser cleaning; (**b**) wet laser cleaning; (**c**) plasma shock wave laser cleaning [23]

etc.) on the surface of the substrate. After laser irradiation, the liquid film heats up sharply, melts and explodes, and the pollutants on the surface are removed by using the energy and shock wave generated by the explosion [21]; laser plasma shock wave cleaning is to focus the laser beam on the surface of the substrate, and use the high energy density of the laser beam to break through the air here to ionize it and generate plasma shock wave to remove pollutants [22]. Schematic diagrams of three laser cleaning methods are shown in Fig. 2.2 [23].

At present, the lasers used for laser cleaning mainly include excimer laser [24], $CO_2$ laser [25] and Nd:YAG laser [26]. When the precision of the surface to be cleaned is very high, excimer laser is preferred; for laser coating cleaning, $CO_2$ laser is the best. Nd:YAG laser is preferred for long-distance laser cleaning because it is transmitted by optical fiber. If scientists already know the nature of pollutants before cleaning, they can choose the appropriate laser according to the absorption of pollutants on the substrate surface to lasers with different wavelengths [27].

## 2.3 Mechanism of Laser Cleaning

Laser cleaning involves many mechanism effects, mainly vibration effect, ablation effect and plasma impact effect [28]. It is of great guiding significance to clarify which effect or effects are playing a major role in laser cleaning. Shi et al. [29] used a 1064 nm Q-switched Nd:YAG laser to clean a steel substrate with a thickness of 55 μm at a scanning speed of 0.55 mm/s. The experimental results showed that the cleaning threshold of the paint layer was 13.5 W and the damage threshold was 24.6 W. The temperature and stress distribution characteristics generated in the paint-substrate system under the action of laser pulses were analyzed by finite element method, and it was found that under the cleaning threshold condition. The main mechanism of laser cleaning is vibration effect. Under the condition of damage threshold, the mechanism of laser cleaning has both vibration effect and ablation effect, which may damage the environment.

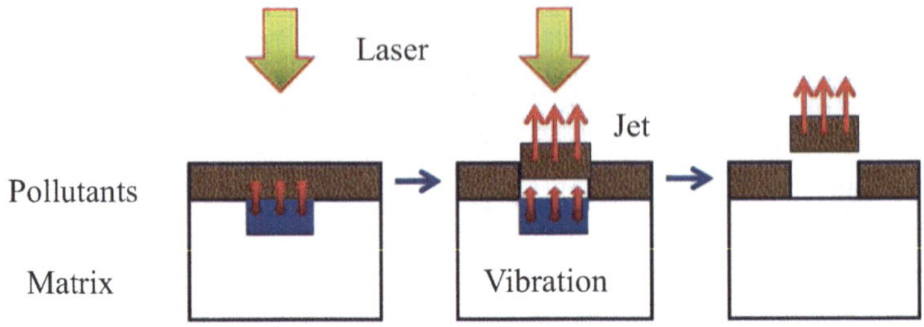


**Fig. 2.3** Schematic diagram of vibration effect during laser cleaning

### *2.3.1 Vibration Effect*

Vibration effect means that when laser irradiates the surface of pollutants, with the extension of irradiation time, not only the surface pollutants will absorb the laser energy, but also the substrate under the pollutants will absorb the laser energy. However, due to their different properties, the absorption rate of laser energy is different and the thermal expansion coefficient is different, so that a temperature gradient will be generated at the interface, which will induce an interfacial stress difference. When the stress difference is greater than the adhesion of pollutants to the substrate surface, the pollutants will be stripped from the substrate surface [30]. The schematic diagram of vibration effect is shown in Fig. 2.3.

### *2.3.2 Ablation Effect*

Ablation effect refers to a series of physical and chemical reactions, such as laser irradiation, interaction between the laser and the contaminated layer, laser energy concentration on the contaminated layer, rapid temperature rise of the contaminated layer, reaching the melting point or even boiling point of the contaminated layer, melting and even gasification of the contaminated layer, so that it can be separated from the surface of the substrate to achieve the cleaning effect [30]. The schematic diagram of ablation effect is shown in Fig. 2.4.

### *2.3.3 Plasma Shock Effect*

The plasma shock effect is that the focused laser beam with high energy density is irradiated to the air above the pollutants, and the gas absorbs the laser energy and ionizes to generate plasma. The plasma has a shielding effect, which prevents the pollutants from continuing to absorb the laser energy, and at the same time produces

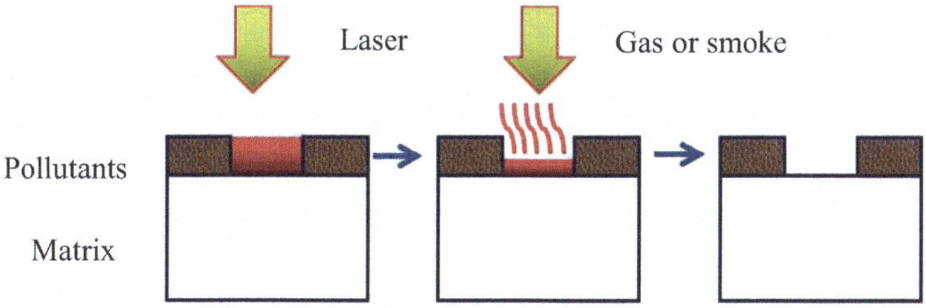


**Fig. 2.4** Schematic diagram of ablation effect during laser cleaning

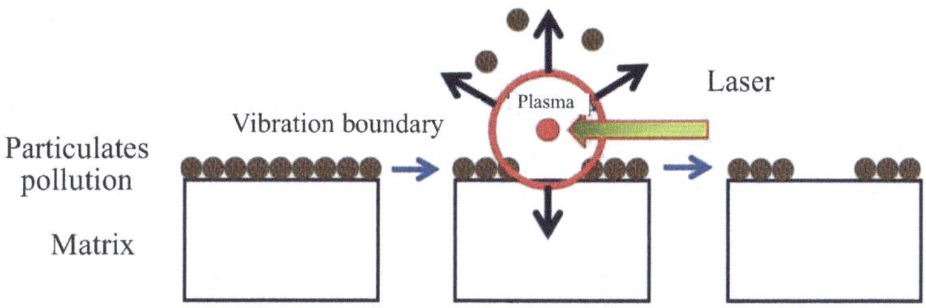


**Fig. 2.5** Schematic diagram of plasma shock effect during laser cleaning

a high-pressure shock wave, which causes the pollutants to vibrate and peel off [31]. The schematic diagram of plasma shock effect is shown in Fig. 2.5.

## 2.4 Typical Application of Laser Cleaning Technology

### *2.4.1 Laser Paint Removal*

Laser paint removal means that the thin layer of paint on the surface of the substrate absorbs laser irradiation. Because of the high laser energy, heat accumulation occurs on the surface of the paint layer instantly, and the temperature rises sharply, so that the surface paint layer dissolves and gasifies, thus peeling the paint layer from the surface of the substrate and achieving the paint removal effect [32, 33].

By optimizing the matching of cleaning parameters, it can be ensured that the paint layer can be removed accurately and effectively without causing any damage to the substrate. Brygo et al. [34] research shows that the greater the laser energy density, the higher the repetition frequency and the longer the pulse duration, the more obvious the ablation effect on the substrate surface will be. Chen et al. [35] used $CO_2$ laser to clean the 50 μm thick paint layer on the surface of low carbon steel, and cleaned it for five times at high power, and the paint layer was completely removed. However, the substrate was seriously damaged. First, most of the paint layer was removed by scanning with high power twice, and then the surface paint

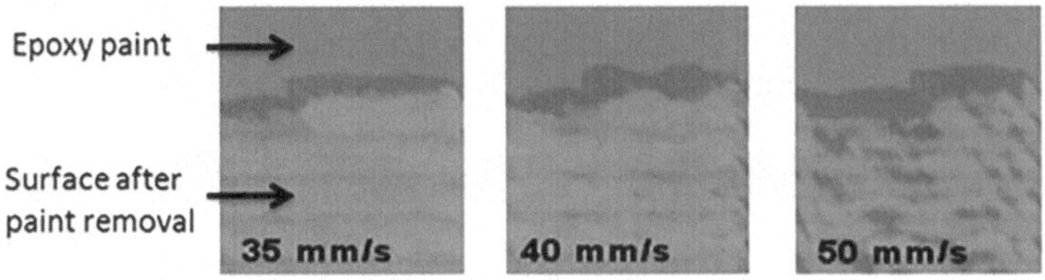


**Fig. 2.6** Surface morphology after paint removal at different scanning speeds [36]

layer was completely removed by cleaning with low power once. Under this cleaning condition, the substrate was not damaged. Kumar et al. [36] used $CO_2$ laser to clean the 40 μm thick epoxy paint layer on the surface of aluminum alloy plate. When the laser density was 7.2 J/cm$^2$ and the pulse frequency was 100 Hz, the scanning speeds of 35 mm/s, 40 mm/s and 50 mm/s were selected, respectively. It was found that the cleaning effect was the best when the scanning speed was 35 mm/s, and the surface paint layer was completely removed, while a certain amount of residual paint layer remained on the surface at the other two scanning speeds. Zheng et al. [37] laser energy density is 7.5 J/cm$^2$, scanning speed is 236 mm/s, pulse frequency is 45 Hz, and spot size is 15.5 mm × 8.2 mm, which is twice as high as dry paint removal efficiency.

In the actual laser paint removal operation, it is necessary to choose the appropriate cleaning process according to the actual situation. In some cases, the wet cleaning method is more worthy of choice. In addition, through the above research, it can be found that the higher the laser power, the more scanning times, the better the paint removal effect. Sometimes too high laser energy density cannot only remove the paint layer well, but even ablate the material, causing damage to the substrate. It is necessary to continuously adjust the process and optimize the laser cleaning parameter matching to obtain satisfactory results (Fig. 2.6).

### 2.4.2 *Laser Derusting*

Laser derusting means that the high-energy beam generated by laser irradiates the surface of the rust layer, so that the local temperature of the surface of the rust layer rises rapidly to above the gasification temperature, causing the rust layer substances to vaporize or the rust layer particles to absorb heat and expand. When this expansion force is greater than the bonding force between the rust layer and the substrate, the rust layer particles will leave the substrate, thus achieving the purpose of laser derusting [38]. The schematic diagram of laser derusting is shown in Fig. 2.7 [39].

The removal of metal surface rust has always been a difficult problem for scholars at home and abroad. In recent years, scholars all over the world have devoted themselves to the research of laser derusting. Compared with traditional derusting, this new derusting method can achieve the purpose of derusting without damaging

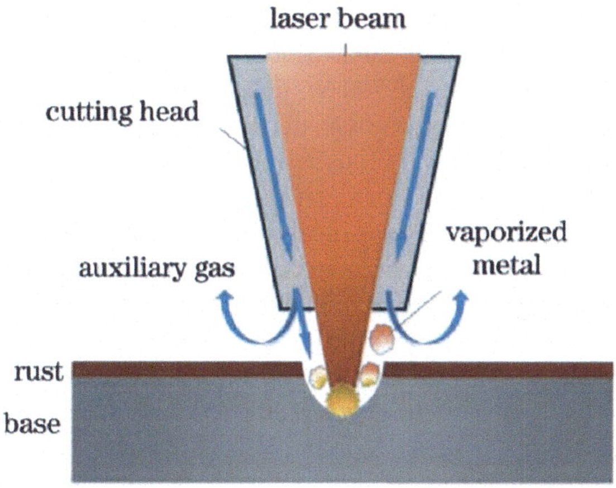


**Fig. 2.7** Schematic diagram of laser derusting [39]

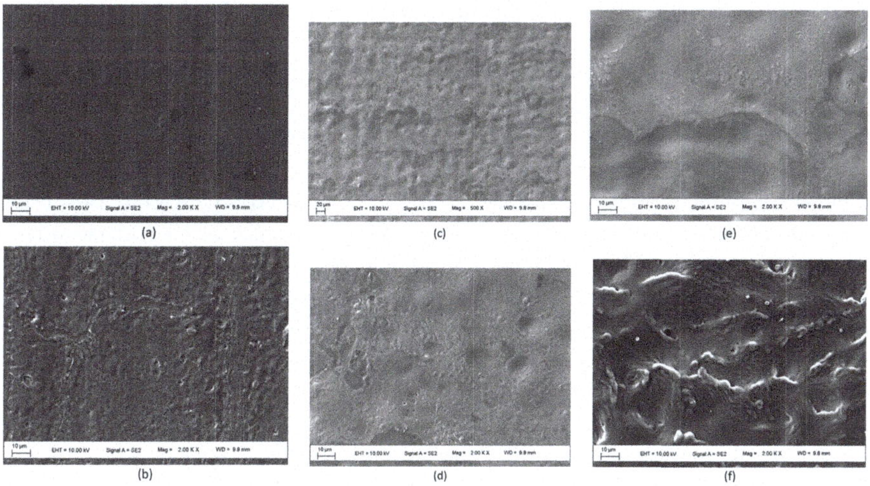


**Fig. 2.8** Microstructure of alloy surface under different energy densities. (**a**) Original sample; (**b**) 3.5 J/cm$^2$; (**c**) 5.7 J/cm$^2$; (**d**) 7.1 J/cm$^2$; (**e**) 8.5 J/cm$^2$; (**f**) 11.3 J/cm$^2$ [44]

the surface. Tang et al. [40] used a 1064 nm nanosecond pulsed fiber laser with a focusing lens, the laser scanning speed was 1000 mm/s, the pulse frequency was 60 kHz, and the pulse width was 200 ns. When the laser energy density was less than 0.41 J/cm$^2$, the cleaning threshold was not reached, and there was no cleaning effect. When the energy density was 0.41–8.25 J/cm$^2$, after scanning for many times, the sulfide on the surface of the sample was completely removed, and the matrix was basically undamaged, and the roughness was reduced from 1.270 to 0.391 μm (Fig. 2.8). At this time, it was at the cleaning threshold but did not reach the damage threshold, and further increasing the energy density would reach the damage threshold and destroy the matrix. Qiao et al. [41] used self-developed high repetition rate and high energy laser cleaning equipment to clean the surface rust layer of 45 carbon steel. Under the matching parameters of laser wavelength of 1064 nm, power of 500 W, pulse width of 80 ns, pulse frequency of 10 kHz and scanning speed of 6 mm/s, the surface rust layer can be completely cleaned without oxidation. Li et al. [42] used pulsed $CO_2$ laser to wet clean the corrosion layer on the surface of carbon steel, and then used acousto-optic Q-switched Nd:YAG laser to heat and remove the residual water film on the surface of carbon steel. By adjusting the laser cleaning parameters, a uniform and dense passivation film was formed on the surface of carbon steel. This double-beam laser derusting method successfully removed the corrosion layer on the surface of carbon steel and prevented the secondary rust return. Wang et al. [43] pointed out that if the action area of the laser beam is equal to the spot size, the scanning speed Vs and the pulse frequency f must satisfy $0 < Vs/f \leq D$ (D is the spot diameter) in order to completely remove the surface rust layer. In the experiment, they used the HGM-50 Q-switched pulse laser to clean the surface rust layer of A3 steel. When the scanning speed was 300 mm/s, Theoretical calculation shows that the laser pulse frequency should be greater than 3.3 kHz to achieve the cleaning purpose. However, in actual cleaning, the laser pulse frequency is only 2.9 kHz, and the rust removal effect can be achieved completely. This is because the actual heating range of the light spot is greater than the diameter of the light spot due to heat conduction, and the corresponding laser density is 26.0 mJ/cm$^2$. After cleaning, the surface hardness of the substrate is slightly improved, and the corrosion resistance is slightly enhanced. Zhang et al. [44] used a pulsed fiber laser with pulse width of 10 ns, pulse frequency of 10 kHz and wavelength of 1064 nm. When the laser cleaning parameters were spot diameter of 0.6 mm, scanning speed of 600 mm/s and pulse number per unit area of 10, different laser energy densities were selected to clean the oxide layer on the surface of AA7024-T4 aluminum alloy. It was found that the newly formed oxide layer after cleaning was better than the original oxide layer on the substrate when the laser energy density was 7.1 J/cm$^2$.

Through the above research, it is found that when laser derusting reaches the cleaning threshold and does not reach the damage threshold, the rust layer on the metal surface can be completely removed without oxidation, and even the secondary rust can be prevented, and the surface of the substrate will not be damaged and its service performance will not be deteriorated, and the surface performance can be improved under the matching of certain parameters.

### *2.4.3 Cultural Relics Restoration*

Removing rust, pollutants, oxides and graffiti from the surface of cultural relics is a heavy and demanding arduous task to repair and protect cultural relics. Laser cleaning technology has great potential in the protection of cultural heritage, from all kinds of stone relics and metal products to paper cultural relics and silk cultural relics [45].

Traditional cleaning technology, whether chemical cleaning method or physical cleaning method, is difficult to avoid damage to cultural relics. Laser cleaning of cultural relics is a widely accepted new cleaning method, which can achieve cleaning effect without basically damaging the surface of cultural relics, and can preserve valuable information in cultural relics more completely [46]. Ye et al. [47] used Q-switched Nd:YAG laser with the wavelength of 1064 nm to clean ink on sandstone surface. The results show that the wet laser cleaning effect is better than the dry laser. When the laser energy is in the range of 17.7–58.8 mJ, the higher the energy, the more obvious the cleaning effect is, and it will not damage the sample surface. Ye et al. [48] used Q-switched Nd:YAG laser with wavelength of 1064 nm and pulse width of 10 ns to irradiate unpolluted sandstone samples, and found that the best energy density range of dry laser cleaning was 0.30–0.96 J/cm$^2$, and that of wet laser cleaning was 0.25–0.96 J/cm$^2$. Under the condition of energy density of 0.65 J/cm$^2$, two cleaning methods were used to clean the samples contaminated by black smoke, respectively. Qi et al. [49] used Q-switched Nd:YAG laser with wavelength of 1064 nm to clean the sandstone in Yungang Grottoes coated with ink by dry and wet laser cleaning methods, respectively. After cleaning, it was found that the wet laser cleaning method was more efficient, mainly due to the fact that ink was soluble in water, which weakened the binding force between ink and substrate. Lee et al. [50] used a Q-switched Nd:YAG laser with a wavelength of 1064 nm, and the laser cleaning parameters were matched to 35 pulses, and the laser energy density was 1.59–2.39 J/cm$^2$. Although most of the pollutants in the area were successfully removed, and the gold-plated layer was exposed, some of the gold-plated layers were destroyed because the bronze ware was not a flat surface, and some of the pollutants in the small concave areas were not removed. Ortiz et al. [51] used Q-switched Nd:YAG pulsed laser to wet clean the surface of 1 cm$^2$ marble. It was found that ultraviolet laser with pulse width of 10 ns, pulse frequency of 10 Hz and wavelength of 266 nm could effectively remove pollutants such as deposits, oxidation spots and graffiti on the surface of the sample. Compared with traditional mechanical and chemical cleaning methods, laser cleaning was considered to be more environmentally friendly and less harmful to the surface of the sample, but the cost was estimated to be 20 times that of traditional cleaning methods.

From the above research at home and abroad, it can be found that although the cleaning effect is not perfect due to the complexity of the surface structure of the cleaned samples, more research results show that laser cleaning can effectively and finely remove all kinds of pollutants on the surface of cultural relics (Fig. 2.9). Compared with traditional cleaning methods, laser cleaning can completely preserve the value of cultural relics on the basis of recreating the past appearance of cultural relics. At present, the cost of laser cleaning cultural relics is still expensive, so it is necessary to comprehensively evaluate the benefits before choosing laser cleaning technology to clean cultural relics.

**Fig. 2.9** Cleaning of graffiti ink on stone wall before and after: (**a**) before laser cleaning, (**b**) after laser cleaning, (**c**) dry laser cleaning, (**d**) wet laser cleaning [48]

### *2.4.4 Laser Degreasing*

Scholars all over the world have been looking for a breakthrough in the removal of oil stains. Crude oil and fuel are extremely difficult to peel off and degrade when adsorbed on the substrate surface. Traditional cleaning methods are mainly chemical solvent cleaning, ultrasonic vibration and manual polishing. These cleaning methods are not only difficult to achieve expectations, but also easy to pollute the environment. The birth of laser cleaning technology provides the possibility for efficient removal of oil stains. Chen et al. [52] used 1064 nm Nd:YAG laser to clean the oil stain on the surface of engine oil filter, and used unidirectional laser to fill it. When the power was 20 W and the scanning speed was 5000 mm/s, the oil stain was completely removed and the best cleaning effect was obtained. Ahn et al. [53] coated 10 μm thick mineral oil films on the surfaces of carbon steel, stainless steel and copper, respectively, and deoiled by Q-switched Nd:YAG laser with wavelength of 1064 nm and pulse width of 25 ns. It was found that the cleaning threshold and damage threshold on carbon steel substrate were

160 ± 10 mJ/cm$^2$ and 410 ± 20 mJ/cm$^2$, respectively. On the stainless steel substrate, it is 130 ± 10 mJ/cm$^2$ and 420 ± 20 mJ/cm$^2$, respectively, while the reflection coefficient of laser light on the copper substrate is much higher than that of the other two substrates, which are 1500 ± 100 mJ/cm$^2$ and 3200 ± 200 mJ/cm$^2$, respectively. All the data show that the oil film can be completely removed on all three substrates without reaching the damage threshold. Ye et al. [54] used $CO_2$ laser for single pulse laser cleaning. The laser power was 10.8 W and the irradiation time was 10 s. The cleaning effect was obvious, and the gold-plated film on the glass surface was not damaged. The micro-morphology before and after laser degreasing is shown in Fig. 2.10. Mateo et al. [55] used Q-switched Nd:YAG laser with laser frequency of 10 Hz and pulse time of 6 ns to clean the residual fuel on the rock surface. It was found that when the laser energy density was 8.0 J/cm$^2$, the residual fuel on the rock surface could be effectively removed and the rock surface was not damaged.

Although there is little research on laser oil removal at present, it can be found that the residual oil on various substrates can be effectively removed by laser cleaning technology without damaging the surface of the substrate to be cleaned.

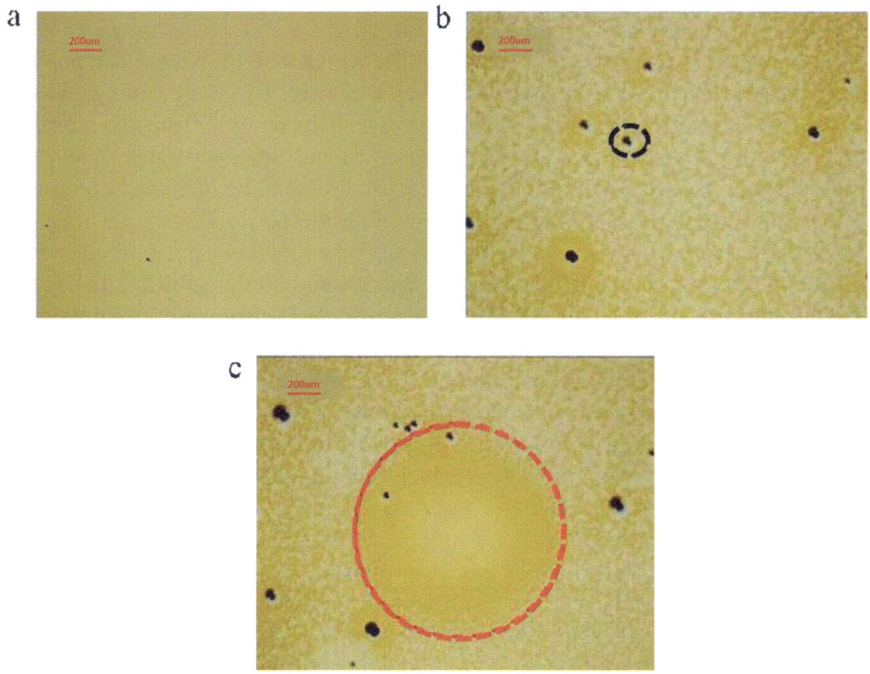


**Fig. 2.10** Microscopic morphology of gold-plated K9 glass before and after laser cleaning: (**a**) surface morphology of gold-coated K9 glass before contamination, (**b**) after contamination, (**c**) gold-coated K9 glass after laser cleaning at laser power of 10.8 W [54]

### 2.4.5 Laser Removing Adhesion of Micron and Submicron Particles

Many precision instruments and glass instruments are easy to be polluted by micron-sized and submicron-sized particles in the process of use, and the performance and reliability of the polluted instruments will deteriorate. The traditional cleaning method is not only inefficient, but also causes damage to instruments. Laser cleaning is a non-contact cleaning method, which can effectively remove pollutants without causing damage to instruments and equipment [56]. For micron and submicron particles, van der Waals force is the main adhesion force. If the particles are assumed to be spherical, when the particles are in contact with the substrate surface, this force can be expressed as:

$$F_v = \frac{hd}{16\pi z^2}$$

where H is the Lifshitz-van der Waals constant, which depends on the properties of the material, D is the particle diameter, and Z is the distance between the particle and the substrate surface. When the distance between them is infinite, this force tends to be infinite [5, 42].

Vatry et al. [57] used an excimer laser with a wavelength of 308 nm and a pulse width of 50 ns. When the laser pulse frequency was 1 Hz, dry laser cleaning method was used to clean carbon particles with an average diameter of 800 nm and tungsten particles with a diameter of 5 μm dispersed on Si and $SiO_2$ substrates. When the laser density exceeded 350 $mJ/cm^2$, carbon particles could be effectively removed on both substrates. However, no matter how high the power density is, tungsten particles can't achieve the ideal cleaning effect. When the laser pulse frequency is 10 Hz and the incident angle between the laser beam and the substrate is 45, the ideal cleaning effect can be achieved on both substrates by using laser plasma shock wave cleaning method. Ye et al. [54] used a 1064 nm Q-switched Nd:YAG laser with a pulse width of 10 ns and a scanning speed of 1 mm/s, and scanned it twice. The cleaning results showed that when the laser energy was 85 mJ, the laser gap was 0.5 mm, and the distance between the $SiO_2$ particles and the focusing point was 1 mm, The clearance rate is as high as 94%. When the laser gap is 1 mm and the distance between $SiO_2$ particles and the focusing point is 1 mm, the energy is increased from 101.14 to 158.41 mJ, and the clearance rate is increased from 84% to 96%, which shows that the laser has a significant cleaning effect on nano $SiO_2$ particles dispersed on the surface of K9 glass under this parameter matching. Ivanova et al. [58] used a semiconductor laser with a wavelength of 1.5 μm and a laser power of 500 mW to clean micron-sized polyethylene, $Al_2O_3$ and talcum powder particles dispersed on welding protective glass and floppy disk by wet laser cleaning. The cleaning effect of polyethylene particles on fully wettable welding protective glass and floppy disk with poor wetting performance was good, while the cleaning effect of the other two particles on two kinds of substrates was poor, and they found that the substrate would not be damaged during laser cleaning. Ye et al. [59] used Q-switched Nd:YAG laser with pulse width of 10 ns and wavelength of 1064 nm to clean the $SiO_2$ particles attached to the surface of gold-plated film K9 glass by laser plasma shock wave. The cleaning effect was the best when the pulse frequency was 1 Hz, the scanning speed was 1 mm/s, the laser focus point was 1 mm from the particle center, and the laser gap was 0.5 mm, and the cleaning effect was twice, and the removal rate of $SiO_2$ particles reached 91% (Fig. 2.11).

At present, many scholars at home and abroad have optimized the matching of various laser cleaning parameters, which can effectively remove the adhesion of micron and submicron particles, and can be applied to the efficient and fine cleaning of various precision instruments and glass instruments, effectively improving the use accuracy and service life of these instruments. In addition, remarkable application progress has been made in the field of nuclear industry, which has improved the stability of nuclear operation equipment.

### 2.4.6 Typical Application of Laser Cleaning in Its Field

As one of the most potential cleaning methods in the twenty-first century, laser cleaning technology has been widely used in various fields of industrial production and life. Sun et al. [60] used $CO_2$ laser to clean the rubber mold. When the laser power was 80 W and the defocus was −3 mm and the scanning speed was 5 mm/s, the cleaning effect was good, and the surface of the mold was not damaged after cleaning. Tian et al. [61] used a nanosecond fiber laser with a wavelength of 1064 nm

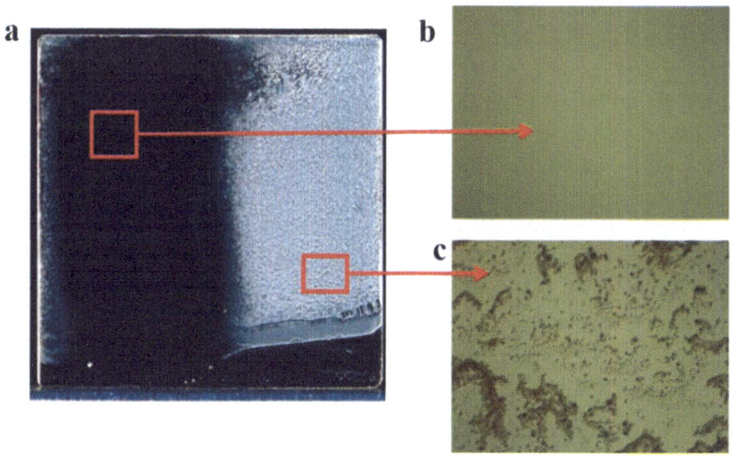


**Fig. 2.11** Surface morphology of gold-plated K9 glass before and after laser cleaning: (**a**) dark field diagram of gold-plated K9 glass after laser cleaning; (**b**) enlarge the surface morphology of the laser cleaned area by 50 times; (**c**) enlarge the surface morphology of the uncleaned area by 50 times [59]

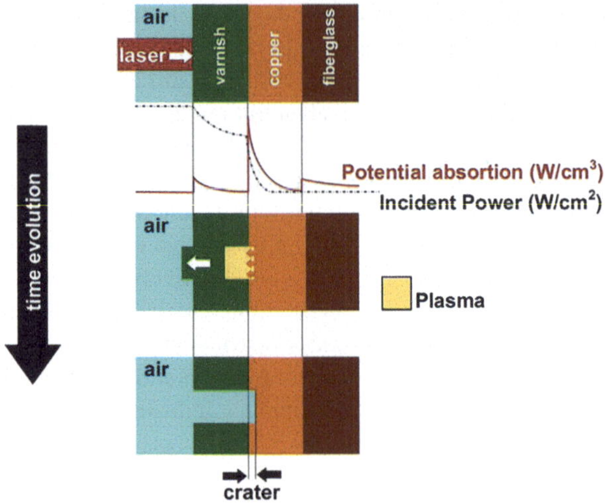


**Fig. 2.12** Schematic diagram of the principle that the solder film is ejected [62]

and a pulse width of 30 ns to clean the natural marine microbial film formed on the surface of aluminum alloy. When the laser density was between 2.76–5.52 J/cm$^2$, the microbial film could be effectively removed, and good surface quality of aluminum alloy was obtained. According to the analysis, the removal mechanism was that the temperature rose sharply after laser irradiation, and the microbial film was gasified and ablated. It is also found that when the density is 5.52 J/cm$^2$, the surface of aluminum alloy will be super-hydrophobic due to laser micromachining, which provides the feasibility for the surface modification of antifouling performance of ships and offshore equipment. Raele et al. [62] used 1064 nm infrared wavelength single pulse Q-switched Nd:YAG laser to irradiate the printed circuit board, so that the thin layer of copper under the solder film on its surface melted, resulting in a lot of thermal stress to completely eject the solder film. The principle is shown in Fig. 2.12, and this method will not have any side effects on the use of the printed circuit board, which proves for the first time that the laser ablation mechanism on the substrate can also be used in the recycling process. Xia et al. [63] used a 1064 nm pulsed laser to clean the oxide film on the surface of 2219 aluminum alloy before welding. When the laser frequency was 100 kHz, the threshold of laser cleaning speed was about 1079 mm/min, and the cleaning effect was good. After welding, the surface was bright and there were no defects such as pores and inclusions. Zhu et al. [64] used a self-developed TEA $CO_2$ laser to clean and remove the adhesive tape stuck on the blue brick. This is a simulation experiment to remove graffiti on buildings. When the laser energy density reaches 5.067 J/cm$^2$, the adhesive tape can be completely removed with 8 pulses, and the substrate is not damaged. Guan et al. [65] used a 1064 nm Nd:YAG laser with a laser power density of $1.37 \times 10^8$ W/cm$^2$ and a scanning speed of 1000 mm/s, and the cleaning effect was good. The micro-morphology before and after cleaning is shown in Fig. 2.13. Qiao et al. [66] used a

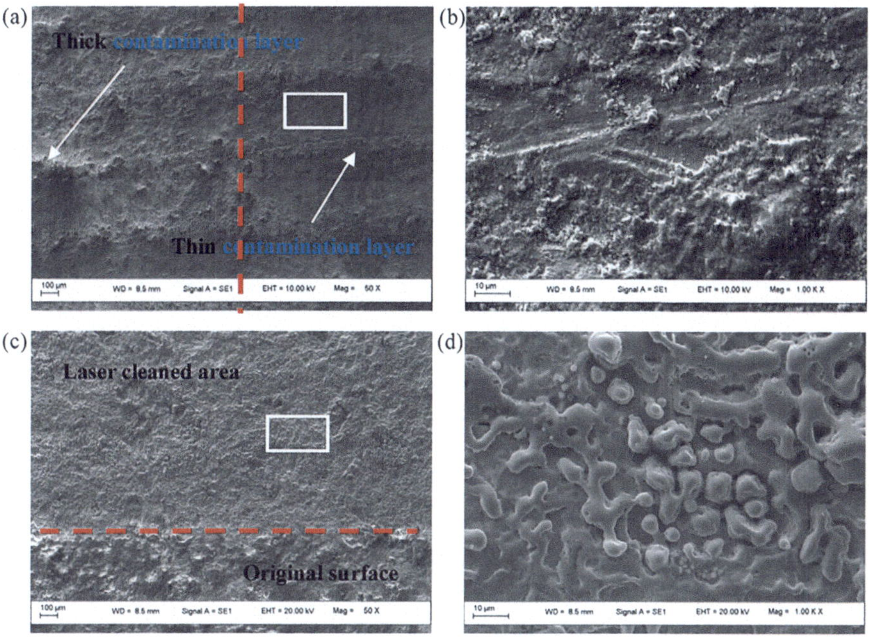


**Fig. 2.13** Scanning images of piston surface before and after laser cleaning: (**a**) pollutants are thin (right) and thick (left); (**b**) the scanned image after enlarging the rectangular area in (**a**); (**c**) laser cleaning the rectangular medium-thick pollutant area; (**d**) the scanned image after enlarging the rectangular area in (**c**) [65]

self-developed 500 W high repetition rate laser to clean the carbon deposit on titanium alloy surface. The experiment showed that the cleaning effect was the best when the laser power was 500 W, the pulse width was 20 ns, the pulse frequency was 18 kHz and the scanning speed was 7 $cm^2/s$.

Practice shows that laser cleaning has excellent effects in cleaning molds, microbial films on ships, soldering films on circuit boards, pre-welding pretreatment, graffiti on buildings and carbon deposition on engines, etc. At present, this technology has developed rapidly and is widely used in various industries. Compared with traditional cleaning methods, laser cleaning has great advantages, and in most fields, laser cleaning is expected to partially or even completely replace traditional cleaning.

## 2.5 Influence of Laser Cleaning on Surface Integrity

Surface integrity mainly includes: mechanical properties such as surface microhardness, tensile properties, residual stress, friction and wear; Surface topography characteristics such as roughness and three-dimensional morphology; Microscopic characteristics such as microstructure, grain size, dislocation change and physical and chemical properties such as wettability and corrosion resistance [67, 68]. Due to the irradiation of laser cleaning energy on the surface of the substrate, interfacial thermal effect and impact effect will be generated, which will lead to the change of the surface integrity of the material, which is a factor that scientists pay special attention to during cleaning operation, because the integrity is directly related to the service life of the cleaned material, and the purpose of regulating the surface integrity of the substrate can be achieved by matching and optimizing the laser cleaning parameters.

### *2.5.1 Effect of Laser Cleaning on Surface Structure and Microstructure of Substrate*

Ensuring the surface integrity of materials means that the surface of the substrate is not damaged or the surface performance of the substrate is enhanced to a certain extent, and the surface integrity of the substrate may be damaged after laser cleaning. Systematic parameter optimization is needed to obtain the best matching of cleaning parameters, realize high-quality cleaning on the premise of ensuring the surface of the substrate is not damaged, and further seek the improvement of surface performance [69].

In recent years, the surface integrity of the substrate after laser cleaning has attracted the attention of many scholars at home and abroad. Jia et al. [70] cleaned micron-sized rubber particles deposited on the surface of chromium-plated die and groove area, and when the laser energy density is 0.59 J/cm$^2$, the particles on the die surface can be effectively removed, while the groove area needs the laser energy density to reach 0.97 J/cm$^2$ to effectively remove the rubber particles. Due to the laser irradiation, the surface temperature rises and the chromium layer is remelted. It can fill the reticular cracks in the chromium layer to a certain extent, and can effectively eliminate the reticular cracks on the surface of the chromium layer. The mechanism of particle removal is mainly thermal expansion effect, and it will not damage the surface structure of the mold during cleaning, which is considered to be an ideal cleaning method. Buccolieri et al. [71] used KrF excimer laser with wavelength of 248 nm and pulse width of 2030 ns to clean the copper rust on the surface of bronze bell, and calculated that the cleaning threshold was 1.8 ± 0.3 J/cm$^2$. They quantitatively analyzed the surface of bronze bell after cleaning, and thought that the laser cleaning method could effectively remove copper sulfate and calcium carbonate pollutants on the surface, and the cleaning mechanism was mainly the ablation effect caused by laser irradiation on the surface of pollutants. At this time, the pollutants absorb heat, so that the temperature reaches the melting point or even the boiling point, and the pollutants melt or even vaporize, thus achieving the cleaning effect. After cleaning, the contents of S and Ca are obviously reduced, and the physical and chemical properties of the bronze bell surface are not changed. Hu et al.

[72] used Nd:YAG pulsed fiber laser to clean the paint layer about 50 μm thick on the surface of Ti17, controlled the scanning speed of 6000 mm/s and the laser pulse frequency of 100 kHz, and changed the laser cleaning power. At 15 W, a good cleaning effect was obtained. Due to the laser shock, dislocations in the surface area slipped, resulting in impact strengthening, and the surface hardness was slightly improved. At the same time, the laser thermal effect caused the surface tissue to remelt and recrystallize, which can be used in metallographic structure. Zhou et al. [73] used a 1064 nm pulsed fiber laser to clean the 200 μm thick rust layer on the surface of AH32 marine steel step by step. Under the condition of scanning speed of 3000 mm/s, the laser energy densities of 30.6 J/cm$^2$ and 10.2 J/cm$^2$ were used for repeated cleaning, respectively. It was found that the surface rust layer was effectively removed. In the first step, the laser energy density was higher, and the cleaning mechanism was mainly hole blasting mechanism. In the second cleaning step, the bottom rust will be further removed due to the accumulation of heat generated by the two cleaning steps. The cleaning mechanism is mainly ablation and evaporation. At the same time, the surface roughness caused by corroded holes is improved due to the laser recoil. After two-step cleaning, the surface roughness reaches the maximum, only 4.517 μm. It is found by scanning electron microscopy that a dendritic mastoid structure is formed on the surface of the sample after cleaning. As shown in Fig. 2.14, the main reason for this structure is that the matrix material is vaporized more slowly than melted. Nilaya et al. [74] used a 1064 nm laser to clean the particles of $CsNO_3$ deposited on the surface of stainless steel. When the pulse duration was 1.1 ns, the cleaning threshold was 62 mJ/cm$^2$. With the density, it was found that the particles were not completely removed at 316 mJ/cm$^2$, but there was pitting corrosion on the surface of the substrate. The main reason for this phenomenon was that the adhesion of the particles to the substrate was very large. With the continuous increase of laser energy density, although the cleaning efficiency is improved, at this time, the heat absorption rate of the substrate is far greater than the

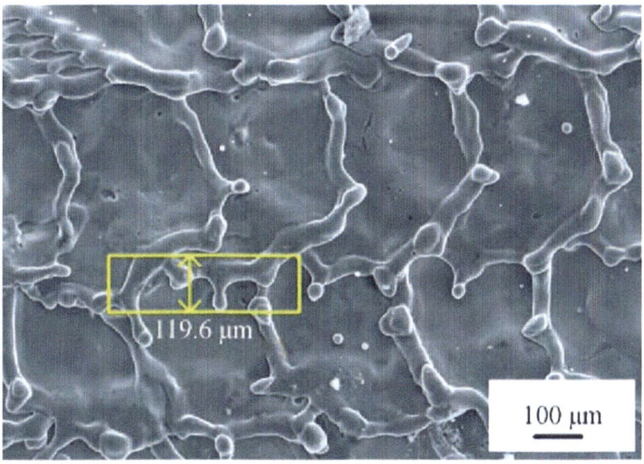


**Fig. 2.14** Dendritic mastoid structure formed after laser cleaning [73]

heat discharge rate, which leads to the rapid increase of the substrate temperature, reaching the melting point or even boiling point of the substrate, which makes the substrate melt or even vaporize, thus forming pits.

Based on the comprehensive analysis of the research results of surface integrity after laser cleaning, as long as the laser cleaning parameters are scientifically controlled and adjusted, the damage of the cleaned substrate can be avoided, the surface roughness changes little, and the physical and chemical properties will not change. The impact effect and thermal effect caused by laser cleaning may lead to dislocation sliding on the surface of the substrate and recrystallization of the microstructure, which may lead to the change of surface hardness and microstructure. If the integrity of the substrate under the premise of cleaning cannot be achieved by systematically adjusting the laser cleaning parameters, it means that the laser cleaning technology is not suitable for cleaning this substrate, or it can be solved by changing the unconventional laser cleaning parameters (such as the incident angle of laser cleaning) or adding other cleaning methods.

### 2.5.2 *Effect of Laser Cleaning on Surface Properties of Substrate*

The interaction between laser and surface pollutants and substrate may lead to the evolution of microstructure and microstructure of substrate surface, cause the change of surface integrity and affect the surface properties of substrate. The results show that different laser cleaning processes have different effects on the surface properties of matrix materials under appropriate parameter matching conditions.

At present, there are few reports about the surface integrity of the substrate based on efficient cleaning by laser cleaning process and the stability of the surface performance of the substrate after laser cleaning. After sporadic reports at home and abroad, it is found that this phenomenon does exist. Shi et al. [75] used a 1064 nm pulsed fiber laser to clean the oxide layer on the surface of 7050-T451 aluminum alloy with a thickness of 2 mm After cleaning, the tensile strength and bending strength of the substrate tended to increase, and the surface of aluminum alloy was plastically deformed to form a hardened layer with a thickness of 930 μm, which improved its surface hardness. The main reason was that the laser irradiated the surface of the substrate to generate heat accumulation, and the surface of the substrate was remelted. When the laser irradiation leaves, the cooling effect of the metal itself causes the surface grains to be refined, and at the same time, the impact of the laser causes the cold plastic deformation of the surface structure of the substrate, and the crystal lattice is distorted, which hinders the dislocation movement and improves the deformation resistance of the surface structure. The surface macromorphology and cross-section micro-morphology after different power cleaning are shown in Fig. 2.15. Rauh et al. [76] used a laser with a wavelength of 266 nm to clean the residual silicone particles on the resin surface. The cleaned resin surface

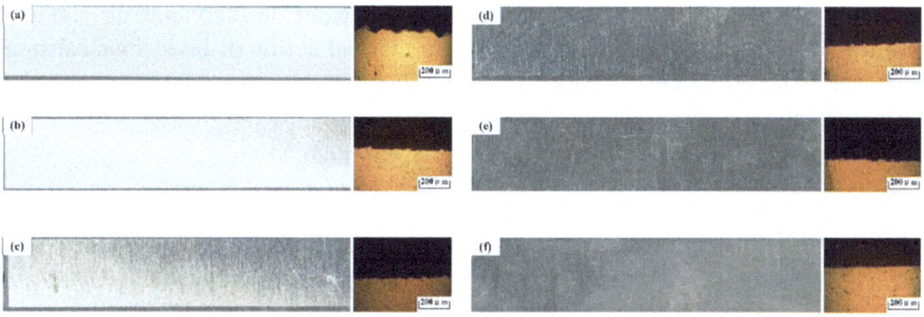

**Fig. 2.15** When the scanning speed v = 10 mm/s, the surface macro-morphology and cross-section micro-morphology under different power: (**a**) P = 120 W; (**b**) P = 108 W; (**c**) P = 96 W; (**d**) P = 84 W; (**e**) P = 72 W; (**f**) P = 60 W [75]

has higher joint strength and smaller standard deviation, and its joint strength is twice as high as that when the carbon fiber surface is exposed after grinding. The main reason for this phenomenon is that a small amount of particles will remain after laser cleaning, which is proved to have higher shear strength than the completely bare fiber. Tian et al. [77] used a nanosecond fiber laser with a wavelength of 1064 nm and a pulse width of 30 ns to clean the 61.7, 26.5 μm marine microbial film naturally formed on the surface of 2 mm thick Al-Mg-Si alloy. When the energy density was 4.14 J/cm$^2$, due to the increase of energy density, heat quickly accumulated on the surface of the substrate and reached the melting point of the substrate. Eventually, the surface of the substrate will be remelted. After laser irradiation, the alloy will be cooled rapidly, and recrystallization will occur. In addition, the particles in plasma smoke will be re-deposited by short pulse laser. The two mechanisms work together to produce an amorphous-nanocrystalline structure layer with a total thickness of about 250 nm on the surface of the substrate, which has strong hydrophobicity, good anti-corrosion and anti-fouling properties, and has important research value for prolonging the life of offshore equipment. From the above research, it can be found that the surface properties of the substrate changed during the laser cleaning process, and even modified and strengthened, resulting in an amorphous-nanocrystalline structure layer. However, some pollutants on the substrate surface are difficult to clean, and even in some cases, the cleaned surface has been damaged when the pollutants are not completely removed. During laser cleaning, laser power, pulse frequency, scanning speed, spot size, spot overlap rate, distance between laser head and substrate, and laser incident angle will all have important effects on the surface performance of substrate after cleaning. In order not to damage the substrate surface during laser cleaning, or even to enhance the performance of the substrate surface, it is necessary to carry out sufficient experimental verification and process optimization, and select the optimal parameter matching. It is worth noting that the modification and strengthening of the substrate surface during laser cleaning is different from laser cladding and melting. Laser melting is to melt a layer of metal on the surface and then rapidly cool it by itself, resulting in grain refinement and thus improving the surface performance of the metal. Laser cladding is to coat a layer of high-quality metal powder on the metal surface and melt it to cover the metal surface through the thermal action of laser, thus causing the surface strengthening of the metal [78], while laser cleaning is not to damage the surface metal.

## 2.6 Conclusion

This paper summarizes the typical research work in the field of laser cleaning technology at present, summarizes the key application direction of laser cleaning, and probes into the surface integrity after cleaning and the mechanism affecting the surface integrity. At present, foreign laser cleaning technology is mature, which is widely used in the fields of aircraft fuselage paint removal, die surface oil removal, engine internal carbon removal and surface cleaning of welded joints. Some universities, research institutes and enterprises in China have shown great interest in laser cleaning technology. Strengthening the research and application of laser cleaning technology and gradually starting independent research and development of laser cleaning equipment have greatly promoted the development of laser cleaning technology. However, there are still some shortcomings, which should be the focus of our future research.

1. Laser cleaning technology has a broad application prospect, which can be used for high quality, high efficiency and green cleaning in the fields of machinery, precision instruments and cultural relics protection. However, there are still many limitations in the application of composite materials and complex structural surfaces, and further research and verification are needed before it can be put into industrial application.
2. Laser cleaning can improve and enhance the surface performance of the substrate while removing the dirt layer on the surface. Compared with the traditional cleaning method, this is a big advantage of laser cleaning, but this aspect has not been paid attention to, and it needs more attention in future research. This research direction has great potential and I believe it will become the focus of our research in the future.
3. The study of laser cleaning mechanism involves complex interdisciplinary integration. On the one hand, it is the mechanism of laser self-cleaning. When thermal effect and vibration effect work together, the dominant mechanism is uncertain. On the other hand, it is the mechanism of surface integrity and performance evolution caused by the interaction between laser and dirt layer and substrate. The mechanism under multiple action mechanisms is complex. In addition, the mechanism of laser interaction with pollutants is different with different types and matrix materials. The parameters of laser cleaning are different, and the interaction with pollutants and matrix will also be different.
4. Laser cleaning equipment is the key to develop laser cleaning technology. The development of high-power equipment and the exploration of laser cleaning

mechanism are urgent problems to be solved. This research should be strengthened in the future. With the development of laser cleaning, there are more and more research on surface integrity, which will be the research focus of laser cleaning.

## References


1. Wu C W. Technical study on laser cleaning process of epoxy resin paint on aluminum alloy surface. Master's Thesis, Soochow University, China, 2019 (in Chinese).
2. Liu P F. Development of high-power all-solid-state laser cleaner. Master's Thesis, Changchun University of Science and Technology, China, 2013 (in Chinese).
3. Koh Y, Sárady I. Journal of Cultural Heritage, 2003, 4, 129s.
4. Wang X. The investigation on the application of laser cleaning on bronze cultural relics. Master's Thesis, East China University of Science and Technology, China, 2019 (in Chinese).
5. Chaoui N, Solis J, Afonso C N, et al. Applied Physics A, 2003, 76, 767.
6. Wang D. The simulation research and experimental verification on high pressure water jet cleaning. Master's Thesis, Harbin Institute of Technology, China, 2014 (in Chinese).
7. Zhang Z Y, Zhang J Y, Wang Y B, et al. Optik, 2018, 174, 46.
8. Hsu S C, Lin J. Optics & Laser Technology, 2006, 38, 544.
9. Marimuthu S, Sezer H K, Kamara A M. Developments in Surface Contamination and Cleaning: Applications of Cleaning Techniques, 2019, 11, 251.
10. Liu P F, Wang S J, Yin F S, et al. Materials Reports, 2020, 34(12), 24121 (in Chinese).
11. Madhukar Y K, Mullick S, Nath A K. Applied Surface Science, 2013, 286, 192.
12. Xie X Z, Huangg Q P, Long J Y. Journal of Materials Processing Technology, 2020, 275, 116321.
13. Kumar A, Biswas D J. Optics and Laser Technology, 2018, 106, 286.
14. Rivas T, Pozo-Antonio J S, López de Silanes M E, et al. Applied Surface Science, 2018, 440, 467.
15. Yu H B. Laser cleaning technology research on metal surface. Master's Thesis, Huazhong University of Science & Technology, China, 2015 (in Chinese).
16. He B L, Deng H P. Surface Technology, 2015, 44(9), 140 (in Chinese).
17. Lu F Y, Song W D, Ang B W, et al. Applied Physics A, 1997, 65, 9.
18. Siano S, Salimbeni R. Accounts of Chemical Research, 2009, 43(6), 739.
19. Kumar A, Prasad M, Bhatt R B, et al. Optics and Lasers in Engineering, 2014, 57, 114.
20. Zapka W, Ziemlich W, Leung W P, et al. Microelectronic Engineering, 1993, 20, 171.
21. Lei Z L, Sun H R, Chen Y B, et al. Chinese Journal of Lasers, 2019, 46(7), 0702003 (in Chinese).
22. Cheng J. Study of laser induced plasma shockwave to clean nanoparticles on surface of silicon wafer. Master's Thesis, Changchun University of Science and Technology, China, 2015 (in Chinese).
23. Ren Z G, Wu C Z, Chen H N, et al. Opto-Electronic Engineering, 2017, 44(12), 1210 (in Chinese).
24. Raza M S, Das S S, Tudu P, et al. Optics and Laser Technology, 2019, 113, 95.
25. Luo H X, Cheng Z G. Laser Journal, 2002, 23(6), 52 (in Chinese).
26. Hua M, Shi X K, Cheung E, et al. Optics & Laser Technology, 2004, 37, 9.
27. Wang Z M, Zeng X Y. Ome Information, 2001(8), 20 (in Chinese).
28. Razab M K A A, Noor Mohaned An'amt, Jaafar M S, et al. Journal of Radiation Research and Applied Sciences, 2018, 11, 393.
29. Shi S D, Du P, Li W, et al. Chinese Journal of Lasers, 2012, 39(9), 0903001 (in Chinese).
30. Chen H. Process of basic research on laser cleaning technology of car body surface paint. Master's Thesis, Harbin Institute of Technology, China, 2018 (in Chinese).

31. Lu S Y. Technical study on laser cleaning of rust on steel surface. Master's Thesis, Huazhong University of Science & Technology, China, 2017 (in Chinese).
32. Jin S, Wang J X, Yuan X D, et al, Aviation manufacturing technology, 2018, 61(17), 63 (in Chinese).
33. Zhang G X, Hua X M, Li F, et al. Hot Working Technology, 2019, 48(18), 1 (in Chinese).
34. Brygo François, Dutouquet C, Le Guern F et al. Applied Surface Science, 2006, 252, 2131.
35. Chen G X, Kwee T J, Tan K P, et al. Applied Physics A, 2010, 101, 249.
36. Kumar M, Bhargava P, BiswasI A K, et al. Optics & Laser Technology, 2013, 46, 29.
37. Zheng G. Study on laser paint stripping of aircraft surface. Master's Thesis, Beijing, Institute of Electronics, Chinese Academy of Sciences, China, 2005 (in Chinese).
38. Zhang G X, Hua X M, Huang Y, et al. Applied Surface Science, 2019, 506, 144666.
39. Xie Y F, Liu H W, Hu Y X. Chinese Journal of Lasers, 2016, 43(4), 0403008 (in Chinese).
40. Tang Q H, Zhou D, Wang Y L, et al. Applied Surface Science, 2015, 355, 334.
41. Qiao Y L, Zhao J X, Wang S J, et al. Laser & Infrared, 2018, 48(3), 299 (in Chinese).
42. Li W. Research on mechanisms of laser rust removal and manufacture of laser cleaning devices. Ph.D. Thesis, Nankai University, China, 2014 (in Chinese).
43. Wang Z M, Zeng X Y, Huang W L. Surface and Coatings Technology, 2003, 166, 10.
44. Zhang F D, Liu H, Suebka C, et al. Applied Surface Science, 2018, 435, 452.
45. CiussaniI B, Monticelli D, Rampazzi L. Analytica Chimica Acta, 2009, 635, 6.
46. Choubey A, Vishwakarma S C, Vachhani D M, et al. Optics and Lasers in Engineering, 2014, 62, 69.
47. Ye Y Y, Qi Y, Qin L, et al. Chinese Journal of Lasers, 2013, 40(9), 0903005 (in Chinese).
48. Ye Y Y, Qi Y, Yuan X D, et al. Optik, 2014, 125, 3093.
49. Qi Y, Ye Y Y, Wang H J, et al. Chinese Journal of Lasers, 2015, 42(6), 0603001 (in Chinese).
50. Lee H, Cho N, Lee J. Applied Surface Science, 2013, 284, 235.
51. Ortiz P, Antúnez V, Ortiz R, et al. Applied Surface Science, 2013, 283, 193.
52. Chen Y K. Application of laser cleaning in aircraft parts repair. Master's Thesis, Huazhong University of Science & Technology, China, 2019 (in Chinese).
53. Ahn D, Jang D, Park T, et al. Surface & Coatings Technology, 2012, 206, 3751.
54. Ye Y Y, Yuan X D, Xiang X, et al. Optik, 2012, 123, 1056.
55. Mateo M P. Applied Surface Science, 2005, 247, 333.
56. Tam A C, Leung W P, Zapka W, et al. Journal of Applied Physics, 1992, 71(7), 3515.
57. Vatry A, Naiimhabib M, Delaporte Ph, et al. Applied Surface Science, 2009, 255, 5569.
58. Lvanova N A, Starov V M, Trybala A, et al. Journal of Colloid and Interface Science, 2016, 473, 120.
59. Ye Y Y, Yuan X D, Xiang X, et al. Optics and Lasers in Engineering, 2011, 49, 536.
60. Sun H Y. Technical study of the laser cleaning rubber products molds. Master's Thesis, Qingdao University of Science & Technology, China, 2009 (in Chinese).
61. Tian Z, Lei Z L, Chen X, et al. Journal of Cleaner Production, 2020, 244, 118724.
62. Raele M P, Pretto L R D, Zezell D M. Waste Management, 2017, 68, 475.
63. Xia P Y, Yin Y H, Cai A J, et, al. Chinese Journal of Lasers, 2019, 46(1), 0102005 (in Chinese).
64. Zhu Y F, Tan R Q. Laser & Infrared, 2011, 41(8), 840 (in Chinese).
65. Guan Y C, Ng G K L, Zheng H Y, et al. Applied Surface Science, 2013, 270, 526.
66. Qiao Y L, Huang K N, Liang X B, et al. Applied Laser, 2017, 37(6), 859 (in Chinese).
67. Gao Y K. Acta Metallurgica Sinica, 2016, 52(8), 915 (in Chinese).
68. Liao Z R, Abdelhafeez A, Li H N, et, al. International Journal of Machine Tools and Manufacture, 2019, 143, 63.
69. Fang S Q, Marro F G, Nuria S, et al. International Journal of Refractory Metals & Hard Materials, 2019, 83, 104982.
70. Jia X S, Zhang Y D, Chen Y Q, et al. Optics and Laser Technology, 2019, 119, 105659.
71. Buccolieri G, Nassisi V, Buccolieri A, et al. Applied Surface Science, 2013, 272, 55.
72. Hu T Y, Qiao H C, Lu Y, et al. Surface Technology, 2018, 47(3), 7 (in Chinese).

73. Zhou J Z, Li H T, Sun Q, et al. Optics and Precision Engineering, 2019, 27(8), 1754 (in Chinese).
74. Nilaya J P, Prasad M B S, Biswas D J. Applied Surface Science, 2012, 263, 25.
75. Shi T Y, Wang C M, Mi G Y, et al. Journal of Manufacturing Processes, 2019, 42, 60.
76. Rauh B, Kreling S, Kolb M, et al. International Journal of Adhesion and Adhesives, 2018, 82, 509.
77. Tian Z, Lei Z L, Chen X, et al. Applied Surface Science, 2020, 499, 144060.
78. Duan Y L, He H, Fu Y C. Materials Reports, 2010, 24(S2), 205 (in Chinese).

# Chapter 3
# Effects of Laser Cleaning on the Surface Morphology of Carbon Fiber Reinforced Polymer

## 3.1 Introduction

Carbon fiber reinforced polymers (CFRP) are lightweight, high strength, good chemical resistance and good processability. They are widely used in various fields such as aerospace, automobile, high-speed rail and wind power equipment. Surface cleaning is required to improve bond strength when CFRP are used as engineered structural components in connection with other materials. The use of conventional cleaning techniques, such as mechanical polishing, chemical cleaning, and ultrasonic cleaning, can result in damage to carbon fibers and deterioration of material properties. Laser cleaning is an efficient, non-destructive, green and safe new cleaning technology, which is an important means to enhance the serviceability of CFRP.

With their good thermal stability, fatigue resistance, corrosion resistance, and excellent mechanical properties [1–5], CFRP are widely used in aerospace [6], automotive [7], construction [8], and medical [9] fields. The manufacture of CFRP is usually accompanied by the use of release agents. During the separation curing process of CFRP, there are release agents left on the surface of the material, which can affect the bonding of CFRP to other materials. In addition, with the widespread use of CFRP structural components in aerospace, the CFRP skin paint layer needs to be cleaned periodically. Traditional cleaning methods include mechanical cleaning and chemical organic solvent cleaning. Among them, the mechanical cleaning efficiency is low, and the working cycle is long [10]; most of the chemical organic solvents such as chlorinated hydrocarbons, fluorinated hydrocarbons, alcohols, and ethers have a certain degree of toxicity and are prone to cause pollution to the environment if not properly handled [11]. The traditional cleaning methods have been unable to meet the cleaning needs of various fields. Laser cleaning has become a highly sought after emerging cleaning method due to its advantages of non-contact, high efficiency, high cleaning quality and compliance with today's green development strategies [12].



S. Wang, *Laser Cleaning*, https://doi.org/10.1007/978-981-95-2935-3_3

Laser cleaning is a highly efficient cleaning cleaning technology that utilizes a high-energy laser beam to irradiate the surface of the object to be cleaned, instantly evaporating and stripping the surface contaminants. Fiber lasers at 1064 nm, $CO_2$ lasers at 10,600 nm, green lasers at 532 nm, and UV lasers at 355 nm are the common laser sources for laser cleaning. The concept of laser cleaning was introduced in the 1960s [13]. It was not until the 1990s that it was gradually applied to industrialized production [14]. Laser cleaning can turn contaminants on the surface of the material into particles that can be collected by a dust collection system. Substances with a lower boiling point will be vaporized directly. This not only reduces environmental pollution but also avoids secondary pollution of the parts surface. Compared with traditional cleaning methods, laser cleaning is non-contact cleaning. Remote cleaning of large equipment facilities can be realized in combination with mechanical automation remote operation. Due to its high precision, good controllability, and many other advantages, this method is widely used in industrialised cleaning for tasks such as paint, rust, oil, carbon, and particle adhesion removal, as well as coating stripping [15–20].

In the early days, laser cleaning targets were mainly metal substrates. Later, as laser technology matured, its application scenarios became more and more numerous. Recently, Nie et al. successfully cleaned a thermally controlled coating on an aluminum alloy surface using a nanosecond pulsed laser [21]. Zou et al. cleaned a ferrite surface varnish layer [22]. Rotondi et al. developed a new laser cleaning tool and accomplished the cleanup of a bronze vessel underwater [23]. Liu et al. used a hybrid laser polishing technique to reduce the surface roughness and improve the surface microhardness and wear resistance of high temperature alloys [24, 25]. Yuan et al. successfully removed bump structures on the surface of C/SiC ceramic matrix composites using femtosecond lasers [26].

Laser cleaning of CFRP effectively removes residual dirt, grease, mold release agent residue and other impurities from the surface, ensuring a clean surface. The cleaned surface helps to improve the bonding effect between CFRP and other materials, reduce the risk of failure of bonded joints, and enhance the durability and reliability of structural components. The specific cleaning schematic is shown in Fig. 3.1 [27]. The output laser becomes a parallel beam through a beam expander, and then the laser beam passes through a scanning galvanometer, is focused by a lens, and finally irradiates the sample surface of the experimental platform.

With the increased use of CFRP, scholars from various countries began to use laser cleaning technology for CFRP surface cleaning, and have achieved preliminary results. However, there are still many problems, such as time-consuming and energy-intensive process trials, difficulty in accurate cleaning, and the need for a more efficient process.

In this paper, the mechanism of laser cleaning CFRP is analyzed, the mechanical properties and bonding properties of laser cleaning CFRP are summarized, and the methods of online monitoring and numerical simulation are discussed to ensure the cleaning quality and efficiency. Finally, the limitations and future research directions of laser cleaning are pointed out.

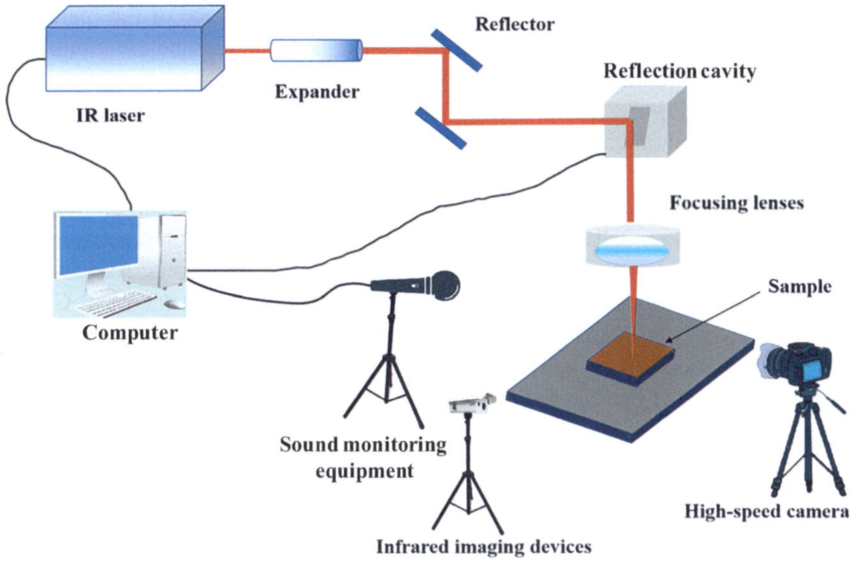


**Fig. 3.1** Schematic diagram of the experimental set-up in laser paint removal

## 3.2 Laser Cleaning Mechanism

With the gradual enhancement of people's environmental awareness, green and safe laser cleaning technology is gradually applied to industrial production. At the same time, in order to ensure the cleaning efficiency and quality, better use of laser characteristics to meet the cleaning requirements, it is very important to understand the relevant laser cleaning mechanism. The mechanism of laser cleaning is mainly based on the interaction between the laser and the material surface, and the nature of the process is the cross-coupling between optics, heat, and mechanics, including a series of complex physical and chemical changes [28, 29]. Laser cleaning mechanisms currently include thermal ablation [30], thermal stress [31], plasma shock waves [32], and photochemical decomposition [33, 34], and the different cleaning processes are competing processes between these mechanisms.

Thermal ablation is essentially when the laser hits the surface of the material to be cleaned and the light energy is absorbed and converted into heat. When the temperature reaches the decomposition temperature of the macromolecular contaminants, they are broken down into small particles that are dislodged from the surface of the substrate. At higher temperatures, the contaminant can even be vaporized directly. The thermal ablation mechanism is more prominent in long-wave lasers such as infrared lasers. The essence of thermal stress is that the contaminant on the surface of the substrate absorbs a large amount of heat for a short period of time to produce thermal expansion. When the expansion force is greater than the adsorption force between the contaminant and the substrate, the contaminant will be dislodged from the substrate. A laser plasma shock wave is generated when the laser irradiation process breaks through the air medium on the surface of the material. The high intensity shock wave can dislodge and remove particulate contaminants from the substrate surface. The coupling schematic of the three cleaning mechanisms is shown in Fig. 3.2a.

The essence of photochemical decomposition is that the single photon energy in the laser is greater than the bond energy of the chemical bond in the pollutant, and the pollutant is broken into small particles due to the destruction of the chemical bond, as shown in Fig. 3.2b. In addition, the photochemical reaction mechanism is a unique cleaning mechanism for short-wave lasers such as UV lasers. Akman [35] used UV (Ultraviolet laser) and IR (infrared laser) lasers to treat identical CFRP sheets. When the sheets were cleaned, the energy density of the IR laser reached 9 J/$cm^2$, while the energy density of the UV laser was only 400 $mJ/cm^2$. The lower energy density means that the temperature during cleaning is lower and damage to

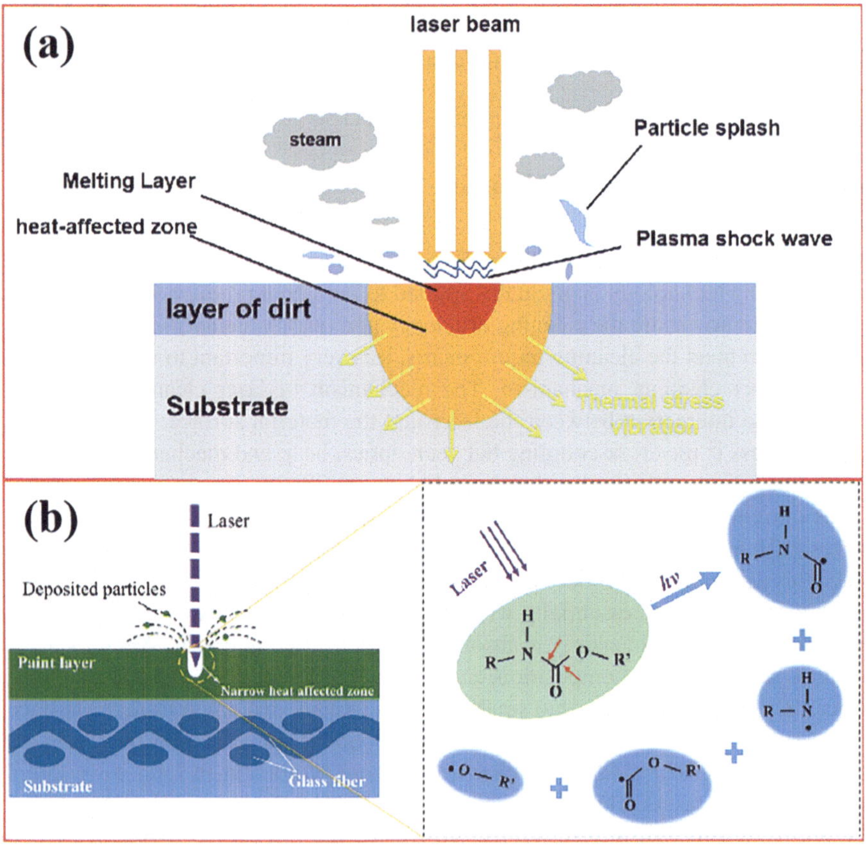


**Fig. 3.2** Schematic diagram of laser cleaning mechanism. (**a**) Schematic diagram of thermal ablation, thermal stress and plasma shock wave cleaning mechanism. (**b**) Schematic diagram of the photochemical-based laser paint removal mechanism

the substrate can be effectively avoided, so UV laser can be considered for the cleaning of precision components, but its cleaning rate is not as high as that of thermal ablation.

In the context of laser cleaning of CFRP, it is essential to consider the composite matrix's absorption characteristics for various laser light wavelengths. Xu et al. demonstrated that CFRP substrates exhibit varying absorption rates across different laser wavelengths [36]. The absorption of laser light by CFRP increases inversely with the laser wavelength; shorter wavelengths result in higher absorption. That is to say, CFRP absorb UV lasers at a wavelength of 355 nm more than an IR laser at a wavelength of 1064 nm. In the application of UV laser cleaning, it is advisable to select a low laser power. This approach capitalizes on the photochemical effects of the UV laser, thereby minimizing the thermal impact on the substrate. When employing an infrared laser as the laser source, it is appropriate to increase the laser power to fully leverage the thermal ablation effect characteristic of the infrared laser.

The cleaning process must account for the interaction between the contaminant and the laser. It is essential to select the laser source and cleaning method based on the morphology of the contaminant. Contaminants requiring removal from CFRP surfaces encompass mold release agents, paint layers, and surface resin layers. The release agent, introduced during the molding phase of CFRP production, leaves residues that can impede the bonding process of CFRP. Therefore, it is crucial to implement cleaning measures to remove these residues prior to bonding. Rauh [37] used a UV laser to successfully clean the mold release agent. Subsequently Schweizer et al. [38] employed UV and IR lasers to effectively eliminate the surface release agent from the material. The experimental findings revealed that both laser types exhibited the capability to remove the release agent. However, noteworthy differences were observed in their effects on the shear strength of the material. Specifically, the IR laser cleaning process resulted in a significant enhancement of the shear strength, which reached 21.2 MPa. In contrast, the UV laser cleaning only led to a comparatively modest increase, raising the shear strength to 15.3 MPa. IR laser is recommended for cleaning release agents. CFRP finds extensive application in aircraft, particularly in the fuselage, wings, and various structural components, owing to its exceptional combination of high strength and low weight. In the course of aircraft maintenance procedures, it becomes necessary to undertake the removal of paint layers from pertinent structural elements, followed by the application of fresh coats of paint. Zhao et al. [39] used an IR laser to treat the paint layer on the CFRP surface, and the experimental results showed that the cleaning mechanism of the paint layer was mainly thermal decomposition. Gu et al. [40] used a nanosecond UV laser for laser paint removal on CFRP aircraft skins to explore the cleaning mechanism and substrate damage mechanism. The experimental results show that the paint removal mechanism of UV laser is thermal stress and photochemical decomposition, while the decomposition of the resin is due to photothermal and photochemical effects, and the fracture of the carbon fibers is due to the effect of thermal shock by the high temperature gas. From the results of Zhao [39] and Gu [40], it can be found that the thermal effect of laser is the prevalent cleaning mechanism. The fracture and shedding of carbon fiber is also due to the thermal shock effect, so it is necessary to reduce the influence of thermal effect on the basis of ensuring the cleaning effect. In order to improve adhesion, the surface resin layer needs to be cleaned to increase the contact area with the adhesive in order to form a mechanical interlocking structure. Gao et al. [41] used a UV laser to clean the resin layer on the CFRP surface and found that the bond strength of partial cleaning was higher than that of complete cleaning, due to the fact that the residual partial resin protects the carbon fibers and increases the contact area with the binder. Tong et al. [42] used IR nanosecond pulsed laser cleaning of resins on CFRP surfaces to improve their surface adhesion properties, and the surface bond strength of the material after laser cleaning was 1.6 times higher than that of the untreated surface. The optimum cleaning parameters for the release agent, paint and resin layers on the CFRP surface are shown in Table 3.1.

Select the laser source or parameters suitable for cleaning the CFRP surface according to the physical and chemical properties of the contaminants to be removed. Since the vaporization temperature of the release agent is lower, the IR laser can be chosen for more efficient cleaning, and a better cleaning surface can be obtained. The vaporization temperature range of the paint layer is larger, for the paint layer with lower vaporization temperature, IR laser can be selected to use laser thermal ablation to remove the paint layer quickly, while for the paint layer with higher vaporization temperature, UV laser can be used to use photochemical decomposition effect to decompose the large paint layer into small particles and then make the particles fall off through vibration effect. For the cleaning of epoxy resin layers, a milder UV laser is recommended. The purpose is to retain a thin layer of epoxy resin on the surface of the material to protect the internal carbon fibers and improve the mechanical properties of the material.

**Table 3.1** Optimal cleaning parameters for different contaminants under different light sources

| Type of pollutant | Type of laser | Optimal cleaning parameters | Ref. |
|---|---|---|---|
| Release agents | UV | $\lambda = 266$ nm, f = 30 kHz, $\tau = 20$ ns<br>P = 3 W, F = 428 mm, V = 20 m/s | [37] |
| Release agents | IR | $\lambda = 1064$ nm, treatment with once IR1 and subsequently three times IR2<br>IR1: P = 19.7 W, $\tau = 200$ ns, f = 27 kHz, V = 270 mm/s<br>IR2: P = 12.7 W, $\tau = 30$ ns, f = 100 kHz, V = 5000 mm/s | [38] |
| Paint layers | IR | $\lambda = 1064$ nm, P = 45 W, f = 20 kHz<br>V = 1600 mm/s, L = 0.03 mm | [39] |
| Paint layers | UV | $\lambda = 355$ nm, P = 15 W, f = 50 kHz<br>V = 1100 mm/s, N = 3 | [40] |
| Surface layer of resin | UV | $\lambda = 355$ nm, $\tau = 12$ ns, P = 13.6 W<br>f = 30 kHz, V = 500 mm/s | [41] |
| Surface layer of resin | IR | $\lambda = 1064$ nm, $\tau = 100$ ns, P = 16 W<br>f = 30 kHz, V = 1600 mm/s, F = 50 μm | [42] |

*Notes*: λ: Wavelength, τ: Pulse duration, f: Repetition frequency, V: Scanning speed, P: Average power, N: Scan times, L: Laser scanning spacing, F: Focal length

## 3.3 Effect of Laser Cleaning on Mechanical Properties of CFRP

Mechanical grinding recovery, thermal decomposition recovery and chemical solvent decomposition recovery are three means of CFRP recovery and reuse [43, 44]. Mechanical grinding recovery can damage carbon fibers and reduce their mechanical properties during grinding or crushing [45]. The gas produced by thermal decomposition contains oily products and granular carbon, which will deposit on the surface of the fiber and reduce the quality of the carbon fiber [46]. The solvents (toluene, acetone, xylene, etc.) and catalysts (hydrogen peroxide, sodium persulfate) used in the chemical solvent method are often toxic and can cause damage to carbon fibers. Toxic gases, such as CO, are generated in the reaction process [47], and the recovery and treatment of solvents, catalysts and toxic gases increase the cost virtually. The structural integrity of CFRP can be maintained to a certain extent through green, efficient and precise laser cleaning. The cleaned CFRP can be reused after repainting. At the same time, the reasonable laser cleaning process can play a certain role in strengthening its surface properties [48–51], which greatly reduces the scrap rate of the material.

### *3.3.1 Effect of Laser Cleaning on the Shear Strength of CFRP*

#### 3.3.1.1 Effect of Surface Physicochemical Properties on Shear Strength

Laser cleaning of CFRP has proven to be feasible when compared to conventional cleaning methods. The cleanliness is guaranteed and at the same time the shear strength of the CFRP is improved when gluing. Belcher et al. [52] used a Nd:YAG laser to etch the surface of CFRP sheets to achieve the effect of cleaning the surface as well as increasing the surface adhesion. It was also analyzed in comparison with the original surface, sandblasted, mechanically polished, and stripped-layer treated surfaces. The "pitting" phenomenon in sandblasting treatment was more serious, the surface uniformity of mechanical grinding was poor, the surface of peel layer treatment had debris and micro-curved surface, and the two surfaces of laser cleaning were flat and regular. At the same time, the single lap shear experiment showed that the shear strength of the laser treatment was slightly better than that of the peeling layer treatment and sandblasting treatment. It can be seen that laser cleaning has the potential to replace physical cleaning such as sandblasting and mechanical grinding.

Laser cleaning results in changes in the functional groups on the CFRP surface and the adhesive used in the bonding process, which can affect the shear strength. Liu et al. [53] investigated the physicochemical properties of laser cleaned CFRP surfaces. Laser treatment with different power (80, 120, 180 W) was compared with alcohol cleaning and sandpaper polishing. It was found that laser treatment can

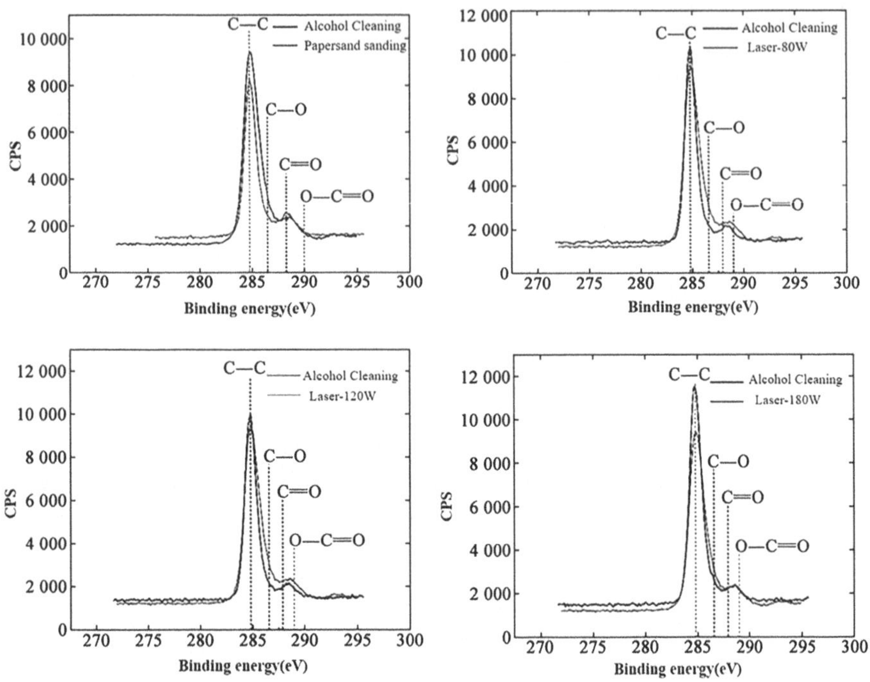


**Fig. 3.3** CFRP C1s peak after different surface treatments: (**a**) sand papering, (**b**) laser-80 W, (**c**) laser-100 W, (**d**) laser-120 W

improve the surface roughness of CFRP, reduce the content of C–C non-polar bonds, and increase the content of C–O, C=O and other polar bonds to improve the surface energy, as shown in Fig. 3.3. The reaction of CFRP with $O_2$ in air under a high-energy laser beam resulted in an increase in the content of oxygen-containing functional groups and a decrease in the number of C–C functional groups [54]. After a series of cleaning tests, it was found that with the increase of laser power, the surface roughness and surface energy first increased and then decreased, and the optimal cleaning power was 120 W. Four different adhesives (BF 2850, PU 6700, Araldite 420, E-120HP) were used in the single lap experiment, as shown in Fig. 3.4. Figure 3.4a, b showed that the joint shear strength was significantly reduced after surface treatment with two polyurethane adhesives, BF 2850 and PU 6700. The reason analyzed was that the polyurethane adhesive with high viscosity and low fluidity was not sensitive to roughness changes and could not fill the gap created by the CFRP surface treatment in time, thus causing the joint to fracture when subjected to shear forces. Araldite 420 and E-120HP in Fig. 3.4c, d were epoxy resin adhesives with low viscosity, high fluidity and high sensitivity to roughness. They could penetrate the CFRP surface in time, increase the contact area, form mechanical interlocking, and substantially increase the shear strength, so epoxy resin

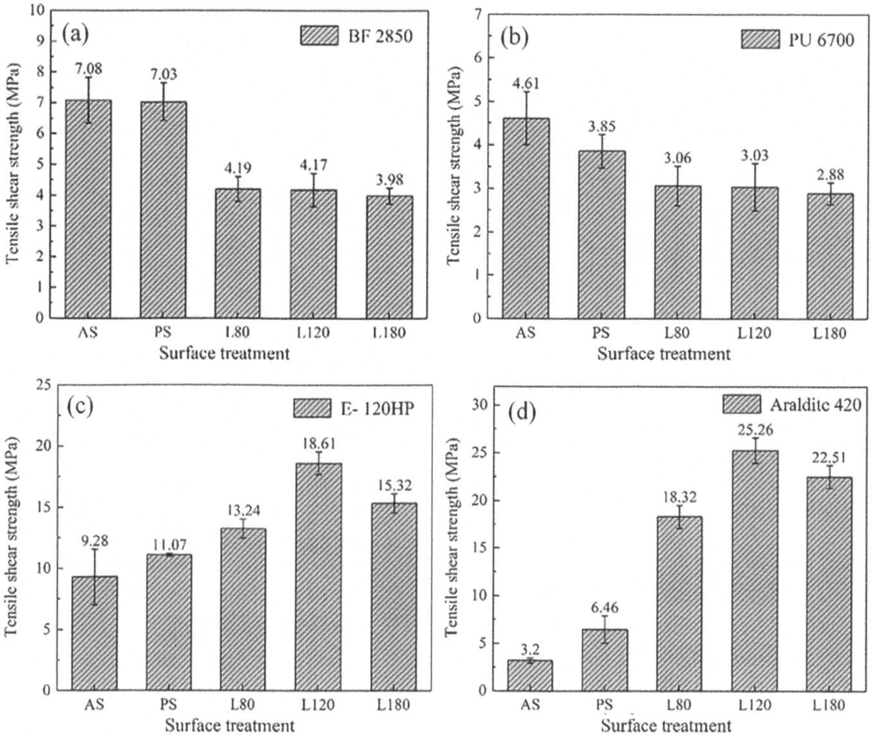


**Fig. 3.4** Lap shear strength of joints after different surface treatments: (**a**) BF2850, (**b**) PU6700, (**c**) E-120HP, (**d**) Araldite420. *AS* mechanical polishing, *PS* chemical soaking, *L80, L120 and L180* three different laser powers

adhesives should be prioritized when gluing. This study also side-steps the fact that laser cleaning can be used as a pretreatment to improve the surface roughness of CFRP to obtain adhesive interfacial bonding.

The laser ablation treatment effectively eliminates surface release agents and contaminants, resulting in the substantial activation of CFRP glued joints. Sun et al. [55] used a fiber laser with a wavelength of 1064 nm for laser ablation of CFRP glued joints. Figure 3.5 showed the SEM images of the as-received and different laser energy density treated CFRP surfaces. From Fig. 3.5a, it can be seen that the surface of the as-received CFRP was relatively uniform and smooth, and the carbon fibers were all covered by epoxy resin. When the energy density was increased to 8.8 mJ/mm$^2$, part of the epoxy resin in Fig. 3.5b was removed and the top carbon fibers were partially exposed. When the energy density was increased to 35.4 mJ/mm$^2$, all carbon fibers in the top layer in Fig. 3.5c were completely exposed. When the laser energy density rose to 70.8 mJ/mm$^2$, the ablation fracture occurred in the top carbon fibers part in Fig. 3.5d, which belongs to excessive cleaning and should be avoided as much as possible. The bonding strength of the joint was increased from 8.5 to 30.4 MPa without damaging the carbon fiber, and the optimal laser

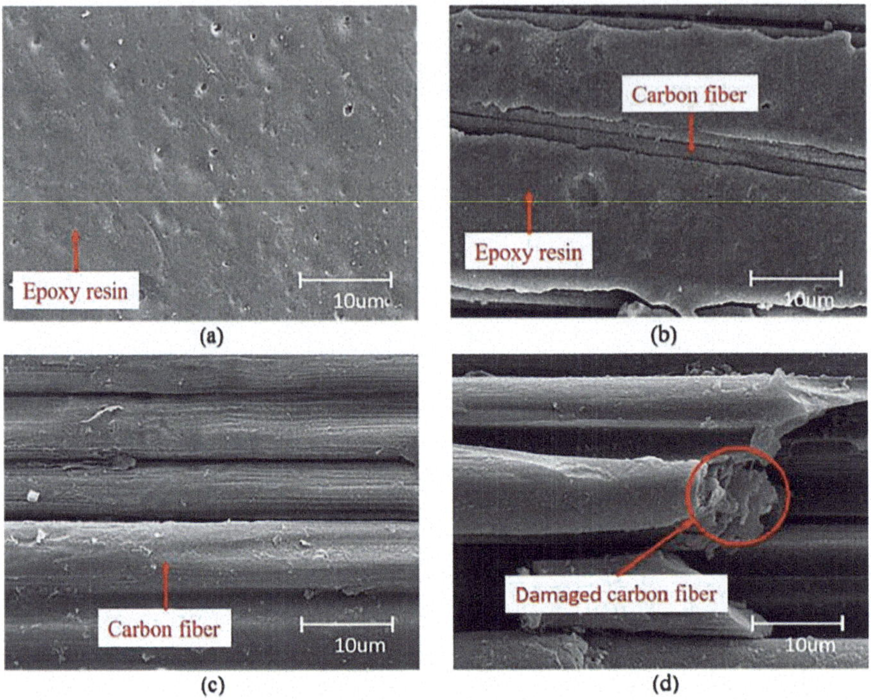


**Fig. 3.5** SEM images of (**a**) as-received CFRP substrate surface and laser treated substrate CFRP surfaces at EDS of (**b**) 8.8 mJ/mm$^2$, (**c**) 35.4 mJ/mm$^2$, and (**d**) 70.8 mJ/mm$^2$

energy density obtained by CFRP was 35.4 mJ/mm$^2$. Zhan et al. [56] used an IPG-50 W pulsed laser with a wavelength of 1064 nm to clean the CFRP surface. The best cleaning effect was achieved when the scanning speed was 6000 mm/s, the laser power density was 22.93 kW/cm$^2$, and the pulse frequency was 100 kHz. At this time, the resin and pollutants on the surface were cleaned, and the exposed carbon fiber was continuous and intact. Single lap shear tested on three groups of samples after original surface, mechanical grinding and laser cleaning showed that the strength of the glued joints was the greatest after laser treatment, reaching 27 MPa, which was 36.5% higher than that of the untreated samples. This shows that laser cleaning can be used not only as a means of CFRP surface cleaning, but also as a means of pre-treatment of the original surface.

It can be seen from the above analysis that reasonable cleaning parameters can not only clean the surface, but even change the physical and chemical properties of the CFRP surface. For the CFRP surface that needs to be recoated, the control of physical and chemical properties of the substrate surface after laser cleaning can significantly improve the interface bonding force between the coating and the CFRP matrix, and obtain high quality coating.

#### 3.3.1.2 Effect of Heat Affected Zone on Shear Strength

The anisotropy [57] inherent in carbon fibers within CFRP results in non-uniform material removal within the laser ablation zone, consequently giving rise to the formation of a heat-affected zone (HAZ). The HAZ of CFRP includes resin charring zones, fiber end swelling, delamination, and cracking. The resin matrix of the CFRP surface layer is composed of polymers. The heat accumulation at the surface of the laser cleaning surface has obvious effect on the surface resin polymer of CFRP, and the instantaneous heat accumulation may cause the damage of the surface resin. Zhan et al. [58] used an IPG-YLP infrared pulsed laser to clean T300 CFRP laminates. The IR laser irradiates the surface of the CFRP laminate, where the heat is absorbed by both the resin and carbon fibers. The surface resin absorbs the laser energy to produce HAZ, as shown in Fig. 3.6a. The laser heat source roughly satisfies the Gaussian distribution of Fig. 3.6b. The carbon fibers are heated initially by the transmitted IR laser radiation during the laser radiation process. The axial thermal conductivity of carbon fibers exceeds that in the radial direction, facilitating heat transfer along the axial orientation of the carbon fiber. Notably, the resin's vaporization temperature is significantly lower than that of the carbon fiber, leading to the removal of surface resin, as illustrated in Fig. 3.6c. Therefore, the energy density is controlled within a certain range during the cleaning process, and the heat accumulation on the laser cleaning surface interface can be reduced to avoid damage to the carbon fiber. The experimental results showed that the maximum CFRP

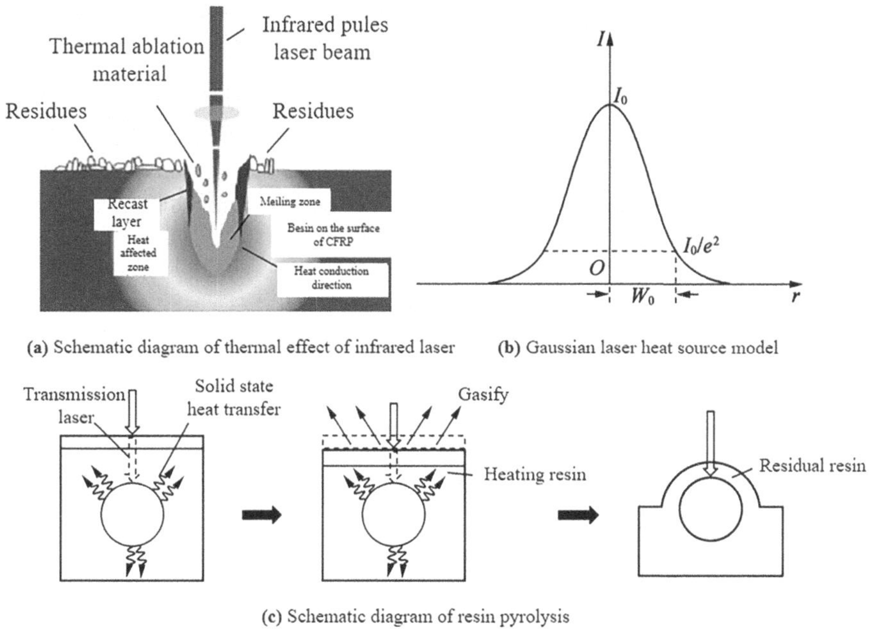


**Fig. 3.6** Schematic diagram of infrared laser irradiation on the surface of CFRP laminates: (**a**) schematic diagram of infrared laser thermal effect; (**b**) Gaussian laser heat source model; (**c**) schematic diagram of resin pyrolysis

shear strength of 27.22 MPa was achieved at a scanning speed of 7000 mm/s and a power of 22.2 W.

Fischer et al. [59] used UV and IR lasers for surface treatment of CFRP sheets, taking into account both the thermal deposition and absorption behaviors generated by the laser radiation. Comparison with the stripped layer and sandblasted samples showed that the UV laser treated samples had the highest shear strength without damaging the carbon fibers, but the treatment efficiency was not as good as that of the infrared laser. But aside from the cleaning efficiency factor, the photon energy of the infrared laser is lower and it is easy to deposit more heat in the material. Meanwhile, CFRP exhibits lower absorption of IR laser compared to UV laser. Consequently, the power of the IR laser is typically higher than that of the UV laser during the cleaning process, thereby elevating the potential risk of ablation. Therefore, when choosing the wavelength of laser equipment for cleaning CFRP surface, the absorption rate of different wavelengths of laser by matrix materials, the thermal deposition effect of laser energy and cleaning efficiency should be considered comprehensively.

Partial melting of the heat affected zone will increase the surface roughness and even reduce the maximum tensile stress of the material, which will affect the main performance of CFRP. Therefore, it is particularly important to avoid the influence of the heat affected zone on CFRP [60]. Li et al. [61] used an infrared nanosecond single-pulse laser to study the effect of laser power density and scanning mode on CFRP surface treatment. The single lap shear test showed that the shear strength of the treated sample after improved scanning method was four times that of the untreated sample. Unlike the continuous pulse laser, the one-dimensional numerical model established by Li showed that the surface gasification depth and heat affected zone were very small after single pulse laser treatment.

Laser as a heat source will have a certain thermal impact on the CFRP surface, and the thermal aggregation at the interface of the laser-cleaned surface will reduce the mechanical properties of the substrate or even the whole specimen. Indeed, the thermal accumulation phenomenon is primarily observed in the utilization of continuous lasers and long-pulse lasers, and it can be mitigated or circumvented through the application of short-pulse lasers. Short-wave lasers are usually UV lasers with a wavelength of 355 nm or shorter. As mentioned in Sect. 3.1, the cleaning mechanism of UV laser is mainly photochemical decomposition, and the heat-affected zone of the cleaning process is small, so the occurrence of thermal ablation can be effectively avoided. Long-wave lasers, such as IR lasers, the cleaning mechanism is mainly thermal ablation, so in the cleaning process should be reasonably adjusted parameters to take into account the cleaning rate and cleaning effect.

#### 3.3.1.3 Effect of Incomplete Cleaning on Shear Strength

By reviewing a large amount of literature, there are studies that characterize an interesting phenomenon, i.e., the existence of partially incompletely cleaned samples with higher shear strengths than fully cleaned samples. Gao et al. [38] used UV laser to treat the surface of CFRP sheets at scanning speeds of 300 mm/s and 500 mm/s, respectively, and conducted bonding experiments on the treated samples. The microstructure of the failed samples after complete UV laser cleaning showed that there were tears on the surface of the carbon fibers, while the failed samples after incomplete UV laser cleaning were intact due to the protection of the resin on the surface of the carbon fibers before bonding. Figure 3.7a showed the single lap shear tensile curves obtained after the samples were cleaned. It is easy to see that the tensile strength of the partially cleaned sample was significantly greater than that of the fully cleaned sample, which was able to carry a shear force of $1 \times 10^5$ N. Figure 3.7b showed the shear strength of the two groups of samples. It can be seen that the strength of the adhesive joint cleaned by UV laser was higher than that of the adhesive joint completely cleaned by UV laser. It is clear that retaining part of the resin while cleaning the sample can improve the bonding performance of CFRP sheets, and the laser incomplete cleaning process for CFRP surfaces is worth digging deeper.

Like Gao's study, Wang found that incomplete exposure of surface fibers after laser cleaning significantly improved tensile shear properties. Wang et al. [62] used nanosecond laser to remove epoxy resins and pollutants on the surface of CFRP composites to improve the bonding strength of CFRP with other materials. It was found that the best cleaning effect was achieved when the laser power density was 8.8 kW cm$^2$ and the scanning speed was 1000 mm/s. Under this parameter, the substrate surface had a high free energy.

Incomplete cleaning without touching the carbon fiber filaments is essential when performing laser cleaning. Zhan et al. [63] used a UV laser to study the effect of the amount of surface residual resin on the shear strength of CFRP joints. The shear test results showed that the shear strength of the sample with part of resin retained on the surface was the highest, but the shear strength of the sample completely cleaned by laser was lower than that of the untreated sample. The analysis of the reason may be that after the carbon fiber was completely exposed, when it was glued again, the fiber surface did not bond well with the binder interface, and the bonding ability with the binder was weakened. Rauh et al. [37] used UV laser to

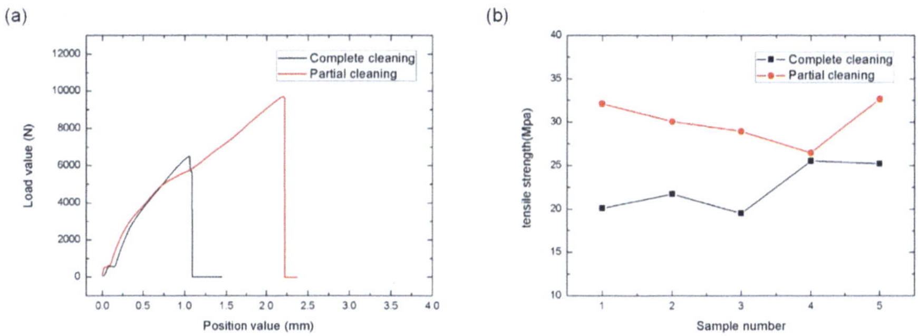


**Fig. 3.7** Cleaning strength of CFRP by different laser scanning speeds: (**a**) tensile curve and (**b**) tensile value (the scanning speed of complete cleaning is 300 mm/s, the scanning speed of incomplete cleaning is 500 mm/s, and samples 1–5 are five experiments under this parameter, respectively)

clean the residue of CFRP mold release agent. Figure 3.8a showed the surface of the CFRP prior to laser treatment, revealing an uneven presence of mold release agent residue. Following the application of the first set of cleaning parameters (single pulse energy of 0.12 J/cm$^2$), the smoothing process resulted in the substantial removal of release agents, concurrently exposing certain carbon fibers, as depicted in Fig. 3.8b. Figure 3.8c showed the surface morphology after treatment using the second set of cleaning parameters (single pulse energy of 0.24 J/cm$^2$), when the contaminants were completely removed and more intact carbon fibers were exposed. The single lap shear experiments of CFRP in original state, mechanical grinding and two sets of UV laser treated samples with different parameters were carried out. The shear strength of the joint was only 7.5 MPa on the original surface because of the silicone residue, and the shear strength was increased to 12.5 MPa after mechanical grinding. After UV laser treatment, the shear strength of the fully exposed carbon fiber reached 24 MPa. After UV laser smoothing treatment, the shear strength of the sample reached 25 MPa. It can be seen that UV laser cleaning of CFRP release agent is better than mechanical grinding, while the shear strength of incompletely cleaned samples is slightly better than that of fully cleaned.

All the above studies show that the shear strength of incompletely cleaned CFRP samples is higher than that of fully cleaned samples. The reason for this may be that

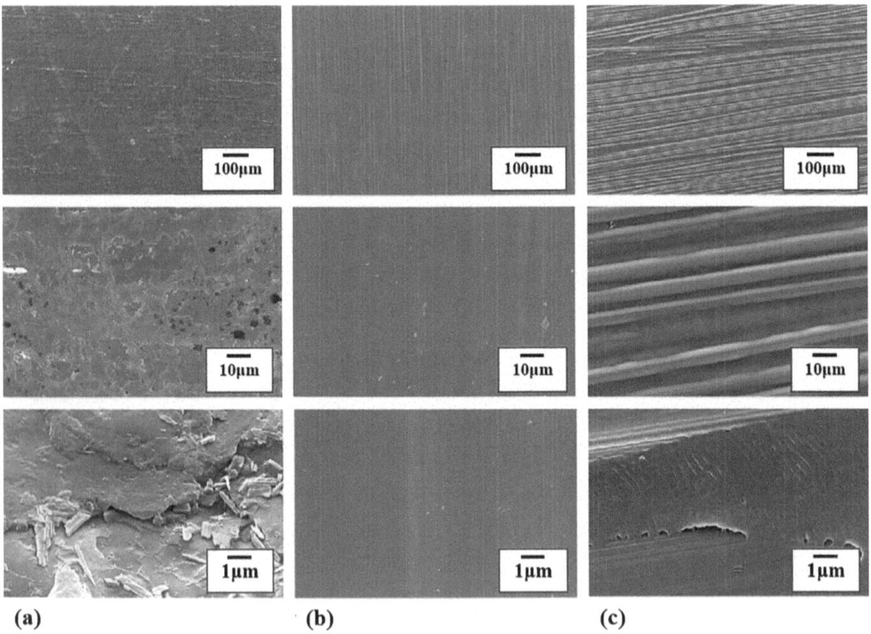


**Fig. 3.8** SEM micrographs of different CFRP surfaces with increasing magnification: (**a**) after demoulding and cutting, without treatment, (**b**) treated with laser parameter set 1 with the purpose of smooth cleaning, (**c**) treated with laser parameter set 2 that exposed the carbon fibres without any observable damage on the fibres

some resin is well bonded to the carbon fiber when it is first coated. Therefore, when laser irradiates the surface of carbon fiber, in the case of not obvious ablative effect, only vibration effect, laser direct impact and other physical effects cannot remove the resin tightly bound with carbon fiber. In the incomplete cleaning state, the residual resin on the substrate surface can increase the contact area with the binder and increase the surface roughness of the carbon fiber sheet substrate during the re-coating process, so that the shear strength of incomplete cleaning will be higher than that of complete cleaning.

The surface morphology of CFRP after laser cleaning can be divided into three stages: incomplete cleaning stage, complete cleaning stage, and excessive cleaning stage. In the incompletely cleaned stage, there are many resin particles or paint layers left on the sample surface. In the fully cleaned stage, the resin particles on the surface of the sample are basically removed, and flat and smooth carbon fibers can be seen. In the over-cleaning stage, localized carbon fibers are fractured and melted. The damaged carbon fibers lead to a reduction in the mechanical properties of the material, while the higher shear strength of the incompletely cleaned samples is due to the fact that the carbon fibers are protected by a thin layer of resin remaining on the surface.

### 3.3.2 Effect of Laser Cleaning on CFRP Tensile Properties

In addition to shear strength, the tensile property of CFRP after laser cleaning is one of the key indicators that scholars and engineers pay attention to [64, 65]. Iwahori [66] removed the paint coating of T300 CFRP by four means: sandpaper sanding, chemical stripping, sandblasting, and TEA-$CO_2$ laser treatment and tested the tensile and compressive strength and modulus before and after paint removal. There was little difference in the tensile properties of CFRP before and after cleaning by the three means, except for a change in the tensile strength of CFRP after chemical stripping. It is shown that the TEA-$CO_2$ laser is a friendly cleaning method for CFRP tensile properties and did not have a deteriorating or strengthening effect on the matrix tensile properties.

The predominant laser sources employed for cleaning CFRP are UV lasers operating at a wavelength of 355 nm and IR lasers at 1064 nm. By comparing the cleaning effect, the properties and characteristics of the two lasers can be fully grasped. It is convenient to select a suitable laser source for cleaning according to different cleaning needs. Li et al. [67] investigated the cleaning characteristics of CFRP sheet surfaces using IR and UV lasers under different process parameters. The strength of the glued joints and the surface free energy of the cleaned samples were comparatively analyzed. It was found that IR laser removed the resin mainly by thermal effect, and UV laser irradiation caused breakage of chemical bonds and decomposition of the resin. Figure 3.9 showed the shear strength graph of the single lap shear test and the tensile curves of the three samples. From Fig. 3.9a, it can be seen that the strength of the joints after UV cleaning was generally better than that of IR. The

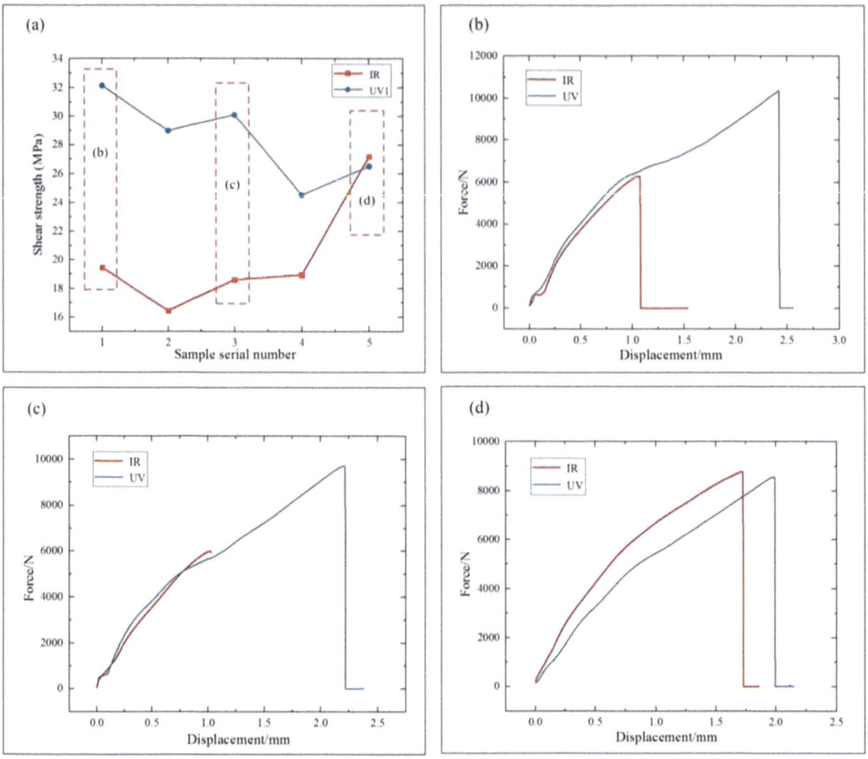


**Fig. 3.9** Uniaxial tensile testing results of single lap joints: (**a**) shear strength summative chart of 5 samples; (**b–d**) the load-displacement curves obtained by tensile tests (The speed of IR laser cleaning is 9000 mm/s, the power is 25.9 W, the speed of UV laser cleaning is 500 mm/s, and the power is 13.6 W. The numbers 1–5 are five samples made under this parameter)

load-displacement curves from Fig. 3.9b to d showed that the tensile strength displacements of the UV-treated samples were greater than those of the IR-treated samples. UV is more suitable for pre-treating CFRP sheets prior to bonding for tensile strength considerations only.

The thermal effect induced by the IR laser surpasses that of the UV laser, potentially causing a reduction in the tensile properties of CFRP following laser cleaning. Noskov et al. [60] investigated the changes in the tensile properties of IR laser irradiated CFRP composites. The experimental results showed that the tensile strength of CFRP decreased slightly after laser irradiation. The tensile properties of CFRP decreased to 68% and 16% at 200 °C and 600 °C, respectively [68]. The reason may be due to the influence of infrared laser heat affected zone. Based on this, Noskov proposed the use of small pulse widths and low frequency lasers in order to achieve a significant reduction in the heat-affected zone, thereby maximizing the retention of the tensile properties of the treated material. Gu et al. [33] used a UV picosecond laser at a wavelength of 355 nm to clean CFRP under different process parameters

(S1: P = 6 W, N = 4, S2: P = 12 W, N = 3, S3: P = 10.5 W, N = 6, S4: P = 12 W, N = 8). The optimum cleaning effect was achieved at a laser power of 12 W and after three cleanings. The "low-temperature cleaning" by the UV laser did not cause any thermal damage to the material, and the tensile properties of the material remained unchanged before and after paint removal. Figure 3.10 showed a schematic diagram of the UV paint removal mechanism, in which it can be seen that the heat-affected zone of the UV laser was very small. UV lasers primarily leverage the photochemical effect between the laser and the material for the removal of the coating layer. Under the irradiation of a UV laser (with a single photon energy of 337 kJ/mol), the bonds on the surface of the paint layer, such as C–C (332 kJ/mol), C–O (326 kJ/mol), and C–N (305 kJ/mol), were broken. Ultimately, the long-chain polymers undergo conversion into monomers or small volatile molecules, which subsequently detach from the material's surface. This cleaning method serves to minimize or entirely prevent damage to the substrate.

In summary, it can be seen that the CFRP tensile properties after laser cleaning are related to the type of laser and the temperature of the cleaning surface interface. In order to maintain the inherent tensile properties of CFRP, the use of $CO_2$ lasers are not recommended, and UV lasers are preferred. In addition, under the premise of considering the cleaning efficiency, smaller pulse width and lower frequency should be selected as far as possible to reduce the range of heat affected zone and improve the tensile property of the material.

### 3.3.3 *Effect of Laser Cleaning on Other Mechanical Properties of CFRP*

Carbon fibers can withstand greater loads than resin, and the bending strength of the samples is increased because the carbon fibers are exposed on the surface of the material after laser cleaning. Wang et al. [62] cleaned CFRP surface resins and contaminants by a nanosecond laser. Three typical samples (laser power densities of

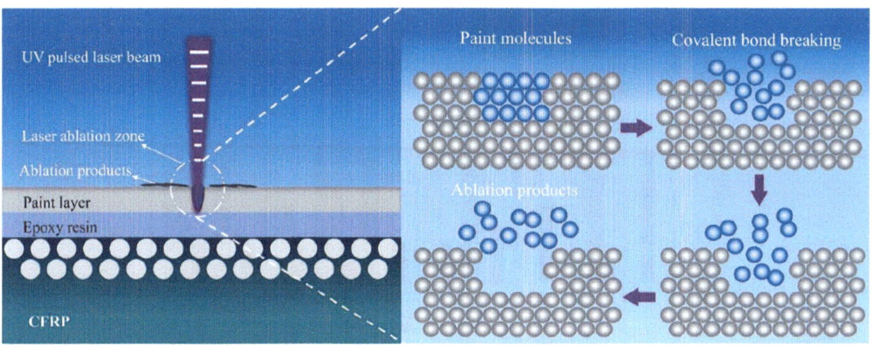


**Fig. 3.10** Schematic diagram of UV laser cleaning mechanism

7.0, 8.8, and 14.1 kW/cm$^2$ corresponding to partial, complete, and excessive cleaning, respectively) and one control sample (not cleaned) were tested for bending after cleaning. The experiments were conducted employing the three-point bending method, as shown in Fig. 3.11. In this configuration, the cleaning surface is positioned on the upper side, with the fiber direction perpendicular to the bending axis. The experimental results showed that: the control sample started to show damage deformation at 1866 N, and the bending strength was 1124 MPa. The fully cleaned samples exhibit a maximum load capacity of 2218 N before initiating deformation, with the optimal bending strength reaching 1294 MPa.

While exposed intact carbon fibers have the potential to enhance the flexural properties of CFRP, an excessive application of laser power can result in the breakage of these fibers, leading to a significant decline in flexural properties. Ji et al. [69] investigated the flexural properties of laser irradiated carbon fiber reinforced polyether ether ketone (CF/PEEK) composites by varying the laser power and scanning speed. The experimental results are shown in Fig. 3.12, as the laser power increases the bending property of the material decreases rapidly and then stabilizes. When the

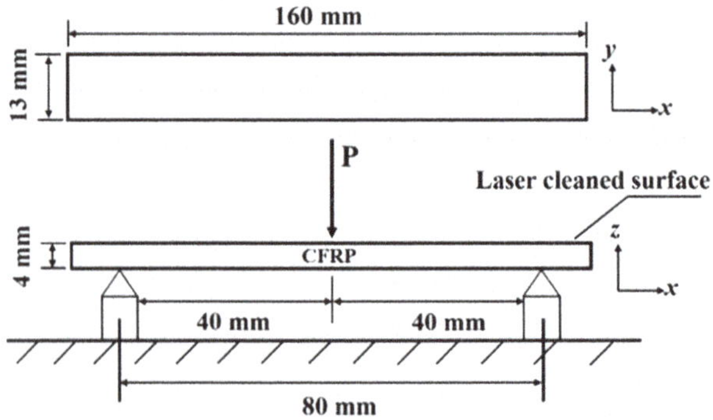


**Fig. 3.11** Schematic diagram of three-point bending test

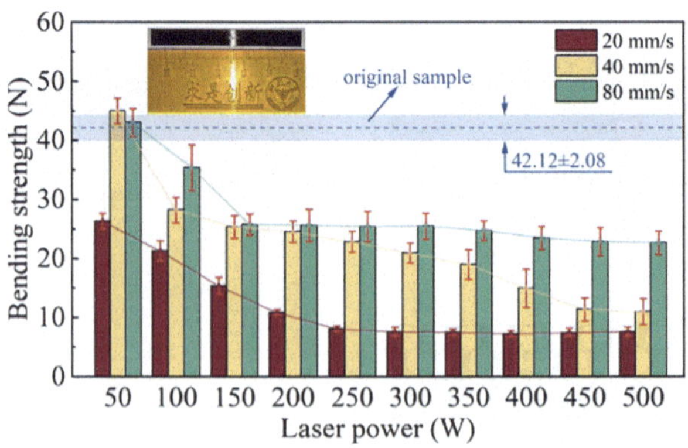


**Fig. 3.12** Bending strength of CF/PEEK composites irradiated by different laser parameters

laser power was 500 W and the scanning speed was 20 mm/s, the material was destroyed and the bending strength was reduced by 81.94%. High-power lasers can cause the resin to evaporate and deteriorate, and local carbon fibers can be exposed to bending or even ablation fracture. Therefore, in the cleaning process to ensure the inherent bending properties of the material, the laser power and scanning speed should be matched.

Laser cleaning has an all-round effect on the physical and chemical effects on the surface of the CFRP matrix, and on the comprehensive mechanical properties of the material. Palmieri et al. [70] used a nanosecond pulsed Nd:YAG laser at a wavelength of 355 nm to laser ablate the CFRP surface and retain the intact carbon fibers, and subsequently prepared single lap shear specimens were stored in different humidity environments for accelerated aging for 3 years. The analysis of the stability of its failure mode showed that the aging treatment did not significantly reduce the properties of the laser-treated interface. In conclusion, the laser bonded pretreated CFRP has a certain stability of comprehensive mechanical properties. Hu et al. [71] used an infrared nanosecond laser with a wavelength of 1064 nm to pretreat the multi-step bonding interface of CFRP composites. The laser was applied to the surface of the material to remove the resin between the carbon fibers, the surface roughness and surface area of the material increased, and the bond strength was improved. The tensile and impact strengths of the material joints after laser cleaning reached 86.29 MPa and 9.43 kJ/m$^2$, respectively, which were increased by 11% and 15%, respectively, compared with the conventional milling treatment process.

Through the above research, it can be seen that laser cleaning CFRP has obvious advantages over traditional cleaning methods. According to the type and nature of the substance to be cleaned, a suitable laser can be selected to achieve the expected cleaning effect. Due to the relatively high energy of the laser beam, attention should be paid to avoiding the negative effects of laser cleaning of the surface heat affected zone during cleaning. Make full use of the thermal effect of laser irradiation to speed up the efficiency of paint removal. Without placing much emphasis on paint removal efficiency, UV laser low temperature paint removal quality is superior to IR laser, which is a more optimized low temperature non-destructive laser cleaning option.

## 3.4 Effect of Laser Cleaning on Bonding Properties of CFRP

### *3.4.1 Effect of Specific Surface Microstructures on Bonding Properties*

In order to increase the bonding properties of repainting, laser-induced specific microstructures on the CFRP surface are used to increase the roughness of the CFRP surface, creating a mechanical interlocking effect and thus increasing the interfacial bonding properties [72, 73]. Oliveira et al. [74] used a near-infrared

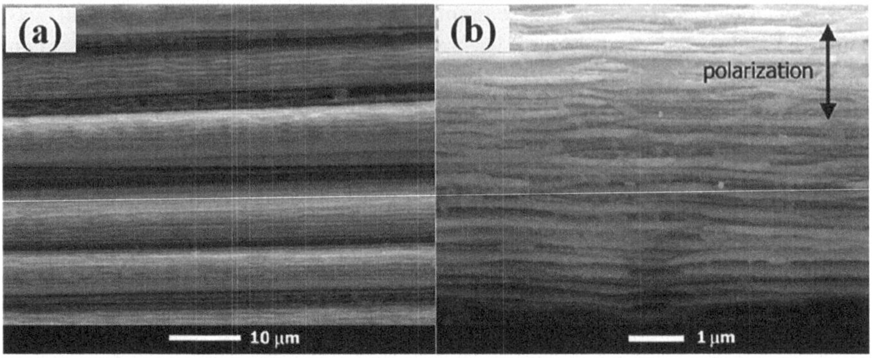


**Fig. 3.13** SEM images of the central region of laser tracks produced with 0.35 mJ and 5 mm/s: (**a**) deep gorges between carbon fibers and (**b**) high magnification image revealing the presence of LIPSS

femtosecond laser to treat CFRP and remove CFRP surface epoxy by modulating the pulse energy and scanning speed. In addition, femtosecond lasers can induce the generation of periodic surface microstructures (LIPSS) on CFRP surfaces within a certain parameter range, as shown in Fig. 3.13b. LIPSS will increase the interfacial contact area and surface roughness, which is favorable to improve the bonding performance when re-coating after laser cleaning. However, looking at the existing research results, it is found that nanosecond laser cleaning is difficult to form LIPSS on CFRP surface.

Similar to Oliveira et al., Chen et al. [75] induced LIPSS on the surface of carbon fiber cloth by femtosecond laser. The structure transforms the surface of CFRP from hydrophobic to hydrophilic. The contact angle of the carbon fiber cloth is reduced from 150.8° to 20.4°, which greatly enhances the wettability of the carbon fiber. Thus, the bonding properties of the material surface interface are improved.

Kunz et al. [76] investigated the formation of LIPSS on carbon fiber surfaces under femtosecond pulsed laser irradiation. LIPSS was induced to form at a pulse width of 300 fs and a wavelength of 1025 nm. The structure disorder and surface roughness of carbon fiber increased after LIPSS formation. At the same time, additional carbon–oxygen bonds are formed, and the surface energy of the material is improved, which can significantly improve the bonding property when joining with other materials. As shown in Table 3.2.

### *3.4.2 Effect of Heterogeneous Joint Connections on Bonding Performance*

In order to meet the requirements of lightweight and high strength in the manufacture of structural components for aerospace equipment, composite structures in which CFRP is connected with other different classes of materials are widely used

**Table 3.2** Detailed parameters and properties of surface microstructure of various carbon fibers

| Substrate | Laser | Scanning parameters | Microstructure shape | Result | Ref. |
|---|---|---|---|---|---|
| $[0]_5$ laminates (CFRP) | $CO_2$ laser | $\lambda = 10{,}600$ nm<br>$\tau = 5$ μs<br>f = 5 kHz<br>V = 200 mm/s | Triangle type, inverted triangle type, frame type, box type | Shear strength increased by 26.8% | [77] |
| $[0_2/90/0_2/90]$ laminates (CFRP) | $CO_2$ laser | $\lambda = 10{,}600$ nm<br>P = 11.3 W<br>f = 25 kHz<br>$\tau = 18$ μs<br>V = 5950 mm/s | Dimples in a square grid | Improve surface roughness and failure load | [78] |
| The CFRP with 0.4–0.5 mm in thickness | Femtosecond laser | $\tau = 500$ fs<br>f = 200 kHz<br>V = 100 mm/s<br>$P_f = 3.8 \times 10^4$ W/cm$^2$, $5.7 \times 10^4$ W/cm$^2$ | Micro-nano structures (100 μm groove, 80 μm grid) | The water contact angle increased from 88° (pristine) to 149° (100 μm groove) and 153° (80 μm grid) | [79] |
| A $[0]_8$ layup with 1.5 mm in thickness | Nanosecond laser | $\lambda = 1060$ nm<br>P = 18 W<br>f = 300 kHZ<br>V = 3850 mm/s | 30 μm cavities | The bond strength can be improved by around 5% | [80] |
| T700 | Nd:YAG laser | P = 10 W<br>f = 20 kHz<br>$\tau = 100$ ns<br>N = 6<br>$\alpha = 0/90°$ | A series of grooves with different patterns | Shear strength increased by 40% | [81] |

*Notes*: $\lambda$: Wavelength, $\tau$: Pulse duration, f: Repetition frequency, V: Scanning speed, P: Average power, $P_f$: Laser power density, N: Scan times, $\alpha$: Scanning angle

[82–86]. The removal of epoxy resin on the surface of CFRP by laser cleaning can ensure the interfacial bond strength when it is connected with heterogeneous joints.

Genna et al. [87] used a 30 W Q-switched Yb:YAG fiber laser for laser cleaning of CFRP laminates and laser-assisted gluing of CFRP to polycarbonate sheets with the help of a 200 W semiconductor continuous laser. The untreated samples were used as a reference for the experiments. Under optimal conditions, the joint strength obtained from the laser pretreatment was more than twice as strong as that of the reference sample.

Zou et al. [88] and Jiao et al. [88, 89] used a picosecond laser to clean carbon fiber reinforced thermoset composites (CFRTS) to improve joint strength when connected to TC4 alloys. After laser cleaning, the epoxy resin on the surface of the CFRTS completely decomposed. The exposed carbon fibers increased their contact area with the binder during the subsequent joining process. The joint strength increased from 0.9 to 14.3 MPa compared to the uncleaned material.

The detailed parameters of the heterogeneous connection between CFRP and other materials are shown in Table 3.3. It can be seen that surface laser cleaning

**Table 3.3** Detailed parameters and properties of heterogeneous connections between CFRP and other materials

| Type | Preprocessing object | Technological parameter | Result | Ref. |
|---|---|---|---|---|
| CFRP/ A7075 | A7075 | $\lambda = 1060$ nm<br>$P = 500$ W<br>$V = 400$ mm/s | Shear strength increased by four times that of untreated | [90] |
| CFRP/ AZ31B | Mg alloy | $\lambda = 355$ nm<br>$\tau = 6$ ps<br>$f = 400$ Hz<br>$E_S = 300$ μJ | The shear strength after laser treatment is 76.55% higher than that after sand blasting | [91] |
| CFRP/ Mg | Mg alloy | $\lambda = 1064$ nm<br>$P = 18$ W<br>$f = 100$ kHz<br>$V = 300$ mm/s | The interlayer stripping strength is 1.5–3 times higher than that of sand blasting | [92] |
| CFRP/ DP590 | DP590 | $\lambda = 1064$ nm<br>$P = 100$ W<br>$f = 120$ kHz<br>$V = 3000$ mm/s | The shear strength increased by 299% from 4.1 to 16.35 MPa | [93] |
| CFRP/ A7075 | A7075, CFRP | A7075: $V = 200$ mm/s<br>CFRP: $V = 600$ mm/s<br>$\lambda = 1064$ nm<br>$P = 20$ W<br>$F = 20$ kHz | The shear strength increased from 4.6 to 17.3 MPa | [94] |
| CFRP/ $AlMg_3$ | $AlMg_3$, CFRP | $AlMg_3$: $\lambda = 1064$ nm, $P = 4.4$ W, $f = 45$ kHz, $V = 117$ mm/s<br>CFRP: $\lambda = 355$ nm, $P = 3.2$ W, $f = 20$ kHz, $V = 500$ mm/s | The shear strength increased from 8.2 to 14.9 MPa | [95] |

*Notes*: *ES* is single pulse energy

pretreatment is necessary before gluing resin-based carbon fiber composites with other heterogeneous materials.

Veltrup et al. [96] used UV nanosecond pulsed lasers with different energy densities to clean CFRP and investigate the interfacial bonding properties after cleaning CFRP sheets at different energy densities. The results showed that lower energy densities lead to the reaccumulation of ablation products as well as corrosion inhibitors, resulting in insufficient adhesion strength. Higher energy densities exposed or ablated carbon fibers, which also led to insufficient bonding between the adhesive and the carbon fibers during the re-bonding process. It can be seen that the selection of appropriate laser parameters is important to obtain high interfacial bond strength of CFRP. Furthermore, to enhance adhesion at the repainting interface, one can explore the laser-induced generation of specific microstructures on the substrate's surface or consider retaining a small amount of resin during the cleaning process.

## 3.5 Online Monitoring of Laser Cleaned CFRP

CFRP is extremely sensitive to the surface interface heat, direct impulse, etc. in laser cleaning and the physicochemical effects caused by them, and its surface quality requirements after laser cleaning are high. In order to obtain a high quality clean substrate, rapid inspection of the surface quality after laser cleaning and real-time monitoring of the cleaning process are particularly important. The real-time monitoring of the cleaning process can not only quickly determine the cleaning quality, but also avoid the damage caused by excessive cleaning of the matrix due to various uncontrollable factors during the cleaning process. The cleaning characteristics and mechanisms of different lasers can also be studied by detecting signals such as temperature, spectrum, image and sound wave during the cleaning process [97–99]. Existing literature on online monitoring of CFRP predominantly concentrates on laser-induced breakdown spectroscopy (LIBS) and temperature monitoring, with other monitoring tools remaining largely unexplored in the context of CFRP. Consequently, this paper specifically addresses laser-induced breakdown spectroscopy and temperature monitoring in the realm of CFRP monitoring.

### *3.5.1 Online Monitoring of Laser-Induced Breakdown Spectroscopy*

LIBS is an emission spectroscopy technique wherein a high-energy pulsed laser interacts with a substance, generating a plasma. Qualitative and quantitative information about the composition of the substance is then acquired by detecting and analyzing the characteristic spectral lines of atoms, ions, or molecular groups within the radiation spectrum [100, 101]. Wang et al. [102] applied the LIBS technique to the online monitoring of laser cleaning. By comparing the elemental composition and LIBS spectrum information of epoxy resin and carbon fiber matrix, the laser cleaning process is analyzed, and the principle of monitoring cleaning quality based on LIBS technology is expounded.

Monitoring experiments proved that the Na atomic line at a wavelength of 588.819 nm can characterize the cleaning quality and the average spectral intensity was positively correlated with the average laser power. The cleaning quality was classified into three stages: not cleaned, optimally cleaned, and over-cleaned based on the LIBS intensity eigenvalues, as shown in Fig. 3.14. Among them, the average power values of each interval of the three stages are [4, 6], [6, 9] and [9, 13] respectively. The optimal threshold for the average power of laser cleaning was determined by a segmented linear fitting method to be 8.2 W, with a corresponding LIBS eigen-spectral line value of 128.22 a.u. This monitoring method provides a new idea for the precise control of the laser cleaning process.

Ledesma et al. [103] used the LIBS technique propelled by pulse energies below 100 μJ to monitor the evolution pattern of trace silicone contaminants during laser

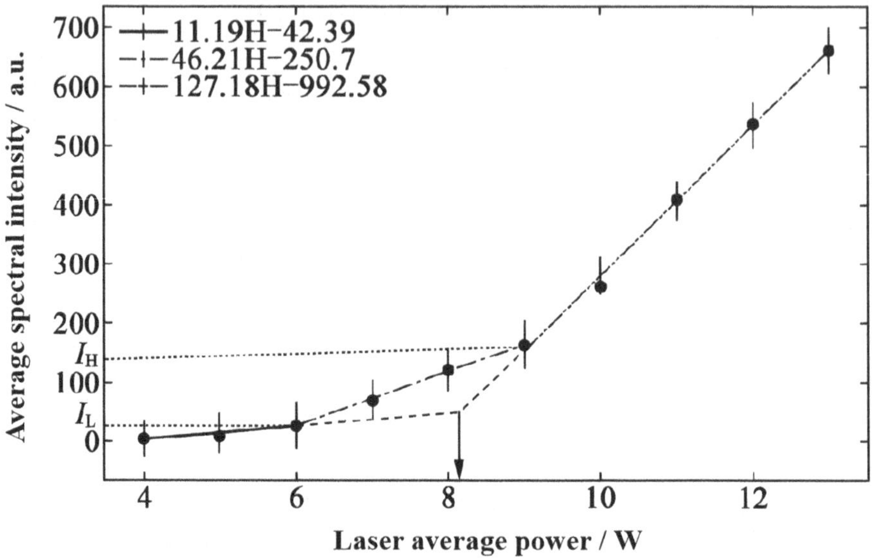


**Fig. 3.14** Piecewise linear fitting graph between average power and the characteristic peak intensity

cleaning of CFRP coated with polydimethylsiloxane (PDMS). The presence of PDMS was detected by Si atomic lines at 251.6 and 288.2 nm before and after laser ablation, and the epoxy resin matrix in CFRP was detected by C atomic lines at 247.9 nm. It was analyzed that the presence of PDMS on the CFRP surface increased the Si emission intensity while suppressing the C emission lines before the laser action. After the laser action, the C emission lines increased and the Si emission lines decreased, as shown in Fig. 3.15. The optimum ablation threshold was 2.4 μJ at a PDMS area density of 2 μg/cm$^2$. The Si/C ratio was reduced by up to 87% after cleaning. It was confirmed that the pulsed laser could effectively remove the silicone from the CFRP surface. This study further illustrates that laser cleaning is an effective means of cleaning the adhesion of particulate matter on the surface of resin matrix composites.

### *3.5.2 Temperature Monitoring*

Laser cleaning stands out as an efficient method, with its thermal effect being the predominant cleaning mechanism. Elevated temperatures during the cleaning process can induce ablation damage to the carbon fibers, thereby diminishing the material's mechanical properties. Process parameters can be dynamically adjusted over time through temperature monitoring to safeguard the integrity of CFRP. Performing cleaning operations at the appropriate temperature can enhance cleaning quality and ensure optimal productivity. Gu et al. [33] used infrared thermography to monitor the temperature of the UV laser cleaning CFRP process as shown in Fig. 3.16.

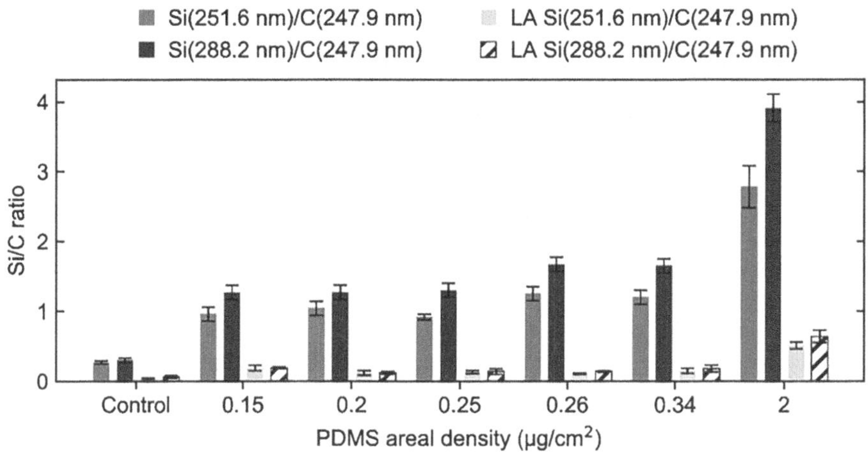


**Fig. 3.15** Si/C ratios for CFRP samples before and after picosecond laser ablation (LA). All the surfaces, except for the control, were coated with PDMS at different concentrations, producing different PDMS areal densities

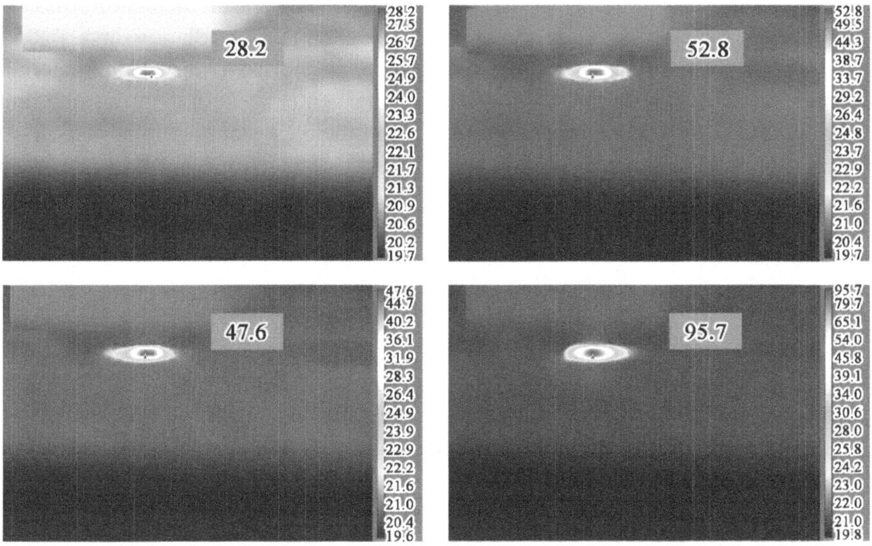


**Fig. 3.16** Maximum temperature of sample surface during laser cleaning: (**a**) S1: P = 6 W, N = 4; (**b**) S2: P = 12 W, N = 3; (**c**) S3: P = 10.5 W, N = 6; (**d**) S4: P = 12 W, N = 8

Among the four sets of samples, only S4 exhibited an elevated temperature, a phenomenon attributed to the higher energy density and an increased number of cleaning cycles under this specific parameter. Consequently, this led to a more pronounced thermal aggregation on the surface. Excessive laser power and an excessive number of cleaning cycles can result in varying degrees of damage to carbon fibers, including fracture, bending, and ablation. Such damage ultimately impacts the substrate's performance. It can be seen that the quality of cleaning can be indirectly assessed by means of temperature monitoring. This improves cleaning efficiency and eliminates the hazards of thermal damage to the CFRP substrate.

Sihn et al. [104] irradiated T650 CFRP composites by continuous laser irradiation in order to study the change in surface temperature of the material under laser irradiation. Two infrared cameras were used to monitor and record the surface temperature of CFRP during laser irradiation, while thermocouples were embedded in the composite to measure the internal temperature changes. Finally, ANSYS nonlinear transient finite element simulation was used to verify the accuracy of the model, and the material damage threshold was obtained. In this method, the damage threshold of laser cleaned CFRP is obtained through the cooperative monitoring of surface interface temperature and matrix internal temperature under laser irradiation. In theory, this method can also help to correct the corresponding effective cleaning threshold, which plays an important role in improving the optimization and optimization efficiency of CFRP laser cleaning process. Zhang et al. [105] investigated the ablation behavior of CFRP under continuous laser irradiation in an atmospheric environment. The temperature of the front and back of the plywood was monitored by a FLIR A655sc infrared camera, while a thermocouple was added to the back of the material for joint monitoring of the temperature. The measurements were shown in Fig. 3.17a, b. From the figure, it can be seen that the laser irradiation time is 20 s, the irradiation time is longer, so the temperature of the front side of the material laminate is higher. The temperature rise rate of the center point of the front side can reach 100 °C/s, while the temperature rise rate of the back side is only 30 °C/s. The large difference between the center temperatures of the front and back surfaces indicates that the thermal conductivity of CFRP laminates in the radial direction is very low, and the surface temperature cannot be transferred to the back in a short time.

Temperature not only affects the cleaning quality, but also the mechanical properties of the material surface. Therefore it is particularly important to monitor the temperature during the actual cleaning process. Temperature monitoring can be used infrared thermal imager, material embedded thermocouples. It should be noted that the installation of thermocouples cannot damage the structure of the material, so the installation process is more difficult, but its temperature monitoring effect is also better.

Current researches have shown that on-line monitoring of CFRP laser cleaning is mainly focused on temperature and spectroscopy. For contaminants with low vaporization temperature, temperature monitoring can ensure the cleaning quality and the integrity of the substrate surface. For the cleaning of multi-coating systems, spectral detection such as LIBS can quickly obtain the elemental changes of the currently acting coating and determine the degree of coating cleaning. In addition, laser cleaning of lacquer, epoxy and even ablative substrates produces wave vibrations of varying amplitude and frequency. Surface interface wave oscillations are important kinetic energy for the effective removal of substrate surface coatings Therefore, acoustic monitoring can also be used in the CFRP laser cleaning process to assess

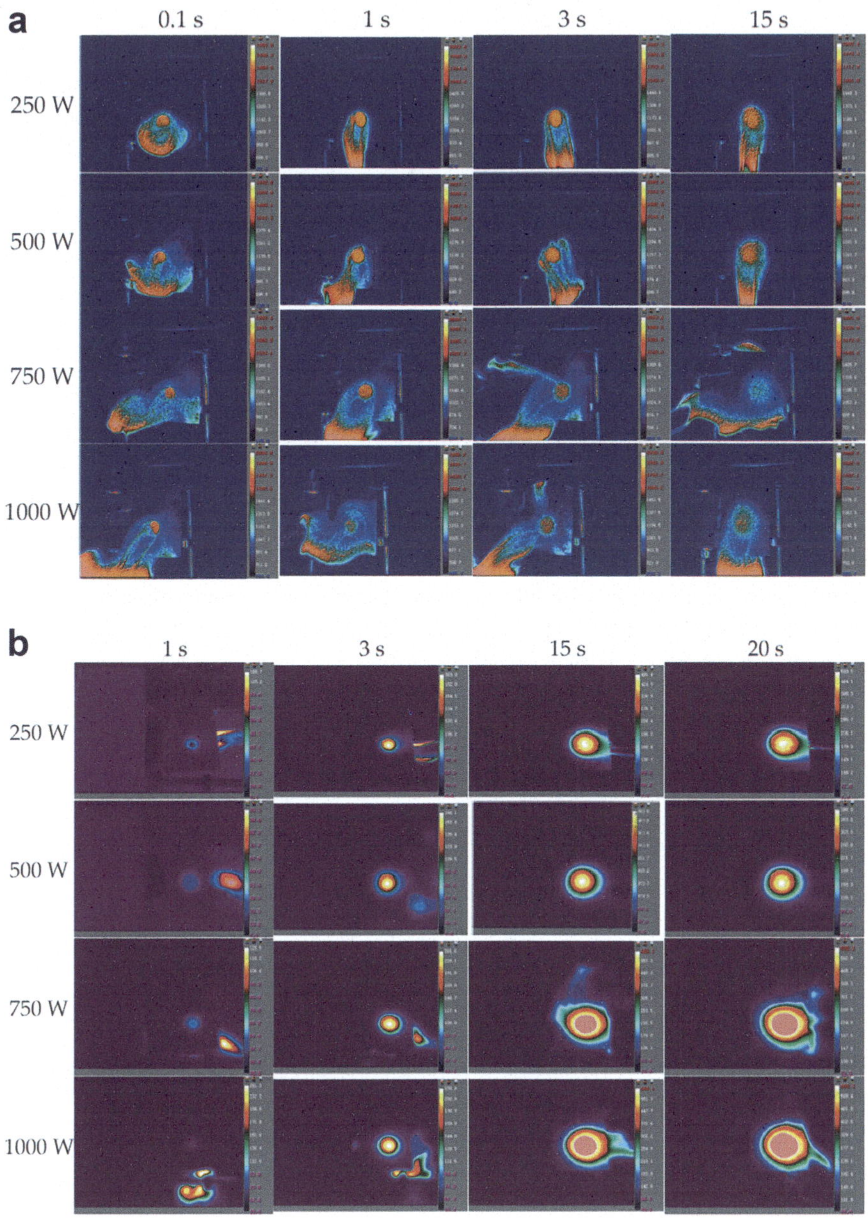


**Fig. 3.17** (**a**) The images of the temperature field on the CFRP laminated front surface irradiated by different laser power in the air environment with different time. (**b**) The images of the temperature field on the CFRP laminated back surface irradiated by different laser power in the air environment at different times

the degree of cleaning and the quality of surface cleanliness through the collection of real-time wave vibration sound. Of course, different monitoring methods should be adopted for different pollutants and surface quality requirements in order to accurately evaluate the cleaning quality on the basis of ensuring high cleaning efficiency. When the surface quality of the substrate is high after cleaning, a variety of monitoring methods can be coupled to enhance the accuracy and sensitivity of monitoring.

## 3.6 Numerical Simulation of Laser Cleaning of CFRP

Based on a large number of CFRP surface laser cleaning process and monitoring experiments, it is an important way to explore the relevant mechanism, obtain high efficiency and high quality cleaning, and regulate the cleaning process. However, experiments need to consume a lot of time, manpower and costs, and often only from a macro perspective to monitor and evaluate the cleaning state. The mechanistic, microscopic and detailed contents embedded in them cannot be clearly characterized. For example, it is difficult to accurately obtain the changes in temperature and stress fields during laser-material interaction. Numerical simulation by computer can effectively solve the above difficulties.

### *3.6.1 Numerical Simulation of the Temperature Field*

In the simulation of the temperature field, the incident laser intensity $I$ can be expressed as

$$I(x,z,t) = \frac{2E_p\omega(t)}{\tau\pi r^2} exp\left(-\frac{2(x^2+z^2)}{r^2}\right) \tag{3.1}$$

where $Ep$, $\omega$, $\tau$ and $r$ are the laser pulse energy, the laser pulse shape function, the pulse duration and the laser beam radius, respectively. The pulse shape function $\omega(t)$ follows a Gaussian distribution:

$$\omega(t) = exp\left(-\frac{t^2}{\tau^2}\right) \tag{3.2}$$

The distribution of laser in the material is expressed as

$$I(x,y,z,t) = \frac{2(1-R_p)\alpha_p \exp(-\alpha_p y) E_p \omega(t)}{\tau\pi r^2} \exp\left(-\frac{2(x^2+z^2)}{r^2}\right) \tag{3.3}$$

where $Rp$ and $\alpha_p$ are the reflectivity and absorption coefficient of the material, respectively.

During the numerical simulation process, it is assumed that all absorbed laser energy transforms into heat. At this point, the heat transfer control equation can be expressed as follows

$$k_x \frac{\partial^2 T}{\partial x^2} + k_y \frac{\partial^2 T}{\partial y^2} + k_z \frac{\partial^2 T}{\partial z^2} + \alpha I(x,y,z,t) = \rho c \frac{\partial T}{\partial t} \tag{3.4}$$

where $T$ is the instantaneous temperature of the material, $I(x, y, z, t)$ is the light intensity, $k_x$, $k_y$, and $k_z$ are the components of thermal conductivity in the $x$, $y$, and $z$ directions, and $\alpha$, $\rho$, and $c$ are the absorption coefficient, density, and specific heat capacity, respectively.

Assuming an ambient temperature of $T_0$, the initial condition is

$$T(x,y,z,0) = T_0 \tag{3.5}$$

The material has a thermally insulating condition on all surfaces except the laser-irradiated upper surface:

$$\frac{\partial T}{\partial n} \mid left, right, front, back, bottom = 0 \tag{3.6}$$

Thermal damage on the upper surface of a workpiece is caused by convection and thermal radiation:

$$k \frac{\partial T}{\partial n} \mid top = h(T - T_0) + \varepsilon\sigma(T^4 - T_0^4) \tag{3.7}$$

where $h$, $\varepsilon$, and $\sigma$ are the heat transfer coefficient, the emissivity of the coating, and the Stefan-Boltzmann constant, respectively [106–108].

Yang et al. [106] established a three-dimensional finite element model based on ANSYS "life and death cell method" for transient temperature field and paint layer removal during laser paint removal on CFRP surface. The effects of laser energy density, spot overlap rate and adjacent line spacing on coating removal and substrate peak temperature were investigated. In the case of single line scanning, two heat affected zones were generated at the resin layer and the internal carbon fiber by means of single pulse laser ablation, as shown in Fig. 3.18. The temperature rise inside the substrate is mainly due to superficial heat conduction and direct absorption of laser energy by deep carbon fibers. The superposition of adjacent pulses increases the ablation depth and raises the substrate temperature. Complete removal of the coating may result in substrate damage. Therefore, a combination of low energy density and high scanning speed is recommended to selectively peel the paint layer during the experiment. This conclusion can be used for reference to reduce the ablative damage of CFRP matrix in laser cleaning.

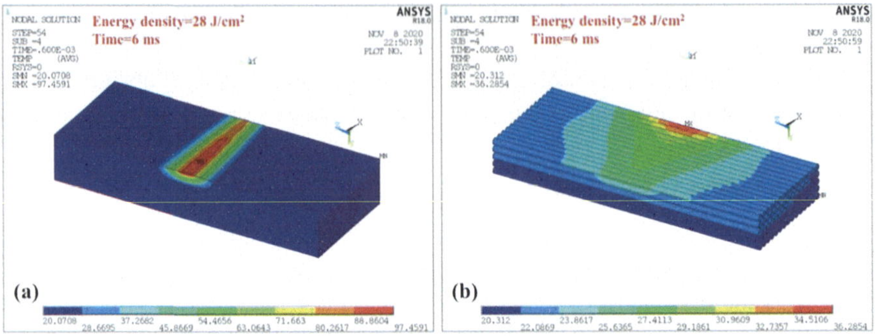


**Fig. 3.18** The HAZ distribution of the substrate and the inner carbon fibers after a single line laser scanning, with a laser energy density of 28 J/cm$^2$ and an overlap ratio of 50%

Nan [109], Ma [110], Wang et al. [107] investigated the effect of continuous laser irradiation on the ablation properties of CFRP. The results of experiments and numerical simulations illustrated that the non-monotonic warming of the material surface and the shape of the heat-affected zone under laser irradiation were related to the laminar structure and fiber orientation. This implies that the nature characteristics of the cleaned CFRP material itself should also be taken into account when controlling the heat accumulation at the laser-cleaned surface interface.

Ding et al. [111] established a finite element model of the temperature field of laser irradiated CFRP based on ABAQUS software and analyzed the temperature field distribution from a three-dimensional perspective. It is found that the temperature field extends along the fiber laying direction due to the fact that the axial thermal conductivity of carbon fibers is larger than the radial thermal conductivity. Therefore, the temperature field distribution in the fiber laying direction and the laser scanning direction produces a large difference. When the fiber direction is parallel to the laser scanning direction, the temperature field distribution of the surface layer is narrow and the heating rate is higher, as shown in Fig. 3.19a. When the fiber direction is perpendicular to the laser scanning direction, the temperature field distribution is wider than when it is parallel, and the peak temperature and high temperature duration are reduced, as shown in Fig. 3.19c. From Figs. 3.19b, d, it can be seen that the temperature field distribution of the lower carbon fibers has little relationship with the layup direction. It indicates that the heat conduction between the composite layers is poor and the laser mainly acts on the surface of the material.

Meanwhile, the heat-affected zones of lasers in different scanning directions have also been studied in more depth [112, 113]. Xu [112] formulated a model for the laser ablation of CFRP to investigate the changes in fiber fracture length ($L$) and heat-affected zone width ($W$) concerning the scanning angle ($\theta$), which represents the angle between the laser scanning direction and the fiber axial direction. The results indicate that the breaking length of carbon fibers initially increases and subsequently decreases with an increase in $\theta$. The difference in fracture length is mainly due to the anisotropy of CFRP. As $\theta$ increases, $W$ increases slowly and then increases

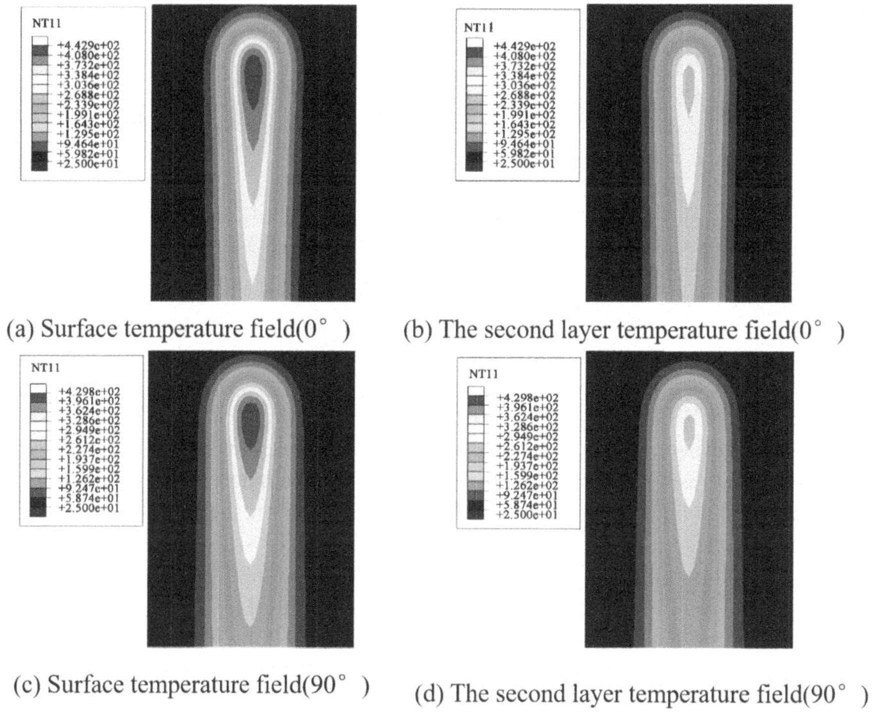


**Fig. 3.19** Temperature field of laminate with 0° fiber: (**a**) surface temperature field, (**b**) second layer temperature field, temperature field of laminate with 90° fiber: (**c**) surface temperature field, (**d**) second layer temperature field

substantially, which is due to the higher thermal conductivity of carbon fiber in axial direction than in radial direction. In the actual processing, both carbon fiber and resin in the ablation area are removed, and the greater the degree of carbon fiber breakage, the better the processing quality. Therefore, when $\theta = 45°$, the fracture is maximum and the heat affected zone is small, which can ensure high machining quality and efficiency. Wang [113] similarly demonstrated that the optimal machining angle is 45°.

### *3.6.2 Numerical Simulation of the Stress Field*

CFRP is a heterogeneous material and the thermal strain due to temperature distribution can be expressed as:

$$\epsilon = \alpha_{ij} \Delta T \quad i = j = 1,2,3 \tag{3.8}$$

where $\alpha_{ij}$ represents the coefficient of thermal expansion of the material in different directions. The stress-strain relationship can be expressed as:

$$[\sigma] = C[\varepsilon] \tag{3.9}$$

$$[\sigma] = [\sigma_{11}\sigma_{22}\sigma_{33}\tau_{12}\tau_{13}\tau_{23}]^T \tag{3.10}$$

$$[\varepsilon] = [\varepsilon_{11}\varepsilon_{22}\varepsilon_{33}\gamma_{12}\gamma_{13}\gamma_{23}]^T \tag{3.11}$$

where $\sigma_{ij}$ (i = j = 1, 2, 3) and $\varepsilon_{ij}$ (i = j = 1, 2, 3) are the normal stress and normal strain along the *x*, *y*, and *z* directions, respectively, $\tau_{12}$, $\tau_{13}$, $\tau_{23}$and $\gamma_{12}$, $\gamma_{13}$, $\gamma_{23}$ are the shear stress and shear strain in the 1–2, 1–3, and 2–3 planes, respectively; C is the stiffness matrix [108, 114].

Liu et al. [114] used optical microscopy and scanning electron microscopy to observe the interlayer damage patterns of CFRP laminates under continuous laser irradiation. The results showed that the continuously irradiated upper surface layer would undergo ablation due to the pyrolysis of the epoxy resin, and the lower surface layer produced large interlayer cracks. The interfacial stresses of the plywood were calculated by finite element analysis and were consistent with the experimental results. The maximum stresses all appeared on the interface around the back of the specimen, and the specimen gradually cracked around the back. The reason for the cracks may be due to the large temperature gradient in the specimen during rapid cooling after the laser irradiation stopped, resulting in cracks caused by the combined action of peak and shear stresses on the back of the specimen.

Li et al. [115] investigated the phenomenon of volumetric ablation of CFRP by a continuous high power density laser. It was found that the peak pressure of the composite appeared at the lower porosity and led to the failure of the composite. This indirectly suggests that the porous structure and honeycomb sandwich structure inside the composites can reduce the internal air pressure and prevent the mechanical damage of the material.

Numerical simulation can not only obtain the temperature distribution on the surface and inside of the material, but also solve the problem of stress distribution inside the material which cannot be obtained due to laser irradiation. For the cleaning of CFRP surface pollutants and paint layers, it is necessary to grasp the gasification temperature of pollutants and paint layers in advance. Then the temperature field of the cleaning process was simulated to find the process parameters that can make the pollutants reach the gasification temperature. For pollutants with high gasification temperature, stress field simulation can be carried out to avoid the cumulative damage to the matrix caused by thermal effect, and the pollutants can be removed by vibration effect. Numerical simulations are conducted within an assumed ideal environment, thus entailing certain limitations. Notably, laser-induced nonlinear effects are omitted. Additionally, with an extended duration of laser irradiation, the laser absorption rate of the composite material undergoes alterations. It is imperative to acknowledge that these limitations may introduce deviations in the results of the numerical simulations. Nonetheless, numerical simulations can still reveal some patterns in the variation of material properties. Through numerical simulation, researchers and engineers can explore the mechanism of laser cleaning, improve the timeliness and efficiency of process optimization, obtain high quality cleaning, and avoid incomplete cleaning and over-cleaning.

## 3.7 Conclusions and Outlook

The need for high-level CFRP surface cleaning and pretreatment technologies is urgent. Laser cleaning shows great potential in CFRP surface cleaning, maintenance, recycling and reuse. As an emerging means of cleaning, laser cleaning has many advantages such as high efficiency, no pollution and easy automation. At the same time, the combination of online monitoring and numerical simulation can better ensure the quality of cleaning and maximize the use of resources. It is expected to become an alternative technology for surface cleaning of CFRP equipment.

Although the cleaning of CFRP has made a lot of progress, there are still many basic theories or engineering applications worthy of attention. Laser cleaning, considered a green and environmentally friendly technology, encounters challenges in cases where the laser energy is excessively high, leading to damage in CFRP. This issue is rooted in the substantial absorption of currently used lasers by CFRP. Consequently, exploring the development of a novel laser type with reduced CFRP absorption emerges as a viable approach. Moreover, the widespread adoption of laser cleaning is hindered by the high cost of lasers, underscoring the need to expedite advancements in laser technology. Simultaneously, efforts should focus on developing new automated cleaning systems to enhance efficiency and bolster productivity. The efficiency of laser cleaning CFRP is the focus of its engineering application. Although laser cleaning and traditional cleaning methods have certain advantages. However, the large number of CFRP makes it particularly important to ensure the high efficiency of laser cleaning. It is urgent to establish a cleaning database with high guidance and universality. The establishment of the database will enable laser cleaning operators to quickly match the corresponding laser and parameters for specific materials, greatly improving cleaning efficiency. Online monitoring of the coupling of laser-cleaned CFRP should be given sufficient attention. Online monitoring of the cleaning process is an important part of intelligence. The current image, spectral and acoustic wave monitoring is more common. A single monitoring technology often fails to achieve the desired effect. The fusion of a variety of online monitoring technologies applied to the laser cleaning process, to enhance the accuracy and intelligence of laser cleaning monitoring is of great significance. With the progress of intelligent algorithms, image recognition and other technologies, this idea will become easy to realize. Multi-beam laser cleaning is expected to become a new direction in the field of laser cleaning. At present, the laser cleaning source is mainly a single laser beam. Cleaning efficiency and substrate surface quality may not be guaranteed in some limited scene cleaning operations. If different kinds of laser beams are coupled together to form a group beam laser, it will have the advantages of many different kinds of lasers at the same time. The laser output beam can be adjusted according to different cleaning materials and different cleaning requirements. In order to apply to more scenarios, it is expected to obtain higher cleaning efficiency and better cleaning quality.

## References

1. A. Bhat, S. Budholiya, S.A. Raj, M.T.H. Sultan, D. Hui, A.U.M. Shah, S.N.A. Safri, Review on nanocomposites based on aerospace applications, Nanotechnol. Rev. 10 (2021) 237–253. https://doi.org/10.1515/ntrev-2021-0018.
2. N.F. Zaaba, H. Ismail, A.M. Saeed, A Review: Metal Filled Thermoplastic Composites, Polym-Plast. Tech. Mat. 60 (2021) 1033–1050. https://doi.org/10.1080/25740881.2021.1882489.
3. X. Rong, Y. Li, X. Chen, X. Zhang, D. Zhao, C. He, C. Shi, N. Zhao, Achieving high mechanical properties and corrosion resistance of Al-Zn-Mg matrix composites via regulating intragranular reinforcements, J. Mater. Sci. Technol. 153 (2023) 1–7. https://doi.org/10.1016/j.jmst.2022.12.066.
4. S. Murugesan, T. Scheibel, Copolymer/clay nanocomposites for biomedical applications, Adv. Funct. Mater. 30 (2020) 1908101. https://doi.org/10.1002/adfm.201908101.
5. W. Guo, Y.C. Lim, C.H. Ong, Atmospheric pressure plasma application on the adhesive bonding improvement of CFRP via surface configuration comparison, Polym. Composite. 45 (2024) 1461–1471. https://doi.org/10.1002/pc.27866.
6. Hueber, G. Fischer, N. Schwingshandl, R. Schledjewski, Production planning optimisation for composite aerospace manufacturing, Int. J. Prod. Res. 57 (2019) 5857–5873. https://doi.org/10.1080/00207543.2018.1554918.
7. V. Khatkar, B.K. Behera, R.N. Manjunath, Textile structural composites for automotive leaf spring application, Compos. Part. B-Eng. 182 (2020) 107662. https://doi.org/10.1016/j.compositesb.2019.107662.
8. Z. Kamble, B.K. Behera, Sustainable hybrid composites reinforced with textile waste for construction and building applications, Constr. Build. Mater. 284 (2021) 122800. https://doi.org/10.1016/j.conbuildmat.2021.122800.
9. S.B. Park, E. Lih, K.S. Park, Y.K. Joung, D.K. Han, Biopolymer-based functional composites for medical applications, Prog. Polym. Sci. 68 (2017) 77–105. https://doi.org/10.1016/j.progpolymsci.2016.12.003.
10. M.K.A.A. Razab, A.M. Noor, M.S. Jaafar, N.H. Abdullah, F.M. Suhaimi, M. Mohamed, N. Adam, N.A.A.N. Yusuf, A review of incorporating Nd: YAG laser cleaning principal in automotive industry, J. Radiat. Res. Appl. Sci. 11 (2018) 393–402. https://doi.org/10.1016/j.jrras.2018.08.002.
11. G. Zhu, Z. Xu, Y. Jin, X. Chen, L. Yang, J. Xu, D. Shan, Y. Chen, B. Guo, Mechanism and application of laser cleaning: A review, Opt. Laser. Eng. 157 (2022) 107130. https://doi.org/10.1016/j.optlaseng.2022.107130.
12. T. Chen, W. Wang, T. Tao, A. Pan, X. Mei, Multi-scale micro-nano structures prepared by laser cleaning assisted laser ablation for broadband ultralow reflectivity silicon surfaces in ambient air, Appl. Surf. Sci. 509 (2020) 145182. https://doi.org/10.1016/j.apsusc.2019.145182.
13. S.M. Bedair, H.P. Smith Jr, Atomically clean surfaces by pulsed laser bombardment, J. Appl. Phys. 40 (1969) 4776–4781. https://doi.org/10.1063/1.1657288.
14. W. Zapka, W. Ziemlich, A.C. Tam, Efficient pulsed laser removal of 0.2 μm sized particles from a solid surface, Appl. Phys. Lett. 58 (1991) 2217–2219. https://doi.org/10.1063/1.104931.

15. Z. Kuang, W. Guo, J. Li, Y. Jin, D. Qian, J. Ouyang, L. Fu, E. Fearon, R. Hardacre, Z. Liu, L. Li, Nanosecond fibre laser paint stripping with suppression of flames and sparks, J. Mater. Process. Tech. 266 (2019) 474–483. https://doi.org/10.1016/j.jmatprotec.2018.11.028.
16. Y.K. Madhukar, S. Mullick, D.K. Shukla, S. Kumar, A.K. Nath, Effect of laser operating mode in paint removal with a fiber laser, Appl. Surf. Sci. 264 (2013) 892–901. https://doi.org/10.1016/j.apsusc.2012.10.193.
17. X. Wang, J. Li, L. Wang, X. Jia, M. Hong, Y. Ren, M. Ma, A novel de-rusting method with molten salt precleaning and laser cleaning for the recycling of steel parts, Clean. Technol. Environ. 23 (2021) 1403–1414. https://doi.org/10.1007/s10098-020-02023-6.
18. K. Ding, K. Zhou, G. Feng, J. Han, N. Xie, Z. Huang, G. Zhou, Mechanism and conditions for laser cleaning of micro and nanoparticles on the surface of transparent substrate, Vacuum 200 (2022) 110987. https://doi.org/10.1016/j.vacuum.2022.110987.
19. Y. Ye, X. Yuan, X. Xiang, X. Cheng, X. Miao, Laser cleaning of particle and grease contaminations on the surface of optics, Optik 123 (2012) 1056–1060. https://doi.org/10.1016/j.ijleo.2011.07.030.
20. X. Ling, G. Wang, C. Zhang, Investigation of laser cleaning on bronze cultural relics, Laser. Phys. 26 (2016) 055603. https://doi.org/10.1088/1054-660x/26/5/055603.
21. J. Nie, H. Zhang, D. Zhang, J. Xu, J. Zhang, D. Shan, B. Guo, Removal mechanism of laser cleaning for inorganic thermal control coatings on aluminum alloys, Appl. Surf. Sci. 633 (2023) 157578. https://doi.org/10.1016/j.apsusc.2023.157578.
22. Zhou, S. Zhao, H. Yu, Z. Zhang, H. Laing, L. Jiang, J. Xu, X. Lin, X. Wang, Plume ejection and product in laser cleaning of ferrite paint, Opt. Laser. Technol. 167 (2023) 109733. https://doi.org/10.1016/j.optlastec.2023.109733.
23. G. Rotondi, A. Brini, F. Cantini, A new underwater laser cleaning tool for large bronze artefacts, J. Cult. Herit. 61 (2023) 188–193. https://doi.org/10.1016/j.culher.2023.03.012.
24. Y. Liu, W. Ouyang, H. Wu, Z. Xu, L. Sheng, Q. Zou, M. Zhang, W. Zhang, J. Jiao, Improving surface quality and superficial microstructure of LDED Inconel 718 superalloy processed by hybrid laser polishing, J Mater Process Tech. 300 (2022) 117428. https://doi.org/10.1016/j.jmatprotec.2021.117428.
25. Y. Liu, S. Sun, J. Wang, W. Ouyang, J. Xu, L. Sheng, Z. Xu, H. Ru, Y. Liu, J. Jiao, Tribological behaviors of LDED Inconel 718 samples polished with a hybrid laser polishing technique, J Mater Res Technol. 65 (2023) 633 633–646. https://doi.org/10.1016/j.jmrt.2023.05.230.
26. C. Yuan, B. Wang, J. Wang, Y. Wang, L. Sheng, J. Jiao, J. Yao, Y. Huang, W. Zhang, Effects of incidence angle and optimization in femtosecond laser polishing of C/SiC composites, Ceram Int. 48 (2022) 32290–32304. https://doi.org/10.1016/j.ceramint.2022.07.171.
27. Y. Song, S. Wang, Y. Pan, Z. Li, L. Yu, Comparative study of infrared nanosecond laser surface paint removal for carbon fiber reinforced polymer and glass fiber reinforced polymer, J Clean Prod. 451 (2024) 142061. https://doi.org/10.1016/j.jclepro.2024.142061
28. Y. Liu, W. Liu, D. Zhang, Z. Tian, X. Sun, Z. Wei, Experimental investigations into cleaning mechanism of ship shell plant surface involved in dry laser cleaning by controlling laser power, Appl. Phys. A-Mater. 126 (2020) 1–17. https://doi.org/10.1007/s00339-020-04050-y.
29. B. Sun, C. Xu, J. He, L. Yu, Y. Zhang, Cleanliness prediction of rusty iron in laser cleaning using convolutional neural networks, Appl. Phys. A-Mater. 126 (2020) 1–9. https://doi.org/10.1007/s00339-020-3363-5.
30. D. Grojo, A. Cros, P. Delaporte, M. Sentis, Experimental investigation of ablation mechanisms involved in dry laser cleaning, Appl. Surf. Sci. 253 (2007) 8309–8315. https://doi.org/10.1016/j.apsusc.2007.02.117.
31. H.T. Hsu, J. Lin, Thermal–mechanical analysis of the surface waves in laser cleaning, Int. J. Mach. Tool. Manu. 45 (2005) 979–985. https://doi.org/10.1016/j.ijmachtools.2004.10.002.
32. A. Kumar, M. Prasad, R.B. Bhatt, P.G. Behere, D.J. Biswas, A comparative study on laser induced shock cleaning of radioactive contaminants in air and water, Opt. Laser. Technol. 100 (2018) 133–138. https://doi.org/10.1016/j.optlastec.2017.10.005.

33. J. Gu, X. Su, Y. Jin, W. Li, Z. Zeng, D. Zhang, J. Xu, B. Guo, Towards low-temperature laser paint stripping by photochemical mechanism on CFRP substrates, J. Manuf. Process. 85 (2023) 272–280. https://doi.org/10.1016/j.jmapro.2022.11.041.
34. J. Gu, X. Su, Y. Jin, W. Li, S. Li, J. Xu, B. Guo, Precise paint stripping based on photochemical effect by UV laser on honeycomb sandwich panel, J. Mater. Res. Technol. 25 (2023) 3768–3779. https://doi.org/10.1016/j.jmrt.2023.06.083.
35. Akman, Y. Erdoğan, M.Ö. Bora, O. Çoban, B.G. Oztoprak, A. Demir, Investigation of the differences between photochemical and photothermal laser ablation on the shear strength of CFRP/CFRP adhesive joints, Int. J. Adhes. Adhes. 98 (2020) 102548. https://doi.org/10.1016/j.ijadhadh.2020.102548.
36. H. Xu, J. Hu, Z. Yu, Absorption behavior analysis of carbon fiber reinforced polymer in laser processing, Opt. Mater. Express. 5 (2015) 2330–2336. https://doi.org/10.1088/1555-6611/aa9d07.
37. B. Rauh, S. Kreling, M. Kolb, M. Geistbeck, S. Boujenfa, M. Suess, K. Dilger, UV-laser cleaning and surface characterization of an aerospace carbon fibre reinforced polymer, Int. J. Adhes. Adhes. 82 (2018) 50–59. https://doi.org/10.1016/j.ijadhadh.2017.12.016.
38. M. Schweizer, D. Meinhard, S. Ruck, H. Riegel, V. Knoblauch, Adhesive bonding of CFRP: a comparison of different surface pre-treatment strategies and their effect on the bonding shear strength, J. Adhes. Sci. Technol. 31 (2017) 2581–2591. https://doi.org/10.1080/01694243.2017.1310695.
39. Z. Zhao, X. Liu, J. Yang, W. Pan, Y. Li, F. Liang, Y. Tian, L. Liu, F. Song, Shape evolution and characteristics of carbon fiber reinforced polymer surface in laser ablation, Vacuum 217 (2023) 112572. https://doi.org/10.1016/j.vacuum.2023.112572.
40. J. Gu, X. Su, W. Li, Y. Jin, J. Xu, B. Guo, Process and mechanism of paint stripping on CFRP by UV nanosecond laser, Opt. Laser. Technol. 171 (2024) 110461. https://doi.org/10.1016/j.optlastec.2023.110461.
41. Q. Gao, Y. Li, H. Wang, W. Liu, H. Shen, X. Zhan, Effect of scanning speed with UV laser cleaning on adhesive bonding tensile properties of CFRP, Appl. Compos. Mater. 26 (2019) 1087–1099. https://doi.org/10.1007/s10443-019-09768-4.
42. Y. Tong, X. Chen, A. Zhang, Y. Fu, X. Ren, Y. Ye, S. Xin, Effect of laser cleaning of carbon fiber-reinforced polymer and surface modification on chemical activity and bonding strength, Appl. Optics 59 (2020) 10149–10159. https://doi.org/10.1364/AO.404846.
43. Oliveux, L.O. Dandy, G.A. Leeke, Current status of recycling of fibre reinforced polymers: Review of technologies, reuse and resulting properties, Prog. Mater. Sci. 72 (2015) 61–99. https://doi.org/10.1016/j.pmatsci.2015.01.004.
44. Y. Liu, M. Farnsworth, A. Tiwari, A review of optimisation techniques used in the composite recycling area: State-of-the-art and steps towards a research agenda, J. Clean. Prod. 140 (2017) 1775–1781. https://doi.org/10.1016/j.jclepro.2016.08.038.
45. K.W. Kim, H.M. Lee, J.H. An, D.C. Chung, K.H. An, B.J. Kim, Recycling and characterization of carbon fibers from carbon fiber reinforced epoxy matrix composites by a novel super-heated-steam method, J. Environ. Manage. 203 (2017) 872–879. https://doi.org/10.1016/j.jenvman.2017.05.015.
46. M.A. Nahil, P.T. Williams, Recycling of carbon fibre reinforced polymeric waste for the production of activated carbon fibres, J. Anal. Appl. Pyrol. 91 (2011) 67–75. https://doi.org/10.1016/j.jaap.2011.01.005.
47. M. Goto, Chemical recycling of plastics using sub-and supercritical fluids, J. Supercrit. Fluid. 47 (2009) 500–507. https://doi.org/10.1016/j.supflu.2008.10.011.
48. Y. Yang, W. Zhou, Z. Tong, L. Chen, X. Ren, Delamination in titanium-based carbon-fibre/epoxy laminates under laser shock peening, Opt. Laser. Technol. 143 (2021) 107280. https://doi.org/10.1016/j.optlastec.2021.107282.
49. J. Nasser, L. Zhang, H. Sodano, Laser induced graphene interlaminar reinforcement for tough carbon fiber/epoxy composites, Compos. Sci. Technol. 201 (2021) 108493. https://doi.org/10.1016/j.compscitech.2020.108493.

50. Y. He, X. Cai, J. Ye, Research on pulse laser cleaning and rust removal technology in power systems, Aip. Adv. 13 (2023) 095101. https://doi.org/10.1063/5.0155263.
51. Y. Li, S. Meng, Q. Gong, Y. Huang, J. Gan, M. Zhao, B. Liu, L. Liu, G. Zou, D. Zhuang, Experimental and theoretical investigation of laser pretreatment on strengthening the heterojunction between carbon fiber-reinforced plastic and aluminum alloy, ACS Appl. Mater. Inter. 11 (2019) 22005–22014. https://doi.org/10.1021/acsami.9b04080.
52. M.A. Belcher, C.J. Wohl, J.W. Hopkins, J.W. Connell, Laser surface preparation for bonding of aerospace composites, Proc. Inst. Civ. En-Eng. 163 (2011) 133–138. https://doi.org/10.1680/eacm.2011.164.3.133.
53. L. Liu, X. Liu, L. Kong, M. Wang, P. Hu, D. Wang, Effect of laser surface treatment on surface and bonding properties of carbon fiber reinforced composites, Int. J. Mater. Form. 13 (2020) 885–895. https://doi.org/10.1007/s12289-019-01509-z.
54. Z. Xu, Y. Huang, C. Zhang, G. Chen, Influence of rare earth treatment on interfacial properties of carbon fiber/epoxy composites, Mat. Sci. Eng. A. 444 (2027) 170–177. https://doi.org/10.1016/j.msea.2006.08.079.
55. C. Sun, J. Min, J. Lin, H. Wan, S. Yang, S. Wang, The effect of laser ablation treatment on the chemistry, morphology and bonding strength of CFRP joints, Int. J. Adhes. Adhes. 84 (2018) 325–334. https://doi.org/10.1016/j.ijadhadh.2018.04.014.
56. X. Zhan, C. Gao, C. Gu, W. Sun, Y. Wei, Laser Cleaning Treatment and its Influence on the Surface Microstructure of CFRP Composite Material, J. Powder. Metall. Min. 6 (2017) 1000165. https://doi.org/10.4172/2168-9806.1000165.
57. Y. Bao, Y. Ma, Y. Yang, J. Wang, D. Cheng, Study on the effect of carbon fiber ply angle on thermal response of CFRP laminate based on the thermal resistance network model, Polym. Composite. 44 (2023) 8792–8804. https://doi.org/10.1002/pc.27737.
58. X. Zhan, Y. Li, C. Gao, H. Wang, Y. Yang, Effect of infrared laser surface treatment on the microstructure and properties of adhesively CFRP bonded joints, Opt. Laser. Technol. 106 (2018) 398–409. https://doi.org/10.1016/j.optlastec.2018.04.023.
59. Fischer, S. Kreling, P. Jäschke, M. Frauenhofer, D. Kracht, K. Dilger, Laser surface pretreatment of CFRP for adhesive bonding in consideration of the absorption behaviour, J. Adhesion. 88 (2012) 350–363. https://doi.org/10.1080/00218464.2012.660042.
60. A.I. Noskov, P.S. Loskutova, R.M. Yanbaev, A.K. Gilmutdinov, The mechanical properties of carbon fiber reinforced polymers after laser processing, Laser. Phys. 29 (2019) 096002. https://doi.org/10.1088/1555-6611/ab344c.
61. X. Li, M. Ehrhardt, P. Lorenz, B. Han, S. Lai, K. Zimmer, L. Xu, P. Nan, X. Ni, Influence of surface treatment with infrared nanosecond laser on adhesion performance of adhesion-bonded carbon fiber/epoxy composite, J. Adhes. Sci. Technol. 34 (2020) 1399–1425. https://doi.org/10.1080/01694243.2019.1710990.
62. J. Wang, Y. Yang, J. Qi, Z. Liu, L. Zhou, Thermodynamic simulation, surface morphology and bending property of carbon fiber reinforced polymer composite material subjected to laser cleaning, Opt. Laser. Technol. 152 (2022) 108099. https://doi.org/10.1016/j.optlastec.2022.108099.
63. X. Zhan, S. Chen, Y. Li, H. Wang, Y. Yang, Effect of surface cold ablation on shear strength of CFRP adhesively bonded joint after UV laser treatment, Int. J. Adhes. Adhes. 94 (2019) 13–23. https://doi.org/10.1016/j.ijadhadh.2019.05.006.
64. M. Li, G. Gan, B. Li, X. Yang, Whole-field strain analysis and strength prediction of fiber laser machined CFRP laminate at elevated temperature, Polym. Composite. 41 (2020) 2662–2672. https://doi.org/10.1002/pc.25565.
65. C. Zhou, Y. Li, G. Zhu, G. Luo, X. Zhao, Q. Yu, Tensile and flexural behavior of metal/CFRP hybrid laminated plates, Polym. Composite. 42 (2021) 2882–2897. https://doi.org/10.1002/pc.26022.
66. Y. Iwahori, T. Hasegawa, K. Nakane, Experimental evaluation for CFRP strength after various paint stripping methods, Aeronautical and Space Sciences Japan 55 (2007) 235–240.

67. Y. Li, X. Zhan, C. Gao, H. Wang, Y. Yang, Comparative study of infrared laser surface treatment and ultraviolet laser surface treatment of CFRP laminates, Int. J. Adv. Manuf. Tech. 102 (2019) 4059–4071. https://doi.org/10.1007/s00170-019-03368-z.
68. Y. Zhang, Y. Li, J. Zhang, J. Pan, L. Zhang, F. Tan, H. Wei, W. Zhang, High-Temperature Effect on the Tensile Mechanical Properties of Unidirectional Carbon Fiber-Reinforced Polymer Plates, Materials 14 (2021) 7214. https://doi.org/10.3390/ma14237214.
69. Y. Ji, C. Luan, X. Yao, Z. Ding, C. Niu, N. Dong, J. Fu, Mechanism and behavior of laser irradiation on carbon fiber reinforced polyetheretherketone (CF/PEEK) during the laser-assisted in-situ consolidation additive manufacturing process, Addit. Manuf. 74 (2023) 103713. https://doi.org/10.1016/j.addma.2023.103713.
70. F.L. Palmieri, M.A. Belcher, C.J. Wohl, K.Y. Blohowiak, J.W. Connell, Laser ablation surface preparation for adhesive bonding of carbon fiber reinforced epoxy composites, Int. J. Adhes. Adhes. 68 (2016) 95–101. https://doi.org/10.1016/j.ijadhadh.2016.02.007.
71. L. Hu, Y. Cai, W. Yang, W. Xue, D. Zhu, Y. Cao, Laser selective ablated multistep interfacing for enhanced adhesive bonding joints of carbon fiber reinforced polymer materials, J. Laser. Appl. 33 (2021) 042005. https://doi.org/10.2351/7.0000378.
72. O. Çoban, E. Akman, M.Ö. Bora, B. Genc. Oztoprak, A. Demir, Laser surface treatment of CFRP composites for a better adhesive bonding owing to the mechanical interlocking mechanism, Polym. Composite. 40 (2019) 3611–3622. https://doi.org/10.1002/pc.25224.
73. B.A. Newcomb, Tailoring thermal expansion of mechanically interlocked carbon fiber composite reinforced metal, Polym. Composite. 43 (2022) 9078–9088. https://doi.org/10.1002/pc.27085.
74. V. Oliveira, S.P. Sharma, M.F.S.F. de Moura, R.D.F. Moreira, R. Vilar, Surface treatment of CFRP composites using femtosecond laser radiation, Opt. Laser. Eng. 94 (2017) 37–43. https://doi.org/10.1016/j.optlaseng.2017.02.011.
75. J. Chen, Z. Liu, F. Su, S. Wu, C. Liang, J. Liu, Surface modification of carbon fiber cloth by femtosecond laser direct writing technology, Mater. Lett. 323 (2022) 132483. https://doi.org/10.1016/j.matlet.2022.132483.
76. C. Kunz, T.N. Büttner, B. Naumann, A.V. Boehm, E. Gnecco, J. Bonse, C. Neumann, A. Turchanin, F.A. Müller, S. Gräf, Large-area fabrication of low-and high-spatial-frequency laser-induced periodic surface structures on carbon fibers, Carbon 133 (2018) 176–185. https://doi.org/10.1016/j.carbon.2018.03.035.
77. M.Ö. Bora, E. Akman, O. Çoban, B. Genc. Oztoprak, A. Demir, The Effect of $CO_2$ Laser-Induced Microhole Formations on Adhesive Bonding Strength of CFRP/CFRP Joints, Polym. Composite. 40 (2019) 2891–2900. https://doi.org/10.1002/pc.25119.
78. L. Sorrentino, G. Parodo, S. Turchetta, CFRP laser texturing to increase the adhesive bonding: morphological analysis of treated surfaces, J. Adhesion 97 (2021) 1322–1335. https://doi.org/10.1080/00218464.2020.1758074.
79. Z. Zhang, J. Zhou, Y. Ren, W. Li, S. Li, N. Chai, Z. Zeng, X. Chen, Y. Yue, L. Zhou, Y. Cheng, S. Li, X. Wang, Passive deicing CFRP surfaces enabled by super-hydrophobic multi-scale micro-nano structures fabricated via femtosecond laser direct writing, Nanomaterials 12 (2022) 2782. https://doi.org/10.3390/nano12162782.
80. S. Harder, H. Schmutzler, P. Hergoss, J. Freese de, J. Holtmannspötter, B. Fiedler, Effect of infrared laser surface treatment on the morphology and adhesive properties of scarfed CFRP surfaces, Compos. Part. A-Appl. S. 121 (2019) 299–307. https://doi.org/10.1016/j.compositesa.2019.02.025.
81. Y. Xie, B. Yang, L. Lu, Z. Wan, X. Liu, Shear strength of bonded joints of carbon fiber reinforced plastic (CFRP) laminates enhanced by a two-step laser surface treatment, Compos. Struct. 232 (2020) 111559. https://doi.org/10.1016/j.compstruct.2019.111559.
82. A. Al-Lami, P. Hilmer, M. Sinapius, Eco-efficiency assessment of manufacturing carbon fiber reinforced polymers (CFRP) in aerospace industry, Aerosp. Sci. Technol. 79 (2018) 669–678. https://doi.org/10.1016/j.ast.2018.06.020.

83. L.Y. Xu, J.R. Lu, K.M. Li, J. Hu, Experimental study of CFRP laser surface modification and bonding characteristics of CFRP/Al6061 heterogeneous joints, Compos. Struct. 283 (2022) 115030. https://doi.org/10.1016/j.compstruct.2021.115030.
84. J. Lin, J. Zhang, J. Min, C. Sun, S. Yang, Laser-assisted conduction joining of carbon fiber reinforced sheet molding compound to dual-phase steel by a polycarbonate interlayer, Opt. Laser. Technol. 133 (2021) 106561. https://doi.org/10.1016/j.optlastec.2020.106561.
85. T. Ishikawa, K. Amaoka, Y. Masubuchi, T. Yamamoto, A. Yamanaka, M. Arai, J. Takahashi, Overview of automotive structural composites technology developments in Japan, Compos. Sci. Technol. 155 (2018) 221–246. https://doi.org/10.1016/j.compscitech.2017.09.015.
86. K.W. Kim, D.K. Kim, B.S. Kim, K.H. An, S.J. Park, K.Y. Rhee, B.J. Kim, Cure behaviors and mechanical properties of carbon fiber-reinforced nylon6/epoxy blended matrix composites, Compos. Part. B-Eng. 112 (2017) 15–21. https://doi.org/10.1016/j.compositesb.2016.12.009.
87. S. Genna, F. Lambiase, C. Leone, Effect of laser cleaning in Laser Assisted Joining of CFRP and PC sheets, Compos. Part. B-Eng. 145 (2018) 206–214. https://doi.org/10.1016/j.compositesb.2018.03.032.
88. Q. Zou, J. Jiao, J. Xu, L. Sheng, Y. Zhang, W. Ouyang, Z. Xu, M. Zhang, H. Xia, R. Tian, X. Cao, Y. Liu, L. Zhao, Effects of laser hybrid interfacial pretreatment on enhancing the carbon fiber reinforced thermosetting composites and TC4 alloy heterogeneous joint, Mater. Today. Commun. 30 (2022) 103142. https://doi.org/10.1016/j.mtcomm.2022.103142.
89. J. Jiao, J. Xu, C. Jing, X. Cheng, D. Wu, H. Ru, K. Zeng, L. Sheng, Effect of Laser Micro-Texturing on Laser Joining of Carbon Fiber Reinforced Thermosetting Composites to TC4 Alloy, Materials 16 (2023) 270. https://doi.org/10.3390/ma16010270.
90. Z. Zhang, J. Shan, X. Tan, J. Zhang, Improvement of the laser joining of CFRP and aluminum via laser pre-treatment, Int. J. Adv. Manuf. Tech. 90 (2017) 3465–3472. https://doi.org/10.1007/s00170-016-9646-5.
91. G. Jiang, X. Zhou, Effect of laser surface texture on the adhesion performance of CFRP/Mg laminates, J. Adhesion (2023) 1–20. https://doi.org/10.1080/00218464.2023.2223547.
92. Y. Pan, G. Wu, Z. Huang, Z. Zhang, H. Ye, M. Li, S. Ji, Improvement in interlaminar strength and galvanic corrosion resistance of CFRP/Mg laminates by laser ablation, Mater. Lett. 207 (2017) 4–7. https://doi.org/10.1016/j.matlet.2017.06.105.
93. X. Zou, L. Liu, T. Chen, L. Wu, K. Chen, L. Kong, M. Wang, Laser surface treatment to enhance the adhesive bonding between steel and CFRP: Effect of laser spot overlapping and pulse fluence. Opt. Laser. Technol. 159 (2023) 109002. https://doi.org/10.1016/j.optlastec.2022.109002.
94. Li, H. Liu, S. Li, Q. Zhao, X. Qin, Influence of high pulse fluence infrared laser surface pretreatment parameters on the mechanical properties of CFRP/aluminium alloy adhesive joints, J. Adhesion 99 (2023) 584–605. https://doi.org/10.1080/00218464.2022.2027242.
95. V. Reitz, D. Meinhard, S. Ruck, H. Riegel, V. Knoblauch, A comparison of IR-and UV-laser pretreatment to increase the bonding strength of adhesively joined aluminum/CFRP components, Compos. Part. A-Appl. S. 96 (2017) 18–27. https://doi.org/10.1016/j.compositesa.2017.02.014.
96. M. Veltrup, T. Lukasczyk, B. Mayer, Effect of re-depositions and fiber exposure on the adhesive bond strength of CFRP after UV excimer laser treatment, Appl. Phys. A-Mater. 128 (2022) 786. https://doi.org/10.1007/s00339-022-05911-4.
97. M. Saeedifar, Impact damage assessment in laminated composites using acoustic emission and finite element methods, Polym. Composite. 45 (2023) 1036–1053. https://doi.org/10.1002/pc.27834.
98. Y. Su, S. Zhang, M. Su, L. Ma, W. Zhou, Failure behavior and crashworthiness characteristic of repaired composites-metal hybrid tubes by acoustic emission, Polym. Composite. 45 (2024) 1632–1648. https://doi.org/10.1002/pc.27878.
99. S. Qiao, M. Huang, Y. Liang, S. Zhang, W. Zhou, Damage mode identification in carbon/epoxy composite via machine learning and acoustic emission, Polym. Composite. 44 (2023) 2427–2440. https://doi.org/10.1002/pc.27254.

100. S. Choi, C. Park, Convolution Neural Network with Laser-Induced Breakdown Spectroscopy as a Monitoring Tool for Laser Cleaning Process, Sensors 23 (2023) 83. https://doi.org/10.3390/s23010083.
101. O. Abdel-Kareem, M.A. Harith, Evaluating the use of laser radiation in cleaning of copper embroidery threads on archaeological Egyptian textiles, Appl. Surf. Sci. 254 (2008) 5854–5860. https://doi.org/10.1016/j.apsusc.2008.03.144.
102. W. Wang, L. Sun, Y. Lu, L. Qi, W. Wang, H. Qiao, Laser induced breakdown spectroscopy online monitoring of laser cleaning quality on carbon fiber reinforced plastic, Opt. Laser. Technol. 145 (2022) 107481. https://doi.org/10.1016/j.optlastec.2021.107481.
103. R. Ledesma, F. Palmieri, J. Connell, W. Yost, J. Fitz-Gerald, Surface characterization of carbon fiber reinforced polymers by picosecond laser induced breakdown spectroscopy, Spectrochim. Acta. B. 140 (2018) 5–12. https://doi.org/10.1016/j.sab.2017.11.014.
104. S. Sihn, J. Pitz, R.H. Gerzeski, A.K. Roy, J.P. Vernon, Experimentally-validated computational model for temperature evolution within laser heated fiber-reinforced polymer matrix composites, Compos. Struct. 207 (2019) 966–973. https://doi.org/10.1016/j.compstruct.2018.09.041.
105. Y. Zhang, J. Pan, S. Zhou, Q. Yin, J. Zhang, W. Xie, F. Tan, W. Zhang, Experimental Investigation on Ablation Behaviors of CFRP Laminates in an Atmospheric Environment Irradiated by Continuous Wave Laser, Polymers 14 (2022) 5082. https://doi.org/10.3390/polym14235082.
106. Yang, H. Liu, R. Gao, X. Liu, X. Yu, F. Song, L. Liu, Numerical simulation of paint stripping on CFRP by pulsed laser, Opt. Laser. Technol. 145 (2022) 107450. https://doi.org/10.1016/j.optlastec.2021.107450.
107. P. Wang, Z. Zhang, B. Hao, S. Wei, W. Qiu, Y. Huang, G. Zhang, Study on anisotropic heat transfer and thermal damage in nanosecond pulsed laser processing of CFRP, Polym. Composite. 44 (2023) 5964–5983. https://doi.org/10.1002/pc.27540.
108. C. Lyu, R. Zhan, Analysis and prediction of interlayer stress and strain of composites such as CFRP irradiated by laser, Mater. Today. Commun. 28 (2021) 102582. https://doi.org/10.1016/j.mtcomm.2021.102582.
109. P. Nan, Z. Shen, B. Han, X. Ni, The influences of laminated structure on the ablation characteristics of carbon fiber composites under CW laser irradiation, Opt. Laser. Technol. 116 (2019) 224–231. https://doi.org/10.1016/j.optlastec.2019.03.015.
110. Y. Ma, Q. Huang, Y. Yu, Y. Dong, H. Li, G. Jin, C. Xin, Theoretical and Experimental Investigation on Temperature Rise of CFRP with Long Pulse Laser Irradiation, Appl. Sci. 11 (2021) 10555. https://doi.org/10.3390/app112210555.
111. Ding, Y. Guo, Z. Wang, Study on laser irradiation temperature field of carbon fiber reinforced plastic composites, Mater. Res. Express. 7 (2020) 035306. https://doi.org/10.1088/2053-1591/ab7dfc.
112. L.Y. Xu, J.R. Lu, K.M. Li, J. Hu, Removal mechanism of CFRP by laser multi direction interaction, Opt. Laser. Technol. 143 (2021) 107281. https://doi.org/10.1016/j.optlastec.2021.107281.
113. P. Wang, Z. Zhang, B. Hao, S. Wei, Y. Huang, G. Zhang, Investigation on heat transfer and ablation mechanism of CFRP by different laser scanning directions, Compos. Part. B-Eng. 262 (2023) 110827. https://doi.org/10.1016/j.compositesb.2023.110827.
114. Y.C. Liu, C.W. Wu, Y.H. Huang, H.W. Song, C.G. Huang, Interlaminar damage of carbon fiber reinforced polymer composite laminate under continuous wave laser irradiation, Opt. Laser. Eng. 88 (2017) 91–101. https://doi.org/10.1016/j.optlaseng.2016.08.001.
115. X. Li, W. Hou, B. Han, L. Xu, Z. Li, P. Nan, X. Ni, Thermal response during volumetric ablation of carbon fiber composites under a high intensity continuous laser irradiation, Surf. Interfaces 23 (2021) 101032. https://doi.org/10.1016/j.surfin.2021.101032.

# Chapter 4
# Laser Cleaning Technology for Surface Pretreatment

## 4.1 Introduction

In the manufacturing and upkeep of aerospace equipment, ships, and various precision instruments, surface cleaning and pretreatment play a crucial role [1]. Traditional methods such as mechanical friction, chemical corrosion, high-pressure water jets, and sandblasting have disadvantages such as environmental pollution, time-consuming material consumption, and substrate damage [2–4]. Surface cleaning and pretreatment are often not carried out continuously, reducing work efficiency [5]. In this context, it is urgent to explore more environmentally friendly and efficient green surface cleaning and pretreatment methods, and the emergence of laser cleaning has brought a new dawn.

The concept of laser cleaning was first put forward by Bedair et al. in 1969 [6]. Its principle is to irradiate the surface of the object to be cleaned with a laser beam with high energy density and make the pollutants on the surface of the substrate fall off through various mechanisms such as vibration, ablation, plasma impact, and photochemistry to achieve surface cleaning. Currently, laser cleaning technology has been widely used in paint removal [7], rust removal [8], oil removal [9], micron (submicron) particle removal [10], marine microorganisms removal [11], and cultural relics restoration [12], and has achieved a good cleaning effect.

According to the different surrounding media when laser cleaning the coating, the laser cleaning technologies mainly include dry laser cleaning, wet laser cleaning, plasma shock wave laser cleaning, etc. After laser cleaning, the residual irradiation will also produce a thermomechanical coupling effect on the substrate, resulting in roughness, texture, polishing, and other phenomena on the surface, which shows the development potential in functional pretreatment such as welding, painting, anti-corrosion, plating, and bonding [13], as shown in Fig. 4.1. Currently, domestic and foreign research on the effect of the laser cleaning process on the surface



S. Wang, *Laser Cleaning*, https://doi.org/10.1007/978-981-95-2935-3_4

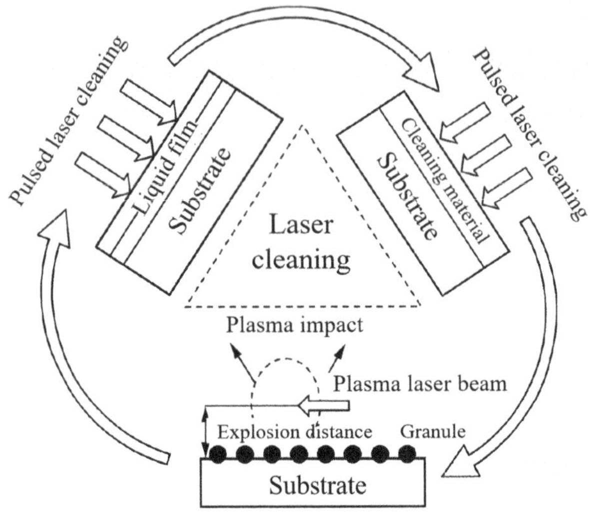


**Fig. 4.1** A schematic diagram of laser cleaning technology types and pretreatment applications

morphology and mechanical properties of the substrate is not mature enough, limiting the development of this technology to high-end equipment manufacturing and maintenance applications.

## 4.2 Laser Cleaning Surface Pretreatment for Welding

Laser welding is a highly efficient and precise weld method utilizing a high-energy-density laser beam as a heat source. During the welding process, the surface condition of the material to weld is decisive for the quality of the weld. To improve the welding quality, it is necessary to completely remove the oxide layer and other impurities from the material's surface before and after welding [14, 15]. Laser cleaning, as an environmentally friendly and highly efficient cleaning technology, cannot only effectively remove oil, rust, accumulated carbon, and other pollutants from the welded parts but can also carry out roughening pretreatment before welding and post-weld stress relief, which has gradually become the most suitable option for high-quality welding operations.

### *4.2.1 Pre-treatment Before Laser Cleaning Welding*

Presently, mechanical polishing, acid washing, and other methods are commonly used for prewelding cleaning, but there are problems such as high cost, low efficiency, and serious pollution. Laser cleaning technology provides a solution to solve these problems. Laser cleaning, which cannot only remove the oxide layer on the welding surface but also change the roughness of the substrate surface and reduce the porosity of the substrate surface, has received wide attention from researchers at home and abroad.

The oxide layer on the surface of welded substrates will hinder the mutual combination between substrates and deteriorate the quality of welded joints [16]. When cleaning 6061 aluminum alloy with a laser before welding, Tong et al. [17] found that under the optimal cleaning parameters of laser power of 15 W and continuous cleaning five times, the mass fraction of surface oxygen decreased to 3.23%, and the lower the surface oxygen content of the aluminum alloy, the higher the welding quality. When Zhang et al. [18] cleaned the 6005A aluminum alloy by laser, they found that the surface oxygen content first decreased, then increased with the increase in laser power, and decreased with the increase in cleaning speed. The surface oxygen content of aluminum alloy was the lowest when the power was 150 W and the cleaning speed was 0.8 m/min. However, Liu et al. [15] found that the laser energy density can adjust the oxygen content. Figure 4.2 shows that when the energy density is lower than the oxidation damage threshold, laser cleaning will make the primary oxide film reach boiling point and vaporize, but the oxidation damage threshold will be exceeded, thermal oxidation will occur, and the oxide content will increase. When the laser energy is 17.5 J/cm$^2$, the surface oxygen content is the lowest, 75% lower than that of the untreated aluminum alloy surface. Afonso et al. [19] also confirmed Liu's view in the real-time detection of UV pulsed laser surface cleaning. This shows that the appropriate laser cleaning energy density, cleaning

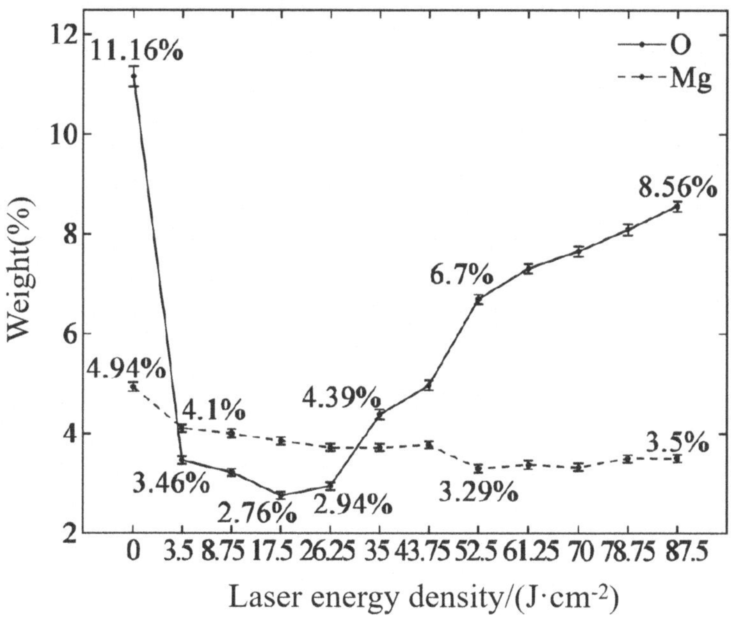


**Fig. 4.2** Changes of surface elements under different laser energy densities [15]

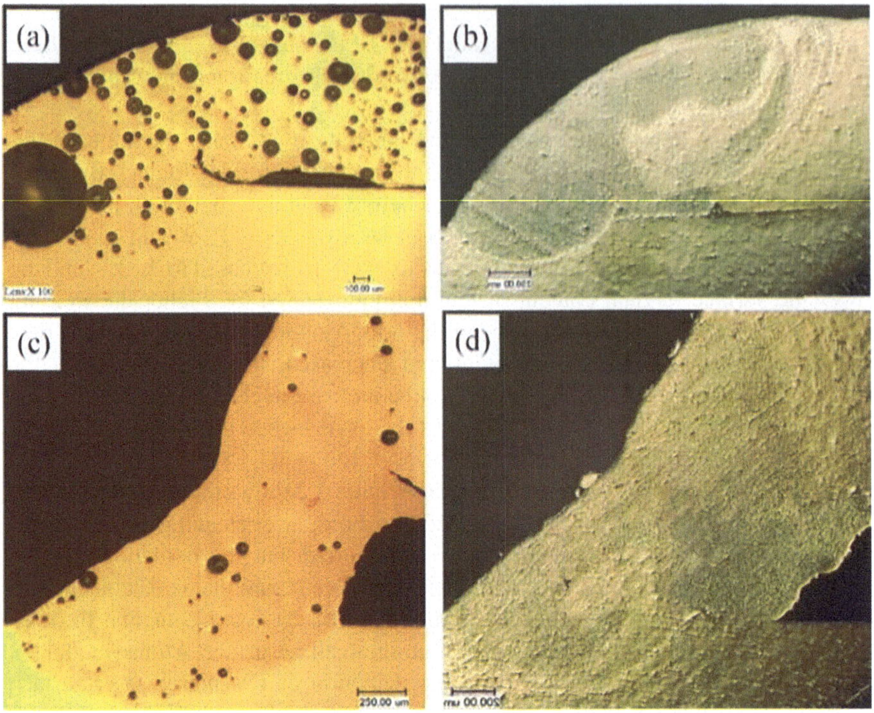

**Fig. 4.3** Comparison in porosity level in laser welded joints: (**a**, **c**) without cleaning area and (**b**, **d**) with laser cleaning area [21]

speed, and scanning times can effectively remove the oxide film, prevent the secondary oxidation of the substrate surface, and improve the welding quality.

In the aspect of reducing porosity, it is found that laser cleaning can remove the original oxide film and its adsorbed water on the alloy surface, thus reducing the hydrogen source during welding and reducing porosity [20]. AlShaer et al. [21] found that in the laser cleaning of the AC-170PX(AA6014) aluminum alloy before welding, weld porosity decreased to less than 1% after the laser removal of about 19 μm of the oxide layer and dirt on its surface. Figure 4.3 shows the cross-section of the samples before and after cleaning, and it shows that there is a significant reduction of porosity in the weld zone. When Zhou [22] and others cleaned the surface of 5A06 aluminum alloy by laser, the porosity of the weld seam was reduced from 2.91% to 1.59% in the argon atmosphere, which was better than that in the air. This is because laser cleaning in the air will re-oxidize the aluminum alloy, and the formed porous oxide film will absorb water again before welding, increasing the source of hydrogen. The argon atmosphere effectively avoids secondary oxidation and significantly reduces the porosity of the weld. Chen et al. [23] chose the process parameters of laser power 600 W and scanning speed 20 mm/s to clean Ti6Al4V titanium alloy. It was found that the main reason for the decrease in porosity on the surface of laser-cleaning titanium alloy was the removal of the oxide layer and the decrease in roughness, which met the requirements of grade I welding in China's aerospace industry standard. Lin et al. [24] carried out laser cleaning on the welded joint of 5083 aluminum alloy, and under the conditions of an average power of 40 W, a scanning speed of 2500 mm/s, and a repetition frequency of 85 kHz, oxides can effectively remove, and weld porosity can reduce. The maximum stress and strain of the joint weld after cleaning are 102.5% and 114.81% higher than before cleaning, respectively. It can be seen that in the atmosphere of inert gas (such as argon), appropriate laser energy density and cleaning speed can significantly reduce the porosity, reduce the introduction of hydrogen source in the welding process, prevent the occurrence of secondary oxidation on the surface of the substrate, and thus improve the welding quality.

### *4.2.2 Laser Cleaning Welded Joints*

A welded joint is an area where two or more parts are welded together, including a weld, a fusion zone, and a heat-affected zone. In recent years, researchers have found that laser cleaning can effectively remove pollutants on the surface of welded joints, strengthen the weld, fusion zone, and heat-affected zone, and then improve the strength of welded joints.

Wei et al. [25] applied laser cleaning to laser welding of SUS310S stainless steel and 6061 aluminum alloy. The research shows that the maximum tensile shear force of welded joint after cleaning reaches 1948N, which is 54% higher than that before cleaning under the laser parameters of wavelength 1064 nm, average rated power 200 W, and scanning speed 5000 mm/s; In the aspect of the strengthening mechanism, it is found that a large amount of easily diffused molten aluminum appears on both sides of the fracture surface after laser cleaning, which fills the fish scale pit during welding to form a relatively firm mechanical bond and improve the strength of the welded joint [26].

Chen et al. [27] found that reducing the thickness of the coating can inhibit the metal vapor generated by laser ablation, eliminate the Fe-Al phase in the welded joint, and improve the stability of the welding process when cleaning the Al-Si coating on the hot-pressed steel welded joint by laser. Figure 4.4 shows the microhardness distribution curve of welded joints with different coating thicknesses, and it can be seen that when the coating thickness is reduced to less than 30 μm, the average microhardness of welded joints is 488.38 $HV_{0.1}$, which is significantly higher than that of the base metal.

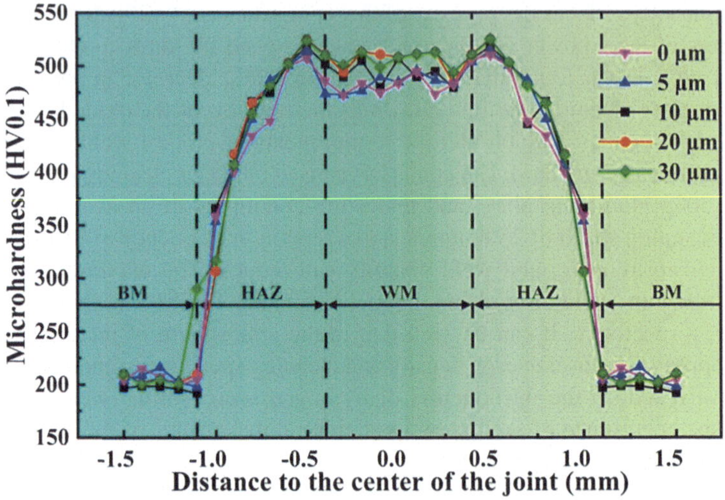


**Fig. 4.4** Microhardness distribution curve of welded joints for different coating thicknesses [27]

The reasons for laser cleaning to enhance the strength of welded joints are multifaceted. Removal of oxides, inhibition of pores, generation of nanoparticles, and formation of strong mechanical bonding are the main mechanisms. By selecting the appropriate laser energy and laser scanning speed, it is found that the tensile strength, stress and strain of the welded joint are improved, which lays a foundation for the continuous coating, reprocessing and long-term service of the welded parts.

### *4.2.3 Laser Cleaning Weld Performance*

The weld seam is the most important area that affects the strength of the welded joint, and the stress magnitude of the weld seam directly affects the performance and stability of the welded structure. At present, the focus of the research is to reduce weld stress by optimizing the welding process parameters. However, some scholars have found that laser cleaning-assisted laser welding can refine the grain size of the substrate surface and change the phase structure, thereby improving the weld stress, which is worthy of further research.

The surface contaminants before welding may cause the welding arc to be unstable, which in turn seriously affects the strength of the weld. Zhu et al. [28] found during laser cleaning of the welding seam of the HSLA steel plate that Under the pulse frequency of 50 kHz, scanning speed of 1500 mm/s, and laser power of 15 W, the average tensile strength of the weld is 120 MPa higher than that of the untreated. It can seen from Fig. 4.5 that some pores with a diameter of about 100 μm can observed on the uncleaned local fracture surface, and a large number of pores appear. However, nanoparticles with a diameter of about 1 μm are found on the

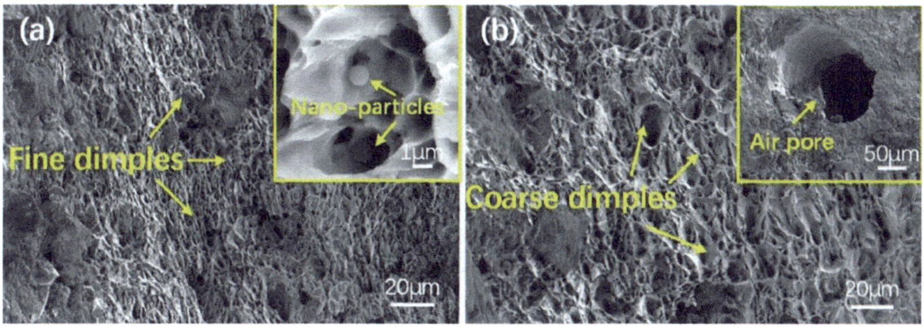


**Fig. 4.5** Scanning electron microscope photo of tensile fracture: (**a**) laser cleaning and (**b**) untreated [28]

fracture surface of the laser-cleaned sample, which will hinder the dislocation movement during tensile loading and lead to dislocation accumulation around the nanoparticles, and the tensile strength performance is improved macroscopically.

Yang et al. [29] innovatively combined laser cleaning with swing laser welding, It was found that the microstructure of the laser cleaned aluminum alloy welding zone changed from coarse columnar dendrites to equiaxed and columnar dendrites under $N_2$ protection, resulting in grain refinement. The tensile strength of the weld can reach 77.4% of the base metal, which is 16% higher than that of the original untreated sample. Zhang et al. [30] designed a " laser cleaning-arc welding " synchronous device to clean the surface of aluminum alloy. It found when the laser power is 200 W, and the cleaning speed is adjustable from 0.5 to 1.1 m/min, the microstructure at the center of the weld is composed of black β ($Mg_2Al_3$) phase and white α aluminum matrix phase. The shear strength of the weld reaches 82.92 MPa, which is 26.4% higher than that of the untreated weld. Zhu [31] used Nd:YAG laser to clean the 5154 aluminum alloy after welding, which proved that the appropriate laser cleaning parameters can eliminate the residual stress of the weld and improve the strength of the welded joint. Zhang et al. [32] performed laser cleaning on the weld of AZ31B magnesium alloy. It was found that the microstructure of the weld surface was evenly distributed along the thickness direction, the columnar crystal zone was narrow, the equiaxed crystal grains were fine, and the defects such as pores and cracks in the weld were significantly reduced, and the hardness of the weld center was significantly improved.

It can be seen that the improvement in weld performance is due to the transformation of phase structure and grain refinement in the laser cleaning area, which improves the mechanical properties of the weld. The laser cleaning-assisted laser welding process cannot only improve the quality of the weld, but also ensure better bonding between the metals during the welding process, and significantly improve the strength and durability of the welded structure.

In summary, laser cleaning technology has great potential in the field of welding. Table 4.1 summarizes the influence of different laser cleaning process parameters on welding performance and its strengthening mechanism. The comparative

**Table 4.1** Effect of different laser cleaning process parameters on welding performance and its strengthening mechanics

| | Substrate | Scanning parameters | Result | Strengthening mechanics | Ref. |
|---|---|---|---|---|---|
| Pre-welding laser cleaning | 6061A aluminum | P = 15 W<br>f = 100 kHz<br>Five times | Oxygen content is 3.23% | Laser cleaning will make the primary oxide film reach boiling point and vaporize, but after exceeding the threshold of oxidation damage, thermal oxidation will occur to increase the oxide content | [17] |
| | 6005A aluminum | P = 150 W<br>f = 100 Hz<br>v = 0.8 m/min | Oxide film disappears | | [18] |
| | 5083 aluminum | λ = 1060 nm<br>v = 2000 mm/s<br>W = 17.5 J/cm$^2$ | The oxygen content has decreased by 75% | | [19] |
| | AC170O4PX aluminum | λ = 1060 nm<br>v = 95 mm/s<br>f = 20 Hz | Porosity reduced to 1% | Elimination of contaminants, lubricants, and oxides | [21] |
| | 5A06 aluminum | f = 400 Hz<br>Scan 1 times | Porosity reduced to 1.59% | Reoxidation is avoided in argon atmosphere | [22] |
| | Ti6Al4V | P = 600 W<br>v = 20 mm/s | Porosity is greatly reduced | Roughness reduction and oxide film removal | [23] |
| Laser cleaning welded joints | SUS310S stainless steel | λ = 1064 nm<br>P = 200 W<br>v = 5000 mm/s | The maximum tensile shear force reaches 1948N | Mechanical bond is formed on the surface after laser cleaning | [25] |
| | Ti/Al dissimilar alloys | P = 120 W<br>v = 6 mm/s<br>f = 30 kHz | Increased tensile strength | Laser cleaning suppresses IMC thickness | [26] |
| | Hot-pressed section steel | W = 30 J/cm$^2$<br>v = 3000 mm/s | The microhardness of weld metal is 488.38 $HV_{0.1}$ | The Fe-Al phase disappears during the welding process. | [27] |
| Laser cleaning weld performance | HSLA steel | f = 50 kHz<br>v = 1500 mm/s<br>P = 15 W | Tensile strength increased to 120 MPa | Dislocation accumulation improves tensile properties | [28] |
| | 2219 aluminum | λ = 1060 nm<br>v = 6000 mm/s | Increased tensile strength of welds | Grain refinement in the welding zone | [29] |
| | 6005A-T6 aluminum | f = 2000 Hz<br>P = 200 W<br>v = 0.5–1.1 m/min | The shear strength of the weld is 82.92 MPa | The phase composition changed | [30] |
| | AZ31B magnesium alloy | f = 100 Hz<br>P = 20 W<br>v = 5000 mm/s | Defects such as pores and cracks are reduced | The surface grain of the welded specimen is fine | [32] |

analysis shows that laser cleaning can effectively remove dirt as oxide film from the surface before welding, improve surface roughness, reduce porosity, improve weld stress, and enhance the strength of the welded joint.

## 4.3 Laser Cleaning Surface Pretreatment for Coating

The purpose of coating is to cover a layer of new material on the surface of the substrate to protect, decorate, and enhance the properties of the material. Coating technology has many forms, including spraying, painting, etc. To improve coating adhesion, cleaning and pretreatment of the substrate surface before coating is essential, and laser cleaning technology ideally agrees with this need before coating. This chapter will introduce the application of laser cleaning in the cleaning and pretreatment of spraying and painting surfaces, focusing on the influence of laser cleaning on the surface integrity of substrates and the improvement of coating adhesion.

### *4.3.1 Surface Cleaning Treatment for Spraying*

Spraying refers to using a spray gun to convert powder into fine, uniform droplets at high temperatures, which are quickly ejected onto the surface of the substrate to form a coating. Coating life mainly depends on the quality of substrate surface pretreatment. Laser cleaning is widely used in pretreatment before spraying because of its high efficiency, cleaning, and surface modification advantages [33, 34].

Laser cleaning can improve the surface state of the substrate, thereby enhancing the adhesion of the sprayed coating. In terms of improving surface roughness, Li et al. [35] found that laser cleaning not only removes surface dirt from the substrate but also forms diffusely distributed micron-sized craters that roughen the surface (Ra < 1.0 μm). The surface roughening effect produced by laser cleaning can improve the flat morphology of particles, inhibit splashing, densify the coating, and improve the coating performance. However, the physical anchoring of the coating (roughness) is not the only factor affecting adhesion, Danlos et al. [36] found that when Ni-Al coatings were prepared by laser cleaning and synchronized thermal spraying processes, Chemical anchoring also plays a role in the interfacial bond between the substrate and the coating. When the substrate surface is ablated, the growth of the oxide layer improves the surface wettability and enhances the coating adhesion. Liu [15] also confirmed that thermal oxidation occurs when the laser energy density exceeds the oxidative damage threshold, thereby increasing the oxide content [18, 19], resulting in the growth of the oxide layer. From Fig. 4.6, compared with the sandblasting pretreatment coating in the left figure, the coating cleaned at the laser energy density of 2.3 J/cm$^2$ is more dense and uniform, and the coating adhesion is significantly better than the sandblasting treatment.

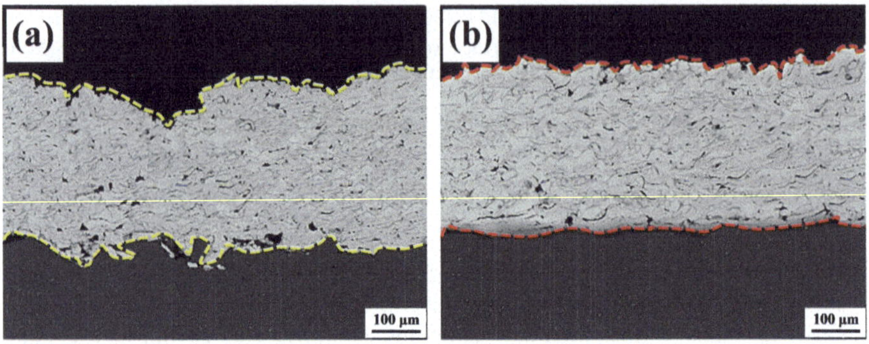


**Fig. 4.6** Cross section of nickel aluminum coating formed on the surface of A2017 aluminum alloy: (**a**) sandblasting treatment and (**b**) laser ablation treatment [36]

The residual heat of laser radiation will cause the substrate temperature to rise, the softened matrix can easily hold the particles, form a mechanical interlock to improve adhesion. Li et al. [37] found that the increase in substrate temperature can promote the atomic diffusion between the substrate material and the coating, increase the possibility of metallurgical bonding, and significantly improve the adhesion when the copper coating is prepared by laser cleaning pretreatment. Wu et al. [38] also proved the above conclusion and found that laser irradiation improved the overall plastic deformation of the deposited particles, promoted the close bonding between the particles, and formed a continuous and dense coating. Figure 4.7 is the image quality and inverse pole figure of copper coating under different laser irradiation power. It can seen that there are more low or un-indexed regions (highlighted by arrows) in the untreated coating (a1), which indicates that the interface area is weak and the defect density is high. In addition, two distinct grain sizes were observed in the coating, showing obvious inhomogeneous deformation. In contrast, With the aid of 700 W laser cleaning power, the boundary defects of copper coating particles gradually decreased, and the interfacial bonding was improved, showing a more homogeneous microstructure and smaller grain size.

At the same time, the ablation effect caused by laser cleaning will cause the surface to appear texture, micro-depressed holes, and other phenomena, which can regarded as surface texture processing [36]. Compared with laser cleaning pretreatment spraying, laser texture processing developed based on laser cleaning has a more significant effect on spraying surface pretreatment. Kromer et al. [39] used laser texture to process the 2017 aluminum alloy substrate and found that under the laser parameters of wavelength 1064 nm, pulse duration 100 ns, and maximum average power of 20 W, the mechanical anchoring effect of texture processing prevents crack propagation, and the bonding strength of the coating can increase by up to 300%. Subsequently, more scholars have also confirmed that laser texture processing as a spraying pretreatment can significantly improve the adhesion and mechanical properties of the coating [40–42].

To sum up, after laser cleaning, on the one hand, ablation causes different degrees of texture and micro-concave holes on the surface of the substrate, which increases the contact area between the coating and the surface of the substrate. On the other

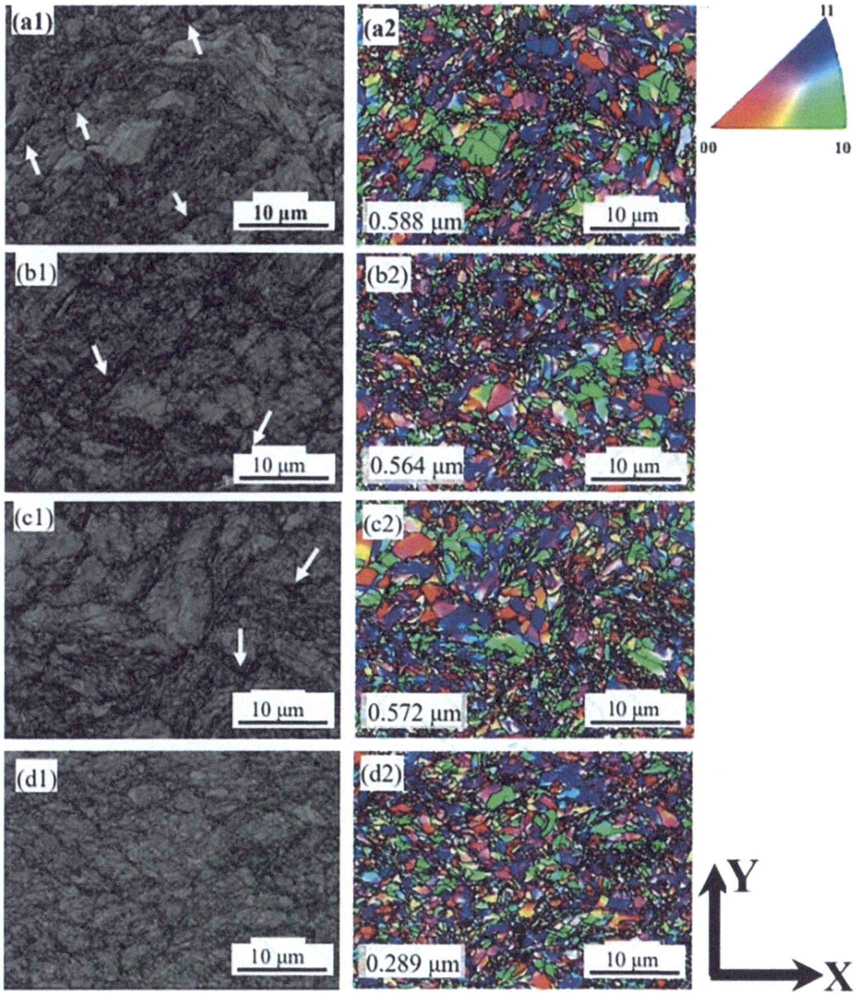


**Fig. 4.7** Image quality (IQ) and inverse pole figure (IPF) maps of laser-assisted Cu coatings under different laser powers of (**a1**–**a2**) 0 W, (**b1**–**b2**) 500 W, (**c1**–**c2**) 600 W, and (**d1**–**d2**) 700 W [38]

hand, the proper choice of laser energy density can make the surface of the substrate oxidize and improve wettability. At the same time, the residual heat of laser radiation softens the substrate, makes the particles stay, forms an interlocking structure, and enhances the adhesion.

### *4.3.2 Surface Cleaning Treatment Before Painting*

In the engineering field, keeping the substrate surface clean is one of the key factors in improving the quality of painting. Laser cleaning, as an advanced cleaning technology, plays a significant role in paint pretreatment. This technology can remove

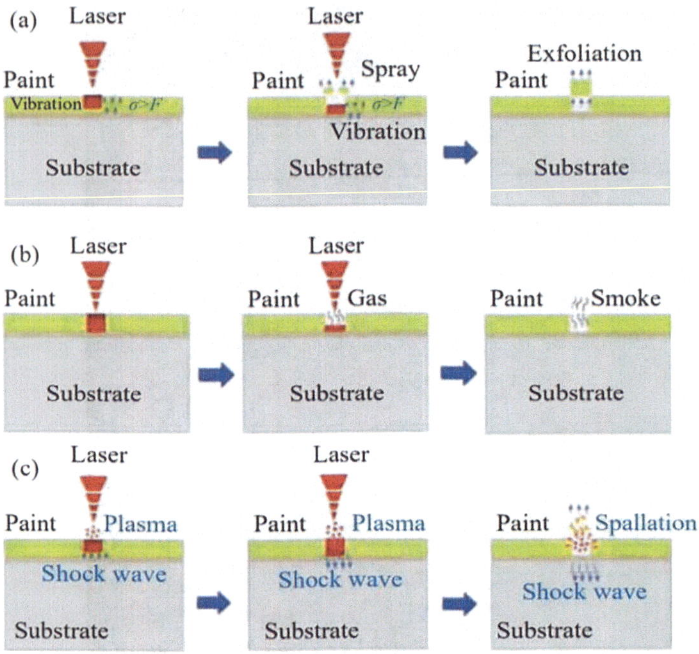


**Fig. 4.8** Schematic diagram of three typical mechanisms in laser paint removal [45]: (**a**) thermoelastic vibration effect; (**b**) ablation effect; (**c**) plasma shock effect

surface pollutants, enhance the adhesion of the paint layer to the substrate, ensure the durability and reliability of the paint layer, and improve the mechanical properties and corrosion resistance of the substrate. Laser cleaning removes the paint layer based on guaranteeing no damage to the substrate through the combined effects of thermoelastic vibration, laser ablation, and plasma shock [43]. This multiple mechanism of action ensures the high efficiency and accuracy of laser cleaning in removing the paint layer [44]. Figure 4.8 is a schematic diagram of the paint removal principle under different mechanisms of laser cleaning [45].

Laser cleaning has made significant progress in removing the paint layer, but the research on the surface state of the substrate after laser cleaning remains to be further studied. The results show that, compared with the traditional method, the mechanical properties of the matrix after laser cleaning are significantly enhanced [46, 47]. For example, Yoo et al. [48, 49] found that the mechanical properties of 304L stainless steel were significantly improved after depainting under the laser parameters of 850 W power, 1064 nm wavelength, and 4 m/s scanning speed, Fig. 4.9 shows the TEM analysis of dislocation arrangement and grain refinement of artificially corroded (CR) and laser cleaned (LSC) pretreated 304L stainless steel, and it can be seen that compared with CR pretreatment, a complex and dense dislocation structure is formed on the surface of LSC treatment, while sub-grains of 100–300 nm in size and nanograins of 20–50 nm in size are generated to refine the

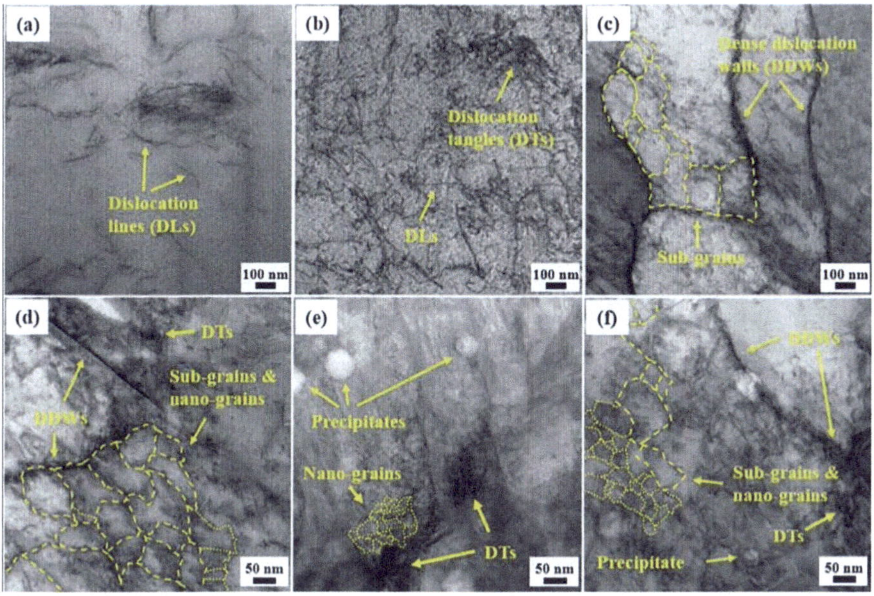


**Fig. 4.9** The TEM analysis of the dislocation arrangements and grain refinement for the CR and LSC specimens [48]

grains. Sun et al. [50] found similar effects in the laser cleaning of 2024-T351 aluminum alloy paint layers, with a significant increase in surface microhardness and tensile strength. In addition, Lu et al. [51] confirmed that laser cleaning can improve the microstructure and dislocation density of the material by rapid melting and quenching in the study of cleaning the paint layer of AH36 steel substrate, thereby improving the microhardness, tensile strength, and bending strength. These findings indicate that laser cleaning has good potential for improving the mechanical properties of matrix materials and has significant application value in future research.

Zhu et al. [52] found that laser cleaning can improve the microhardness of the substrate surface. At the same time, the paint layer absorbs laser energy to vibrate, break, and evaporate, resulting in a shock wave acting on the substrate surface, which will produce residual tensile stress, resulting in plastic deformation and cracks [53]. Therefore, although laser cleaning enhances surface hardness, the impact of residual tensile stress it causes on component stability is an urgent technical challenge that needs to be addressed.

At the same time as laser cleaning to remove paint pollutants, the cleaned substrate surface cannot only form a dense oxide film but also refine the grains and produce a specific surface texture, thus improving the corrosion resistance of the substrate. Li et al. [54] found that the formation of passivation film significantly improved the corrosion resistance of 5083 aluminum alloy during laser paint removal, which was 78% higher than that of untreated samples. Figure 4.10 shows the grain size statistics of the aluminum alloy matrix after laser cleaning. It is found

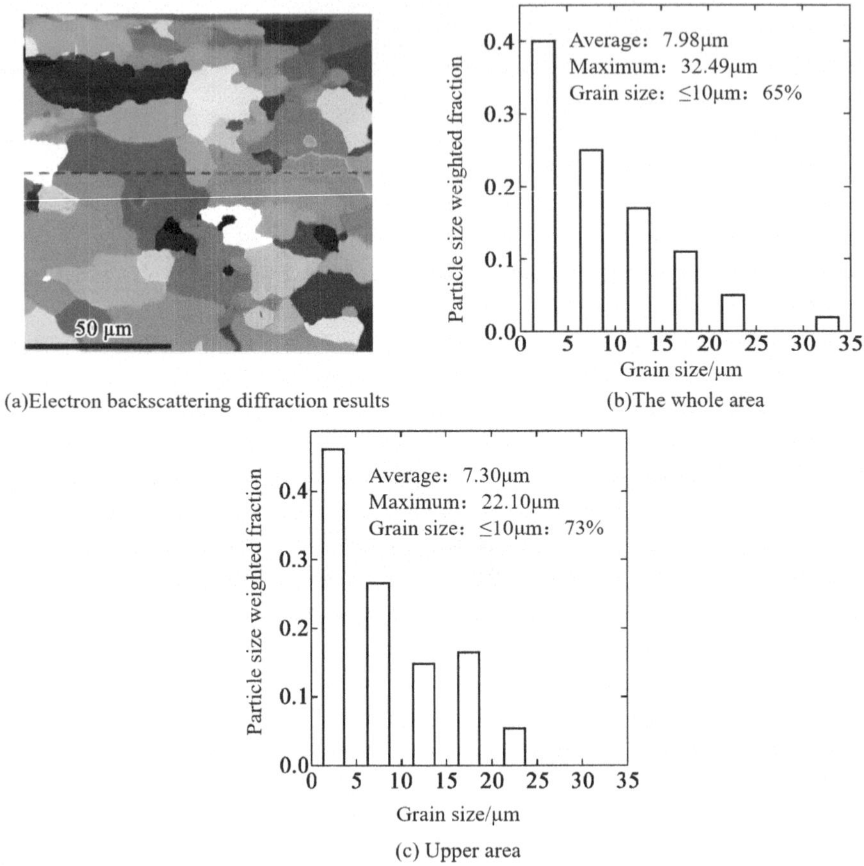


**Fig. 4.10** Statistics of grain size of aluminum alloy substrate after laser cleaning: (**a**) electron backscatter diffraction results; (**b**) entire region; (**c**) upper region [54]

that the average grain size on the surface of the cleaned aluminum alloy is 7.98 µm, with the percentage of grains smaller than 10 µm reaching 65%. This effectively reduces the variability between the grain boundaries and the intercrystalline components, consequently reducing the corrosion current density. Lu et al. [55] conducted continuous laser cleaning on the surface of the 5A06 aluminum alloy hull and found that when the laser intensity reached 11.9 W/cm$^2$, the dense micro-nano structure formed on the surface of the aluminum alloy helped to improve corrosion resistance. These results together indicate that laser cleaning is of great significance in removing the paint layer and improving the corrosion resistance of the substrate.

In terms of improving the adhesion of the paint layer, the surface of the cleaned substrate is affected by laser residual irradiation, the roughness changes, and the surface activity is enhanced, thereby obtaining better surface wettability and adhesion. Li et al. [54] found in the laser cleaning of 5083 aluminum alloy that the surface roughness was lower and coating adhesion was improved by about 22% compared to

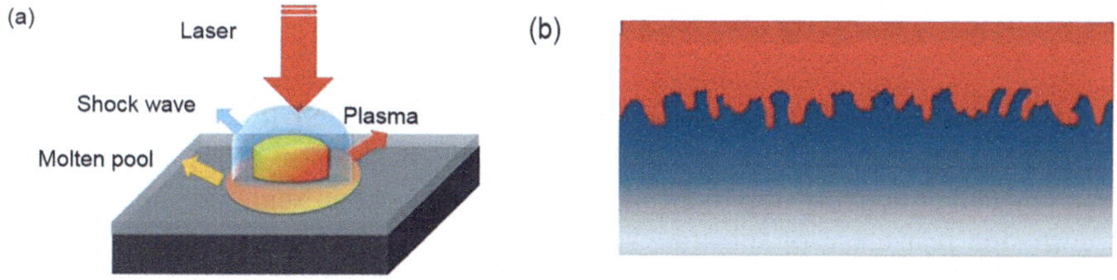


**Fig. 4.11** Laser cleaning pre-treatment spray painting: (**a**) schematic diagram of laser ablation process and (**b**) schematic diagram of recoating substrate bonding [59]

mechanical grinding. Ding et al. [56] regulated the surface adhesion of 5754 aluminum-magnesium alloy by optimizing the laser cleaning process parameters and found that at a power of 100 W and a frequency of 100–200 kHz, the wettability was better with an increase in surface roughness, which significantly improved the paint adhesion. However, some scholars have pointed out that it is not the higher the surface roughness, the stronger the adhesion of the paint layer [57]. It is necessary to select appropriate laser cleaning parameters for different substrate materials to obtain optimal roughness and then make the adhesion of the paint layer the best.

In engineering applications, laser cleaning plays an important role in improving the adhesion of engine, hull, and aircraft skin surface coatings. Kim et al. [58] conducted laser cleaning on the surface of the dual-fuel engine and found that the optimal laser energy density was 5 J/cm$^2$. After cleaning, the surface roughness of the substrate was about 5 μm, and the adhesion of the paint layer was significantly enhanced. Li et al. [59] laser-cleaned the paint layer on the steel surface of a ship's hull and found that the adhesion strength of the repainted paint layer reached 20 MPa, far exceeding the requirements of the shipyard. Figure 4.11 is a schematic diagram of the principle of laser ablation cleaning and the combination of repainting interfaces. The ablation caused by cleaning will make the surface contain a large number of micropores, which enhances the spread ability of the paint and its mechanical embedding with the substrate and improves the adhesion of the paint layer [53]. Shuangjiang et al. [60] found in laser cleaning of aircraft skin paint layers that at a laser energy density of 47.54 J/cm$^2$ and a scanning speed of 2800 mm/s, the surface roughness of the substrate was at a minimum of 1.801 μm, the surface-free energy reached 45.93 mJ/m$^2$, the adhesion of the repainted paint layer was further enhanced compared to the untreated one.

To sum up, laser cleaning cannot only effectively remove the surface dirt before coating, but also has the potential to improve the coating quality. On the one hand, the improvement of mechanical properties and corrosion resistance is due to the increasing dislocation density and micro-nano structure, the formation of oxide film, and grain refinement. On the other hand, the improved adhesion of the repainted layers is due to the influence of surface texturing, micro texturing, and wettability after the laser treatment. However, while pursuing the high performance of the substrate surface after cleaning, it is necessary to pay close attention to the deterioration problems such as plastic deformation caused by surface residual stress.

## 4.4 Laser Cleaning Surface Pretreatment for Anticorrosion

Metal corrosion will cause economic losses to society and threaten life safety. Therefore, effective surface anti-corrosion treatment is necessary [61]. Surface pretreatment can effectively improve corrosion resistance. At present, the improvement of corrosion resistance of substrate surface after laser cleaning has attracted wide attention from scholars in various countries. It has been found that appropriate laser cleaning process parameters can effectively improve the corrosion resistance of metal material surfaces. This chapter expounds on the potential of laser cleaning in improving the surface corrosion resistance of aluminum alloy, ferroalloy, copper alloy, and titanium alloy, and analyzes the mechanism of strengthening the surface corrosion resistance of different alloy substrates after laser cleaning.

### *4.4.1 Laser Surface Cleaning of Aluminum Alloy*

Aluminum alloy is easy to form a dense oxide film with corrosion resistance in the air, but it cannot serve in a long-term harsh environment [62]. Compared with the traditional pretreatment method, laser cleaning has a significant effect on improving the corrosion resistance of the aluminum alloy. Figure 4.12 is a schematic diagram of the mechanism of laser cleaning of pollutants on the surface of aluminum alloy. This technology cannot only remove pollutants such as oil stains and rust on the surface of aluminum alloy but also cause ablation, melting, recrystallization, quenching, and other physical and chemical reactions during the cleaning process, so that the molten aluminum atoms are in contact with the air to form a more corrosion-resistant aluminum-magnesium oxide layer [63, 64]. In addition, relevant scholars have carried out laser cleaning on A7N01-T4 aluminum alloy [65] and hot-rolled AA5083-O aluminum alloy [66]. It is found that the new aluminum magnesium oxide layer formed after cleaning will increase the impedance, reduce the capacitance, and then improve the corrosion resistance. Some scholars have also found that even under extremely short laser pulse width (10 ns), Al-Mg alloy will undergo a thermal oxidation process to form a continuous MgO coating, and its mechanism is similar to that of steady-state thermal oxidation [67].

It is well known that the laser cleaning residual irradiation will cause the substrate surface to melt and condense to form uniform pits and emulsion protrusions. This surface micro-texture hinders the contact between the liquid and the surface of

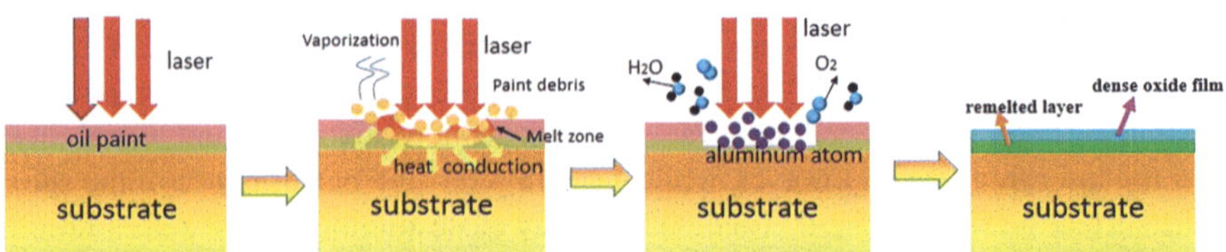


**Fig. 4.12** Schematic diagram of mechanism of laser cleaning aluminum alloy surface pollutants

the aluminum alloy substrate, thereby improving the corrosion resistance [68]. When Tian et al. [69, 70] cleaned the marine biological film on the surface of the aluminum alloy substrate by laser, at the laser energy of 4.14 J/cm$^2$, After cleaning, a nano-structure layer composed of amorphous nanocrystalline phase is formed on the surface of aluminum alloy, including nano-convex layer with a thickness of about 200 nm and bright band layer with a thickness of about 50 nm. Through element scanning, it is further observed that these nanostructures are mainly composed of aluminum and oxygen elements. This micro-nano structure has good potential in anti-adhesion, antifouling, and corrosion resistance, which is helpful for the subsequent recycling of aluminum alloys.

Rajan et al. [71] observed the decrease of grain size and the increase of dislocation density during laser cleaning of 6061 aluminum alloy, which led to the transformation of the material from crystalline to amorphous state, thus improving the corrosion resistance of the matrix [72]. Wang et al. [73] conducted laser cleaning on 5083-111H aluminum alloy and found that the uniform distribution of elements and the fine grain size helped to slow down the diffusion of oxygen into the substrate and enhance its corrosion resistance [74, 75].

When laser cleaning 7075-T651 aluminum alloy in a nitrogen atmosphere, Yue et al. [76] found that laser cleaning cannot only remove the coarse particles in the melting layer but also form fine honeycomb and dendritic nanostructures. More importantly, an AlN phase electrical insulator is formed in the melting layer, which greatly improves the corrosion resistance of the aluminum alloy

Table 4.2 summarizes the comparison of the corrosion resistance mechanism of different process parameters on the aluminum alloy matrix. It is found that the appropriate laser cleaning parameters will reduce the grain size of the aluminum alloy matrix surface after cleaning, increase the dislocation, and generate the oxide layer, micro-nano structure, and AlN phase, thereby improving the corrosion resistance. This shows that laser cleaning has great potential in improving the corrosion resistance of aluminum alloy matrix.

**Table 4.2** Comparison of corrosion resistance mechanism of magnesium alloy matrix with different process parameters

| | Material type | Processing parameter | Mechanism | Ref. |
|---|---|---|---|---|
| Al alloy | AA7024-T4<br>A7N01-T4<br>Al-Mg | λ = 1064 nm, v = 600 mm/s, W = 11.3 J/cm$^2$<br>λ = 1064 nm, v = 8000 mm/s, P = 30 W<br>λ = 1064 nm, W = 1 J/cm$^2$, t = 10 ns | The oxide layer is composed of $Al_2O_3$ and MgO | [63]<br>[65]<br>[67] |
| | 7075Al | λ = 1064 nm, v = 2500 mm/s, W = 4.44 J/cm$^2$ | Micro-nano structure | [68] |
| | 6063 Al<br>6061 Al<br>7050 Al | λ = 1064 nm, W = 5.52 J/cm$^2$, v = 4000 mm/s<br>λ = 800 nm, W = 15.3 J/cm$^2$, v = 0.4 mm/s<br>λ = 1064 nm, v = 10 mm/s, P = 120 W | The grain size decreases and the dislocation increases | [69]<br>[71]<br>[74] |
| | 7075-T651 Al | λ = 1064 nm, v = 30 mm/s, P = 500 W | AlN phase | [76] |

### 4.4.2 Laser Surface Cleaning of Ferroalloy

Ferroalloy has become the focus of attention due to its wide range of uses and huge consumption. However, how to effectively remove pollutants from the surface of ferroalloy and prolong its service life has been plaguing researchers. Compared with the traditional pretreatment method, laser cleaning cannot only effectively clean the surface contaminants without damaging the substrate but also change the surface roughness of the substrate and produce surface texture, thereby improving the corrosion resistance of the ferroalloy.

After laser cleaning, the depth of pits on the surface of ferroalloy can be reduced, and the surface roughness can be reduced so that the corrosion solution often has no time to diffuse to the depth of pits, which hinders the corrosion [77]. Lu et al. [8] used a nanosecond laser to clean the surface of AH36 steel and found that under the laser energy of 3.17 J/cm$^2$, the roughness was reduced to one-third of the original, and the corrosion resistance was improved by about five times.

Some researchers have also found that the oxide film formed on the surface of iron alloy by laser cleaning can reduce the corrosion current density and improve the corrosion resistance at a laser energy density of 2.546 J/cm$^2$ and a scanning speed of 3000 mm/s [78]. Yansong et al. [79] innovatively used continuous-pulse laser two-step continuous operation to clean the pollutants on the surface of the steel substrate and found that the surface would form new $\alpha$-$Fe_2O_3$, which played a role in corrosion protection. At the same time, the above conclusions are studied under the premise of not damaging the substrate.

The surface texture phenomenon caused by laser cleaning also has a positive guiding effect on the corrosion resistance of ferroalloy [80]. Guo et al. [81] cleaned iron-based alloys under specific laser process parameters. As can be seen from Fig. 4.13a, b, its bright-field image shows almost completely disordered structure, which belongs to a typical amorphous structure. With the increase of power, $\alpha$-Fe(Si) nanocrystalline and $Fe_2B$ crystalline phases are found in Fig. 4.13c, h. At the same time, it can be seen in figure (e) that the glass ribbon of $Fe_{78}Si_9B_{13}$ forms a new amorphous structure, which makes it have excellent pitting corrosion resistance.

After cleaning the SMA490BW weathering steel, Zhang et al. [82] found that the surface of the sample after cleaning formed micron-level, tightly packed columnar particles, which improved the corrosion resistance. However, the local thermal effect caused by the rapid and slight melting and solidification of the surface increased the residual tensile stress on the surface of the sample by 130–200 MPa. The generation of this residual tensile stress is one of the urgent problems to solve.

It can be seen from Table 4.3 that the main factors for improving the corrosion resistance of the surface of the ferroalloy substrate after laser cleaning are the reduction of roughness, the formation of oxide film, and micro-nano structure, which provides a new development direction for the surface cleaning and anti-corrosion pretreatment of the ferroalloy substrate. However, the local thermal effect caused by laser cleaning will produce plastic deformation caused by surface residual stress and deteriorate the matrix. This problem needs to be focused on.

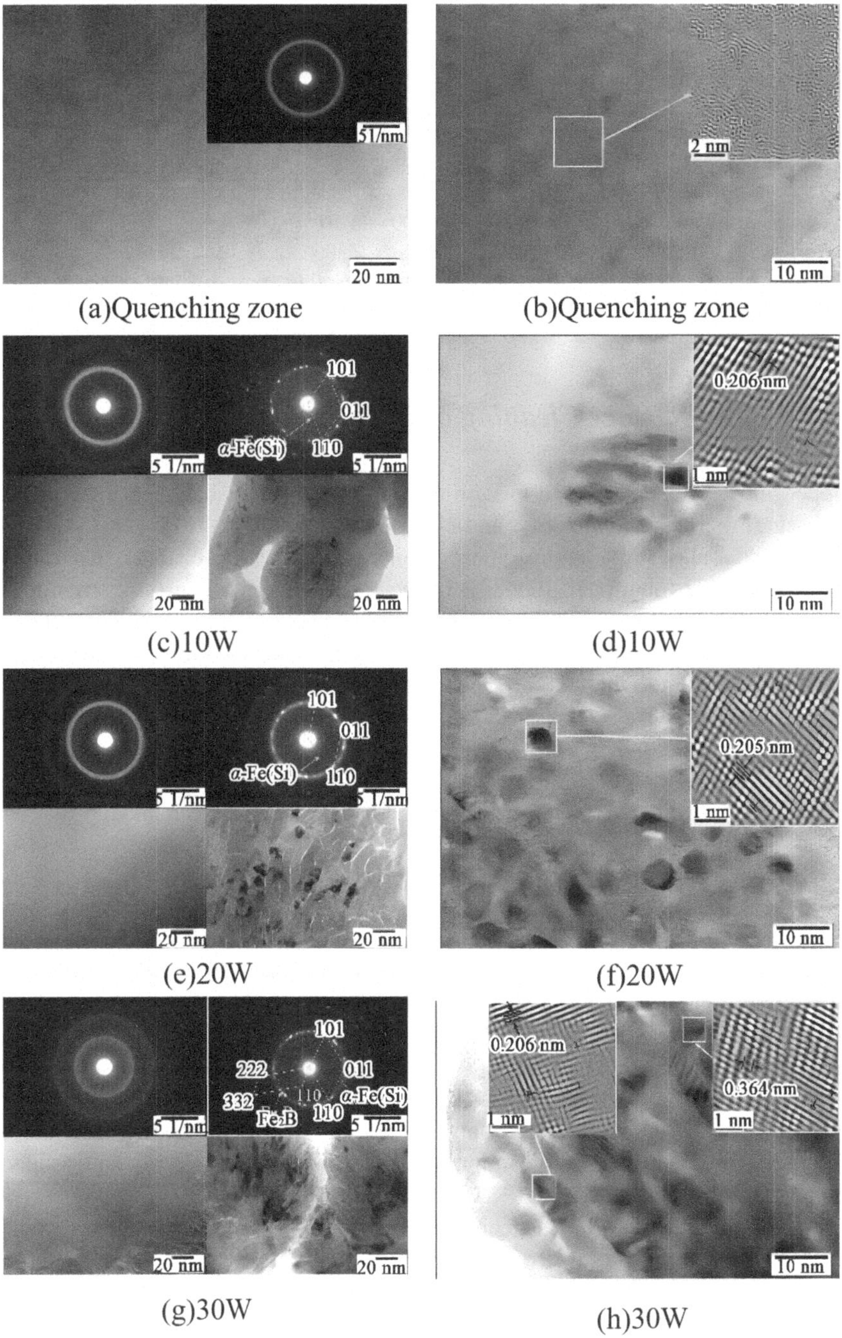


**Fig. 4.13** Shows the TEM results of quenched ferroalloys without and with different laser power treatment: (**a**, **b**) quenching belt, (**c**, **d**) 10 W, (**e**, **f**) 20 W and (**g**, **h**) 30 W of ribbons post-pulse laser processing. The illustrations in (**b**), (**d**), (**f**) and (**h**) are Fourier transform images of selected areas in HRTEM images

**Table 4.3** Comparison of corrosion resistance mechanisms of ferroalloy materials under different process parameters

| | Material type | Processing parameter | Mechanism | Ref. |
|---|---|---|---|---|
| Fe alloy | AH32 steel<br>AH36 steel | W = 20 J/cm$^2$, the overlap rate is 50%<br>$\lambda$ = 355 nm, P = 20 W, W = 3.17 J/cm$^2$ | Roughness reduction | [77]<br>[8] |
| | EH36 marine steel<br>Q235B steel | t = 100 ns, v = 3000 mm/s, W = 2.54 J/cm$^2$<br>$\lambda$ = 1064 nm, v = 1000 mm/s, P = 15 W | Oxide film formation | [78]<br>[79] |
| | SMA490BW steel<br>AH36 steel | P = 28.5 W, d = 70 μm, v = 8000 mm/s<br>t = 15 ns, $\lambda$ = 355 nm, W = 5.30 J/cm$^2$ | Micro-nano structure | [82]<br>[80] |
| | $Fe_{78}Si_9B_{13}$ steel | f = 20 kHz, v = 3000 mm/s, P = 20 W | | [81] |

### 4.4.3 *Laser Surface Cleaning of Copper Alloy*

Copper alloys are widely used in various fields due to their excellent mechanical processing, thermal conductivity, and electrical conductivity. However, the corrosion of copper alloys in complex environments is still one of the challenges to their performance and functional life [83]. In the aspect of removing the corrosion of bronze products, Di Francia [84] and others used various techniques to characterize the laser cleaned and unwashed areas, and found that laser cleaning technology can effectively reduce the thickness of the corrosion layer without causing damage to the original alloy surface. Baek's et al. [85] studied the influence of laser cleaning on brass, and found that under the conditions of laser energy density of 1.62 J/cm$^2$ and spacing distance of 0.05 mm, the corrosion products on brass surface can be effectively removed, and the influence on microstructure and mechanical properties of brass is minimal.

When laser cleaning the contaminants on the surface of copper alloy, it is found that the roughening and texture on the surface of the substrate after cleaning is the main reason for improving its corrosion resistance. Shu et al. [86] filled the modified $SiO_2$ powder into the micro-pit array prepared by nanosecond laser cleaning to form a micro-nano layered structure. It was found that the $SiO_2$ protective layer helps to reduce the contact area between the corrosive medium and the substrate and improve the corrosion performance [87]. Dong et al. [88] used a femtosecond laser to clean the surface of copper mesh. Under the laser frequency of 1 kHz and scanning speed of 50 um/s, superhydrophobic non-uniform nano-ripples and microstructures were formed after cleaning, and the corrosion resistance of the copper mesh surface was improved.

In addition, when DiFrancia cleaned the copper alloy by laser [89], it was found that the pollutants on the surface of the substrate could be effectively removed after cleaning. By using the method of $^{18}O$ isotope and laser phase identification, it can be observed that the ablation area has merged with the blue $^{18}O$ block isotope after laser cleaning, causing a slight re-oxidation of the substrate surface. The interaction process between laser and copper alloy in air is shown in Fig. 4.14, and the formation of an oxide film improves the corrosion resistance of the copper alloy surface.

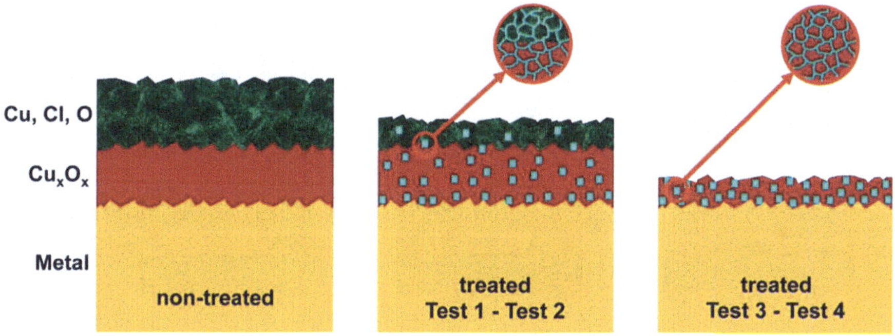


**Fig. 4.14** The schematic diagram of the interaction between laser-copper alloy and oxygen in the air during laser treatment

In summary, laser cleaning of the surface of the copper alloy substrate can re-oxidize the surface, form an oxide film and a micro-nano-scale structure, and improve the corrosion resistance of the copper alloy. This provides a new idea for laser cleaning to clean and pre-treat the surface of the copper alloy substrate.

### *4.4.4 Laser Surface Cleaning of Titanium Alloy*

Titanium alloy is widely used in various fields because of its high specific strength and high heat resistance [90], but it is easy to form an oxide layer and pollutants in harsh environments, which affects its subsequent use. In terms of laser cleaning of titanium alloy oxide film and pollutants, it was found that the removal mechanism changed from the ablation effect to the combination of gasification and explosion [90–92]. At the same time, the heat generated by high energy output at the moment of cleaning caused the surface Ti atoms to react with oxygen in the air, forming an extremely thin and dense oxide film on the surface of the remelted layer to enhance corrosion resistance [93]. Liu et al. [94] found that when laser-cleaned TA15 alloy, a uniform, and flat oxide layer was found on the cross-section at an energy density of 8.75 J cm$^2$, and the surface corrosion rate was 20.43 times lower than that of the untreated sample.

In the aspect of corrosion resistance of laser cleaning titanium alloy, Liu et al. [95] found that the decrease of surface roughness and the formation of an oxide layer after laser cleaning of TA15 alloy hindered further contact between the substrate and the external environment, thus improving the corrosion resistance. However, it was found that the huge residual stress generated by pulse frequency would lead to dense cracks and reduce the mechanical properties of the substrate. Li et al. [96] found that the surface corrosion resistance of titanium alloy is closely related to the laser energy density when cleaning titanium alloy with specific laser parameters. Figure 4.15 shows the process of cleaning titanium alloys with different laser energies. At a low laser flux of 1.99 J/cm$^2$, the surface roughness decreases,

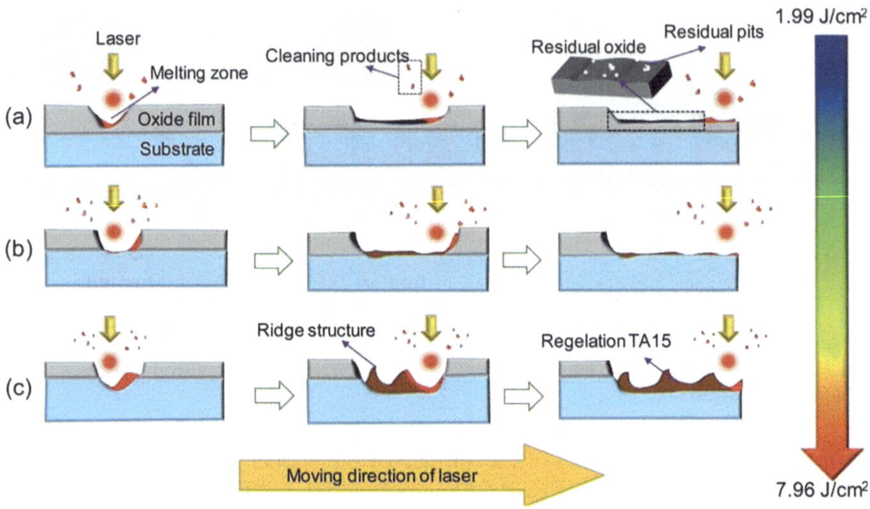


**Fig. 4.15** Evolution behavior of oxide film surface structure of TA15 titanium alloy after laser cleaning: (**a**) low energy density, (**b**) appropriate energy density, and (**c**) high energy density

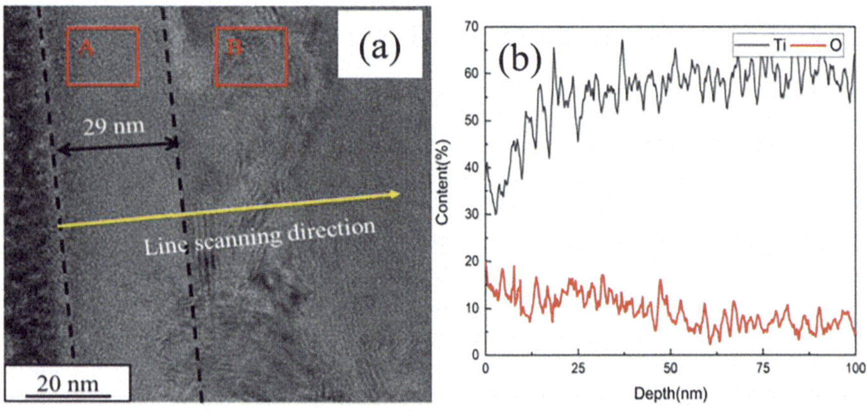


**Fig. 4.16** HRTEM images and SAED images of cross section (31.85 MW/cm$^2$ and 80%)

and the surface oxide film melts slightly and then oxidizes again. As the energy continues to increase, the oxide film is completely removed, and the laser impact force causes the surface to form a periodic 'ridge' structure. This microstructure has great advantages, such as superhydrophobicity, anti-adhesion, anti-fouling, and anti-corrosion.

Li et al. [97] used nanosecond pulse technology to clean titanium alloy oxide film. It was found that when the maximum laser power density was 31.85 MW/cm$^2$ and the spot overlap rate was 80%, the surface became smoother and no cracks appeared. Under this parameter, Fig. 4.16 shows the TEM and SAED image analysis of the cleaned titanium alloy cross section. It can be seen that under strong laser radiation, oxides with different valence states are formed on the surface in a short time. From the line scanning element, the thickness of the newly formed dense oxide film on the titanium alloy surface is 29 nm, which has a certain positive effect on improving the corrosion of the titanium alloy matrix.

In summary, the laser cleaning process parameters are closely related to the improvement of the corrosion resistance of titanium alloy. Choosing the appropriate laser cleaning energy density, the reoxidation layer and micro-nanostructure will appear on the surface of the substrate after cleaning, which will prevent the corrosion of titanium alloy. It is of great significance for the subsequent application of laser-cleaning titanium alloy materials.

## 4.5 Laser Cleaning Surface Pretreatment for Plating

Surface plating refers to the technology of surface modification of the substrate by plating a layer of other substances on the surface of the product to cover the original surface through electroplating, brush plating, chemical plating, coating, and other processes. However, in the process of processing, handling, and preserving the substrate, there will be oxides, dust, and other contaminants attached to the surface of the plating. If these contaminants are not removed and directly plated, the adhesion between the product surface and the coating will decrease, affecting the surface properties of the plating. Laser cleaning technology cannot only effectively clean the pollutants in the coating but also enhance the adhesion of the coating and optimize the surface characteristics of the coating. At present, laser cleaning technology mainly focuses on electroplating and surface pretreatment of film, so this chapter focuses on electroplating and filming.

### *4.5.1 Electroplating Surface Cleaning Treatment*

Electroplating refers to the use of the electrolysis principle to plate a layer of metal film on the surface of the substrate, which plays a role in decoration and anti-oxidation. A good surface state of the substrate before electroplating is a prerequisite for ensuring the quality of the coating. If the surface of the substrate is rough, rusty, or oily, the adhesion of the coating will be affected [98]. Therefore, it is very important to clean and pretreat the substrate surface before electroplating.

It has been found that laser cleaning can effectively remove the pollutants from the coating by using a high-energy beam without damaging the substrate, compared with traditional cleaning. When Lee et al. [99] cleaned the surface contaminants of gold-plated bronze products, the surface corrosion products were successfully removed at a laser energy density of 1.59–2.39 $J/cm^2$, but in some damaged groove areas, the corrosion products still existed after cleaning due to the rough and inhomogeneous surface morphology, which could not respond effectively to the laser beam. Therefore, corrosion products are still present after cleaning. Jia et al. [100] used laser cleaning technology to remove the surface rubber contaminants from the grooved chromium plating mold and found that it could not only clean the gap efficiently but also eliminate the micro-cracks on the chromium layer. Figure 4.17 shows the cleaning results of the rubber layer of the chromium-plated mold with grooves. Compared with a single laser cleaning, there is no rubber residue on the surface after three scans, showing an obvious pit structure, and only a small amount of iron oxide residue in the damaged chromium layer.

In addition, the cleaning degree and damage degree of the electroplating substrate are also different when using different wavelength lasers to pretreat the electroplating substrate. When Kumar et al. [101] removed the powder coating on the surface of galvanized steel, it was found that the 1064 nm pulsed laser removed the 60 μm thick coating at a rate of about 35 $cm^2$/min. Although the 532 and 355 nm laser wavelengths can effectively remove the powder coating, the galvanized layer is damaged. Mustafa et al. [102] found that the 1030 nm laser wavelength can effectively remove the contaminants on the surface of the coating, the 515 nm laser wavelength can be used for texture processing, and the 343 nm laser wavelength is most suitable for laser drilling of galvanized steel. This shows that selecting the appropriate laser pulse wavelength cannot only realize the functionalization of the metal surface of the plating layer but also effectively reprocess the substrate surface.

In terms of improving the bonding strength of the plating layer, laser cleaning will produce rough, texture, micro-nano structure on the surface of the substrate, so that a close interlock is formed between the plating layer and the substrate to enhance the adhesion [103]. When Miguel et al. [104] cleaned the contaminants on the galvanized stainless steel substrate by laser, they not only effectively removed the fouling particles with an average size of 50 μm but also prepared a micronano structure on the surface, and the adhesion of the coating was improved. Zhu et al. [105] found in nanosecond laser cleaning of contaminants on the surface of hot-dip galvanized coatings of HSLA steel sheets that when the coating thickness was the same, the appearance of $\zeta$ and $\delta$ phases in the zinc coatings on the cleaned samples

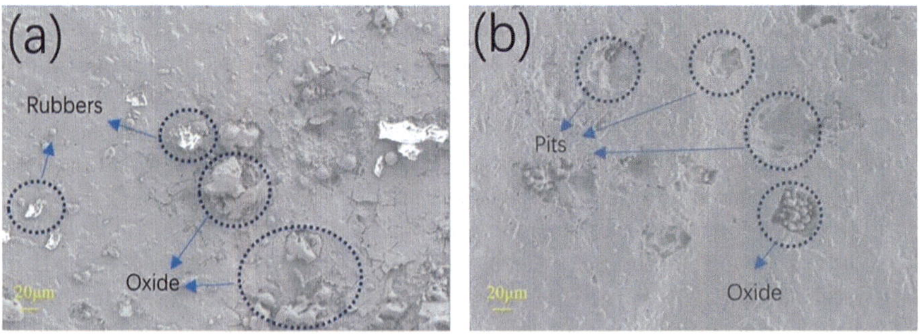


**Fig. 4.17** Cleaning results of the rubber layer of the chrome-plated mold with grooves: (**a**) laser single scan results and (**b**) laser three scan results [100]

contributed to the preferential expansion of cracks from the coating depressions to the surface and enhanced the resistance of the coatings to chalking, As shown in Fig. 4.18.

In summary, laser cleaning before electroplating can effectively remove the pollutants on the electroplating surface, improve the bonding strength of the coating, and improve the comprehensive performance of the plating parts. Laser cleaning shows a vigorous development trend in electroplating pretreatment. However, for metal samples with complex morphology, such as bronze products and chromium-plated molds, the effective removal ability of pollutants by laser cleaning is limited, which needs further improvement in electroplating pretreatment.

## 4.5.2 Coating Surface Cleaning Treatment

Coating is a process of covering a thin film on the surface of an object, which is used to improve the comprehensive performance of the object and is widely used in electronics, automobiles, aerospace, and other industries. Physical vapor deposition (PVD) and chemical vapor deposition (CVD) are common techniques in the coating process. The pretreatment steps before coating are crucial to ensure the formation of high-quality films. To effectively remove pollutants without damaging the surface of the substrate material, researchers have been exploring more efficient and environmentally friendly pretreatment methods. The emergence of laser cleaning provides a feasible solution for surface pretreatment in coating applications.

It has been found that the main contaminants on the surface of coated optical elements are micron and submicron $SiO_2$ particles. In the aspect of laser cleaning contaminants on the surface of coated optical components, Ye et al. [106] found that

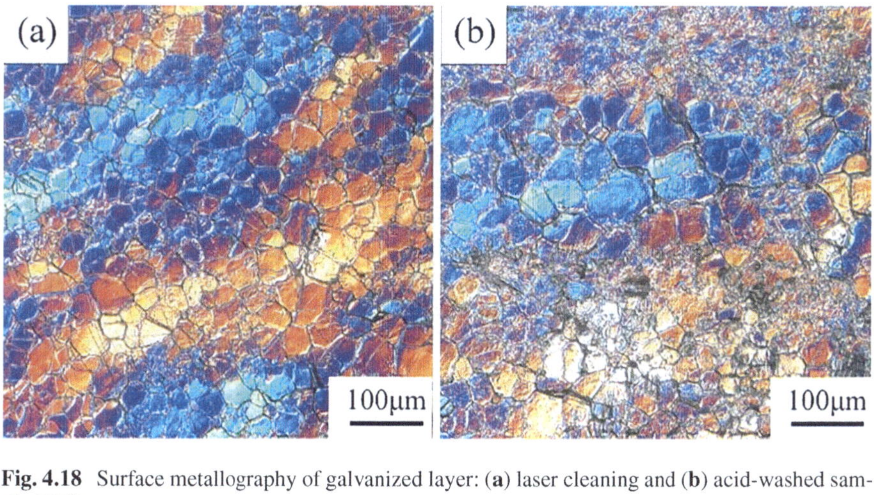


**Fig. 4.18** Surface metallography of galvanized layer: (**a**) laser cleaning and (**b**) acid-washed sample [105]

when laser cleaning dimethyl silicone oil on the surface of gold-plated K9 glass, the gold film did not cause damage under the $CO_2$ laser with a power of 10.8 W and a frequency of 100 Hz, and the cleaning effect was remarkable. The next year, Ye et al. [107] carried out laser cleaning of $SiO_2$ particles attached to the surface of gold-plated K9 glass and found that the removal rate of $SiO_2$ particles could reach 91%. Guo et al. [108] also found that the laser parameters with a wavelength of 355 nm and a pulse width of 10 ns can effectively remove the pollutants of optical components without damaging the substrate. In terms of equipment improvement, Wang et al. [109] innovatively proposed laser cleaning assisted by an airflow replacement system. After cleaning, the total particle removal rate on the surface of the optical element reached 82.21%, which could completely remove the contaminated particles with a radius of more than 5 μm and reduce the phenomenon of particle reattachment [110].

Laser cleaning cannot only effectively clean the pollutants on the coating surface but also improve the film preparation efficiency and adhesion and achieve a high-performance film layer [111]. In terms of physical vapor deposition (PVD), Zhang et al. [112] found that laser cleaning of cemented carbide TiAlN coatings after the PVD process can effectively reduce the defects in the coating at a laser scanning speed of 10 mm/s, densify the coating, and increase the critical load of the coating by 32%. At the same time, the obvious microgroove structure generated on the surface of the substrate after cleaning forms a mechanical interlock with the TiAlN coating, which effectively improves the adhesion of the coating [113]. Before using the PVD process, Bobzin et al. [114] performed laser cleaning pretreatment on the CrAl based hard coating deposited on AISI420 steel. It was found that the increase in laser pulse energy can significantly expand the depth of the coating structure and improve the deposition efficiency. However, beyond a certain laser pulse energy, the laser-induced thermal effect will increase the proportion of amorphous and crystalline regions in the PVD coating, resulting in a decrease in the hardness of the coating.

In chemical vapor deposition (CVD), Jeong et al. [115] used CVD technology to perform laser cleaning pretreatment before cobalt deposition on silica substrates. Under the laser wavelength of 355 nm, scanning speed of 4 μm/s, and laser power of 3.33 W, it was found that the grain size of the coating increased, the degree of densification increased, and the resistivity of the film decreased from 58 to 19 Ω/cm, which improved the crystallinity and function of the Co film. Matsumoto et al. [116] used a 976 nm semiconductor laser for cleaning pretreatment in the CVD process and found that a highly oriented $HfO_2$ film was produced on a polycrystalline AlN substrate. Compared with the traditional CVD method, the deposition rate of the film after cleaning increased by 3–5 times.

Practice shows that laser cleaning can effectively remove the pollutants on the coating surface without damaging the substrate, improve the compactness of the coating, improve the deposition efficiency of the film, and enhance the adhesion between the film and the substrate. Laser cleaning technology provides a green and efficient solution for the functional pretreatment of coatings in physical vapor deposition (PVD), chemical vapor deposition (CVD), and other fields. It can be predicted that with the maturity of laser cleaning technology, its importance in improving the quality of industrial products and environmental sustainability will be further highlighted.

## 4.6 Laser Cleaning Surface Pretreatment for Adhesive Bonding

Adhesive bonding is a process method to connect two bonded parts by using the mechanical bonding force, physical adsorption force, and chemical bonding force generated by the adhesive on the joint surface. Effective connections are needed between engineering structural components to ensure the stability of the structure. Compared with other methods such as welding and riveting, bonding has the advantages of a simple process and uniform stress distribution. To improve the bonding quality of various materials, especially the bonding of carbon fibers and metals commonly used in modern manufacturing, the cleaning of material surfaces has become particularly important. As a non-contact cleaning method, laser cleaning not only avoids the problems of substrate damage or chemical residue that may be caused by traditional physical grinding or chemical cleaning but also improves the bonding strength.

### *4.6.1 Surface Cleaning Treatment of Carbon Fiber Reinforced Composites*

Carbon fiber reinforced polymer (CFRP) has the characteristics of low density, high specific strength, and excellent thermal stability [117]. However, the surface of CFRP is easily polluted by oil, adhesives, and release agents, which reduces the bonding strength between CFRP and bonded parts. Therefore, it is necessary to clean the surface pretreatment. Laser cleaning as a surface pretreatment technology, the use of instantaneous high temperature generated by thermal stress stripping, thermal vibration, and other mechanisms, can in the case of no damage to the substrate, not only effectively clean the surface contaminants on the surface of the CFRP substrate, and at the same time improve the surface properties of the CFRP [117, 118].

In terms of the mechanism of laser cleaning to enhance the bonding strength of the CFRP surface, Fig. 4.19 shows the schematic diagrams of CFRP surface morphology and fracture modes with different laser energy densities, respectively [119]. At low energy density, the residual epoxy resin on the surface of the cleaned CFRP substrate will break with the adhesive during the bonding process and reduce the bonding strength. As the laser energy density increases, the epoxy resin is completely removed, and the exposed and undamaged carbon fiber will increase the bonding area of the adhesive. At the same time, the surface of the carbon fiber is

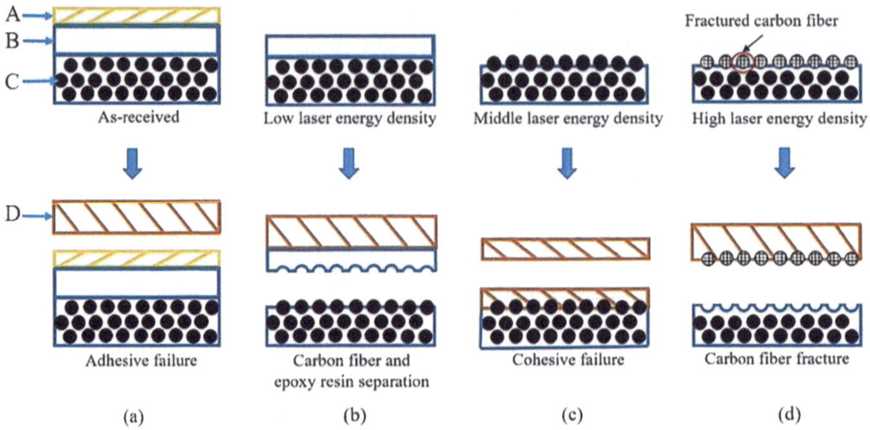


**Fig. 4.19** Effect of laser energy density on the surface morphology and fracture mode of the joint [119]: (**a**) receive, (**b**) ED < 17.7 mJ/mm$^2$, (**c**) 17.7 < ED < 35.4 mJ/mm$^2$, (**d**) ED > 35.4 mJ/mm$^2$. A release agent, B epoxy resin, C carbon fiber, D adhesive. *ED* refers to laser energy density

oxidized to increase the surface wettability, and the bonding strength is enhanced. However, as the laser density further increases, the surface of the CFRP substrate is damaged, and Joint fracture occurs in the carbon fiber layer [119–122].

Therefore, the appropriate laser cleaning process parameters are very important to improve the bonding strength of the CFRP surface. Gao et al. [123] found that when cleaning the CFRP surface at a scanning speed of 500 mm/s and an average power of 13.6 W, they found that the surface energy after local cleaning increased from 56.67 to 83.35 mJ/m$^2$ compared to complete cleaning, and the surface wettability and bonding strength increased. Leone et al. [121] at the wavelength of 1064 nm, pulse power of 20 kW, and scanning speed of 4000 mm/s, laser cleaning effectively removes the dirt on the surface of the CFRP substrate, so that the exposed carbon fiber is directly combined with the new substrate. Compared with the sandpaper polishing treatment, the shear strength after cleaning is increased to 21 MPa. When Tong et al. [124] laser cleaned the surface of the CFRP substrate, under the scanning speed of 1500 mm/s and the average power of 16 W, there was no resin residue left after cleaning, and the surface exposed the complete carbon fiber surface. The bonding strength was 1.6 times higher than that of untreated.

In addition, finite element simulation technology can predict material response, optimize processing parameters, reduce trial and error process, and reveal the performance changes of materials under different conditions, providing important data support for future material design and application. Yang et al. [125] used finite element software to establish a three-dimensional model of the transient temperature field and paint layer peeling during laser depainting of the CFRP surface. It was found that the superposition of adjacent laser pulses would increase the ablation depth. If the coating is completely removed, it may cause damage to the substrate. Therefore, a combination of low energy density and high scanning speed is recommended for selective stripping of the paint layer during the experiments. Ponticelli et al. [126] combined a fuzzy algorithm and a genetic algorithm to simulate the laser cleaning of the surface of CFRP laminates. It was found that the bonding strength of CFRP was significantly improved under the power of 100–160 W and the energy of 3–3.3 J/mm$^2$ [127]. Similarly, Xu et al. [128] found that keeping the angle θ at 45° is beneficial to improve the processing quality and efficiency when simulating the scanning angle of laser cleaning CFRP parts by finite element method.

Selecting the appropriate laser cleaning process parameters, laser cleaning can effectively remove the dirt on the surface of the CFRP matrix, increase the contact area between the exposed and undamaged carbon fiber and the adhesive, and improve the surface wettability, thereby achieving high-strength bonding of carbon fiber and other composite materials. In addition, numerical simulation and simulation technology can be used to optimize laser cleaning process parameters and guide process control, which is likely to become the focus of future green and efficient cleaning pretreatment research.

### 4.6.2 Surface Cleaning Treatment of Metal Bonding Joints

Metal bonding is an important connection method. It can replace the traditional welding method to achieve a solid connection between metal materials and is widely used in various fields. To improve the bonding strength of metals, many surface cleaning methods based on physical or chemical modification have been developed, but there are some problems, such as matrix damage and environmental pollution. Laser cleaning cannot only remove the dirt on the substrate surface to ensure integrity but also modify the substrate surface, which is widely recognized in the fields of aluminum alloy and titanium alloy bonding.

In terms of the bonding strength of laser-cleaning aluminum alloy, Gacs et al. [129] found that after laser cleaning of $AlMg_{4.5}Mn_{0.4}$ aluminum sheets, the microstructure of the cleaned substrate surface can affect the bonding strength of aluminum alloys. When laser cleaning AA7075-T6 bonded joints, Guo et al. [130] found that the strength and toughness of bonded joints increased with the increase of surface roughness under low energy input and decreased with the increase of surface roughness under high energy input. At the same time, the surface after cleaning has an excellent topological structure, and the tensile shear strength and toughness are increased by 5.9% and 13.5%, respectively. Zhang [131] and others innovatively used crater array-multi-groove laser cleaning to improve the shear strength of 7075-T6 aluminum alloy adhesive joint. It was found that when the groove pattern was parallel grooves, the scanning interval was less than 100 μm and the scanning times were 2–5 times, the shear strength reached the maximum value of about 29.34 MPa.

Zhu et al. [132] studied the effect of laser cleaning energy on the lap shear strength of bonded aluminum alloy joints. It can be seen from Fig. 4.20 that the laser ablation of 26.3 J/cm$^2$ increased the oxygen content on the surface of the aluminum

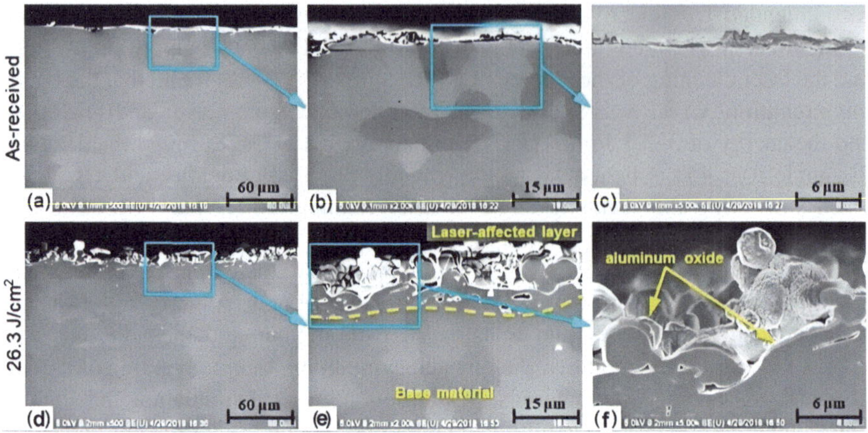


**Fig. 4.20** Cross-sectional view of aluminum alloy matrix: (**a–c**) Received aluminum alloy matrix. (**d–f**) In laser ablation of aluminum alloy substrate, the laser flux is 26.3 J/cm$^2$ [128]

alloy and formed a new oxide layer of about 0.5 μm thickness, which made the surface more active and improved the wettability of the adhesive on the surface of the aluminum alloy, thereby improving the strength of the bonded joint [133]. Wu et al. [134] studied the effect of laser cleaning on the properties of AA6022-T4 aluminum alloy joints. It was found that a more uniform and thicker alumina layer was formed on the surface after laser cleaning, which increased the joint strength by 25%. Therefore, the changes in microstructure, oxide film, and roughness on the surface of the aluminum alloy after laser cleaning are the key factors in improving the strength of aluminum alloy bonded joints [135].

In terms of laser cleaning to improve the strength of titanium alloy bonded joints, Rotella et al. [136] studied the effect of pulsed laser cleaning on the surface of Ti6Al4V and found that laser irradiation can improve the average roughness, wettability, and surface hardness under specific parameters, thereby enhancing the delamination resistance of titanium alloy. Kurtovic et al. [137] found that a uniform nano-scale oxide layer was formed on the surface of the Ti6Al4V substrate after laser cleaning, which effectively improved the bonding performance. Palavra et al. [138] studied the effect of laser cleaning on the adhesion between CFRP composites and titanium alloy. It was found that when the surface of the titanium alloy was treated with a $5.03 \times 10^{-3}$ J pulse, the tensile strength reached 8.2 MPa, and the bonding strength significantly improved. Zhang C et al. 139] fabricated the microstructure of crater array with different parameters on the surface of Ti6Al4V titanium alloy plate by nanosecond fiber laser. It is found that with the increase of overlap ratio, the increase of pit depth and bump height leads to the increase of surface roughness. When the overlap ratio is 75%, the surface area is the largest, and the maximum shear strength is 29.2 MPa (Table 4.4).

In summary, the increase of surface bonding area, the change in roughness, and the generation of micro-nanostructure and oxide film are the main factors for laser

**Table 4.4** Effect of different laser cleaning process parameters on bonding properties and mechanism of aluminum and titanium alloys

| Substrate | Scanning parameters | Result | Mechanics | Ref. |
|---|---|---|---|---|
| $AlMg_{4.5}Mn_{0.4}$ | $\Phi \le 33\ J/cm^2$ | The adhesive strength is improved by three times | The change of surface roughness affects the bonding strength, but too much roughness will reduce the bonding strength | [129] |
| AA7075-T6 | Laser pulse energy is 0.31 mJ | The strength and fracture toughness of the joint can be improved by 5.9% and 13.5%, respectively | | [130] |
| 7075-T6 | Parallel slot $L \le 100\ \mu m$ $N = 2–5$ | The maximum shear strength is about 29.34 MPa | | [131] |
| AA6022-T4 | $\Phi = 26.3\ J/cm^2$ | A new oxide layer about 0.5 μm thick | The surface has higher cleanliness, wettability and chemical modification | [132] |
| AA6022-T4 | $\Phi = 19.01\ J/cm^2$ | The alumina layer still exists at least at a depth of 900 nm | | [134] |
| Ti6Al4V | $\lambda = 1064$ nm $f = 10$ kHz $P = 5$ W | The contact angle decreases and the wettability increases | Formation of nanostructures and oxide films | [136] |
| Ti-6Al-4V | $\lambda = 1064$ nm $v = 800$ mm/s $f = 10$ kHz | Presenting a uniform nanostructured oxide layer with a thickness of about 16–47 nm | | [137] |
| Titanium alloy | $\lambda = 532$ (nm) $v = 500$ (mm/s) $f = 5$ kHz | The tensile strength reached the highest level of 8.2 MPa | | [138] |
| Ti6Al4V | $\Phi = 6.5\ J/cm^2$ $L = 75\%$ | The maximum shear strength is 29.2 MPa | | [139] |

cleaning to improve the strength of metal alloy bonded joints. As an efficient and environmentally friendly surface pretreatment method, laser cleaning can effectively improve the strength of metal alloy joints by selecting appropriate cleaning process parameters. In future technological development, accurately grasping the appropriate process parameters in the laser cleaning process and deeply exploring the different process enhancement mechanisms will become the core focus of further enhancing the bonding strength.

## 4.7 Conclusion and Prospect

Laser cleaning technology cannot only efficiently clean the substrate without damaging it but also obtain the effect of pretreatment on the surface of the cleaned substrate. In this paper, the application progress of laser cleaning technology in welding, coating, anticorrosion, plating, and bonding is systematically expounded, and the mechanism of laser cleaning technology with surface cleaning and pretreatment ability in different application scenarios is discussed. Although laser cleaning technology shows great pretreatment potential in many application scenarios such as welding and coating, there are still many problems to be solved.

1. At present, most scholars and engineers at home and abroad pay attention to the improvement of laser cleaning efficiency and the control of surface cleaning degree and often ignore the surface integrity of the substrate after cleaning, as well as excessive cleaning problems such as large, slight burns on the surface of the substrate after laser cleaning. The pursuit of high cleaning efficiency and cleaning degree based on maintaining the integrity of laser cleaning surface is the development direction of laser cleaning technology to become an accurate, green, and high-quality cleaning technology in the future.
2. Laser cleaning can effectively remove the oxide film, corrosion, and other attachments on the surface of the substrate before welding, improve the surface roughness of the substrate, and enhance the welding strength. In terms of weld stress relief, laser cleaning can refine the weld grains, alleviate the tensile stress concentration of the weld after welding, and improve the fatigue resistance of the weld. Many studies have shown that laser cleaning technology has a good application prospect in cleaning texturing before welding and weld stress relief.
3. Laser cleaning residual irradiation will induce regular micro-texture on the surface of the substrate, change the surface roughness, enhance the surface activity, and make the surface of the substrate have better wettability and adhesion. After cleaning, the interface adhesion of the substrate recoating (or plating) is improved. If the ability of laser cleaning technology to roughen pretreatment before coating (or plating) can be fully exploited, it will be expected to make cleaning, roughening, and coating (or plating) coherently implemented, greatly improving the efficiency of the entire coating (or plating) operation.
4. Laser surface anti-corrosion pretreatment technology is widely used on the surface of aluminum, iron, copper, and titanium alloys. By analyzing the corrosion resistance characteristics and mechanism of the substrate surface after laser cleaning, it is found that the formation of the oxide layer, micro-nano structure hydrophobicity, and grain refinement are the main reasons. However, it is difficult to control the process of improving the corrosion resistance of the substrate surface after laser cleaning. If the laser residual irradiation parameters cannot be accurately controlled, the substrate surface may produce residual tensile stress, thereby reducing the fatigue resistance of metal materials. How to eliminate the residual tensile stress after laser cleaning surface pretreatment will become one of the key directions of future research.
5. High-power laser cleaning equipment is an important means to improve the efficiency of laser cleaning and obtain a good pretreatment effect. However, after cleaning the surface coating of the substrate, the high laser energy density is easy to causes the ablation of the substrate surface, which not only fails to achieve the strengthening effect of surface texturing and micro-structure but also degrades the surface performance of the substrate. Studies have shown that the layer-by-

layer peeling method using continuous-nanosecond, nanosecond-picosecond, and other composite laser cleaning can protect the substrate to the greatest extent and obtain the strengthening effect based on rapid cleaning of the substrate surface coating, which may be an important direction for the future development of laser cleaning technology.

6. Laser cleaning is essentially a kind of laser processing technology. At present, there are many simulations of laser cladding, laser welding, laser shock strengthening, and other technologies, and there are few professional simulations for laser cleaning. Through simulation, the mechanism of laser cleaning can be explained more clearly, the control of the laser cleaning process can be optimized, and the impact-strengthening function of the substrate surface of laser cleaning can be maximized. It can predicted that the combination of comprehensive and systematic laser cleaning computer simulation and experiment will become a research hotspot of surface cleaning pretreatment in the future.

## References


1. J.M. Sánchez-Amaya, M.R. Amaya-Vázquez, L. González-Rovira, M. Botana-Galvin, F.J. Botana, Influence of surface pre-treatments on laser welding of Ti6Al4V alloy, J. Mater. Eng. Perform. 23 (2014) 1568–1575. https://doi.org/10.1007/s11665-014-0967-y.
2. L. Zhong, Z. Li, G. Wang, H. He, G. Wei, S. Zheng, G. Feng, N. Xie, R. Zhang, Erosion resistance of valve core surface combined with WC-10Co-4Cr coating process under different pretreatments, Materials 15 (2022). https://doi.org/10.3390/ma15228140.
3. S.T. Auwal, S. Ramesh, F. Yusof, S.M. Manladan, A review on laser beam welding of titanium alloys, Int. J. Adv. Manuf. Technol. 97 (2018) 1071–1098. https://doi.org/10.1007/s00170-018-2030-x.
4. G. Baiocco, S. Genna, D. Salvi, N. Ucciardello, Laser texturing to increase the wear resistance of an electrophoretic graphene coating on copper substrates, Materials 16 (2023). https://doi.org/10.3390/ma16155359.
5. Y. Song, S. Wang, Y. Pan, Z. Li, L. Yu, Current research status of laser cleaning of carbon fiber reinforced polymers, Mater. Today Commun. 39 (2024) 109312. https://doi.org/10.1016/j.mtcomm.2024.109312.
6. S.M. Bedair, Jr., Smith Harold P., Atomically clean surfaces by pulsed laser bombardment, J. Appl. Phys. 40 (1969) 4776–4781. https://doi.org/10.1063/1.1657288.
7. Z. Zhao, X. Liu, Z. Chen, Y. Tian, M. Chen, L. Liu, F. Song, Evaluation of laser cleaning effect for the removal of paint on aluminum alloys, Int. J. Adv. Manuf. Technol. 126 (2023) 3193–3203. https://doi.org/10.1007/s00170-023-11224-4.
8. Y. Lu, Y. Ding, G. Wang, L. Yang, M. Wang, Y. Wang, B. Guo, Ultraviolet laser cleaning and surface characterization of AH36 steel for rust removal, J. Laser Appl. 32 (2020) 032023. https://doi.org/10.2351/1.5127569.
9. L. Hou, F. Yin, S. Wang, J. Sun, H. Yin, A review of thermal effects and substrate damage control in laser cleaning, Opt. Laser Technol. 174 (2024) 110613. https://doi.org/10.1016/j.optlastec.2024.110613.
10. T. Chen, W. Wang, T. Tao, A. Pan, X. Mei, Multi-scale micro-nano structures prepared by laser cleaning assisted laser ablation for broadband ultralow reflectivity silicon surfaces in ambient air, Appl. Surf. Sci. 509 (2020) 145182. https://doi.org/10.1016/j.apsusc.2019.145182.
11. J. Li, D. Zhang, X. Chen, D. Xu, D. Qiu, F. Wang, M. Easton, Laser directed energy deposited, ultrafine-grained functional titanium-copper alloys tailored for marine environments:

Antibacterial and anti-microbial corrosion studies, J. Mater. Sci. Technol. 166 (2023) 21–33. https://doi.org/10.1016/j.jmst.2023.05.020.
12. R. Xiang, J. Wang, Y. Wang, X. Chao, B. Mai, Y. Zhang, J. Cao, Study on the application of laser technology in the restoring and transformation of PbS on painted cultural relics, J. Cult. Herit. 66 (2024) 455–463. https://doi.org/10.1016/j.culher.2024.01.002.
13. G. Zhu, Z. Xu, Y. Jin, X. Chen, L. Yang, J. Xu, D. Shan, Y. Chen, B. Guo, Mechanism and application of laser cleaning: A review, Opt. Lasers Eng. 157 (2022) 107130. https://doi.org/10.1016/j.optlaseng.2022.107130.
14. M. Bertasa, C. Korenberg, Successes and challenges in laser cleaning metal artefacts: A review, J. Cult. Herit. 53 (2022) 100–117. https://doi.org/10.1016/j.culher.2021.10.010.
15. B. Liu, C. Wang, G. Mi, J. Wang, W. Zhang, X. Zhang, Oxygen content and morphology of laser cleaned 5083 aluminum alloy and its influences on weld porosity, Opt. Laser Technol. 140 (2021) 107031. https://doi.org/10.1016/j.optlastec.2021.107031.
16. M. Harooni, B. Carlson, R. Kovacevic, Detection of defects in laser welding of AZ31B magnesium alloy in zero-gap lap joint configuration by a real-time spectroscopic analysis, Opt. Lasers Eng. 56 (2014) 54–66. https://doi.org/10.1016/j.optlaseng.2013.11.015.
17. Y. Tong, A. Zhang, J. Huang, R. Li, J. Zhou, X. Chen, X. Ren, Distribution and influence of oxygen content of laser pretreated aluminum alloy surface before welding, Chinese Journal of Lasers-Zhongguo Jiguang 46 (2019). https://doi.org/10.3788/CJL201946.0202003.
18. Y. Zhang, Q. Yao, W. Long, C. Wang, J. Lin, Z. Liu, Welding defect and mechanical properties of nanosecond laser cleaning 6005A aluminum alloy, Materials 15 (2022). https://doi.org/10.3390/ma15217841.
19. C.N. Afonso, F. Vega, J. Solís, Real time optical diagnostics during pulsed uv laser induced surface cleaning and oxidation, Microelectron. Eng. 25 (1994) 223–228. https://doi.org/10.1016/0167-9317(94)90019-1.
20. A. Kumar, M. Sapp, J. Vincelli, M.C. Gupta, A study on laser cleaning and pulsed gas tungsten arc welding of ti–3Al–2.5V alloy tubes, J. Mater. Process. Technol. 210 (2010) 64–71. https://doi.org/10.1016/j.jmatprotec.2009.08.017.
21. A.W. AlShaer, L. Li, A. Mistry, The effects of short pulse laser surface cleaning on porosity formation and reduction in laser welding of aluminium alloy for automotive component manufacture, Opt. Laser Technol. 64 (2014) 162–171. https://doi.org/10.1016/j.optlastec.2014.05.010.
22. C. Zhou, H. Li, G. Chen, G. Wang, Z. Shan, Effect of single pulsed picosecond and 100 nanosecond laser cleaning on surface morphology and welding quality of aluminium alloy, Opt. Laser Technol. 127 (2020) 106197. https://doi.org/10.1016/j.optlastec.2020.106197.
23. Chen J, Wen P, Laser cleaning of titanium alloy and its effect on laser welding porosity, China Mechanical Engineering 31 (2020) 379. https://doi.org/10.3969/i.issn.1004-132X.2020.04.001.
24. J. Lin, C. Wang, A study of surface quality and mechanical strength in 5083 aluminum alloy plates using pulsed laser cleaning, J. Mech. Sci. Technol. 36 (2022) 5223–5229. https://doi.org/10.1007/s12206-022-0935-7.
25. P. Wei, Z. Chen, D. Wang, R. Zhang, X. Li, F. Zhang, K. Sun, Y. Lei, Effect of laser cleaning on mechanical properties of laser lap welded joint of SUS310S stainless steel and 6061 aluminum alloy, Mater. Lett. 291 (2021) 129549. https://doi.org/10.1016/j.matlet.2021.129549.
26. J. Zhao, P. Jiang, S. Geng, Y. Wang, B. Xu, Effect of pulsed laser pretreatment induced pit-structure on the formation of intermetallic compounds in titanium-aluminum dissimilar welded joints, Opt. Laser Technol. 167 (2023) 109589. https://doi.org/10.1016/j.optlastec.2023.109589.
27. X. Chen, S. Zhang, M. Jiang, Y. Chen, Y. Jiang, Z. Wang, N. Jiang, A. Chen, B. Li, Z. Lei, Y. Chen, Microstructure and mechanical properties of laser welded hot-press-formed steel with varying thicknesses of al–si coatings cleaned by nanosecond pulsed laser, Journal of Materials Research and Technology 22 (2023) 2576–2588. https://doi.org/10.1016/j.jmrt.2022.12.124.

28. L. Zhu, B. Sun, Z. Li, X. Pan, Y. Chen, Y. Cao, The weld quality improvement via laser cleaning pre-treatment for laser butt welding of the HSLA steel plates, Welding in the World 64 (2020) 1715–1723. https://doi.org/10.1007/s40194-020-00949-5.
29. F. Yang, Study on surface pretreatment of 2219 aluminum alloy and swing laser welding technology in negative pressure environment, Guangxi University (2022). https://doi.org/10.27034/d.cnki.ggxiu.2022.001387.
30. Y. Zhang, W. Long, Q. Yao, Study on welding synchronous laser cleaning technology of aluminum alloy car body, Infrared and Laser Engineering 52 (2023). https://doi.org/10.3788/IRLA20220534.
31. G. Zhu, S. Wang, M. Zhang, X. Yang, W. Liu, G. Wang, Application of laser cleaning in postwelding treatment of aluminum alloy, Appl. Opt. 59 (2020) 10967–10972. https://doi.org/10.1364/AO.406171.
32. M. Zhang, xiong Li, heqing Li, Influence of Nanosecond Pulsed Laser Cleaning on Laser Welding Quality of AZ31B Magnesium Alloy, Automotive Engineering 44 (2022) 1786–1796. https://doi.org/10.19562/j.chinasae.qcgc.2022.11.017.
33. M. Jones, A. Cockburn, R. Lupoi, M. Sparkes, W. O'Neill, Solid-state manufacturing of tungsten deposits onto molybdenum substrates with supersonic laser deposition, Mater. Lett. 134 (2014) 295–297. https://doi.org/10.1016/j.matlet.2014.07.091.
34. R. Lupoi, M. Sparkes, A. Cockburn, W. O'Neill, High speed titanium coatings by supersonic laser deposition, Mater. Lett. 65 (2011) 3205–3207. https://doi.org/10.1016/j.matlet.2011.07.014.
35. H. Li, S. Costil, H.-L. Liao, C. Coddet, Role of the laser surface preparation on the adhesion of ni−5%al coatings deposited using the PROTAL process, J. Therm. Spray Technol. 15 (2006) 191–197. https://doi.org/10.1361/105996306X108011.
36. Y. Danlos, S. Costil, H. Liao, C. Coddet, Influence of ti–6Al–4V and al 2017 substrate morphology on ni–al coating adhesion — impacts of laser treatments, Surf. Coat. Technol. 205 (2011) 2702–2708. https://doi.org/10.1016/j.surfcoat.2010.08.147.
37. B. Li, L. Yang, Z. Li, J. Yao, Q. Zhang, Z. Chen, G. Dong, L. Wang, Beneficial effects of synchronous laser irradiation on the characteristics of cold-sprayed copper coatings, J. Therm. Spray Technol. 24 (2015) 836–847. https://doi.org/10.1007/s11666-015-0246-y.
38. L. Wu, Z. Lin, Z. Luo, B. Li, J. Liu, Q. Zhang, J. Yao, Microstructure and corrosion properties of cu coatings deposited via laser-assisted low-pressure cold spray, T. Nonferr. Metal. Soc. 34 (2024) 604–617. https://doi.org/10.1016/S1003-6326(23)66421-0.
39. R. Kromer, S. Costil, J. Cormier, D. Courapied, L. Berthe, P. Peyre, M. Boustie, Laser surface patterning to enhance adhesion of plasma sprayed coatings, Surf. Coat. Technol. 278 (2015) 171–182. https://doi.org/10.1016/j.surfcoat.2015.07.022.
40. V. Kumar, R. Verma, K. Chauhan, R. Kumar, Influence of laser texturing pre-treatment on HVOF-sprayed WC-10Co-4Cr+GNP coatings on AISI 304, Surf. Topogr.: Metrol. Prop. 10 (2022) 015052. https://doi.org/10.1088/2051-672X/ac5c68.
41. P. Maressa, L. Anodio, A. Bernasconi, A.G. Demir, B. Previtali, Effect of surface texture on the adhesion performance of laser treated Ti6Al4V alloy, J. Adhes. 91 (2015) 518–537. https://doi.org/10.1080/00218464.2014.933809.
42. D. Jia, D. Zhou, P. Yi, S. Zhang, X. Zhan, Y. Liu, Effect of laser-textured substrate on adhesion and microstructure of deposited mo–NiCrBSiCFe coating, Journal of Materials Research and Technology 29 (2024) 2072–2082. https://doi.org/10.1016/j.jmrt.2024.01.243.
43. H.C. Zhao, Y.L. Qiao, Q. Zhang, X. Du, Y. Zang, X.T. Liu, B.Y. Han, Study on the characteristics and mechanism of pulsed laser cleaning of polyacrylate resin coating on aluminum alloy substrates, Appl. Opt. 59 (2020) 7053–7065. https://doi.org/10.1364/AO.387532.
44. J. Ouyang, P. Mativenga, N. Goffin, W. Liu, Z. Liu, N. Mirhosseini, L. Jones, E. Woolley, L. Li, Energy consumption and performance optimisation of laser cleaning for coating removal, Cirp J. Manuf. Sci. Technol. 37 (2022) 245–257. https://doi.org/10.1016/j.cirpj.2022.02.001.

45. M. Jiang, J. Xu, K. Liu, Current status and prospect of engineering application of laser paint removal technology for aircraft skin (Invited), Infrared and Laser Engineering 52 (2023). https://doi.org/10.3788/IRLA20220780.
46. T. Shan, F. Yin, S. Wang, Y. Qiao, P. Liu, Surface integrity control of laser cleaning of an aluminum alloy surface paint layer, Appl. Opt. 59 (2020) 9313–9319. https://doi.org/10.1364/AO.404030.
47. T. Shan, Study on thermal effect of interface cleaning by high energy pulsed laser and its influence on surface integrity of substrate, Shandong University of Technology (2021). https://doi.org/10.27276/d.cnki.gsdgc.2021.000252.
48. H.J. Yoo, S. Baek, J.H. Kim, J. Choi, Y.-J. Kim, C. Park, Effect of laser surface cleaning of corroded 304L stainless steel on microstructure and mechanical properties, Journal of Materials Research and Technology 16 (2022) 373–385. https://doi.org/10.1016/j.jmrt.2021.11.147.
49. H.J. Yoo, H. Park, J.H. Kim, C. Park, Effect of paint removal with kilowatt level nanosecond pulsed laser on microstructure and mechanical properties of stainless steel, Opt. Laser Technol. 163 (2023) 109434. https://doi.org/10.1016/j.optlastec.2023.109434.
50. Q. Sun, J. Zhou, X. Meng, P. Li, Mechanical properties and microstructure characteristics of 2024-T351 aluminum alloy specimen subjected to paint removal by laser cleaning, Vacuum 211 (2023) 111927. https://doi.org/10.1016/j.vacuum.2023.111927.
51. Y. Lu, Y. Ding, M. Wang, L. Yang, Y. Wang, A characterization of laser cleaning painting layer from steel surface based on thermodynamic model, Int. J. Adv. Manuf. Technol. 116 (2021) 1989–2002. https://doi.org/10.1007/s00170-021-07566-6.
52. G. Zhu, S. Wang, W. Cheng, Y. Ren, D. Wen, Corrosion and wear performance of aircraft skin after laser cleaning, Opt. Laser Technol. 132 (2020) 106475. https://doi.org/10.1016/j.optlastec.2020.106475.
53. Md. Shamsujjoha, S.R. Agnew, M.A. Melia, J.R. Brooks, T.J. Tyler, J.M. Fitz-Gerald, Effects of laser ablation coating removal (LACR) on a steel substrate: Part 1: surface profile, microstructure, hardness, and adhesion, Surf. Coat. Technol. 281 (2015) 193–205. https://doi.org/10.1016/j.surfcoat.2015.01.071.
54. Y. Li, C. Wang, G. Mi, Influence of cleaning modes on the microstructure and performance of 5083 alloy substrate, J. Mech. Sci. Technol. 35 (2021) 3943–3949. https://doi.org/10.1007/s12206-021-0807-6.
55. Y. Lu, L. Yang, Y. Wang, H. Chen, B. Guo, Z. Tian, Paint removal on the 5A06 aluminum alloy using a continuous wave fiber laser, Coatings 9 (2019). https://doi.org/10.3390/coatings9080488.
56. Y. Ding, M. Wang, G. Luo, Study on the influence of laser cleaning on the subsequent gluing effect of aluminum alloy surface, Surface Technology (2022) 1–14. https://doi.org/10.16490/j.cnki.issn.1001-3660.2024.13.019.
57. T. Yuan, Y. Wang, J. Gong, Effect of laser cleaning on adhesion of aluminum conductor insulation coating, Insulation Material 55 (2022) 103–108. https://doi.org/10.16790/j.cnki.1009-9239.im.2022.04.015.
58. J.-E. Kim, P.-S. Kim, J.-M. Lee, H.-S. Choe, J.-D. Kim, A study on the application of laser cleaning on the weld of the gas fuel supply pipe for DF engine, Applied Sciences 11 (2021). https://doi.org/10.3390/app112411832.
59. X. Li, T. Huang, A.W. Chong, R. Zhou, Y.S. Choo, M. Hong, Laser cleaning of steel structure surface for paint removal and repaint adhesion, 44 (2017).
60. H. Shuangjiang, Y. Wenfeng, Study on the effect of infrared pulse laser paint removal on the recoatability of CFRP skin spray paint, Study on the Effect of Infrared Pulse Laser Paint Removal on the Recoatability of CFRP Skin Spray Paint (2023). https://doi.org/10.27722/d.cnki.gzgmh.2023.000039.
61. N. Sim, H. Lee, N. Goyal, J.D. Cramer, Surgical site antiseptic preparations for otolaryngology – head and neck surgery: A current review, Am. J. Otolaryngol. 45 (2024) 104280. https://doi.org/10.1016/j.amjoto.2024.104280.

62. W. Wang, J. Shen, W. Liu, H. Bian, Q. Li, Effect of laser energy density on surface physical characteristics and corrosion resistance of 7075 aluminum alloy in laser cleaning, Opt. Laser Technol. 148 (2022) 107742. https://doi.org/10.1016/j.optlastec.2021.107742.
63. F.D. Zhang, H. Liu, C. Suebka, Y.X. Liu, Z. Liu, W. Guo, Y.M. Cheng, S.L. Zhang, L. Li, Corrosion behaviour of laser-cleaned AA7024 aluminium alloy, Appl. Surf. Sci. 435 (2018) 452–461. https://doi.org/10.1016/j.apsusc.2017.11.141.
64. D. Zhang, J. Xu, Z. Li, K. Li, C. Wang, D. Shan, B. Guo, Removal mechanism of blue paint on aluminum alloy substrate during surface cleaning using nanosecond pulsed laser, Opt. Laser Technol. 149 (2022) 107882. https://doi.org/10.1016/j.optlastec.2022.107882.
65. D. Xie, H. Chen, S. Yin, F. Wang, J. Chen, S. Ai, Effect of different laser energy densities on the corrosion resistance of aluminum alloys, Anti-Corrosion Methods and Materials 67 (2020) 437–444. https://doi.org/10.1108/ACMM-02-2020-2263.
66. S.L. Zhang, C. Suebka, H. Liu, Y.X. Liu, Z. Liu, W. Guo, Y.M. Cheng, F.D. Zhang, Y.D. Peng, L. Li, Mechanisms of laser cleaning induced oxidation and corrosion property changes in AA5083 aluminum alloy, J. Laser Appl. 31 (2019) 012001. https://doi.org/10.2351/1.5046470.
67. Th. Dimogerontakis, R. Oltra, O. Heintz, Thermal oxidation induced during laser cleaning of an aluminium-magnesium alloy, Appl. Phys. A 81 (2005) 1173–1179. https://doi.org/10.1007/s00339-004-3143-7.
68. W. Wang, X. Li, W. Liu, F. Xing, J. Wang, K. Zhang, Experimental study on hydrophobic properties and corrosivity of laser cleaned 7075 aluminum alloy anodized film surface, Opt. Laser Technol. 166 (2023) 109615. https://doi.org/10.1016/j.optlastec.2023.109615.
69. Z. Tian, Z. Lei, X. Chen, Y. Chen, Evaluation of laser cleaning for defouling of marine biofilm contamination on aluminum alloys, Appl. Surf. Sci. 499 (2020) 144060. https://doi.org/10.1016/j.apsusc.2019.144060.
70. Z. Tian, Z. Lei, X. Chen, Y. Chen, L.-C. Zhang, J. Bi, J. Liang, Nanosecond pulsed fiber laser cleaning of natural marine micro-biofoulings from the surface of aluminum alloy, J. Cleaner Prod. 244 (2020) 118724. https://doi.org/10.1016/j.jclepro.2019.118724.
71. R.A. Rajan, S. Rao Konda, C. Sajed Saraj, Y. Hang Lai, G. Verma, Z. Yu, W. Yu, D. Yan, J. Yang, Long-term seawater anti-corrosion properties of al alloy triggered by femtosecond laser structuring with phase change, Appl. Surf. Sci. 573 (2022) 151612. https://doi.org/10.1016/j.apsusc.2021.151612.
72. C.A.C. Souza, F.S. Politi, C.S. Kiminami, Influence of structural relaxation and partial devitrification on the corrosion resistance of Fe78B13Si9 amorphous alloy, Scr. Mater. 39 (1998) 329–334. https://doi.org/10.1016/S1359-6462(98)00159-6.
73. F. Wang, S. Ai, Q. Wang, Y. Cheng, H. Huang, Effect of narrow pulse width laser cleaning on corrosion resistance of A5083-H111 aluminum alloy, Anti-Corrosion Methods and Materials 68 (2021) 192–201. https://doi.org/10.1108/ACMM-12-2020-2414.
74. S. Meng, Y. Yu, X. Zhang, L. Zhou, X. Liang, P. Liu, Investigations on electrochemical corrosion behavior of 7075 aluminum alloy with femtosecond laser modification, Vacuum 221 (2024) 112911. https://doi.org/10.1016/j.vacuum.2023.112911.
75. T. Shi, C. Wang, G. Mi, F. Yan, A study of microstructure and mechanical properties of aluminum alloy using laser cleaning, J. Manuf. Processes 42 (2019) 60–66. https://doi.org/10.1016/j.jmapro.2019.04.015.
76. T.M. Yue, L.J. Yan, C.P. Chan, Stress corrosion cracking behavior of nd:YAG laser-treated aluminum alloy 7075, Appl. Surf. Sci. 252 (2006) 5026–5034. https://doi.org/10.1016/j.apsusc.2005.07.052.
77. J. Nie, D. Zhang, X. Su, H. Zhang, J. Zhang, J. Xu, D. Shan, B. Guo, Laser cleaning for inorganic thermal control coatings on aluminum alloys: Microstructure evolution and mechanical properties of substrate, Appl. Surf. Sci. 659 (2024) 159939. https://doi.org/10.1016/j.apsusc.2024.159939.
78. Z. Wang, Z. Yang, jianfeng ShangGuan, Effect of laser derusting process on corrosion resistance of EH36 marine steel, Progress in Laser and Optoelectronics 60 (2023) 0514011. https://doi.org/10.3788/LOP220916.

79. Y. Wang, S. Wang, Laser cleaning and protection of industrial heritage steel structure, Journal of Shenzhen University 37 (2020) 466–473. https://doi.org/10.3724/SP.J.1249.2020.05466.
80. Y. Lu, Y. Ding, M. Wang, L. Yang, Y. Wang, An environmentally friendly laser cleaning method to remove oceanic micro-biofoulings from AH36 steel substrate and corrosion protection, J. Cleaner Prod. 314 (2021) 127961. https://doi.org/10.1016/j.jclepro.2021.127961.
81. L.Y. Guo, S.N. Geng, J. Pang, Y.H. Hu, S. Lan, C.M. Wang, W.M. Wang, Structural transformation and property improvement of Fe78Si9B13 amorphous ribbon by pulsed laser processing, Mater. Design 160 (2018) 538–548. https://doi.org/10.1016/j.matdes.2018.09.045.
82. C. ZHANG, H. CHEN, C. CAI, Effect of laser cleaning on surface stress and corrosion of SMA490BW steel welded Joint, Transactions of the China Welding Institution 41 (2020) 89–96. https://doi.org/10.12073/i.hixb.20200618001.
83. W. Dou, R. Jia, P. Jin, J. Liu, S. Chen, T. Gu, Investigation of the mechanism and characteristics of copper corrosion by sulfate reducing bacteria, Corros. Sci. 144 (2018) 237–248. https://doi.org/10.1016/j.corsci.2018.08.055.
84. Di Francia E., Lahoz R., Neff D., De Caro T., Angelini E., Grassini S., Laser-cleaning effects induced on different types of bronze archaeological corrosion products: Chemical-physical surface characterisation, Appl. Surf. Sci. 573 (2022) 150884. https://doi.org/10.1016/j.apsusc.2021.150884.
85. Baek S., Choi S., Kim J.O., Kim Y.-J., Park C., Effect of laser surface cleaning on yellow brass: Application for cartridge case maintenance, Optics & Laser Technology 174 (2024) 110638. https://doi.org/10.1016/j.optlastec.2024.110638.
86. Y. Shu, X. Lu, W. Lu, W. Su, Y. Wu, H. Wei, D. Xu, J. Liang, Y. Xie, Mechanically robust superhydrophobic copper surface with self-cleaning, anti-icing, and corrosion resistance, Surf. Coat. Technol. 455 (2023) 129216. https://doi.org/10.1016/j.surfcoat.2022.129216.
87. B. Wei, J. Xu, L. Gao, H. Feng, J. Wu, C. Sun, Z. Wang, W. Ke, Nanosecond pulsed laser-assisted modified copper surface structure: Enhanced surface microhardness and microbial corrosion resistance, J. Mater. Sci. Technol. 107 (2022) 111–123. https://doi.org/10.1016/j.jmst.2021.08.023.
88. Z. Dong, X. Sun, D. Kong, D. Chu, Y. Hu, J.-A. Duan, Spatial light modulated femtosecond laser ablated durable superhydrophobic copper mesh for oil-water separation and self-cleaning, Surf. Coat. Technol. 402 (2020) 126254. https://doi.org/10.1016/j.surfcoat.2020.126254.
89. E. Di Francia, R. Lahoz, D. Neff, V. Rico, N. Nuns, E. Angelini, S. Grassini, Novel procedure for studying laser-surface material interactions during scanning laser ablation cleaning processes on cu-based alloys, Appl. Surf. Sci. 544 (2021) 148820. https://doi.org/10.1016/j.apsusc.2020.148820.
90. M. Najafizadeh, S. Yazdi, M. Bozorg, Mehran. Ghasempour-Mouziraji, M. Hosseinzadeh, M. Zarrabian, P. Cavaliere, Classification and applications of titanium and its alloys: A review, Journal of Alloys and Compounds Communications 3 (2024) 100019. https://doi.org/10.1016/j.jacomc.2024.100019.
91. Z. Dong, Y. Tian, L. Zhang, T. Jiang, D. Wang, Y. Chang, D. Chen, Research status of high efficiency deep penetration welding of medium-thick plate titanium alloy: A review, Defence Technology (2024). https://doi.org/10.1016/j.dt.2024.08.004.
92. W. Wang, Q. Jiang, W. Liu, H. Bian, J. Wang, Effect of galvanometer scanning speed on the cleaning quality of anodic oxide film on TC2 titanium alloy surface in laser cleaning, Optik 287 (2023) 171055. https://doi.org/10.1016/j.ijleo.2023.171055.
93. Dasong L, Qian W, Feisen W, Effect of Nanosecond Pulsed Laser Cleaning Scanning Speed on Cleaning Quality of Oxide Films on TC4 Titanium Alloy Surface, Chinese Journal of Lasers 50 (2023). https://doi.org/10.3788/CJL220819.
94. B.-W. Liu, G.-Y. Mi, C.-M. Wang, Reoxidation process and corrosion behavior of TA15 alloy by laser ablation, Rare Met. 40 (2021) 865–876. https://doi.org/10.1007/s12598-020-01553-8.
95. B. Liu, G. Mi, C. Wang, Modification of TA15 alloy surface by high-pulse-frequency laser cleaning, J. Laser Appl. 32 (2020) 032019. https://doi.org/10.2351/7.0000139.

96. Z. Li, D. Zhang, X. Su, S. Yang, J. Xu, R. Ma, D. Shan, B. Guo, Removal mechanism of surface cleaning on TA15 titanium alloy using nanosecond pulsed laser, Opt. Laser Technol. 139 (2021) 106998. https://doi.org/10.1016/j.optlastec.2021.106998.
97. Li Z., Xu J., Zhang D., Xu Z., Su X., Jin Y., Shan D., Chen Y., Guo B., Nanosecond pulsed laser cleaning of titanium alloy oxide films: Modeling and experiments, Journal of Manufacturing Processes 82 (2022) 665–677. https://doi.org/10.1016/j.jmapro.2022.08.033.
98. Y. Zhao, X. Yue, P. Zhang, J. Liu, X. Cao, D. Yu, Removal mechanism and cleaning method of electroplating masking paint with a microwave plasma jet, J. Manuf. Processes 124 (2024) 1285–1293. https://doi.org/10.1016/j.jmapro.2024.07.020.
99. H. Lee, N. Cho, J. Lee, Study on surface properties of gilt-bronze artifacts, after nd:YAG laser cleaning, Appl. Surf. Sci. 284 (2013) 235–241. https://doi.org/10.1016/j.apsusc.2013.07.087.
100. X. Jia, Y. Zhang, Y. Chen, H. Wang, G. Zhu, X. Zhu, Laser cleaning of slots of chrome-plated die, Opt. Laser Technol. 119 (2019) 105659. https://doi.org/10.1016/j.optlastec.2019.105659.
101. A. Kumar, M. Prasad, S. Shail, R.B. Bhatt, P.G. Behere, Md. Afzal, A. Kumar, R. Kar, J.P. Nilaya, D.J. Biswas, Pulsed laser-assisted removal of powder coating from galvanised steel surface: A characterisation study, Appl. Phys. A 119 (2015) 853–858. https://doi.org/10.1007/s00339-015-9050-2.
102. H. Mustafa, D.T.A. Matthews, G.R.B.E. Römer, Wavelength dependence of picosecond-pulsed laser ablation of hot-dip galvanized steel, Appl. Phys. A 128 (2022) 296. https://doi.org/10.1007/s00339-022-05393-4.
103. X. Wang, M. Xu, Z. Wang, L. Shen, M. Qiu, Z. Tian, M.N. Ahsan, C. Wang, Properties of jet-plated ni coating on ti alloy (Ti6Al4V) with laser cleaning pretreatment, Metals 9 (2019). https://doi.org/10.3390/met9020248.
104. M. Martinez-Calderon, T.A. Haase, N.I. Novikova, F.S. Wells, J. Low, G.R. Willmott, N.G.R. Broderick, C. Aguergaray, Turning industrial paints superhydrophobic via femtosecond laser surface hierarchical structuring, Prog. Org. Coat. 163 (2022) 106625. https://doi.org/10.1016/j.porgcoat.2021.106625.
105. L. Zhu, Q. Gao, B. Sun, Y. Ke, Y. Tan, Y. Cao, Nanosecond laser cleaning for enhanced zinc coating quality of HSLA steel, Opt. Laser Technol. 143 (2021) 107311. https://doi.org/10.1016/j.optlastec.2021.107311.
106. Ye Y, Yuan X, Xiang X, CO2 Laser-cleaning of dimethylsilicone contamination on gilded film Surface, High Power Laser and Particle Beams 22 (2010) 968–972. https://doi.org/10.3788/HPLPB20102205.0968.
107. Y. Ye, X. Yuan, X. Xiang, W. Dai, M. Chen, X. Miao, H. Lv, H. Wang, W. Zheng, Laser plasma shockwave cleaning of SiO2 particles on gold film, Opt. Lasers Eng. 49 (2011) 536–541. https://doi.org/10.1016/j.optlaseng.2010.12.006.
108. Guo Naihao, Study on Laser and Multi-means Assisted Cleaning Technology of Optical Elements with Chemical Coating, University of Electronic Science and Technology of China (2020). https://doi.org/10.27005/d.cnki.gdzku.2020.002688
109. Wang J, Zhang J, Li Y, In-situ laser cleaning of large-aperture optical components with sol-gel film (Invited), Infrared and Laser Engineering 52 (2023). https://doi.org/10.3788/IRLA20220739.
110. M. Basso, E. Colusso, C. Carraro, C. Kalha, A.A. Riaz, G. Bombardelli, E. Napolitani, Y. Chen, J. Jasieniak, L.E. Ratcliff, P.K. Thakur, T.-L. Lee, A. Regoutz, A. Martucci, Rapid laser-induced low temperature crystallization of thermochromic VO2 sol-gel thin films, Appl. Surf. Sci. 631 (2023) 157507. https://doi.org/10.1016/j.apsusc.2023.157507.
111. E. Korzeniewska, M. Tomczyk, M. Walczak, The influence of laser modification on a composite substrate and the resistance of thin layers created using the PVD process, Sens. 20 (2020). https://doi.org/10.3390/s20071920.
112. Zhang K, Deng J, Xing Y, INFLUENCE of POST-LASER PROCESSING on the MECHANICAL and TRIBOLOGICAL PROPERTIES of PVD TiAlN COATINGS, Surface Review and Letter 27 (2020). https://doi.org/10.1142/S0218625X19501373.

113. K. Zhang, X. Guo, C. Wang, F. Liu, L. Sun, Effect of plasma-assisted laser pretreatment of hard coatings surface on the physical and chemical bonding between PVD soft and hard coatings and its resulting properties, Appl. Surf. Sci. 509 (2020) 145342. https://doi.org/10.1016/j.apsusc.2020.145342.
114. K. Bobzin, T. Brögelmann, A. Gillner, N.C. Kruppe, C. He, M. Naderi, Laser-structured high performance PVD coatings, Surf. Coat. Technol. 352 (2018) 302–312. https://doi.org/10.1016/j.surfcoat.2018.07.094.
115. K. Jeong, J. Lee, I. Byun, M. Seong, J. Park, H.W. Kim, M.J. Kim, J.-H. Kim, J. Lee, Synthesis of highly conductive cobalt thin films by LCVD at atmospheric pressure, Mater. Sci. Semicond. Process. 68 (2017) 245–251. https://doi.org/10.1016/j.mssp.2017.06.032.
116. S. Matsumoto, Y. Kaneda, A. Ito, Highly self-oriented growth of (020) and (002) monoclinic HfO2 thick films using laser chemical vapor deposition, Ceram. Int. 46 (2020) 1810–1815. https://doi.org/10.1016/j.ceramint.2019.09.156.
117. F.L. Palmieri, M.A. Belcher, C.J. Wohl, K.Y. Blohowiak, J.W. Connell, Laser ablation surface preparation for adhesive bonding of carbon fiber reinforced epoxy composites, Int. J. Adhes. Adhes. 68 (2016) 95–101. https://doi.org/10.1016/j.ijadhadh.2016.02.007.
118. Y. Song, S. Wang, Y. Pan, Z. Li, L. Yu, Comparative study of infrared nanosecond laser surface paint removal for carbon fiber reinforced polymer and glass fiber reinforced polymer, J. Cleaner Prod. 451 (2024) 142061. https://doi.org/10.1016/j.jclepro.2024.142061.
119. C. Sun, J. Min, J. Lin, H. Wan, S. Yang, S. Wang, The effect of laser ablation treatment on the chemistry, morphology and bonding strength of CFRP joints, Int. J. Adhes. Adhes. 84 (2018) 325–334. https://doi.org/10.1016/j.ijadhadh.2018.04.014.
120. B. Rauh, S. Kreling, M. Kolb, M. Geistbeck, S. Boujenfa, M. Suess, K. Dilger, UV-laser cleaning and surface characterization of an aerospace carbon fibre reinforced polymer, Int. J. Adhes. Adhes. 82 (2018) 50–59. https://doi.org/10.1016/j.ijadhadh.2017.12.016.
121. C. Leone, S. Genna, Effects of surface laser treatment on direct co-bonding strength of CFRP laminates, Compos. Struct. 194 (2018) 240–251. https://doi.org/10.1016/j.compstruct.2018.03.096.
122. W. Yang, D. Wang, Y. Cao, S. Li, L. Wang, H. Zhang, Implementation and precision of CFRP geometric configurations before adhesive bonding repair based on 1064 nm infrared pulsed fiber laser ablation, Opt. Laser Technol. 175 (2024) 110765. https://doi.org/10.1016/j.optlastec.2024.110765.
123. Q. Gao, Y. Li, H. Wang, W. Liu, H. Shen, X. Zhan, Effect of scanning speed with UV laser cleaning on adhesive bonding tensile properties of CFRP, Appl. Compos. Mater. 26 (2019) 1087–1099. https://doi.org/10.1007/s10443-019-09768-4.
124. Y. Tong, X. Chen, A. Zhang, Y. Fu, X. Ren, Y. Ye, S. Xin, Effect of laser cleaning of carbon fiber-reinforced polymer and surface modification on chemical activity and bonding strength, Appl. Opt. 59 (2020) 10149–10159. https://doi.org/10.1364/AO.404846.
125. H. Yang, H. Liu, R. Gao, X. Liu, X. Yu, F. Song, L. Liu, Numerical simulation of paint stripping on CFRP by pulsed laser, Opt. Laser Technol. 145 (2022) 107450. https://doi.org/10.1016/j.optlastec.2021.107450.
126. G.S. Ponticelli, F. Lambiase, C. Leone, S. Genna, Combined fuzzy and genetic algorithm for the optimisation of hybrid composite-polymer joints obtained by two-step laser joining process, Materials 13 (2020). https://doi.org/10.3390/ma13020283.
127. P. Wang, Z. Zhang, B. Hao, S. Wei, Y. Huang, G. Zhang, Investigation on heat transfer and ablation mechanism of CFRP by different laser scanning directions, Composites Part B: Engineering 262 (2023) 110827. https://doi.org/10.1016/j.compositesb.2023.110827.
128. L.Y. Xu, J.R. Lu, K.M. Li, J. Hu, Removal mechanism of CFRP by laser multi direction interaction, Opt. Laser Technol. 143 (2021) 107281. https://doi.org/10.1016/j.optlastec.2021.107281.
129. J. Gacs, E. Sára Bogya, L. Kocsis, T. Jacob, Application of laser cleaning on AlMg4.5Mn0.4 sheets for adhesive bonding, Int. J. Adhes. Adhes. 115 (2022) 103132. https://doi.org/10.1016/j.ijadhadh.2022.103132.

130. S. Guo, B.E. Carlson, L.G. Hector, Y. Xia, Q. Zhou, Increasing strength and fracture toughness of AA7075-T6 adhesively-bonded joints with laser ablation, J. Mater. Process. Technol. 259 (2018) 368–379. https://doi.org/10.1016/j.jmatprotec.2018.05.010.
131. C. Zhang, Y. Zhang, L. Chen, Y. Chen, Effect of laser micromachining crater-array–multi-grooves on the bonding strength and failure mode of aluminum alloy adhesive joints, Opt. Laser Technol. 175 (2024) 110803. https://doi.org/10.1016/j.optlastec.2024.110803.
132. C. Zhu, H. Wan, J. Min, Y. Mei, J. Lin, B.E. Carlson, S. Maddela, Application of pulsed yb: Fiber laser to surface treatment of al alloys for improved adhesive bonded performance, Optics and Lasers in Engineering 119 (2019) 65–76. https://doi.org/10.1016/j.optlaseng.2019.03.017.
133. W. Qiang, G. Yingchun, C. Baoqiang, Q. Bojin, Laser cleaning of commercial al alloy surface for tungsten inert gas welding, J. Laser Appl. 28 (2016) 022507. https://doi.org/10.2351/1.4943909.
134. Y. Wu, J. Lin, B.E. Carlson, P. Lu, M.P. Balogh, N.P. Irish, Y. Mei, Effect of laser ablation surface treatment on performance of adhesive-bonded aluminum alloys, Surf. Coat. Technol. 304 (2016) 340–347. https://doi.org/10.1016/j.surfcoat.2016.04.051.
135. C. Mandolfino, E. Lertora, S. Genna, C. Leone, C. Gambaro, Effect of laser and plasma surface cleaning on mechanical properties of adhesive bonded joints, 9th CIRP Conference on Intelligent Computation in Manufacturing Engineering—CIRP ICME'14 33 (2015) 458–463. https://doi.org/10.1016/j.procir.2015.06.054.
136. G. Rotella, M. Alfano, S. Candamano, Surface modification of Ti6Al4V alloy by pulsed yb-laser irradiation for enhanced adhesive bonding, Cirp Ann. 64 (2015) 527–530. https://doi.org/10.1016/j.cirp.2015.04.042.
137. A. Kurtovic, E. Brandl, T. Mertens, H.J. Maier, Laser induced surface nano-structuring of ti–6Al–4V for adhesive bonding, Int. J. Adhes. Adhes. 45 (2013) 112–117. https://doi.org/10.1016/j.ijadhadh.2013.05.004.
138. A. Palavra, B.N. Coelho, J.Th.M. de Hosson, M.S.F. Lima, S.M. Carvalho, A.R. Costa, Laser surface treatment for enhanced titanium to carbon fiber-reinforced polymer adhesion, Journal of the Brazilian Society of Mechanical Sciences and Engineering 39 (2017) 2917–2924. https://doi.org/10.1007/s40430-017-0826-5.
139. C. Zhang, L. Chen, G. Wang, J. Jin, Y. Zhang, Effects of laser processing crater array microstructure on the surface characteristics and bonding strength of Ti6Al4V adhesive joints, Opt. Laser Technol. 141 (2021) 107125. https://doi.org/10.1016/j.optlastec.2021.107125.

# Chapter 5
# Thermal Effects on the Surface-Interface During Laser Cleaning and Substrate Damage Control

## 5.1 Introduction

Laser cleaning technology removes contaminants from the surface of an object by irradiating a high-energy laser beam [1]. Laser cleaning technology, in comparison to conventional cleaning methods like mechanical friction, high-pressure water jet cleaning, chemical reagent cleaning, and ultrasonic cleaning, offers superior advantages such as high efficiency, non-contact operation, and environmental friendliness. This technology is primarily utilized for various applications including the removal of oxides from metal surfaces [2], the remanufacturing of coatings [3], the eradication of microorganisms [4], and the cleaning of micron and nanoscale particles [5]. When the laser irradiates on the surface of cleaned material, the laser energy can be divided into three parts according to the law of conservation of energy [6]: reflection from the material's surface, absorption within the material, and conduction between the materials. The absorption of laser energy by materials results in interactions where thermal effects are indispensable for the efficacy of dry laser cleaning in removing contaminants. However, the aggregation of thermal effects may result in damage to the surface of the substrate, and non-destructive laser cleaning has always been a focus of research. The non-uniformity in the thickness of the adherent material, the irregularity of the object's surface, and the high laser energy density employed to ensure cleaning efficiency, inevitably subject the substrate's surface to thermal effects during the laser cleaning process. However, some studies have indicated that the slight remelting of the substrate after laser cleaning may improve the surface properties of the substrate under thermal effects, which is beneficial for secondary processing and utilization.

The temperature field generated coupled with the temperature-induced stress field when the laser energy is absorbed by the contaminants and the substrate is the key to the effectiveness of contaminants removal and the surface quality of the substrate. However, obtaining accurate data on the distribution of temperature and stress fields through direct experimental measurements is challenging. Consequently,



S. Wang, *Laser Cleaning*, https://doi.org/10.1007/978-981-95-2935-3_5

the finite element analysis method has emerged as an effective tool for studying the interactions between the laser and the material during the laser cleaning process [7]. The finite element analysis method enables the determination of optimal laser cleaning process parameters, thereby enhancing efficiency and reducing production costs. To further streamline the efficiency of laser cleaning, real-time monitoring has become a crucial aspect of the laser cleaning process. It can quickly and accurately respond to the state of contaminants removal during the laser cleaning process, and then optimize the adjustment of laser parameters to achieve closed-loop control of laser cleaning and reduce the risk of substrate damage.

## 5.2 Thermal Effects of Laser Cleaning

Laser cleaning methods can be categorized into three major groups: Dry laser cleaning, wet laser cleaning and laser-induced plasma shock wave cleaning, as shown in Fig. 5.1 [8]. Dry laser cleaning involves the direct interaction of the laser beam with the substrate's surface, where both the substrate and the contaminants absorb the laser energy. The primary mechanism for the removal of surface contaminants is the combined effects of heat and impact. Wet laser cleaning requires a liquid film coated on the surface to be cleaned. This film absorbs the laser energy resulting in evaporation explosion that generates a pressure wave to remove contaminants from the substrate. Additionally, the presence of the liquid film significantly reduces the cleaning threshold for contaminants and minimizes heat accumulation on the substrate surface compared to dry laser cleaning [9]. Laser-induced plasma shock wave cleaning is the laser beam acting on the air above the substrate surface, ionizing it to produce a plasma shock wave. This shock wave, as it diffuses, impacts the surface of the substrate. Removal of adherent particles occurs when the lateral pressure generated by the shock wave, parallel to the substrate, surpasses the bonding force between the particles and the substrate [10]. In summary, wet laser cleaning and laser-induced plasma shockwave cleaning weaken the influence of thermal effects, while dry laser cleaning is dominated by thermal effects in the cleaning process. Notably, the thermal effects in laser cleaning are instrumental in the removal of contaminants and the enhancement of the substrate's surface quality. Consequently,

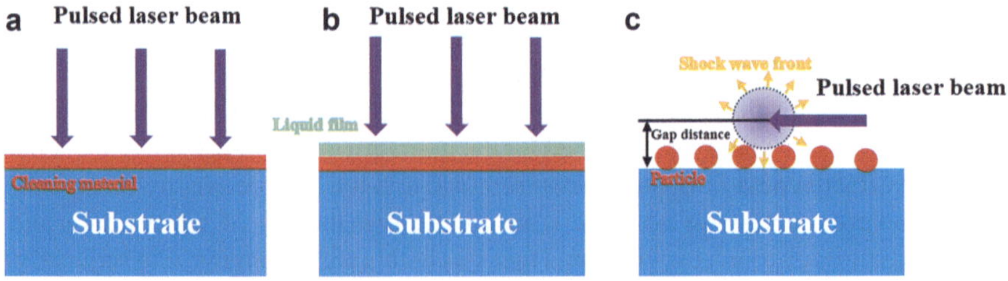


**Fig. 5.1** Schematic diagram of laser cleaning methods: (**a**) dry, (**b**) wet, (**c**) plasma shock wave

the discussion of the thermal effects in this context is mainly applicable to dry laser cleaning.

### 5.2.1 *Thermal Effects in Contaminants*

Laser cleaning of contaminants is a thermal process where the laser beam transfers energy to a very small surface area within a brief time span. During this process, the material absorbs the beam energy, resulting in rapid heating. In this context, the predominant factor is not a solitary mechanism; instead, there is a simultaneous competition among multiple removal mechanisms for influence [11]. Laser cleaning of contaminants is mainly thermal and impact effects, where thermal effects include ablation mechanism, thermal stress vibration mechanism, crack extension mechanism; impact effects include plasma shock, steam recoil pressure, photon impact. The process of laser-material interaction can be summarized as follows: The substrate and its surface contaminants absorb laser energy, leading to the generation of thermal stresses due to the differing coefficients of thermal expansion between the contaminants and the substrate. This differential expansion overcomes the adhesion forces on the material's surface, facilitating the removal of the contaminants [12]. As heat accumulates, the temperature of the contaminants reaches the melting point and begins melting to form a molten pool. Here, the ablation mechanism becomes the primary driver of laser cleaning, characterized by processes such as vaporization, cavitation, and phase explosion of the contaminants [13]. When the temperature of the molten pool surpasses the boiling point, metal vapors are emitted from the surface, forming a plasma through the absorption of laser energy. With the input of laser energy, the plasma undergoes continuous expansion. Subsequently, both the contaminants and the substrate experience the action of recoil pressure. This action facilitates the expulsion of the molten pool's liquid into the surrounding area, thereby aggravating the removal of the contaminants from the substrate's surface [14]. The interplay of thermal and impact effects is collaboratively illustrated in Fig. 5.2.

During the laser cleaning process, the predominant cleaning mechanism varies according to the type of laser and the nature of the materials being cleaned. Zhang

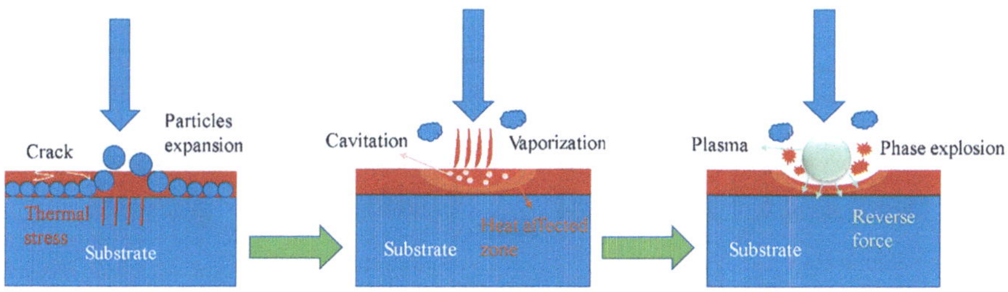


**Fig. 5.2** Schematic diagram of pulsed laser cleaning mechanism

[15] used a nanosecond pulsed laser with a wavelength of 1064 nm to remove red and blue paints. Under identical laser energy conditions, the blue paint layer has a higher absorption coefficient, absorbing most of the laser energy, which results in a significantly higher temperature in the blue paint layer compared to the red. When the temperature reaches the boiling point of the blue paint, the layer evaporates and is removed. As for the red paint layer cleaning, the laser energy penetrates into the substrate, resulting in raised temperature of the substrate and non-evaporation of the red paint layer. The raised temperature of the substrate generates thermal stresses that cause the paint to peel off. The schematic diagram of the removal mechanism is shown in Fig. 5.3. Gu [16] employed ultraviolet picosecond lasers to clean the paint layers on the surfaces of honeycomb sandwich structures. Under the combined influence of single and two-photon absorption, the chemical bonds on the paint surface directly break, resulting in the formation of small molecular fragments or gases. The higher the material's absorption coefficient for the laser, the quicker the intensity of the laser diminishes in the z-direction, leading to a smaller heat affected zone. This enables the gradual delamination of the paint layer, a process that can be described by the generalized Beer-Lambert's law: $I(z) = I_0E_u(-\alpha z^u)$ , where $\alpha$ is the absorption coefficient, $I(z)$ is the laser intensity in the laser irradiation direction, $I_0$ is the laser heat source intensity, $u$ is the fractional order $(0 < u \leq 2)$, $E_u\left(X\right) = \sum_{k=0}^{\infty} \frac{x^k}{\Gamma\left(uk+1\right)}$.

Since various materials exhibit distinct absorption coefficients for different laser wavelengths, they consequently generate disparate thermal effects. Therefore, coupling two or more wavelengths presents an advantageous strategy for enhancing cleaning outcomes. This approach leverages the unique absorption characteristics of each material, optimizing the cleaning process by effectively targeting the specific

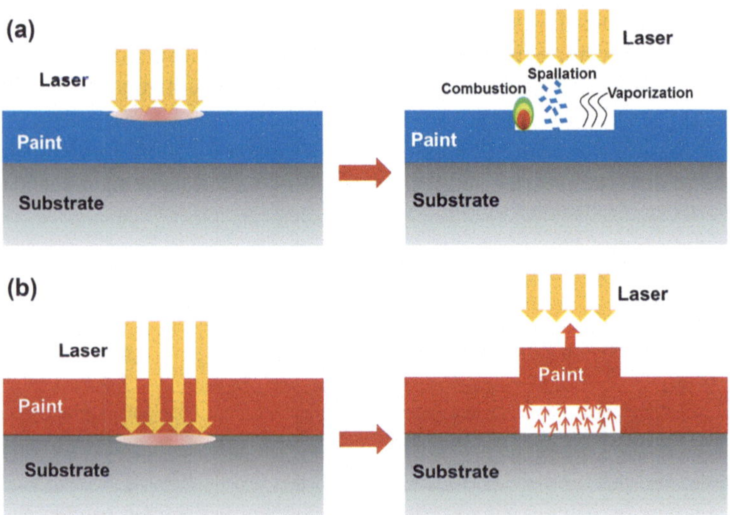


**Fig. 5.3** Schematic diagram of the paint removal principle: (**a**) blue and (**b**) red

thermal responses induced by different wavelengths. Pozo-Antonio [17] carried out a comparative analysis of cleaning methods employing either single wavelengths (1064 or 355 nm) or a dual-wavelength approach (simultaneous application of 1064 and 355 nm). The study found that the concurrent use of dual-wavelength beams markedly enhanced the removal of blue and black graffiti compared to the single-wavelength cleaning. This method's superiority in cleaning efficiency was clearly evidenced on surfaces irradiated with both beams at a $F_{IR}/F_{UV}$ ratio of 1/3. Additionally, this approach demonstrated a significant advantage in minimizing potential damage to the substrates being cleaned. Pouli [18] conducted research on the issue of discoloration on the surfaces of cultural heritage artifacts due to laser cleaning. The study explored the effectiveness of using either a 1064 nm IR laser or a 355 nm UV laser independently, as well as the sequential application of UV and IR lasers. However, these methods did not yield optimal results. In contrast, the simultaneous spatial and temporal overlapping use of both wavelengths demonstrated significantly improved outcomes, as shown in Fig. 5.4. This approach effectively combines the distinct cleaning capacities of both UV and IR lasers, offering a more comprehensive solution for the delicate task of cleaning cultural heritage surfaces without causing undesirable discoloration.

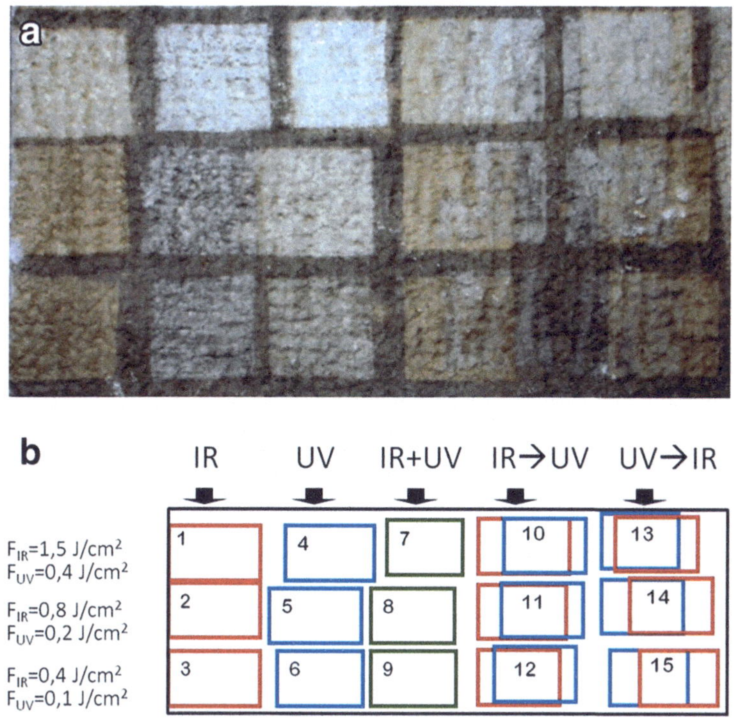


**Fig. 5.4** Demonstrates laser cleaning using two wavelengths (infrared at 1064 nm and ultraviolet at 355 nm) applied through three distinct methods: individually, sequentially, and synchronously. (**a**) Surface morphology of the sample after cleaning (**b**) Schematic diagram of three different cleaning parameters

Laser cleaning takes into account not only the characteristics of the contaminants but also the characteristics of the substrate. When the melting point and boiling point of the attached material is lower than the substrate or the energy absorption rate of the attached material to the laser is greater than that of the substrate, the attached material is removed mainly by its absorption of laser energy to produce expansion, ablation, evaporation, plasma shock phenomenon. When the melting point and boiling point of the attached material is higher than the substrate or the substrate strongly absorbs the laser energy, the attached material is removed mainly by the substrate absorbing the laser energy [19]. The removal of relatively thick contaminants through thermal stressing of the substrate is challenging. It is inevitable that the substrate will be damaged by ablation under excess heat accumulation. In this case, contaminants wavelengths with high absorption coefficients should be considered for cleaning [20].

### 5.2.2 *Thermal Effects in Substrates*

In the process of laser cleaning, the laser energy is often increased to obtain higher cleaning efficiency [21]. Thermal effects are one of the most important causes of damage to the substrate. The thermal effects of the substrate surface is generated mainly from two aspects: on the one hand, with the removal of the substrate surface contaminants, the laser acts directly on the substrate surface and the substrate absorbs laser energy; on the other hand, the laser acts on the surface contaminants of the substrate and the laser energy is transferred to the substrate [22]. Thermal damage on the surface of the substrate comes from the stress and strain produced by thermal effects, leading to the evolution of surface morphology, organizational structure and chemical composition. Eventually the surface integrity is destroyed resulting in surface defects, material deformation and so on. Li [23] used a nanosecond laser to clean the oxide film on the surface of TA15 titanium alloy. Under certain high laser energy densities, the substrate surface undergoes melting. This process, influenced by rapid heating and cooling along with surface tension, results in the formation of microcracks on the surface. These microcracks extend further as the laser energy density increases, correlating with the increased depth of surface melting.

Given the strong interfacial bonding between the contaminants and the substrate, it is imperative to enhance the physicochemical effects of the laser on the contaminants to ensure their complete removal. Irradiation of the substrate with a high-energy laser beam induces an intense thermal input. Excessive surface heat input inevitably causes micro-damage to the cleaned substrate surface. However, some studies have shown that the surface properties of the substrate after laser cleaning under thermal effects have been improved. Tian [24] used additional nanosecond laser energy to form a micro-nanostructured oxide layer on the surface of an aluminum alloy. To further analyze the crystal structure of the nano-oxide layer, high-resolution electron microscopy was employed to observe the atomic arrangement and provide selected area electron diffraction results, as shown in Figs. 5.5 and 5.6. The nanostructured layer was primarily composed of the $Al_2O_3$ phase and the surrounding amorphous phase. Liu [25] investigated the surface modification after

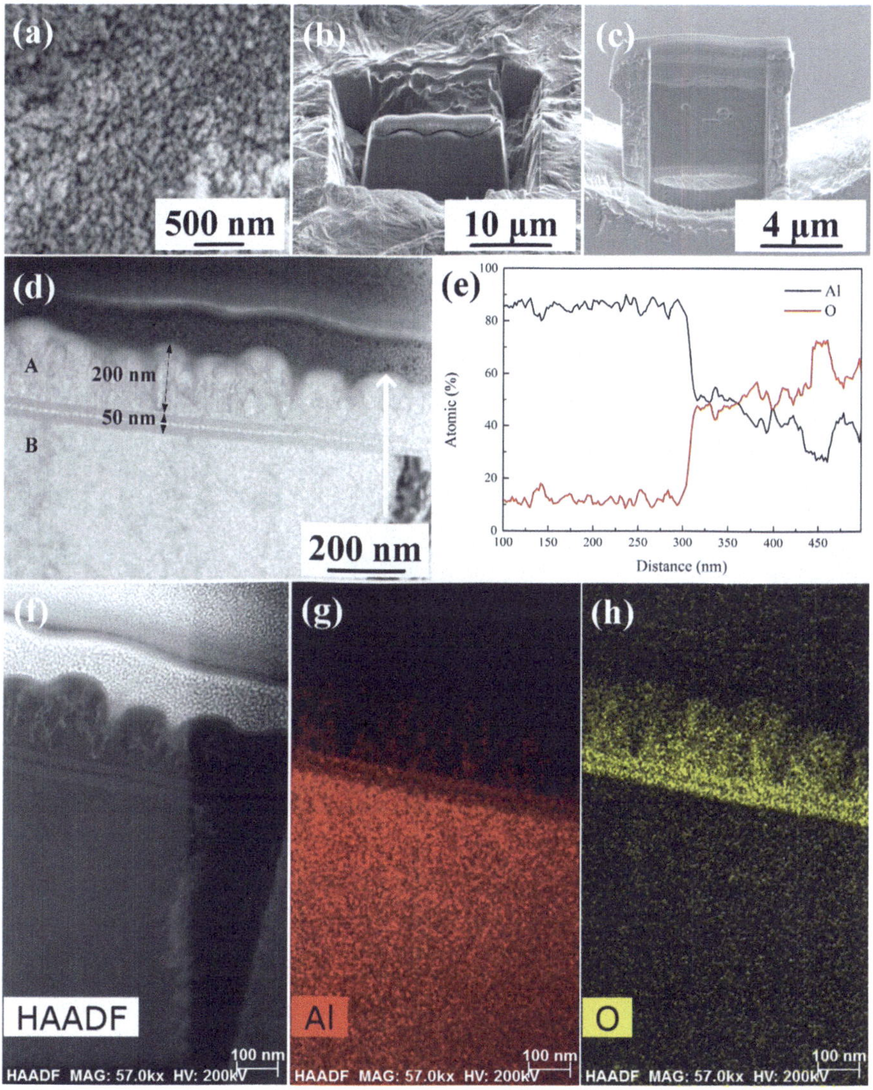


**Fig. 5.5** TEM analysis of the nanostructured layer formed on the surface of aluminum alloy substrate at a laser fluence of 4.14 J/cm$^2$: (**a**) a selected sample surface position for FIB preparation under the view of SEM (**b**) the cross-sectional morphology of the sample before ion beam polishing, (**c**) the cross-sectional morphology of the sample after ion beam polishing, (**d**) TEM image of nanostructured layer in cross section prepared by FIB, (**e**) line scanning energy spectrum analysis result along the white arrow in (**d**), (**f**) HAADF image of the nanostructured layer, (**g**, **h**) chemical elemental distribution (**Al** and **O**) images of the nanostructured layer

laser cleaning at different energy densities. The results show that the large energy input from laser cleaning increases the nucleation rate of the substrate while the high solidification cooling rate prevents the full growth of the grains. The rapid heating and cooling results in the refinement of the grains on the substrate surface.

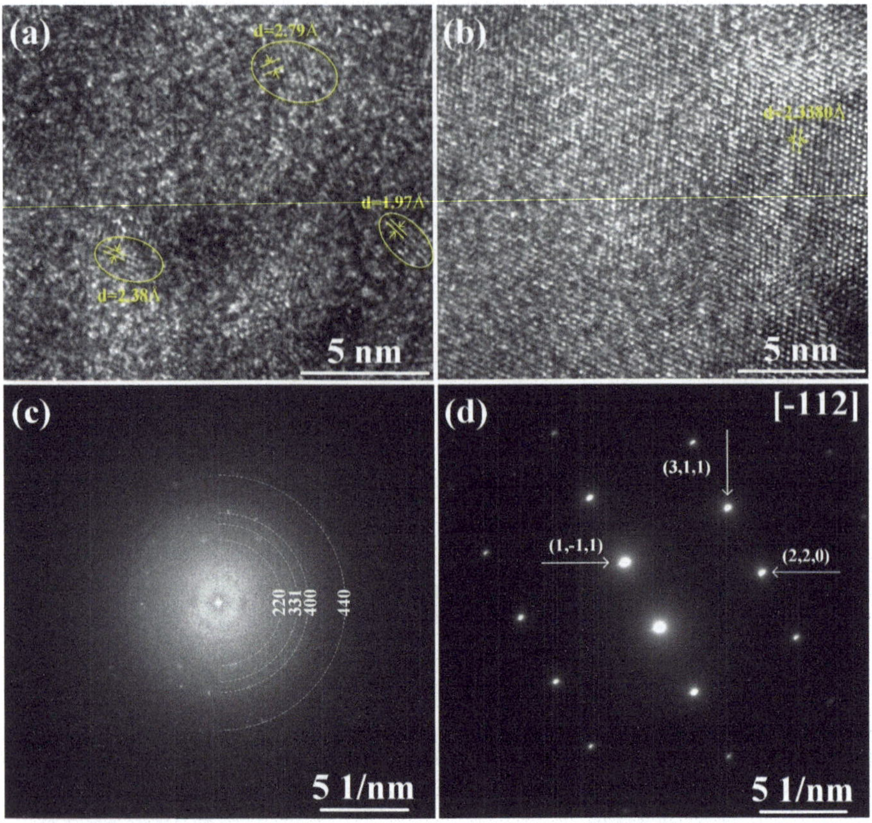


**Fig. 5.6** HRTEM image and SAED image: (**a**) atomic arrangement of the nanostructured layer, (**b**) atomic arrangement of the aluminum substrate, (**c**) SAED of the nanostructured layer, (**d**) SAED of the aluminum substrate

As the laser energy density increases, the extent of the grain refinement region increases and the ratio of large-angle grain boundaries increases, as shown in Fig. 5.7. He [26] compared the influence of laser-induced thermal and force effect on the microstructure and mechanical properties of aluminum alloy surfaces as shown in Fig. 5.8. Under thermal action, extreme temperature changes cause significant undercooling, leading to the formation of dendrites and protrusions on the solidifying aluminum alloy surface. At high temperatures, magnesium and copper dissolve more in the aluminum alloy but precipitate out as the cooling rate increases, forming precipitated phases. Under the laser-induced force effect, dislocations on the aluminum alloy surface begin to proliferate and move along slip planes, forming dislocation walls. These dislocations propagate through movement and interaction, leading to the formation of dislocation tangles. The dislocation tangles further evolve into sub-grain boundaries through accumulation and rearrangement, resulting in the formation of smaller-sized crystals. A new substance and microstructure

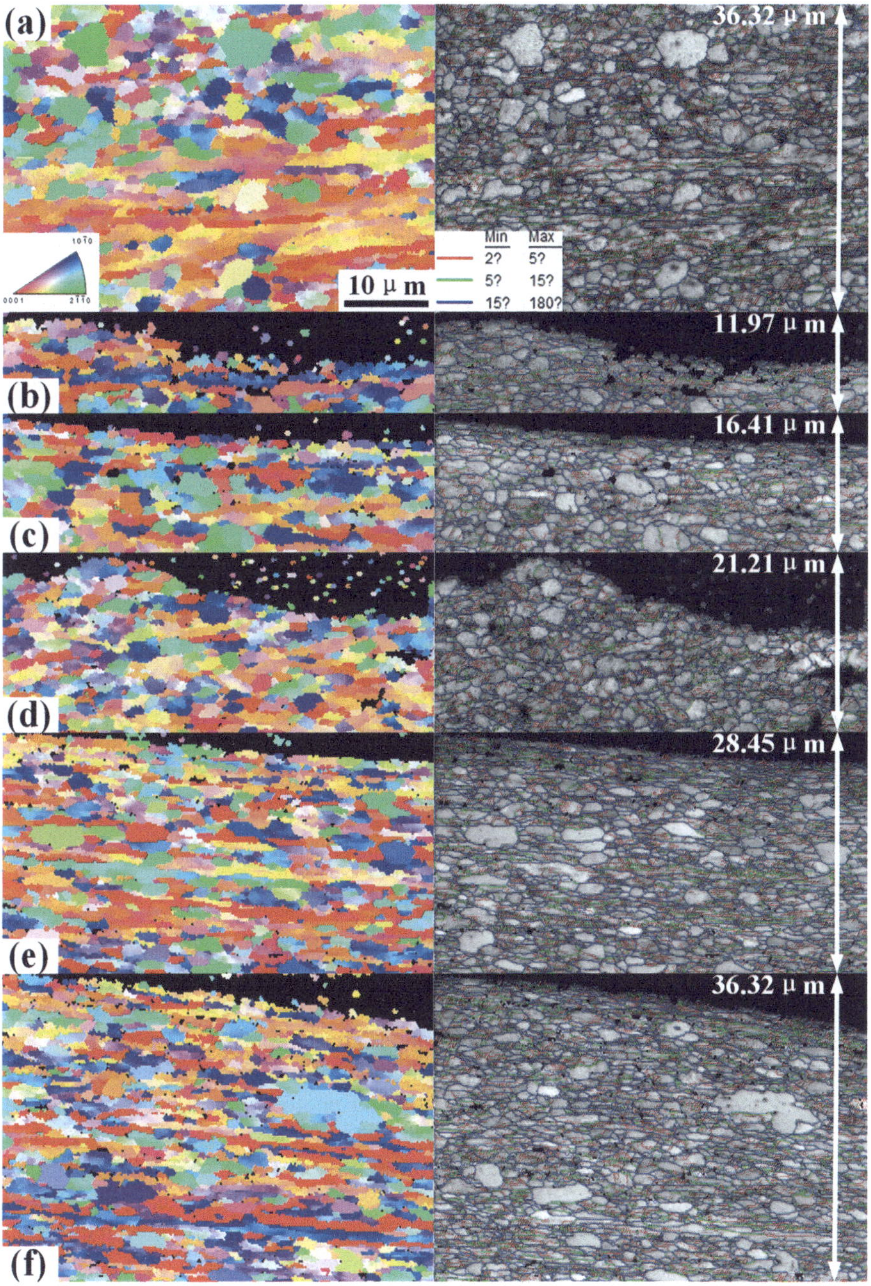


**Fig. 5.7** EBSD generated inverse pole figure (IPF) and grain boundary (GB) micrographs of the TA15 alloy, (**a**) untreated and laser cleaning at (**b**) $10^7$ W/cm$^2$, (**c**) $2.5 \times 10^7$ W/cm$^2$, (**d**) $5 \times 10^7$ W/cm$^2$, (**e**) $7.5 \times 10^7$ W/cm$^2$ and (**f**) $10^8$ W/cm$^2$

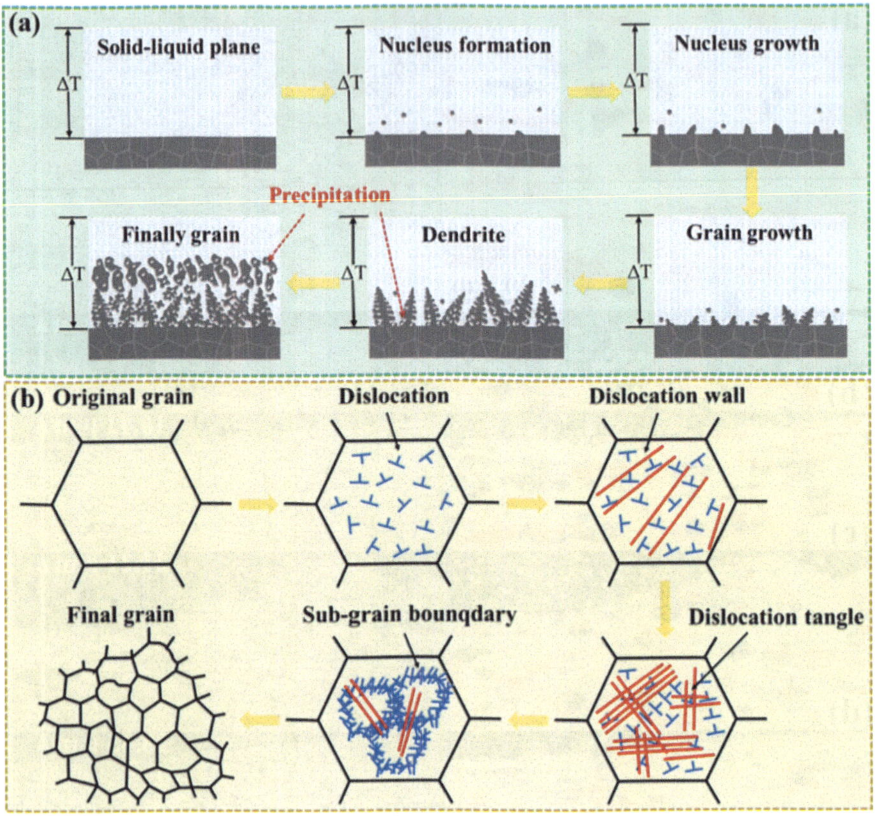


**Fig. 5.8** Surface microstructure evolution mechanism: (**a**) laser-induced thermal effects and (**b**) laser-induced force effect

are created on the surface of the material after laser cleaning, resulting in improved properties of the cleaned material, as summarized in Table 5.1.

The thermal effects of laser cleaning result in a morphological evolution of the substrate surface, such as fine crystallization and micro-texture. These laser cleaning "side effects" with enhanced performance characteristics can be applied in a variety of aspects. In the field of solar cells and photodetectors, Chen [45] used femtosecond laser ablation to remove the oxide deposition on the silicon surface and fabricated multi-scale micro-nano structures on the surface of the substrate, which resulted in a decrease in the reflectivity of the silicon material from 23.98% to 5.3%. Furthermore, it was confirmed that elliptical spots are more effective than circular spots in enhancing the anti-reflective properties of the material, as illustrated in Fig. 5.9. In the field of coatings, Gao [46] applied laser layered cleaning to aircraft skins. This process was shown to create numerous "craters" on the surface of the residual paint layer following the cleaning. This texturing proved beneficial in increasing the interfacial contact area of the paint layer, thereby enhancing the

**Table 5.1** Improved surface properties of substrate after laser cleaning with different cleaning materials

| Cleaning object | Parameter | Improved performance |
|---|---|---|
| Surface paint layer of aluminum alloy [27] | Wavelength 1064 nm, laser energy density 5.3 J/cm$^2$, frequency 10 kHz, scan speed 5000 mm/s | Microhardness, ultimate tensile strength and yield strength increased by 9.7%, 6.6% and 13.2%, respectively |
| SUS310S stainless steel [28] | Wavelength 1064 nm, defocused length 20 mm, scanning speed 5000 mm/s | The maximum tensile shear force achieved 1948 N and increased 54% |
| Epoxy resin of CFRP [29] | Frequency 100 kHz, pulse width 100 ns, scanning speeds 1000 mm/s, laser beam diameter 0.85 mm, cleaning times 1, laser power density 8.8 kW/cm$^2$ | Improved the tensile shear strength and bending strength |
| Rusty stainless steel [30] | Laser power density 19.1 × 10$^5$ W/cm$^2$, repetition rate 80 kHz, scanning speed 1900 mm/s, scanning spacing 0.03 mm | Surface microhardness enhance and surface roughness reduce |
| Oceanic micro-biofoulings of AH36 steel [31] | Wavelength 355 nm, laser pulse duration 15 ns, repetition rate 100 kHz, laser fluences 5.30 J/cm$^2$ | Corrosion resistance improved by three times |
| Bare carbon steel plates [32] | CW fiber laser, wavelength of 1070 nm | Corrosion resistance improved |
| Oxide layer on the surface of aluminum alloy [33] | Laser power 120 W, spot diameter 60 μm, pulse width 250 ns, pulse frequency 90 kHz, scanning speed 10 mm/s, scanning frequency 250 Hz | Bending strength of the substrate increases by 11.23% and improved tensile |
| Aluminum alloy 6061-T6 [34] | Wavelength 1064 nm, pulse duration 50 ns, average power 30 W, pulse frequency 30 kHz, focal length 160 mm | Joint strength up to the double compared to the untreated |
| Oxidized layer of a Q235 surface [35] | Scanning speed 7000 mm/s, pulse frequency 350 kHz, pulse width 340 ns, spot diameter 50 μm | Eliminate the welded joint's residual stress and improve its strength slightly |
| Corroded grey cast iron brake disc [36] | Wavelength 10.6 μm, continuous wave Rofin multi-scan $CO_2$ laser, laser power 60 W, transverse speed 900 mm/s, spot size 722 μm | Surface microhardness of laser cleaned specimen was 37% compare to the rusted specimen surface |
| Primer and topcoat obtained from the CFRP [37] | Pulse frequency 10 kHz, pulse width 70 ns, spot size 1.5 mm, laser energy density 2.22–4.44 J/cm$^2$, scanning speed 2500 mm/s | Tensile strength improved |

(continued)

**Table 5.1** (continued)

| Cleaning object | Parameter | Improved performance |
|---|---|---|
| Corroded 304L stainless steel [38] | Power (W) 1140, pulse width 89 ns, frequency 15 kHz, beam size 2.1 mm, laser energy density 2.2 $J/cm^2$, scan speed 1000 m/s, scan hatch distance 0.1, 1.8 mm | Tensile strength and surface microhardness improved |
| T300 unidirectional carbon fiber PR prepreg [39] | Wavelength 355 nm, pulse duration 12 ns, spot diameter size 50 μm, power 13.6 W, frequency 30 kHz, velocity 500 mm/s | Adhesive strength of bonding joints and CFRP surface energy enhanced |
| 5083 aluminum alloy [40] | Wavelength 1064 nm, average power 40 W, scanning speed 2500 mm/s, repetitive frequency 85 kHz | The stress and strain of joints were 277.9 MPa and 13.88% |
| Oxide layers of AA5083-O aluminum alloy [8] | Wavelength 1064 nm, pulse duration 10 ns, frequency 10 kHz, scanning velocity 600 mm/s, spot diameter 0.6 mm, number of pulses per unit area 10, laser fluence 3.5, 7.1, 8.5 $J/cm^2$ | Corrosion resistance improved |
| Surface paint layer of 5A06 aluminum alloy [41] | Wavelength 1070 nm, spot size 5.0 mm, laser power density 11.9 $W/cm^2$ | Corrosion resistance improved |
| Surface painting layer of AH36 steel [42] | Wavelength 355 nm, laser pulse width 15 ns, repetition rate 100 kHz, laser fluences 2, 5, 7 $J/cm^2$ | Microhardness, tensile strength, and bending strength dramatically improved |
| Boeing aircraft skins: first layer is a BMS10-11primer, the second layer is an aluminum-clad layer, and the third layer is 2024 aluminum alloy [43] | Wavelength 1064 nm, cleaning speed 5 mm/s, spot diameter 0.5 mm, spot overlap rate 0.3, the repetition frequency 10 kHz, energy density 2, 3, 4, 5, 6 $J/cm^2$ | Friction and wear properties, microhardness, corrosion resistance improved |
| Surface paint layer of 2024 aluminum alloy [44] | Wavelength 1064 nm, pulse width 100 ns, frequency 30 kHz, scanning speed 500 mm/s, spot diameter 60 μm | Microhardness and young's modulus were improved by 6% and 25% |

adhesion of any subsequent layers of paint applied. However, it was observed that when the thickness of the residual paint layer was reduced to 11.2 μm, the disparity in the coefficients of thermal expansion between the paint layer and the substrate negatively impacted the adhesion, necessitating a re-application of the primer. Wang [47] used a combination of laser cleaning and jet electrodeposition process to improve the mechanical properties of the surface of $Ti_6Al_4V$ samples, and investigated the effect of laser cleaning on the morphology and adhesion of nickel coatings under different power conditions. The shallow pits produced by laser ablation increase the adhesion of the paint layer when the laser energy density is 4 $J/cm^2$. As

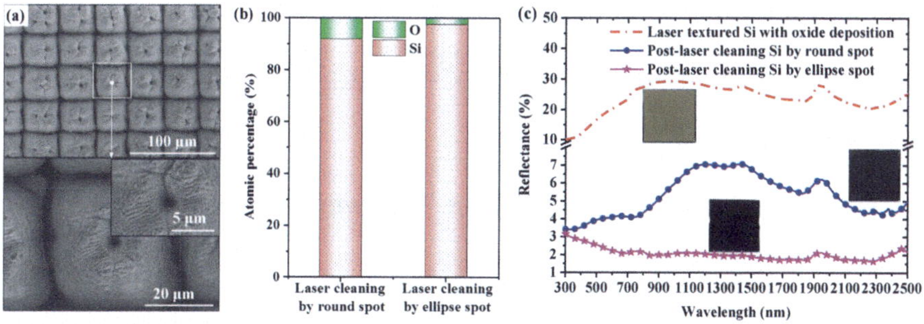


**Fig. 5.9** Laser cleaning at 500 mW, a scanning speed of 8 mm/s, and a scanning pitch of 60 μm. (**a**) SEM images of the surface structural morphology. (**b**, **c**) Chemical composition analysis and reflection spectra of the laser textured Si surfaces with and without laser cleaning by round spot or ellipse spot, respectively

the laser energy density increases, the substrate surface is over-cleaned and the substrate re-oxidizes with adhesion beginning to decline.

The aforementioned studies collectively demonstrate that excessive accumulation of thermal effects during laser cleaning can compromise the surface integrity of the substrate. This leads to alterations in the properties and morphology of the substrate surface post-cleaning. With appropriate calibration of laser cleaning parameters, the thermal effects of laser cleaning can positively contribute to enhancing the surface properties of the substrate, such as corrosion resistance, abrasion resistance, surface hardness, tensile strength, etc. Notably, the cleaned substrate surface maintains the material's secondary use properties without degradation. However, research on the micro-texture of surfaces formed after laser cleaning remains limited, highlighting a need for further investigation to fully comprehend the underlying patterns and principles.

## 5.3 Influence of Laser Cleaning Parameters on Thermal Effects

During laser cleaning, the thermal effects produced by the laser's interaction with the target material are influenced by both the laser characteristics and the material properties [48]. These thermal effects play a crucial role in the laser cleaning process, impacting both the cleaning efficiency and the properties of the cleaned substrate surface. High heat input can enhance cleaning efficiency, but excessive heat may damage the substrate surface. Conversely, insufficient heat input can compromise cleaning efficiency, failing to fully utilize the capabilities of laser cleaning. By adjusting the laser cleaning process parameters, the thermal effects can be effectively controlled, directly influencing the laser cleaning mechanism. Moreover, optimizing laser cleaning parameters and precisely regulating thermal effects are vital to avoid thermal ablation damage and maintain the integrity of the substrate [49, 50]. Key parameters in laser cleaning operations that require careful regulation include pulse width, cleaning speed, and energy density.

### 5.3.1 *Influence of Pulse Width on Thermal Effects*

The pulse width refers to the period during which the laser power is sustained at a specific value. The magnitude of the pulse width influences the duration of a single laser pulse's action, thereby determining the varying thermal effects of laser cleaning, as depicted in Fig. 5.10. For pulse widths in the nanosecond range or longer, a heat-affected zone is generated in the area yet to be cleaned, owing to continuous heat transfer. With longer pulse widths, the phenomenon of heat transfer becomes more pronounced, and the volume of the heat-affected zone increases [51, 52]. In this case, the removal of the contaminants is carried out by rapid thermal expansion of the contaminants and the substrate or if the temperature of the contaminants is high enough it may also evaporate. For pulse widths in the picosecond range or shorter, the laser pulse duration is shorter than the electron-phonon coupling time during the interaction between the laser and the material. In such instances, the ablation rate of the contaminants exceeds the rate of thermal diffusion on their surface, preventing the accumulation and transfer of absorbed laser energy via thermal conduction. Consequently, the temperature of the substrate surface does not increase during laser cleaning, a phenomenon termed "cold ablation" [53]. The thermal penetration depth depends on the thermal diffusivity and pulse width, and can be expressed as $L_{th} = 2\sqrt{k \cdot \tau_H}$, where $k$ is the material's thermal diffusivity, $\tau_H$ is the pulse width [54].

There are different cleaning results at different pulse widths due to different laser irradiation times and peak powers. Zhou [55] used picosecond and nanosecond lasers to clean the oxide film on the surface of aluminum alloys. In the case of nanosecond lasers used for cleaning aluminum alloys, the material's relaxation time is shorter than the laser's pulse width. This discrepancy allows for the energy, conducted by the laser, to be absorbed more deeply into the material. At the end of the

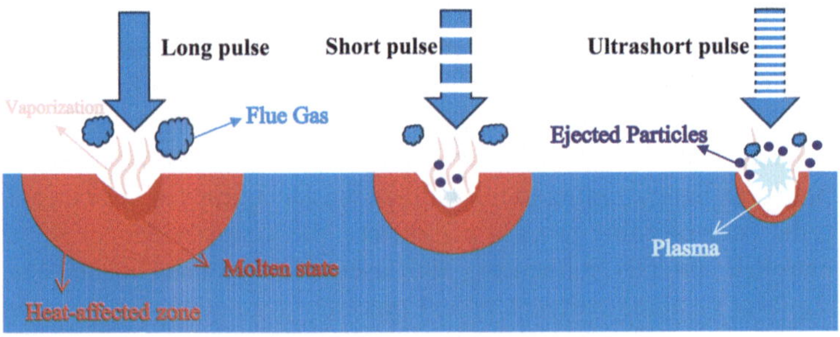


**Fig. 5.10** Schematic diagram of laser cleaning with different pulse widths

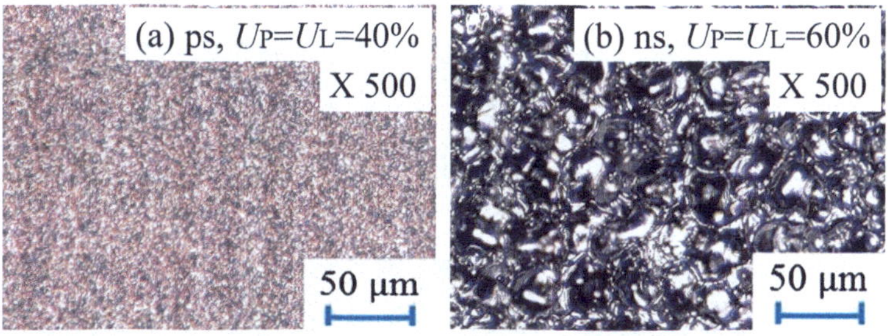


**Fig. 5.11** Comparison of results between nanosecond laser cleaning and picosecond laser cleaning. (**a**) Picosecond laser (**b**) Nanosecond lase

pulse, the molten material flows under the reaction force of the vaporized material and again solidifies to form the molten phase. For picosecond lasers, heat conduction is no longer a diffusion process, but is transmitted in the form of heat flow waves, showing a hyperbolic heat conduction pattern. Only a small fraction of the laser energy absorbed by the aluminum surface contaminants is transmitted to the depth and the thermal penetration depth is very small. The laser energy accumulates in the superficial layer where the surface molten material is in the form of a superheated liquid. At the end of the pulse, the superheated liquid evaporates, so that the surface of aluminum alloys cleaned with picosecond lasers is virtually free of molten phases. It is also the reason why nanosecond laser cleaning has greater surface roughness than picosecond laser cleaning of aluminum alloy substrate as shown in Fig. 5.11. Seo [56] used laser cleaning with a pulse width of 4–200 ns to remove the oxide layer and slag formed after arc welding as shown in Fig. 5.12. When the pulse duration was between 8 and 20 ns, the surface was slightly charred by the delayed residual irradiation from the laser cleaning and macroscopically appeared brown or light black. For pulse durations of more than 30 ns, the laser-cleaned substrate surface is more oxidized than the original surface due to the absorption of excess laser energy into the contaminants or substrate. During slag removal, the absorption of the laser beam by the slag is reduced due to the presence of silica in the slag. The interaction time between the laser beam and the material should be extended by increasing the pulse duration. Ling [57] applied a picosecond laser and an excimer laser for laser cleaning of metal orthodontic brackets, respectively. The results show that the intermittent light pulse action will make the irradiated part of the material repeatedly thermal expansion and contraction to form surface thermoelastic waves, causing very strong mechanical vibration of the material surface, to achieve the removal of the material without thermal ablation damage. The main reason for this is that the material is characterized by the presence of electronic state absorption of picosecond photon energy, which significantly reduces the thermal accumulation of laser energy and decreases the chance of producing evaporative removal. Boynukara [58] used femtosecond and nanosecond lasers to clean contaminated paper, respectively. It was found that femtosecond laser ablation occurring only on the surface of the paper could achieve high-precision cleaning. However, nanosecond laser cleaning is thermally ablated on both the front and back of the paper.

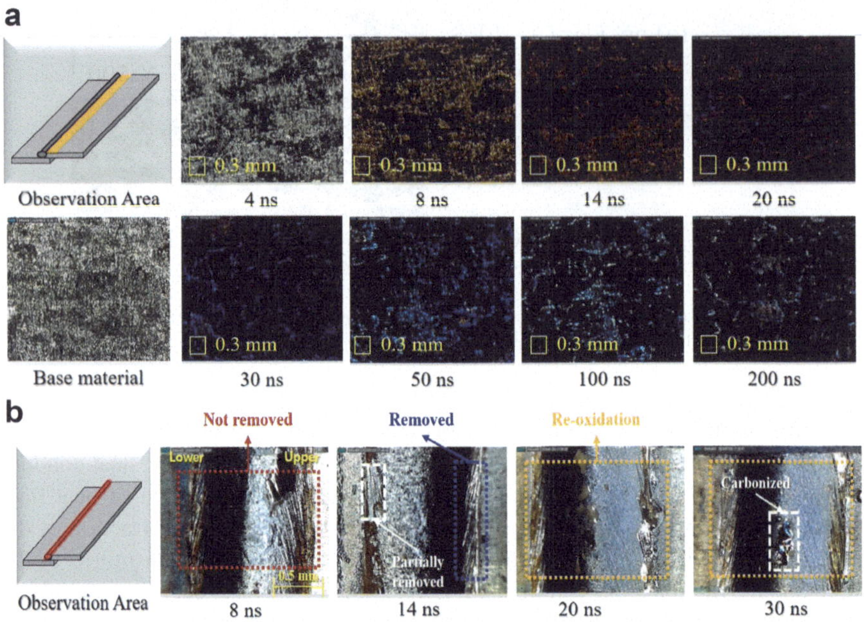


**Fig. 5.12** Post-weld by-products of laser cleaning with different pulse widths: (**a**) oxide layer and (**b**) welding slag

Obviously, pulse width is one of the key factors affecting the thermal effects of laser cleaning, which is related to the surface quality and cleaning efficiency of laser cleaning. The use of multi-pulse width composite, namely multi-laser beam composite cleaning method is expected to obtain high surface quality and fast cleaning speed. Wang [59] used two steps of continuous-pulsed laser operation to achieve high efficiency and high precision cleaning of thick corrosion layers of steel, while also actively forming an oxidized layer on the surface of the steel substrate to provide corrosion protection. Further studies have confirmed that continuous lasers have the highest cleaning efficiency but are prone to ablative damage to the substrate surface and nanosecond lasers cause more controlled damage to the substrate but the cleaning efficiency is too low in Fig. 5.12. As a result, high efficiency and substrate surface integrity retention cleaning of thick corrosion layers of steel is achieved by utilizing the combined advantages of high energy and high efficiency of a continuous laser as well as the high precision of a pulsed laser. Bi [60] conducted cleaning experiments on primer coatings and composite coatings using millisecond pulsed $CO_2$ laser and nanosecond pulsed fiber laser. The use of millisecond pulsed $CO_2$ lasers to clean paint layers produces black combustion products and can easily thermally deform the substrate. When using a nanosecond pulsed fiber laser to clean primer coatings, the laser energy can be precisely controlled to avoid damage on the substrate but it is difficult to completely remove the composite coating in one scan. Based on the above conclusions, the authors proposed a multi-laser combined cleaning method for composite coatings, i.e., millisecond pulsed $CO_2$ laser is used to thin the coating and nanosecond pulsed laser is used to remove the thinner primer from the substrate surface, which is a flexible cleaning method that can effectively avoid the damage of the substrate and at the same time obtain a very high cleaning efficiency.

It follows that laser cleaning processes with nanosecond or longer (e.g., millisecond) pulse widths have high cleaning efficiency. However, the low peak power, long thermal action time and heavy heat accumulation lead to a high susceptibility of the substrate to ablation damage, which is suitable for materials with low melting and boiling points of the contaminants or where the substrate is less affected by the heat accumulation zone. However, the thermal aggregation zone of the substrate tends to cause surface damage of the substrate and impaired surface integrity. The thermal aggregation zone expansion is inhibited by limiting the thermal diffusion effect by varying the laser pulse width in order to modulate the time of laser-material interaction. The ultrashort laser (e.g., picosecond, femtosecond) has a high peak power, short thermal action time and basically does not receive heat inside the lattice of the substrate surface, which makes the remelting phenomenon of the substrate surface effectively suppressed and is more suitable for the precision cleaning of the substrate surface quality with stringent requirements [61]. Of course, there are obvious bottlenecks in the large-scale application of ultrashort pulsed laser. On the one hand, the equipment is expensive to use and maintain; on the other hand, there are constraints in the research and development of related equipment and difficulties in obtaining larger power cleaning equipment and low cleaning efficiency.

### 5.3.2 *Effect of Cleaning Speed on Thermal Effects*

The cleaning speed primarily affects the distance between adjacent pulse spots in laser cleaning, known as the overlap rate. At a specific laser output energy, a low cleaning speed results in a high overlap rate of the spots. Under these conditions, the heat generated by the preceding pulse does not dissipate before the next pulse commences, leading to an accumulation of temperature. If the temperature substantially exceeds the substrate's melting point, it can result in the substrate's melting and vaporization, causing surface damage. Conversely, an excessively high laser cleaning speed leads to a significantly reduced or even zero overlap rate, thereby hindering complete cleaning. The cleaning speed includes the speed in the cleaning direction and the speed in the scanning direction, as shown in Fig. 5.13 [62]. The lap rate $U_p$ in the direction of scanning speed and $U_L$ in the direction of cleaning speed are expressed by the following equation: $U_p = \left(1 - \frac{V}{Df}\right) \times 100\%, U_L = \left(1 - \frac{L}{D}\right) \times 100\%$, where $V$ is the scanning speed of the galvanometer, $D$ is the spot diameter of the laser, $f$ is the pulse repetition frequency of the laser, $L$ is the centerline spacing between two adjacent laser scanning tracks.

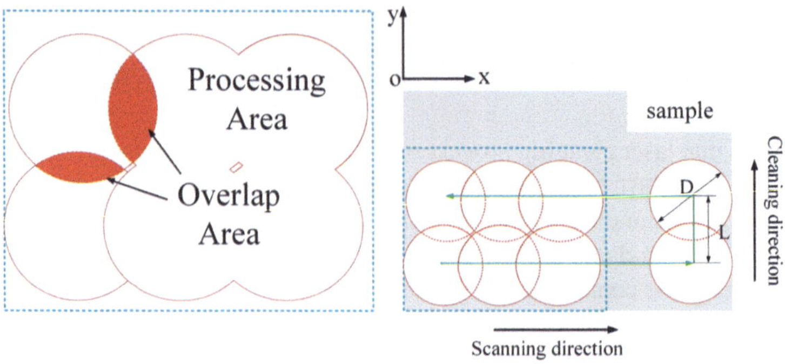


**Fig. 5.13** Schematic of laser cleaning speed

Under the premise of ensuring that the substrate surface contaminants are completely removed without damage to the substrate, the cleaning speed is increased appropriately in order to obtain a higher cleaning efficiency. Song [63] investigated the effect of overlap rate in laser rust removal on the surface properties of 7B04 aluminum alloy substrate for aerospace applications, as shown in Fig. 5.14. At an overlap rate of 40%, the small overlap area between neighboring spots leads to a short spot action time per unit area only sufficient to effectively remove the outer layer of corrosion products. Areas (1) and (2) show a large amount of rust residue on the surface of the sample without ablation and remelting of the substrate surface in an incompletely cleaned state. When the overlap rate increases to 60%, the rust layer absorbs more laser energy and the temperature rises instantaneously to reach the vaporization point where rust is almost completely removed by ablation-induced vaporization and phase explosion. When the overlap increases to 80% and 90%, significant ablation and remelting of the substrate surface occurs. As shown in regions (3) and (4), many inhomogeneous "craters" are formed on the surface of the substrate due to the low scanning speed and high spot overlap, which raise the surface temperature of the sample by heat accumulation effects. As the rust melts and sublimates, the excessive temperature causes an explosion of the liquid phase on the surface of the substrate forming craters and bumps. Zhang [64] investigated the relationship between the overlap rate and the surface roughness of the substrate at a laser energy density of 8.5 J/cm$^2$, as shown in Fig. 5.15. When the overlap was 0%, neighboring ablation craters did not interact with each other. When the overlap is 25%, the laser energy affects the formation of ablation craters and their edges. As the overlap area increases when the overlap rate is 50%, part of the laser energy is absorbed by the ablation crater A. The ablation crater becomes deeper and there is enough substrate material to participate in the formation of the elevated edge of the ablation crater. When the overlap rate is 75%, the overlap region almost covers the ablation crater A. The next pulse mainly irradiates the previous ablation crater A. The ablation crater A becomes deeper after absorbing the laser energy but the laser-irradiated surface of the original substrate is reduced without

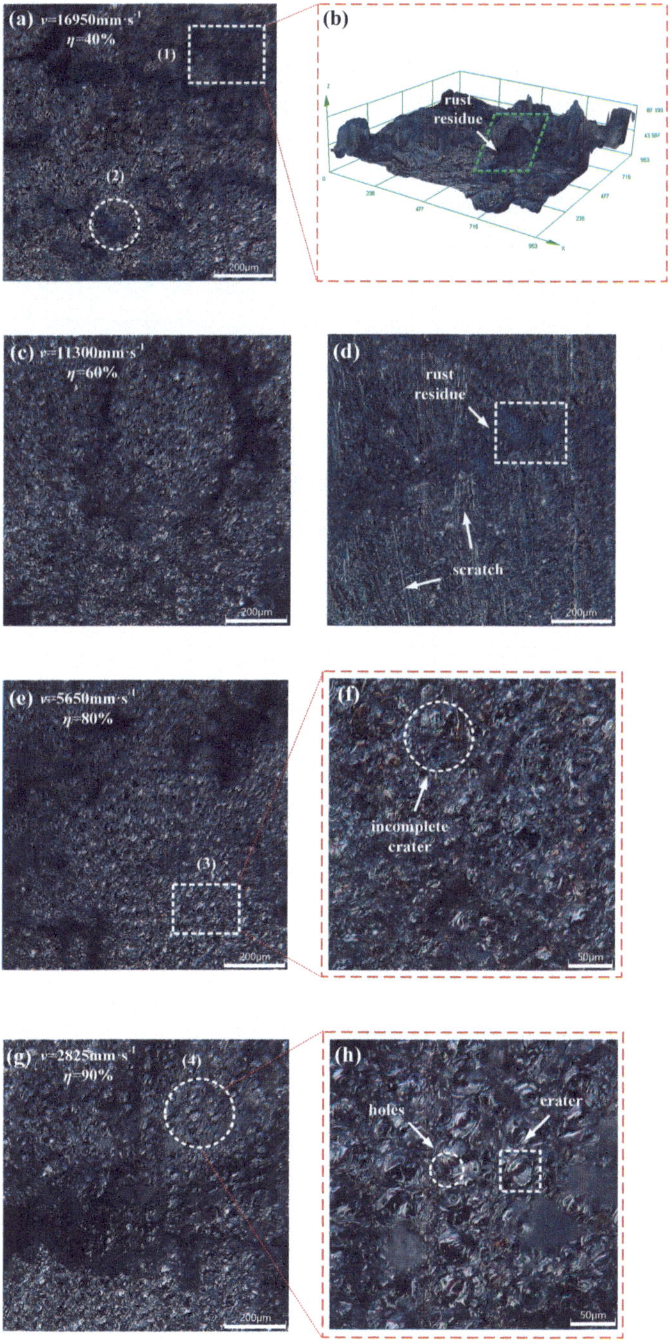


**Fig. 5.14** Microscopic morphology of rust removal in different regions. (**a**) and (**b**) at ascanning speed of 16,950 mm $s^{-1}$ (**c**) at a scanningspeed of 11,300 mm $s^{-1}$, (**d**) mechanical grinding, (**e**) and (**f**) at a scanning speed 5650 mm $s^{-1}$, (**g**) and (**h**) at ascanning speed of 2825 mm

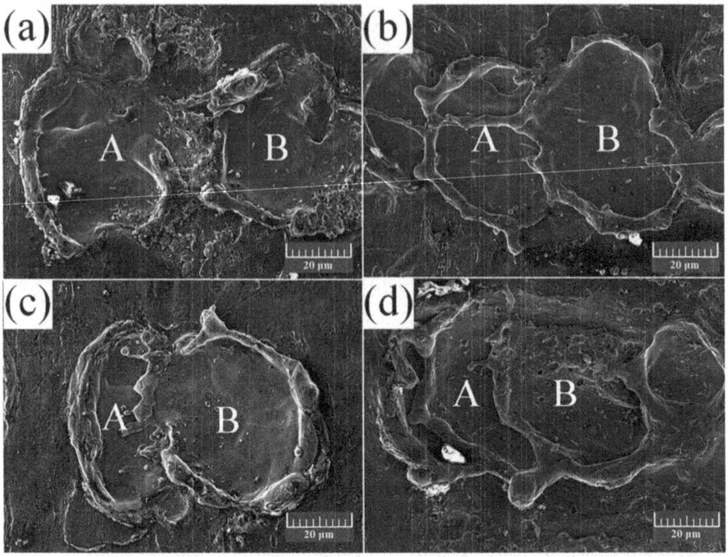


**Fig. 5.15** SEM images of the sample surface with different overlap rates: (**a**) 0%, (**b**) 25%, (**c**) 50%, (**d**) 75% (A is the ablation crater of the previous pulse and B is the ablation crater of the next pulse)

enough substrate material to form a higher ablation crater edge. In general, there is a close correlation between the overlap rate and surface roughness as the overlap rate increases the roughness tends to increase gradually. Xia [65] verified the relationship between repetition frequency and cleaning speed during laser cleaning of oxide films and found that the threshold of cleaning speed depends on the pulse repetition frequency. With higher repetition frequencies, the cleaning speed threshold becomes larger. When the actual cleaning speed is greater than the cleaning speed threshold, too low a spot stacking rate results in incomplete removal of the oxide film. Kim [66] studied laser removal of epoxy coatings from boat surfaces. The number of pulses in the spot increased when the laser beam overlap rate increased, which increased the heat input of laser energy to the epoxy coating. Thermal ablation played an important cleaning role, achieving effective removal of the epoxy paint layer in fewer scans.

The above studies show that the heat accumulation at the laser cleaned has been significantly altered at different cleaning speeds and the cleaning mechanism has evolved accordingly. Under the same cleaning conditions, the slower the cleaning speed or the greater the thickness of the cleaned contaminants, the more serious the heat accumulation leading to the substrate surface thermal ablation damage is more likely. In order to maximally weaken the thermal effects on the substrate surface and avoid secondary damage of the substrate, a fast but multiple scanning method can be used [67].

### 5.3.3 *Influence of Laser Energy Density on Thermal Effects*

A beam of laser irradiation to the surface of the material to form a certain area of the spot, the ratio of the laser pulse energy and the area of the spot reflects the size of the energy density of this beam of light, namely, the average energy density F. The specific relationship can be expressed by the following formula: $F = \frac{4P}{\pi D^2 f}$, where $P$ is the average laser power, $D$ is the spot diameter, $f$ is the repetition frequency. The cleaning threshold and the damage threshold are the critical values of two important parameters in the laser cleaning process. If the energy density is lower than the cleaning threshold, the substrate surface contaminants cannot be effectively cleaned. On the contrary, if the energy density is higher than the damage threshold, the substrate surface will be damaged by ablation. Due to the varying sensitivity of the substrate and contaminants to laser, the cleaning and damage thresholds differ for different substrates and contaminants. These thresholds can be determined by fitting the relationship between the material's ablation depth and the single pulse laser energy density [68–70], as represented by the equation $h = \frac{1}{\alpha} \ln\left(\frac{F}{F_1}\right)$, where $h$ is the material's ablation depth, $\alpha$ is the material's absorption coefficient, and $F_1$ is the cleaning threshold. If the properties of the contaminants significantly differ from those of the substrate, the removal process can be self-limiting. This means that the energy density deposited on the sample surface by the focused laser is sufficient to remove the contaminants, yet insufficient to damage the substrate material. While this represents an ideal removal process, practical applications are often more complex [71]. Gaussian and flat-top beams exhibit significant differences in the distribution of laser energy density, as shown in Fig. 5.16. The single-pulse energy density of a Gaussian beam is high, with considerable radial variance in energy distribution. The energy density at the pulse center exceeds the substrate damage threshold, leading to damage on the substrate surface. In contrast, a flat-top beam has a lower single-pulse energy density, and its energy is uniformly distributed radially. The laser energy density is ideally positioned between the cleaning threshold and the

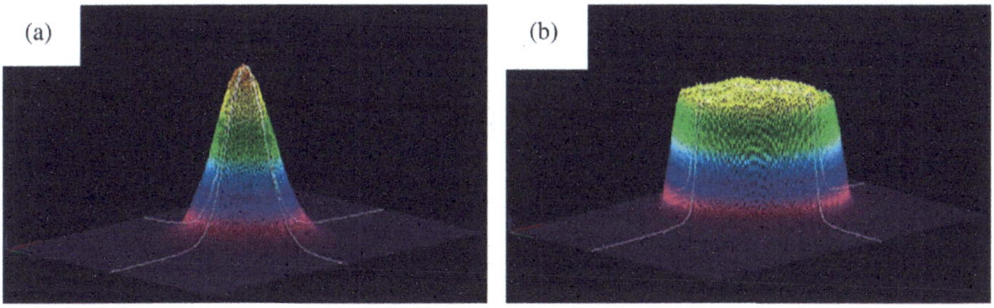


**Fig. 5.16** Three-dimensional shape of energy distribution: (**a**) Gaussian beam and (**b**) flat-topped beam

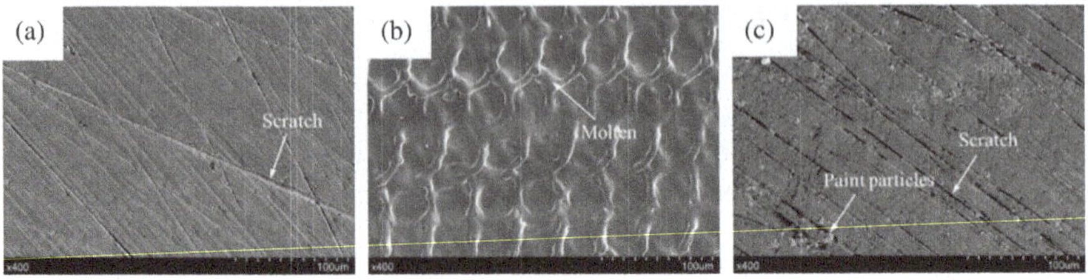


**Fig. 5.17** Surface micromorphology: (**a**) original substrate, (**b**) Gaussian beam, and (**c**) flat-top beam

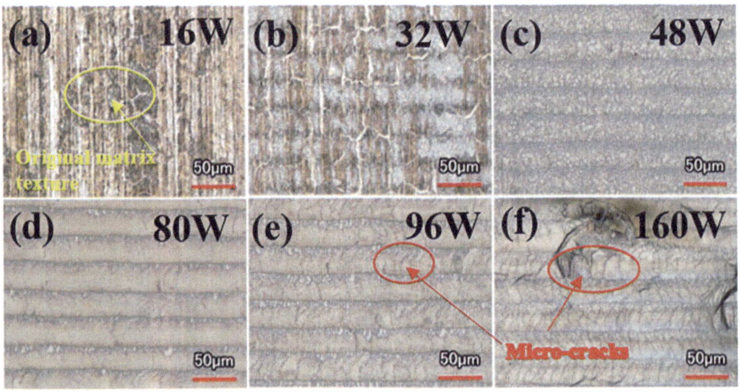


**Fig. 5.18** The morphology of 7075 aluminum alloy after laser cleaning with varying laser power. (**a**)16W (**b**) 32W (**c**) 48W (**d**) 80W (**e**) 96W (**f**) 160W

damage threshold, facilitating non-destructive cleaning of the substrate surface. However, due to the lower energy density of each pulse, its ability to remove stubborn paint from scour marks is weaker, as illustrated in Fig. 5.17 [72].

If the damage threshold of the substrate is lower than the cleaning threshold of the contaminants, the contaminants cannot be removed in a self-limiting manner during the cleaning process, inevitably leading to the substrate surface being affected by thermal effects. Jing [73] given that the melting point of the oxide layer is significantly higher than that of the aluminum alloy, higher laser energy is required for cleaning, which inevitably affects the aluminum alloy surface, as shown in Fig. 5.18. When the laser power was below 32 W, a considerable amount of oxide layer residue was still observable. At 48 W, the oxide layer was significantly reduced, but discoloration began to appear on the aluminum alloy surface. The best surface morphology quality, with the oxide layer completely removed, was achieved at a laser power of 80 W. However, when the laser power exceeded 96 W, cracks appeared on the surface of the aluminum alloy. In a similar study, Wang [74] utilized pulsed lasers with varying energy densities to remove the oxide film from the surface of 7075 aluminum alloy, as shown in Fig. 5.19. The microscopic morphology post-cleaning at a laser energy density of 1.82 J/cm$^2$ indicated a substantial reduction in striation defects, resulting in a smoother surface with no apparent damage to

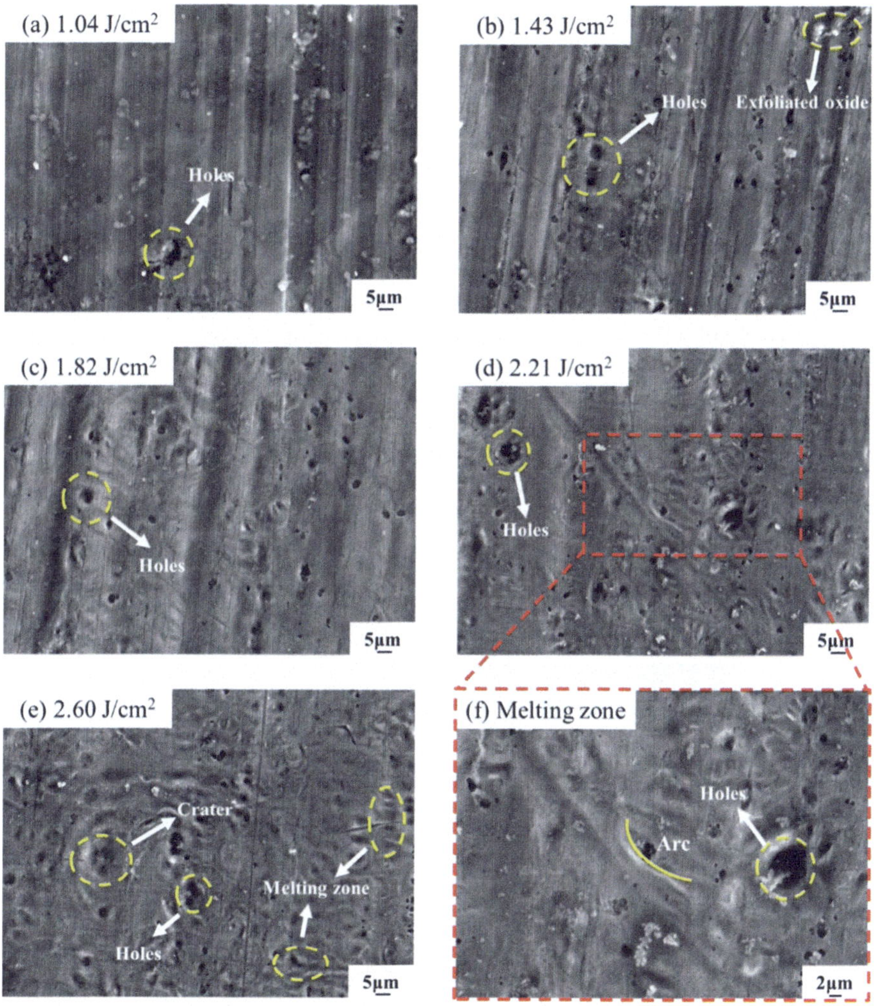


**Fig. 5.19** Surface morphology of 7075 aluminum alloy after laser cleaning with different energy density. (**a**) 1.04 J/cm$^2$ (**b**) 1.43 J/cm$^2$ (**c**) 1.82 J/cm$^2$ (**d**) 2.21 J/cm$^2$ (**e**) 2.60 J/cm$^2$ (**f**) Melting zone

the aluminum alloy substrate. However, at laser energy densities of 2.21 J/cm$^2$ or higher, melting phenomena were observed on the surface of the aluminum alloy. This melting can be attributed to the laser energy partly being used for oxide film removal and partly absorbed by the substrate, causing the substrate's surface temperature to exceed its melting point.

From the aforementioned studies, it was found that the surface morphology of the substrate changes with the increase in laser energy density. To investigate the underlying patterns, Li [75] explored the relationship between laser energy density and the evolution of the surface morphology of TA15 titanium alloy after laser cleaning. Under high laser energy density, extensive remelting occurred on the surface of TA15 titanium alloy. The periodic ridge-like structure was formed under the effect of laser impact. Similarly, Zhang [76] used a nanosecond pulsed laser to remove the blue paint layer on the surface of an aluminum alloy substrate at energy densities ranging from 1.2 to 8.0 J/cm². Excessively high laser energy creates a molten layer on the substrate, which becomes more pronounced as the laser energy density increases. Due to variations in melt thickness and duration, this leads to three different surface morphologies: At a laser energy density of 4.8 J/cm², tiny craters and bumps are formed due to evaporative recoil pressure; when the laser energy density exceeds 6.4 J/cm², the molten substrate material flows into surrounding scratches, resulting in a smoother surface with reduced roughness; at a laser energy density of 8.0 J/cm², ridge-like structures develop on the surface of the substrate, caused by the temperature difference between the laser irradiation zone and the anterior solidification zone, as shown in Fig. 5.20.

In situations with thick contaminants or to prevent damage to the substrate surface due to high laser energy density, a multi-pass cleaning method is likely to yield better results. Liu [77] employed varying laser powers for repeated cleaning of the external surface paint layer on a ship's hull, performing the process twice. Their

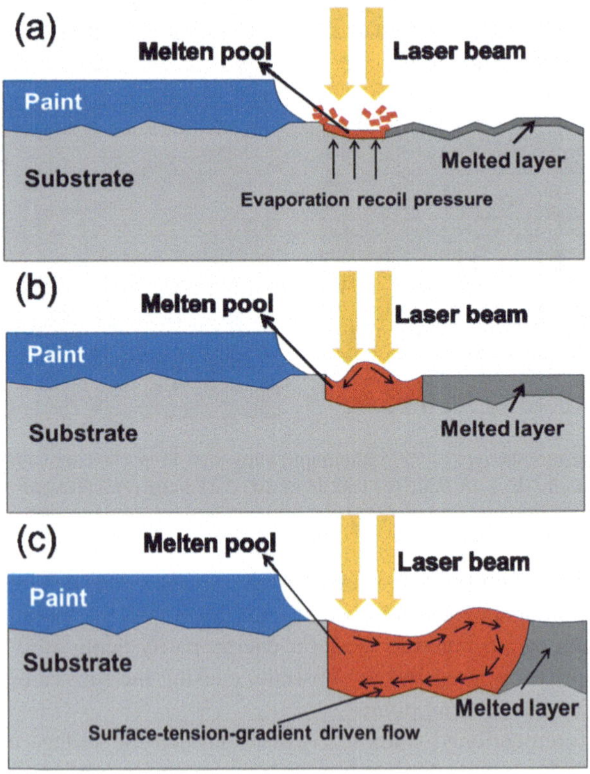


**Fig. 5.20** Schematic of the aluminum alloy surface behavior during laser cleaning: laser fluence of (**a**) 4.8 J/cm², (**b**) 6.4 J/cm², (**c**) 8.0 J/cm²

findings indicate that higher laser power increases the efficiency of removing polyurethane paint from the ship's exterior. A strategy of two or more repeated cleanings can thoroughly remove the polyurethane layer, demonstrating that the effectiveness of two cleaning sessions far surpasses that of a single session.

In summary, the cleaning mechanisms demonstrate distinct characteristics under varying laser energy densities [78]. Generally, higher laser energy densities are associated with enhanced cleaning efficiency. However, with the growth of plasma, the substrate surface becomes prone to secondary damage and secondary oxidation phenomena [14]. The varied energy densities directly influence the substrate surface's microstructure, crystallography, and compositional makeup [79]. At high energy densities, the substantial thermal input from the laser can cause temperatures to exceed the substrate's melting point, leading to remelting. The surface morphology post-remelting is closely associated with the energy density. Moreover, employing a method of multiple scans for cleaning can achieve better surface morphology while preventing damage to the substrate due to high energy densities.

## 5.4 Numerical Simulation of Thermal Effects

During the laser cleaning process, the instantaneous interaction between the laser and the material occurs within nanoseconds to microseconds. The temperature rise at the laser cleaning interface is both rapid and localized. Characterizing this temperature change through experimental methods requires a significant investment of time, labor, and materials. Moreover, these methods are unable to depict the variations in the temperature field or thermal stress field over time. Numerical analysis and simulation can more effectively replicate and assist in interpreting this process, providing important guidance for monitoring changes in temperature, stress, and surface morphology of the contaminants and substrate surfaces during laser cleaning operations.

When a laser is irradiated onto the surface of a material, a portion of the energy is absorbed and conducted by the material, while another portion is reflected. Since most substrates and contaminants are opaque, their transmittance is zero. In the transmission process within a material, the original laser intensity $I_0$ diminishes exponentially. The intensity at a distance $z$ from the surface is expressed as $I(z) = (1 - R)I_0e^{-\alpha z}$, where $\alpha$ is the absorption coefficient of the material, and $R$ represents its reflectivity [80]. The Gaussian distribution of a laser heat source represents a common form of laser energy distribution, with the distribution function of laser power density being given by: $P(x,y,t) = \frac{2P_0}{\pi r^2}\exp\left(-2\frac{x^2+y^2}{r^2}\right)\frac{t}{l^2}\exp\left(-\frac{t}{l}\right)^2$, where $P_0$ is the power of the incident laser, $r$ is the radius of the laser spot, and $l$ is the laser pulse width. During laser cleaning, the thermodynamic properties of the contaminants and the substrate material differ. The thermal conduction equation for the material is expressed as $k\left(\frac{\partial^2 T_e}{\partial x^2}+\frac{\partial^2 T_e}{\partial y^2}+\frac{\partial^2 T_e}{\partial z^2}\right) = \rho c\frac{\partial T}{\partial t}$, where $\rho$, $c$, $k$ is the

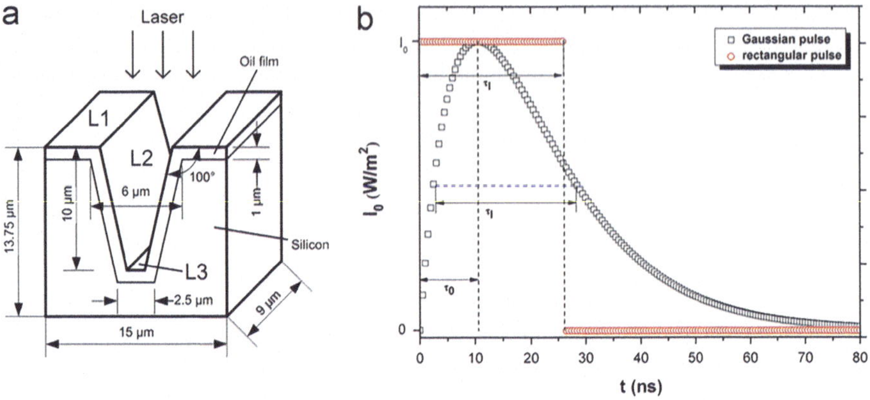


**Fig. 5.21** (**a**) The dimension of micro-slot structure and (**b**) temporal variation of the intensity for rectangular and Gaussian pulses

density, specific heat capacity, and thermal conductivity coefficient of the material, respectively; $T_e$ is the material's instantaneous temperature; and $t$ is the duration of laser exposure. According to Newton's Law of Cooling, the heat convection equation is given by $Q_c = \alpha S_c(T - T_w)$, where $\alpha$ is the convective heat transfer coefficient, $S_c$ is the contact surface area between the solid and the fluid, and $T$ and $T_w$ are the temperatures of the solid and the fluid, respectively. According to Stefan-Boltzmann's Law, the heat radiation equation is formulated as $Q_{rc} = w\sigma S_{rc}\left(T^4 - T_0^4\right)$, where $w$ is the emissivity of the material, $\sigma$ is the Stefan-Boltzmann constant, and $S_{rc}$ is the radiative surface area, $T$ and $T_0$ are the temperatures inside and outside the radiative zone, respectively [81–84]. The temperature field, as a key parameter characterizing the laser-material interaction, is an important aspect of monitoring the entire laser cleaning process. Yue [85] studied rectangular and Gaussian shapes with different time pulses based on finite element method, as shown in Fig. 5.21. The finite element model was validated with a planar analytical solution and applied to a conical microgroove structure finding higher cleaning depth than the Gaussian pulse. Yang [86] used ANSYS to establish a finite element model to investigate the effects of process parameters such as laser scanning speed and scanning interval on the paint removal process. In addition, the surface morphology after planar scanning laser paint removal was simulated and the numerical simulation results were experimentally verified. Figure 5.22 shows a single trench simulation at 50 J/cm$^2$ and 100 kHz using various scanning speeds (4000–8000 mm/s), which introduces the phenomenon of thermal accumulation during the laser cleaning process, where the lower scanning speed makes the transient process longer. Ai [87] analyzed the removal of oxides from the surface of titanium alloys at different scanning speeds using finite elements. It was shown that the increase or decrease of scanning speed has little effect on the peak temperature and the minimum temperature after the first pulse, but has an effect on the thermal equilibrium temperature and the time required to reach the thermal equilibrium state. As can be seen in Fig. 5.23, when the

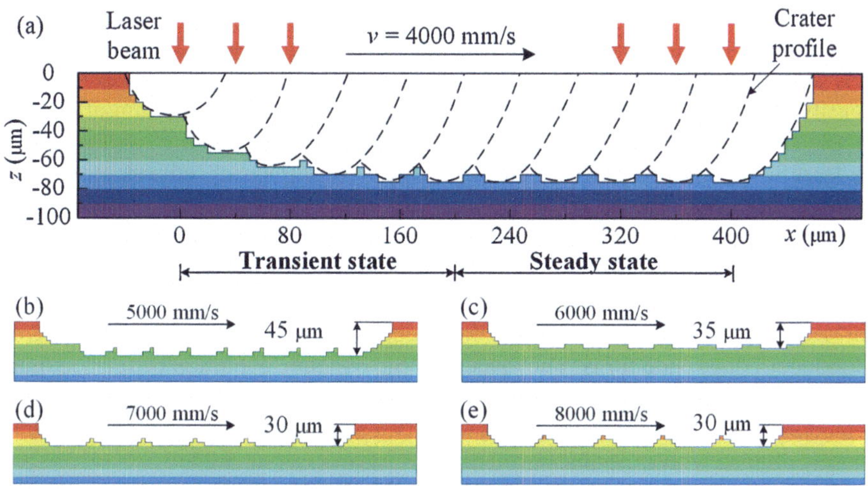


**Fig. 5.22** Cross-sectional morphology of ablation craters at different scanning speeds. (**a**) 4000mm/s (**b**) 5000mm/s (**c**) 6000mm/s (**d**) 7000mm/s (**e**) 8000mm/s

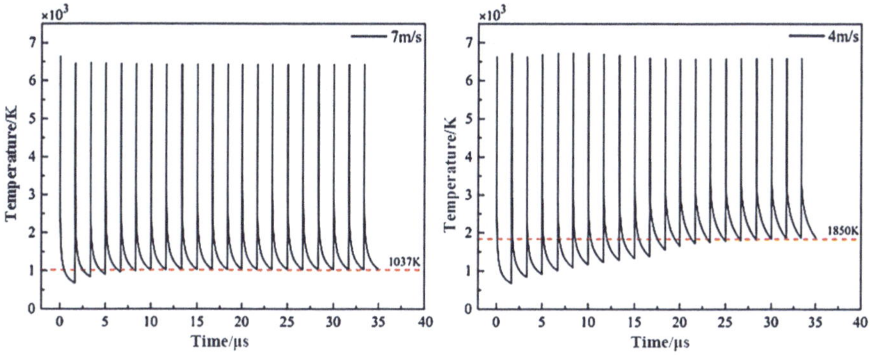


**Fig. 5.23** Variation of surface temperature of $TiO_2$ oxide layer under different scanning speeds

scanning speed is 7 m/s, the surface of the oxide layer reaches thermal equilibrium after only about 3 pulses. When the scanning speed is reduced to 4 m/s, it needs to experience 13 pulses to reach the thermal equilibrium state. The thermal equilibrium temperature increases dramatically with the decrease of scanning speed.

When a material is heated, it undergoes thermal expansion, which can be described by the equation $\epsilon = \gamma\Delta T$. Here, $\epsilon$ is the thermal strain induced by temperature changes, $\gamma$ is the coefficient of thermal expansion, and $\Delta T$ is the change in temperature. Employing Hooke's Law, the relationship between stress and strain in the material can be established. The thermal stress can thus be expressed as $\sigma = Y\epsilon = Y\gamma\Delta T$, where $Y$ is the elastic modulus. Due to the differing thermodynamic properties of the contaminants and the substrate, the force causing the detachment of the contaminants from the substrate can be expressed as $\Delta\sigma = \sigma_c - \sigma_s > f_A$, $\sigma_c$ is the thermal stress in the contaminants, $\sigma_s$ is the thermal stress in the substrate, and $f_A$ is the adhesion force [88–90]. Miao [7] investigated the influence of laser energy density and scanning speed on the paint removal from 7075 aluminum alloy surfaces. They analyzed the surface micro-morphology and related cleaning effects under different laser parameters. Using COMSOL Multiphysics finite element analysis software, they established a three-dimensional thermodynamic model of laser cleaning. This model calculated the variations in temperature and thermal stress at different depths of the coating and metal substrate over time. To further study the surface morphology of the substrate post-laser cleaning. Guo [91] developed a three-dimensional heat transfer and flow coupled model for the laser cleaning of aluminum alloy oxide films. They analyzed the dynamic evolution of surface morphology under different laser powers, examined the transient distributions of the temperature and fluid fields in detail, and demonstrated that the simulation results closely aligned with experimental findings, as shown in Fig. 5.24.

Through the above research, it is found that the extensive use of finite element analysis software can visualize and analyze the temperature, stress and fluid fields in the laser cleaning process to achieve a scientific and reasonable prediction of different laser cleaning parameters and different material systems. This is facilitated to analyze the changes in temperature, stress, morphology and even properties of the surface during the laser cleaning process. However, the simulation model has some limitations and needs to be explored in developing a generalized numerical simulation for laser cleaning and further validated with actual laser cleaning operations.

## 5.5 Online Monitoring of Laser Cleaned Surface

For different cleaning objects, different cleaning quality requirements, the inhomogeneity of the contaminants thickness and components of the laser cleaning process control is more complex. There are many parameters that need to be matched and regulated, such as laser power, frequency, pulse width, scanning speed, scanning angle, additional external field and so on. In the process of laser cleaning operations, real-time adjustment of cleaning parameters is the inevitable requirement for improving cleaning efficiency and preventing incomplete cleaning and over-cleaning. The online monitoring of the state of the cleaned surface is an effective means of support. According to the ability to adjust the cleaning parameters in real time and evaluate the cleaning effect while the laser cleaning is in progress, it can be categorized into offline monitoring and online monitoring. Offline monitoring using first laser cleaning operations and then static detection and analysis, namely, cleaning and monitoring is carried out step by step. The disadvantages of this method are poor timeliness and inefficiency. Online monitoring can monitor the whole process of laser cleaning operations in real time without damaging the substrate. The basic principle is to monitor the thermal radiation, acoustic or optical signals of laser cleaning, and intelligently evaluate or assist the laser cleaning operator to determine whether the surface contaminants has been cleaned or the substrate has been damaged.

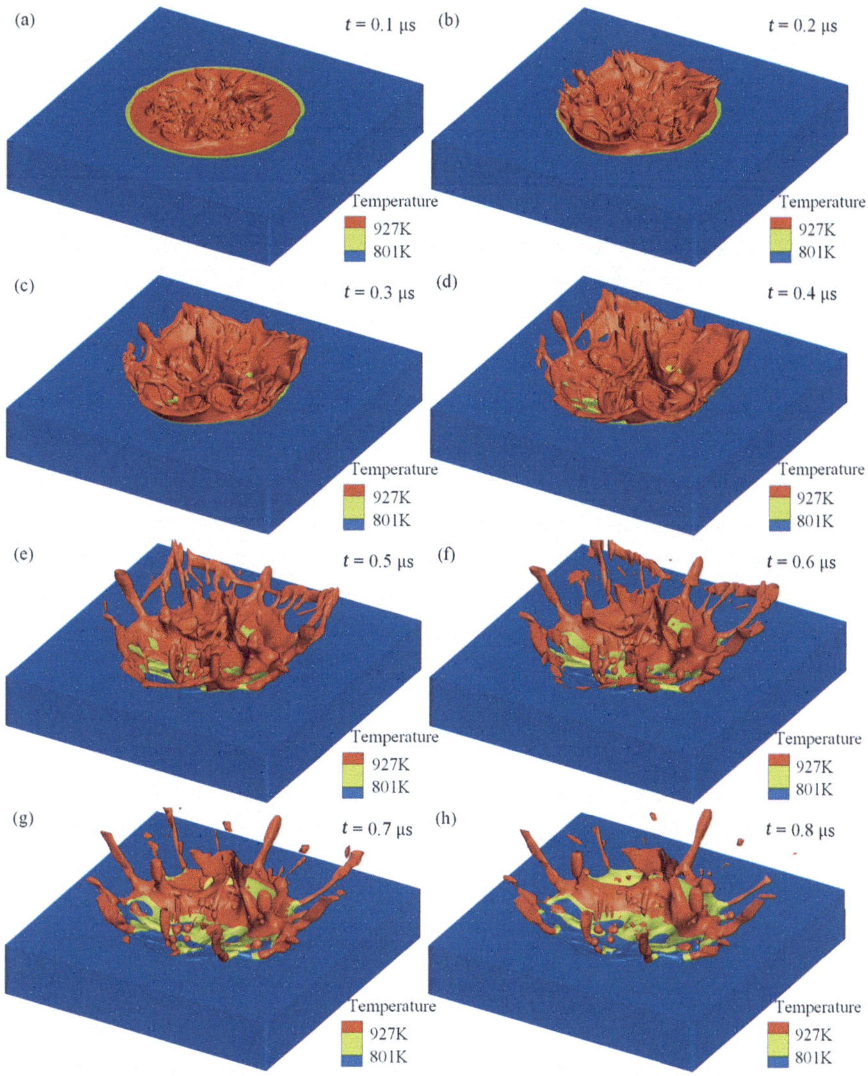


**Fig. 5.24** The evolution process of the surface morphology of 6061 aluminum alloy cleaned by 60 W at different time points: (**a**) 0.1 μs (**b**) 0.2 μs (**c**) 0.3 μs (**d**) 0.4 μs (**e**) 0.5 μs f 0.6 μs (**g**) 0.7 μs (**h**) 0.8 μs

### *5.5.1 Laser-Induced Breakdown Spectroscopy*

Laser-induced breakdown spectroscopy (LIBS) is a cutting-edge technology for online monitoring and intelligent feedback in the field of laser cleaning. The principle is a high energy density laser irradiation to the surface of the sample and the atoms are excited by the laser energy from the ground state to the excited state to produce a large number of high-temperature plasma. Subsequently, the plasma

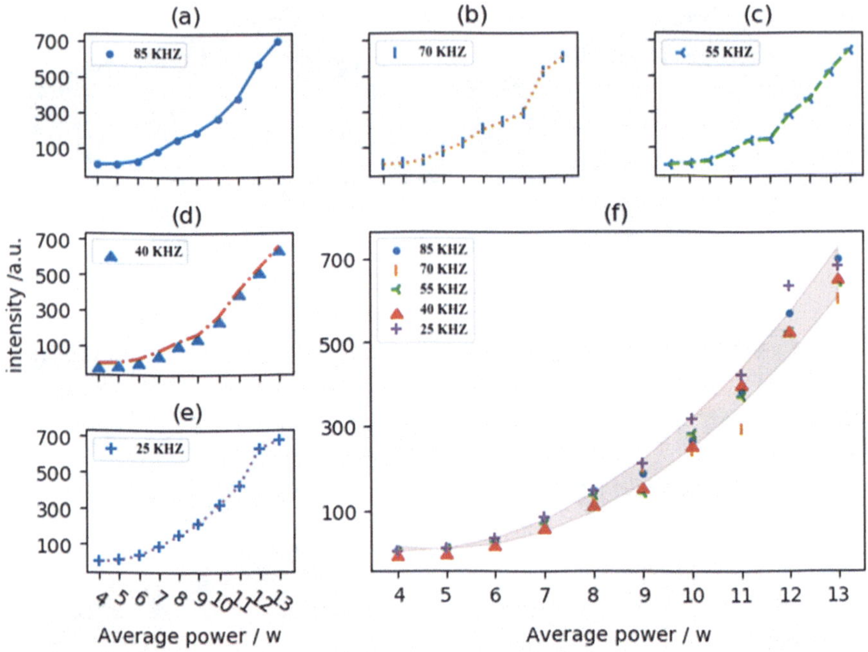


**Fig. 5.25** The curves of the average spectral intensity change under different parameter combinations. (**a**)-(**e**) the relationship between the line intensity and the average laser power at different frequencies (**f**) Contrast diagram of spectral line strength at different frequencies

temperature decreases at the end of the laser pulse causing a relaxation phenomenon when the excited state atoms in the plasma return to the ground state. Some of the energy is radiated in the form of light with distinctive elemental properties to produce a characteristic spectrum [92]. Wang [93] designed an online monitoring system for laser cleaning using LIBS to monitor the quality of laser cleaning in real time. By comparing the differences in the LIBS spectra of epoxy resin and carbon fiber substrates, it was found that the Na atomic spectral line with a wavelength of 588.819 nm in the LIBS spectrum could characterize the cleaning quality. The effect of laser parameters on the characteristic spectrum of LIBS was tested experimentally proving that the average spectral intensity is positively correlated with the average laser power as shown in Fig. 5.25, and the system is built as shown in Fig. 5.26. Yang [94] used the LIBS technique to measure the spectral signal changes of aluminum alloy, primer, and topcoat surfaces at different pulse numbers to demonstrate that the LIBS characteristic peaks of the characteristic elements in different paint layers change with the laser paint removal process which could be realized as a layered paint removal. Zhou [95] investigated the time-resolved characteristics of the elemental peaks in LIBS of paint layers and substrates. When the paint layer is removed, the peak intensity of the elemental spectrum of the paint layer suddenly decreases while the peak intensity of the elemental spectrum of the substrate suddenly increases. The time-resolved signal can be fitted with a double exponential function, and the

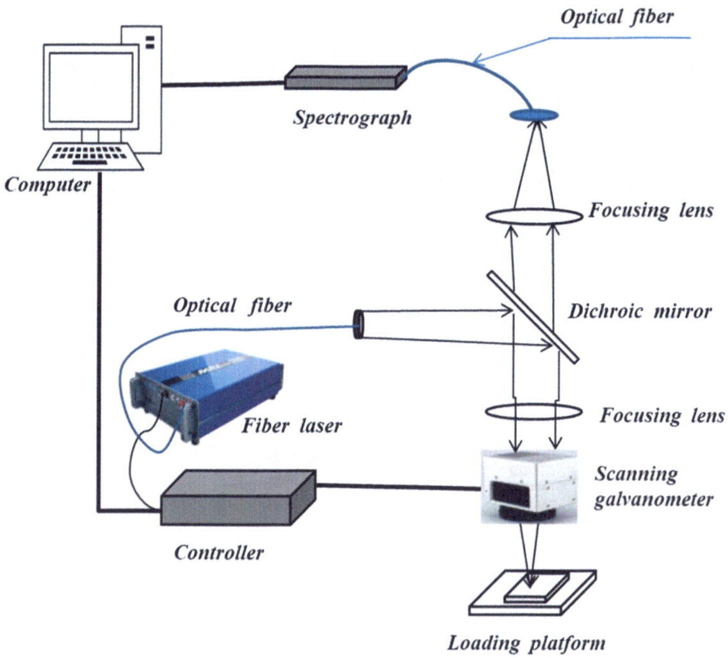


**Fig. 5.26** Laser cleaning and LIBS monitoring system

ratio of the peak intensities of the elemental spectra of the paint and the substrate and the double exponential function coefficients ($A_{short}/A_{long}$) can be used for real-time monitoring of the laser paint removal process as shown in Fig. 5.27.

By fully utilizing the plasma excited during the laser cleaning process, LIBS is a non-contact, real-time, high-accuracy on-line monitoring method with the laser cleaning process, which avoids secondary damage to the material surface by off-line monitoring, and can be used to monitor the emission spectra of the laser cleaned contaminants, substrate, and even ambient splash fumes. However, this monitoring method requires plasma generation hence is not suitable for low energy density cleaning. In addition, LIBS is suitable for layer cleaning of heterogeneous " contaminants–substrate" materials, where the spectra of different materials are used to determine whether or not the contaminants have been removed, but it is difficult to work with homogeneous "contaminants–substrate" material systems.

## *5.5.2 Photoacoustic Monitoring Techniques*

Acoustic signal characteristics are closely related to the laser cleaning state. Photoacoustic monitoring techniques has unique advantages for monitoring the laser cleaning process. The photoacoustic effect can be simply described as the

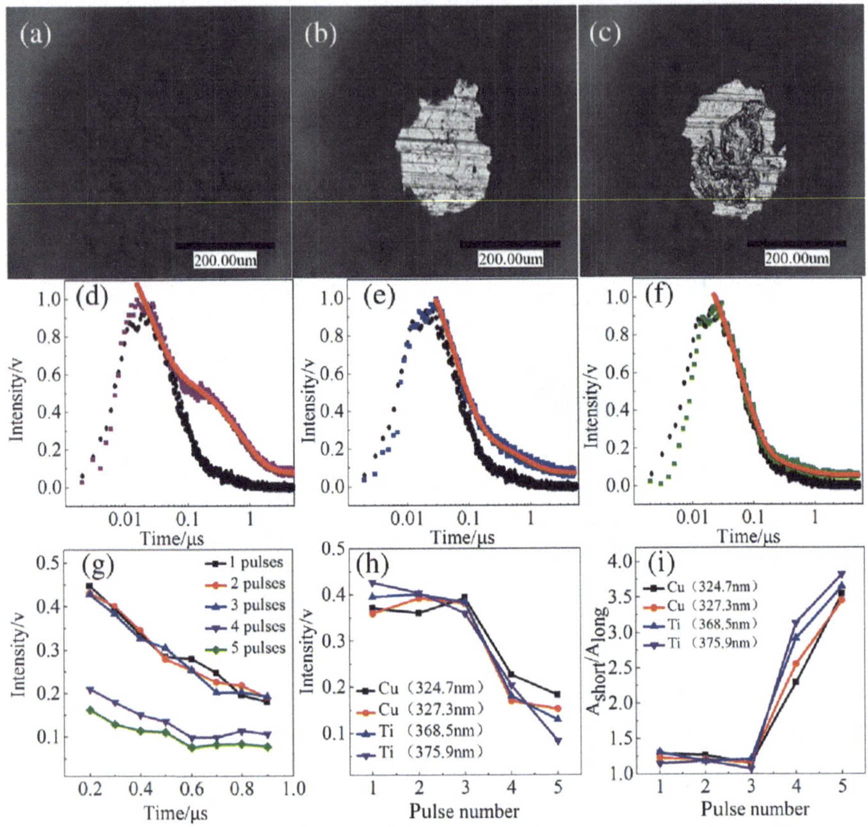


**Fig. 5.27** Surface morphology of the thin blue paint after being irradiated by (**a**) three, (**b**) four, and (**c**) five laser pulses; TRS signal at 368.5 nm and its double exponential function fitting (solid line) of (**d**) third, (**e**) fourth, and (**f**) fifth pulses on thin blue paint and substrate (black dotted line). (**g**) Detailed TRS signal at 368.5 nm (Ti). (**h**) Intensity of the TRS signal versus pulse number at 0.3 μs. (**i**) Ratio of the coefficient ($A_{short}/A_{long}$) of the TRS signal of blue paint as a function of pulse numbers

generation of sound waves by a material absorbing laser energy radiation, and the degree of photoacoustic effect presentation is proportional to the applied laser energy and the effective light absorption coefficient of the interacting materials [96]. Since the photoacoustic properties of the contaminants and the substrate are different, the amplitude of the photoacoustic signal will change during the process of laser cleaning of the contaminants until the laser touches the substrate. Through the ultra-sensitive monitoring of the photoacoustic changes and pulse-by-pulse recording of the average photoacoustic amplitude, real-time photoacoustic information of the surface of the laser cleaning process can be obtained to monitor the real-time state of laser cleaning. Zou [97] analyzed different time-domain, frequency-domain and time-frequency-domain spectral features for three sets of states: laser energy below the cleaning threshold, above the cleaning threshold with undamaged substrate and above the cleaning threshold with damaged substrate. The audible acoustic signal has a weakly correlated time-domain characteristic. From the spectrogram, it can be seen that the paint layer is completely removed when the frequency amplitude of 0.04 kHz reaches its maximum value. From the time-frequency spectrum, it can be seen that the amplitude of the power spectral density increases with increasing laser energy. Chen [98] investigated the characteristics of acoustic signals at various stages of laser paint removal and found that with the continuous removal of the paint layer, the amplitude and the local standard deviation of the acoustic signals in the frequency range of 7–10 kHz were gradually reduced to a stable value, which were more realistically reflecting the state of the laser paint removal than the time-domain signals.

Papanikolaou [99] developed a hybrid photoacoustic and optical system for online monitoring of the laser cleaning process, and the system is built as shown in Fig. 5.28. By comparing the photoacoustic signals of two different wavelengths of laser cleaning and relating them to the laser ablation mechanism, it was confirmed that each wavelength maps a different ablation mechanism, and that 1064 nm IR laser radiation has stronger thermal effects than 355 nm UV. Xie [100] utilized a laser to remove different thicknesses of rust layers on the surface of Q235 steel, and acoustic emission signals and high-speed photography were used to monitor the laser rust removal process. The frequency band of pulsed laser interaction with the rust layer is mainly concentrated in 15–35 kHz, which can assist the quantitative analysis of the characteristics of rust layer removal under different laser energies. The laser-substrate interaction bands of 65–85 and 140–160 kHz can be used to monitor the thermal damage characteristics of the substrate surface.

The acoustic amplitude is related to multiple factors such as laser energy density, number of pulses, pulse width, and material properties of the contaminants and substrate. Based on the laser cleaning process of photoacoustic signal variability of

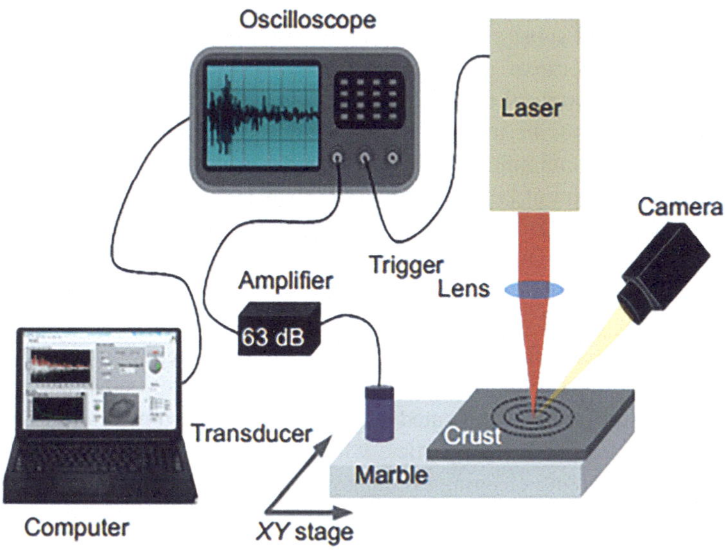


**Fig. 5.28** Schematic diagram of the photoacoustic and optical experimental setup

the online monitoring is used to determine whether the laser cleaning contaminants is clean or whether the substrate is damaged and autonomous or auxiliary laser cleaning operator real-time regulation of laser cleaning process parameters. The photoacoustic signal monitoring method is widely used, convenient and low-cost, however, it is susceptible to the noise of the laser cleaning atmosphere and is not very accurate. For certain scenarios where the surface cleanliness of the substrate is required and the substrate must not be damaged at all, the photoacoustic technique has obvious unreliability.

## 5.6 Conclusion and Prospects

This paper clarifies the origins of thermal effects and discusses the thermal effects within both the contaminants and the substrate during laser cleaning, as well as strategies to prevent thermal ablation of the substrate surface. It comprehensively elucidates the influence of laser cleaning parameters on thermal effects, the simulation of thermal effects in laser cleaning, and monitoring methods such as laser-induced breakdown spectroscopy and photoacoustic techniques. The paper also focuses on proposals for process optimization, aimed at enhancing laser cleaning efficiency, ensuring the integrity of the substrate surface, and improving the surface properties of the substrate. Despite the increasing maturity and rapid application of laser cleaning technology in engineering, there are still some issues that require focused effort.

1. The scientific control of the thermal effects at the laser cleaning is essential for optimizing cleaning efficiency. Thermal effects serve a dual role: it facilitates the effective removal of contaminants and simultaneously can cause thermal ablation of the cleaned substrate. Consequently, precise control of this thermal phenomenon is imperative for achieving high-quality and efficient cleaning outcomes. Accordingly, the selection of suitable laser cleaning equipment should be based on the specific material systems of the "contaminants–substrate", the thickness of the contaminants, and the desired surface quality standards. A systematic optimization of laser parameters such as power, pulse width, frequency, scanning speed, angle of incidence, and the auxiliary atmosphere is necessary. This approach aims to maximize cleaning efficiency while ensuring the integrity of the substrate surface, necessitating extensive foundational process experiments for exploration.
2. Balancing the relationship between laser cleaning efficiency and the quality of the substrate surface is critical for achieving superior substrate cleaning. Numerous studies have indicated that post-laser cleaning, substrates often exhibit enhanced surface properties, which facilitate their reuse. However, in pursuit of maximizing cleaning efficiency, these enhancements in substrate surface performance are frequently overlooked. It has been observed that achieving a notable improvement in substrate surface properties through laser cleaning

requires substantial adjustments to the laser cleaning parameters, often at the expense of reduced cleaning efficiency. Therefore, navigating the balance between laser cleaning efficiency and substrate surface quality is a pivotal challenge in the advancement of high-quality laser cleaning technology. Moreover, the current understanding of these enhancement characteristics is sporadic and unpredictable. Investigating methods to transform the irregular enhancement of substrate surface properties post-laser cleaning into a controlled and predictable process is a valuable avenue for future research.

3. Advancements in more precise and reliable intelligent online monitoring technologies are crucial for enhancing the autonomous matching and optimization of laser cleaning parameters. The inability to monitor the quality state of the cleaning interface in real-time during laser cleaning operations leads to pervasive issues of low cleanliness and over-cleaning. These issues largely stem from the inhomogeneity of the contaminants being cleaned and the rudimentary nature of manual labor. On one hand, there is an imperative need to develop accurate and reliable online monitoring technologies, along with corresponding cleaning effect evaluation and feedback mechanisms, to enable independent dynamic adjustment of laser cleaning process parameters. On the other hand, intensifying research in intelligent laser cleaning technology is essential. This involves integrating laser cleaning sensors and robotic arms to achieve high-precision positioning and automation in the cleaning process, which also facilitates the seamless processing of the cleaned substrate.

## References


1. Rivas T, Pozo-Antonio J S, de Silanes M E L, et al. Laser versus scalpel cleaning of crustose lichens on granite. Applied Surface Science, 2018, 440: 467–476.
2. Ma M, Wang L, Li J, et al. Investigation of the surface integrity of Q345 steel after Nd: YAG laser cleaning of oxidized mining parts. Coatings, 2020, 10(8): 716.
3. Zheng Z, Wang C, Huang G, et al. Effect of Defocused Nanosecond Laser Paint Removal on Mild Steel Substrate in Ambient Atmosphere. Materials, 2021, 14(20): 5969.
4. Tian Z, Lei Z, Chen X, et al. Nanosecond pulsed fiber laser cleaning of natural marine micro-biofoulings from the surface of aluminum alloy. Journal of Cleaner Production, 2020, 244: 118724.
5. Ding K, Zhou K, Feng G, et al. Mechanism and conditions for laser cleaning of micro and nanoparticles on the surface of transparent substrate. Vacuum, 2022, 200: 110987.
6. Nie J, Zhang H, Zhang D, et al. Removal mechanism of laser cleaning for inorganic thermal control coatings on aluminum alloys. Applied Surface Science, 2023, 633: 157578.
7. Miao R, Wang T, Yao T, et al. Experimental and numerical simulation analysis of laser paint removal of aluminum alloy. Journal of Laser Applications, 2022, 34(1): 012002.
8. Zhang S L, Suebka C, Liu H, et al. Mechanisms of laser cleaning induced oxidation and corrosion property changes in AA5083 aluminum alloy. Journal of Laser Applications, 2019, 31(1): 012001.
9. Li G, Gao W, Zhang L, et al. The quality improvement of laser rubber removal for laminated metal valves. Optics & Laser Technology, 2021, 139: 106785.

10. Zhang S, Yan Q, Lin J, et al. Elimination of blind zone in nanoparticle removal on silicon wafers using a double-beam laser shockwave cleaning process. Applied Surface Science, 2021, 539: 148057.
11. Zhao H, Qiao Y, Du X, et al. Laser cleaning performance and mechanism in stripping of Polyacrylate resin paint. Applied Physics A, 2020, 126(5): 1-14.
12. Zhang Ruo-nan, et al. Research on Laser derusting Mechanism and Cleaning Threshold of EH36 Marine Steel Plate. 2021:22–32.
13. Wang Y, Shen H, Liu C, et al. Evolution of nanosecond laser-induced phase explosion based on a high-speed continuous imaging system. Results in Physics, 2021, 29: 104782.
14. Zhang G, Hua X, Huang Y, et al. Investigation on mechanism of oxide removal and plasma behavior during laser cleaning on aluminum alloy. Applied Surface Science, 2020, 506: 144666.
15. Zhang D, Xu J, Li Z, et al. Removal mechanisms of nanosecond pulsed laser cleaning of blue and red polyurethane paint. Applied Physics A, 2022, 128(2): 1–14.
16. Gu J, Su X, Jin Y, et al. Precise paint stripping based on photochemical effect by UV laser on honeycomb sandwich panel [J]. Journal of Materials Research and Technology, 2023.
17. Pozo-Antonio J S, Papanikolaou A, Melessanaki K, et al. Laser-assisted removal of graffiti from granite: Advantages of the simultaneous use of two wavelengths. Coatings, 2018, 8(4): 124.
18. Pouli P, Papakonstantinou E, Frantzikinaki K, et al. The two-wavelength laser cleaning methodology; theoretical background and examples from its application on CH objects and monuments with emphasis to the Athens Acropolis sculptures [J]. Heritage Science, 2016, 4(1): 1–11.
19. Hu C, He G, Chen J, et al. Research on Cleaning Mechanism of Anti-Erosion Coating Based on Thermal and Force Effects of Laser Shock. Coatings, 2020, 10(7): 683.
20. Roberts D E. Pulsed laser coating removal by detachment and ejection. Applied Physics A, 2004, 79(4–6): 1067–1070.
21. Zhou R, Hong M. Laser Cleaning of Contaminated Substrate Surfaces. Handbook of Laser Micro-and Nano-Engineering, 2020: 1–46.
22. Labuschagne K, Pityana S L. Laser induced damage threshold on metallic surfaces during laser cleaning. 2005.
23. Li Z, Xu J, Zhang D, et al. Nanosecond pulsed laser cleaning of titanium alloy oxide films: Modeling and experiments. Journal of Manufacturing Processes, 2022, 82: 665–677.
24. Tian Z, Lei Z, Chen X, et al. Evaluation of laser cleaning for defouling of marine biofilm contamination on aluminum alloys. Applied Surface Science, 2020, 499: 144060.
25. Liu B, Mi G, Wang C. Research on grain refinement and wear behavior of micro-remelted Ta15 alloy surface by laser cleaning. Materials Chemistry and Physics, 2021, 259: 124022.
26. He Z, Shen Y, Xiong W, et al. Uncovering the roles of laser action modes in surface mechanical properties of 2024 aluminum alloy [J]. Applied Surface Science, 2023, 613: 156032.
27. Sun Q, Zhou J, Meng X, et al. Mechanical properties and microstructure characteristics of 2024-T351 aluminum alloy specimen subjected to paint removal by laser cleaning [J]. Vacuum, 2023, 211: 111927.
28. Wei P, Chen Z, Wang D, et al. Effect of laser cleaning on mechanical properties of laser lap welded joint of SUS310S stainless steel and 6061 aluminum alloy [J]. Materials Letters, 2021, 291: 129549.
29. Wang T, Bai Q, Li Y, et al. Investigation of the Multimechanism Laser Cleaning Dynamics for Rough Fused Silica Surfaces with Organic Contaminants: A Computational Simulation and Atomic Analysis [J]. Langmuir, 2023, 39(42): 15095–15106.
30. Wang A, Feng A, Gu X, et al. Effect of picosecond laser cleaning on surface morphology and properties of stainless steel [J]. Optics & Laser Technology, 2023, 159: 109041.
31. Lu Y, Ding Y, Wang M L, et al. An environmentally friendly laser cleaning method to remove oceanic micro-biofoulings from AH36 steel substrate and corrosion protection [J]. Journal of Cleaner Production, 2021, 314: 127961.

32. Zhuang S, Kainuma S, Yang M, et al. Characterizing corrosion properties of carbon steel affected by high-power laser cleaning [J]. Construction and Building Materials, 2021, 274: 122085.
33. Shi T, Wang C, Mi G, et al. A study of microstructure and mechanical properties of aluminum alloy using laser cleaning [J]. Journal of Manufacturing Processes, 2019, 42: 60–66.
34. Mandolfino C, Lertora E, Genna S, et al. Effect of laser and plasma surface cleaning on mechanical properties of adhesive bonded joints [J]. Procedia Cirp, 2015, 33: 458–463.
35. Zhu G, Wang S, Zhang M, et al. Application of laser cleaning in postwelding treatment of aluminum alloy [J]. Applied optics, 2020, 59(34): 10967–10972.
36. Ogbekene Y F, Shukla P, Zhang Y, et al. Laser cleaning of grey cast iron automotive brake disc: Rust removal and improvement in surface integrity [J]. 2018.
37. Liu W, Ji D, Wang W, et al. Effect of laser cleaning on adhesive properties of carbon fiber composite materials [J]. Modern Physics Letters B, 2023: 2450045.
38. Yoo H J, Baek S, Kim J H, et al. Effect of laser surface cleaning of corroded 304L stainless steel on microstructure and mechanical properties [J]. Journal of Materials Research and Technology, 2022, 16: 373–385.
39. Gao Q, Li Y, Wang H, et al. Effect of scanning speed with UV laser cleaning on adhesive bonding tensile properties of CFRP [J]. Applied Composite Materials, 2019, 26: 1087–1099.
40. Lin J, Wang C. A study of surface quality and mechanical strength in 5083 aluminum alloy plates using pulsed laser cleaning [J]. Journal of Mechanical Science and Technology, 2022, 36(10): 5223–5229.
41. Lu Y, Yang L J, Wang Y, et al. Paint removal on the 5A06 aluminum alloy using a continuous wave fiber laser [J]. Coatings, 2019, 9(8): 488.
42. Lu Y, Ding Y, Wang M L, et al. A characterization of laser cleaning painting layer from steel surface based on thermodynamic model [J]. The International Journal of Advanced Manufacturing Technology, 2021, 116: 1989–2002.
43. Zhu G, Wang S, Cheng W, et al. Corrosion and wear performance of aircraft skin after laser cleaning [J]. Optics & Laser Technology, 2020, 132: 106475.
44. Shan T, Yin F, Wang S, et al. Surface integrity control of laser cleaning of an aluminum alloy surface paint layer [J]. Applied Optics, 2020, 59(30): 9313–9319.
45. Chen T, Wang W, Tao T, et al. Multi-scale micro-nano structures prepared by laser cleaning assisted laser ablation for broadband ultralow reflectivity silicon surfaces in ambient air. Applied Surface Science, 2020, 509: 145182.
46. Gao K et al. Adhesion of Residual Primer Paint After Laser Cleaning Aircraft Aluminum Alloy Skin. Laser & Optoelectronics Progress, 2021, 58(9): 0914006.
47. Wang X, Xu M, Wang Z, et al. Properties of Jet-Plated Ni Coating on Ti Alloy (Ti6Al4V) with Laser Cleaning Pretreatment. Metals, 2019, 9(2): 248.
48. Sun X, Yu Q, Bai X, et al. Substrate Cleaning Threshold for Various Coated Al Alloys Using a Continuous-Wave Laser//Photonics. MDPI, 2021, 8(9): 395.
49. Wang Z, Zhang L, Jiang J, et al. Mechanism and parametric control of laser-layered paint removal for aircraft skin. Optics & Laser Technology, 2023, 165: 109612.
50. Wen J, Li Y, Fan J, et al. Pulsed laser cleaning of resin-based surface coating on the titanium alloy substrate. AIP Advances, 2021, 11(11): 115212.
51. Maharjan N, Zhou W, Zhou Y, et al. Ablation morphology and ablation threshold of Ti-6Al-4V alloy during femtosecond laser processing. Applied Physics A, 2018, 124(8): 1–10.
52. Seo C, Ahn D, Kim D. Removal of oxides from copper surface using femtosecond and nanosecond pulsed lasers. Applied Surface Science, 2015, 349: 361–367.
53. Zhu G, Xu Z, Jin Y, et al. Mechanism and application of laser cleaning: A review [J]. Optics and Lasers in Engineering, 2022, 157: 107130.
54. Dausinger F, Hugel H, Konov V I. Micromachining with ultrashort laser pulses: from basic understanding to technical applications[C]//ALT'02 International Conference on Advanced Laser Technologies. SPIE, 2003, 5147: 106–115.

55. Zhou C, Li H, Chen G, et al. Effect of single pulsed picosecond and 100 nanosecond laser cleaning on surface morphology and welding quality of aluminium alloy. Optics & Laser Technology, 2020, 127: 106197.
56. Seo Y, Son S, Lee D. A study of laser cleaning to remove by-products occurring after arc welding. Materials Chemistry and Physics, 2022: 126375.
57. Ling C, et al. Research and mechanism study on refurbishing stainless bracket using picosecond laser based on laser cleaning technology. Chinese Journal of Lasers 41.4 (2014): 403005-1.
58. Boynukara C Y, Uguryol M, Mavili G, et al. An investigation of the cleaning performances of femtosecond and nanosecond laser pulses for artificially soiled papers with sizing. Applied Physics A, 2021, 127(4): 1–10.
59. Wang Y et al. Laser cleaning and protection of industrial heritage steel structure. Journal of Shenzhen University: Science and Engineering Edition, 2020, 37(5): 466–473.
60. Bi S, Lei Z, Feng J. Comparison and Improvement of Millisecond Pulsed Laser and Nanosecond Pulsed Laser for Cleaning to Remove Paint Coating from Aluminum Alloy Surface. Available at SSRN 4229851.
61. Peng W, Lei F, et al. Effect of Energy Density on Surface Morphology and Properties of H13 Mold Steel via Picosecond Laser Cleaning. Laser & Optoelectronics Progress, 2021, 58(17): 1714002.
62. Zhu L, Gao Q, Sun B, et al. Nanosecond laser cleaning for enhanced zinc coating quality of HSLA steel. Optics & Laser Technology, 2021, 143: 107311.
63. Song Y, Zhang T, Fan W, et al. Effect of the overlap ratio on surface properties of 7B04 aluminum alloy for aviation during laser derusting. Journal of Materials Research and Technology, 2022, 20: 1495–1511.
64. Zhang G, Hua X, Li F, et al. Effect of laser cleaning process parameters on the surface roughness of 5754-grade aluminum alloy. The International Journal of Advanced Manufacturing Technology, 2019, 105(5): 2481–2490.
65. Xia P, Yin Y, Cai A, et al. Laser Cleaning Process of 2219 Aluminum Alloy Anodic Oxide Film Before Welding. Chinese Journal of Lasers, 2019, 46(1): 102005--1.
66. Kim J E, Lee J M, Hyun J H, et al. A Study on the Laser Removal of Epoxy Coatings on SS400 Surface by Beam Scanning Patterns. Coatings, 2021, 11(12): 1510.
67. Di Francia E, Lahoz R, Neff D, et al. Laser-cleaning effects induced on different types of bronze archaeological corrosion products: Chemical-physical surface characterisation. Applied Surface Science, 2022, 573: 150884.
68. Sun Q, Zhou J Z, Meng X K, et al. Mechanism and threshold fluence of nanosecond pulsed laser paint removal [J]. Rare Metals, 2022: 1–10.
69. Abdul Razab M K A, Jaafar M S, Rahman A A, et al. Estimation of threshold fluence, absorption coefficient and thermal loading of car coated substrate in laser paint removal [J]. applied mechanics and materials, 2014, 554: 439–443.
70. Greifzu G, Kahl T, Herrmann M, et al. Laser-based decontamination of metal surfaces. Optics & Laser Technology, 2019, 117: 293–298.
71. Mateo M P, Nicolas G, Piñon V, et al. Laser cleaning: an alternative method for removing oil-spill fuel residues [J]. Applied surface science, 2005, 247(1–4): 333–339.
72. Tong Y, ShangGuan J, Ren X, et al. Surface quality of laser paint removal of marine steel: a comparative study using a Gaussian beam and a flat-top beam. Applied Optics, 2022, 61(9): 2237–2246.
73. Shao J, Zhang R, Dong H, et al. The energy density threshold evaluation for high-precision laser cleaning of the oxidized layer on 7075 aluminum alloy with 100 ns laser [J]. The International Journal of Advanced Manufacturing Technology, 2022, 120(3–4): 1901–1907.
74. Wang W, Shen J, Liu W, et al. Effect of laser energy density on surface physical characteristics and corrosion resistance of 7075 aluminum alloy in laser cleaning. Optics & Laser Technology, 2022, 148: 107742.
75. Li Z, Zhang D, Su X, et al. Removal mechanism of surface cleaning on TA15 titanium alloy using nanosecond pulsed laser. Optics & Laser Technology, 2021, 139: 106998.

76. Zhang D, Xu J, Li Z, et al. Removal mechanism of blue paint on aluminum alloy substrate during surface cleaning using nanosecond pulsed laser. Optics & Laser Technology, 2022, 149: 107882.
77. Liu Y, Liu W, Zhang D, et al. Experimental investigations into cleaning mechanism of ship shell plant surface involved in dry laser cleaning by controlling laser power. Applied Physics A, 2020, 126(11): 1–17.
78. Zhao H Chao, Qiao Y Lin, Du Xian, et al. Effect and Mechanism of Energy Density on the Aluminum Alloy Paint Cleaned by Laser. Laser & Optoelectronics Progress, 2020, 57(13): 131403.
79. Xie D, Chen H, Yin S, et al. Effect of different laser energy densities on the corrosion resistance of aluminum alloys [J]. Anti-Corrosion Methods and Materials, 2020, 67(5): 437–444.
80. Li X, Qin Y, Tian J, et al. Simulation and experimental research on laser cleaning of glass insulators [J]. Optics & Laser Technology, 2024, 170: 110198.
81. Wang J, Yang Y, Qi J, et al. Thermodynamic simulation, surface morphology and bending property of carbon fiber reinforced polymer composite material subjected to laser cleaning [J]. Optics & Laser Technology, 2022, 152: 108099.
82. Peng Z, Yin J, Cui Y, et al. Numerical and experimental investigation on pulsed nanosecond laser ablation processing of aluminum alloy [J]. Journal of Materials Research and Technology, 2022, 19: 4708–4720.
83. Ma Z, Li X, Tian J, et al. Temperature Field Distribution Characteristics of Ceramic Insulator during Laser Cleaning//2021 3rd Asia Energy and Electrical Engineering Symposium (AEEES). IEEE, 2021: 179–183.
84. Marimuthu S, Mhich A, Molchan I S, et al. Numerical simulation of excimer laser cleaning of film and particle contaminants. Journal of heat transfer, 2013, 135(12).
85. Yue L, Wang Z, Li L. Modeling and simulation of laser cleaning of tapered micro-slots with different temporal pulses. Optics & Laser Technology, 2013, 45: 533–539.
86. Yang J, Zhou J, Sun Q, et al. Digital analysis and prediction of the topography after pulsed laser paint stripping. Journal of Manufacturing Processes, 2021, 62: 685–694.
87. Ai Sifei, et al. Numerical Simulation and Experimental Study of Laser Cleaning of High Temperature Oxide Layer on TC4 Titanium Alloy Surface. Southwest Jiaotong University,2021.
88. Lu Y, Yang L J, Wang M L, et al. Improved thermal stress model and its application in ultraviolet nanosecond laser cleaning of paint [J]. Applied Optics, 2020, 59(25): 7652–7659.
89. Dong H, Li J, Li Y, et al. Thermal stress analysis of laser cleaning of aluminum alloy oxide film [J]. Applied Optics, 2023, 62(29): 7805–7811.
90. Lu Y, Yang L J, Wang M L, et al. Simulation of nanosecond laser cleaning the paint based on the thermal stress. Optik, 2021, 227: 165589.
91. Guo L, Li Y, Geng S, et al. Numerical and experimental analysis for morphology evolution of 6061 aluminum alloy during nanosecond pulsed laser cleaning. Surface and Coatings Technology, 2022: 128056.
92. Yuan D, et al. The Theory and the Influential Factors of Laser Induced Breakdown Spectroscopy. Spectroscopy and Spectral Analysis, 2008, 28(9): 2019–2023.
93. Wang W, Sun L, Lu Y, et al. Laser induced breakdown spectroscopy online monitoring of laser cleaning quality on carbon fiber reinforced plastic. Optics & Laser Technology, 2022, 145: 107481.
94. Yang W, Qian Z, Cao Y, et al. LIBS Monitoring and Analysis of Laser-Based Layered Controlled Paint Removal from Aircraft Skin. Journal of Spectroscopy, 2021, 2021.
95. Zhou Q, Deng G, Chen Y, et al. Laser paint removal monitoring based on time-resolved spectroscopy. Applied Optics, 2019, 58(34): 9421–9425.
96. Tserevelakis G J, Pozo-Antonio J S, Siozos P, et al. On-line photoacoustic monitoring of laser cleaning on stone: Evaluation of cleaning effectiveness and detection of potential damage to the substrate. Journal of Cultural Heritage, 2019, 35: 108–115.

97. Zou W, Song F, Luo Y. Characteristics of audible acoustic signal in the process of laser cleaning of paint on metal surface. Optics & Laser Technology, 2021, 144: 107388.
98. Chen Y, Deng G L, Zhou Q, et al. Acoustic signal monitoring in laser paint cleaning. Laser Physics, 2020, 30(6): 066001.
99. Papanikolaou A, Tserevelakis G J, Melessanaki K, et al. Development of a hybrid photoacoustic and optical monitoring system for the study of laser ablation processes upon the removal of encrustation from stonework. Opto-Electronic Advances, 2020, 3(2): 190037-1–190037-11.
100. Xie X, Huang Q, Long J, et al. A new monitoring method for metal rust removal states in pulsed laser derusting via acoustic emission techniques. Journal of Materials Processing Technology, 2020, 275: 116321.

# Chapter 6
# Acoustic-Optical-Thermal Dynamic Monitoring of Surface-Interface in Laser Cleaning

## 6.1 Introduction

As a non-invasive and non-contact cleaning technology, laser cleaning technology can efficiently clean the surface of materials without using harmful chemicals, and has little impact on the environment, so it has been widely concerned and developed rapidly [1]. However, the laser beam is a high-energy beam, and the laser cleaning of the interface is bound to have a complex thermal-mechanical coupling effect. The real-time monitoring of the cleanliness and integrity of the substrate surface during the laser scanning process is very important to ensure the effective cleaning of the matrix surface and avoid excessive cleaning damage to the substrate [2]. The research on accurate dynamic monitoring of the laser cleaning interface has become a hot issue widely used in laser cleaning technology [3].

In the process of laser cleaning, it involves overcoming the forces between the substrate and pollutants (such as van der Waals force, electrostatic force, capillary force, etc.) [4] and using the photothermal, photochemical and opto-mechanical effects between the laser and the material to efficiently remove the dirt on the surface of the material. The laser beam interacts with the surface of the material, and the laser energy is absorbed by the material, which leads to the rapid increase of local temperature, causing thermal expansion and producing phenomena such as ablation, vibration and plasma shock wave (Fig. 6.1a). These phenomena are accompanied by the generation of physical signals such as sound waves, spectra, images, and thermal radiation, which can be detected by corresponding sensors and converted into analyzable data [5]. This provides crucial information for monitoring and assessing the cleaning process (Fig. 6.1b). The function of online monitoring technology is to quickly collect the signals generated in the cleaning process and feed them back in real time, so as to achieving precise control to prevent over-cleaning and potential damage, while ensuring the safety and effectiveness of the cleaning operation. Therefore, this technology plays a significant role in various fields, including the cleaning of artworks [6] and cultural relics [7], the maintenance



S. Wang, *Laser Cleaning*, https://doi.org/10.1007/978-981-95-2935-3_6

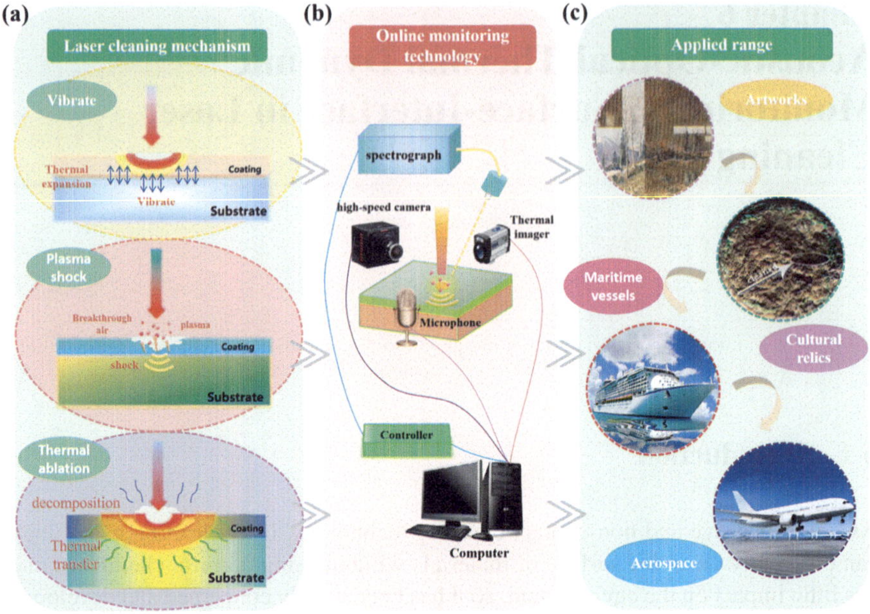


**Fig. 6.1** Schematic diagram of laser cleaning online monitoring and its application field (**a**) Laser cleaning mechanism (**b**) Online monitoring technology (**c**) Applied range

of aerospace components [8], and the surface treatment of maritime vessels [9] (Fig. 6.1c).

Compared with traditional offline monitoring, online monitoring technology has obvious advantages. First of all, online monitoring can provide real-time data feedback of surface cleanliness and substrate integrity in the cleaning process. This real-time ensures the continuity of cleaning operation, can quickly identify and solve potential problems, and avoids the problem that offline monitoring needs to be detected after cleaning stops [10]. Secondly, online monitoring can help operators to adjust the cleaning parameters in real time, and prevent the substrate from being damaged by excessive cleaning, thus improving the cleaning efficiency and quality. In addition, online monitoring technology can also integrate a variety of monitoring means, such as sound wave, spectrum, image recognition and thermal radiation, which can comprehensively evaluate the surface condition and ensure the stability and consistency of the cleaning process. These advantages make online monitoring widely used in laser cleaning technology, which greatly improves the control accuracy and reliability of cleaning process.

How to extract the physical signal of laser cleaning interface in time and help to judge the cleanliness and integrity of the surface is an important topic in the field of laser cleaning. It is of significance for adjust the laser cleaning process, avoid substrate damage, maintain the consistency of cleaning operation, and then improve the cleaning efficiency and quality. This paper aims to review the latest research progress of the dynamic monitoring technology of laser cleaning interface, and deeply explore the properties of wave vibration, spectroscopy, image recognition and thermal radiation and their typical applications in the field of laser cleaning. By analyzing the principles, advantages and challenges of these monitoring technologies, the feasibility of dynamic monitoring technology to assist laser cleaning coherent operation is explored. In addition, this paper also prospects the future development trend of laser cleaning monitoring technology, and predicts the potential of integrating multi-monitoring means and intelligent algorithm in improving the control accuracy of cleaning process control, in order to provide reference and guidance for the research and application of laser cleaning monitoring technology.

## 6.2 Wave Vibration Monitoring Technology

Laser cleaning essentially involves overcoming the forces between the substrate and the contaminants. During the process of overcoming these forces, a series of phenomena such as ablation, vibration, and plasma shock will occur [11]. The principle of wave vibration generation is that when a laser beam irradiates the surface of a material, the energy of the laser is absorbed by the material. As a result of the interaction between the laser photons and the electrons within the material, the electrons are excited from a lower energy level to a higher one. The laser energy is absorbed by the surface or the near-surface region of the material, leading to a rapid increase in temperature of that area. The increase in temperature causes thermal expansion of the material at the area irradiated by the laser. Since the thermal expansion is instantaneous, it generates a pressure wave within the material [12], The mechanical waves generated by the accelerated motion of internal particles under the influence of pressure propagate within the material in the form of acoustic wave vibrations due to rapid changes in pressure. Similarly, there are vibrational sound waves and plasma shock waves generated during the cleaning process that attenuate as they propagate and become impact sound waves [13], These sound waves spread from the point of laser irradiation to the surrounding medium (such as air), producing a multitude of acoustic signals. Researchers collect these acoustic signals and transform them into understandable spectrograms. By finding out the correlation between the spectrogram and the cleaning progress, the interface monitoring of the laser cleaning process is realized. To put it simply, the monitoring of acoustic signals involves the fact that when laser irradiation is applied to materials with different optical sensitivities, the resulting acoustic wave vibrations will change accordingly. Therefore, by analyzing the spectrum of the acoustic signals, it is possible to identify different stages of the cleaning process, predict the cleaning effect, determine the onset of substrate damage, and assess the surface smoothness of the substrate [2].

Generally, during the laser shock cleaning process, when the laser beam is focused above the cleaning surface, plasma is produced through laser-induced air breakdown, which is followed by the generation of a strong shock wave. Audible sound waves are produced in this process, which can be detected by acoustic sensors, such as microphones. The acquired acoustic emission signals are displayed digitally on a computer and are subjected to Fast Fourier Transform (FFT) analysis. Through this analysis, researchers can evaluate the frequency characteristics of the sound waves and their relationships with changes in laser power density, type of gas, surface flatness, number of pulses, type of contaminant, and material type [14–18].

Zou et al. [17] employed time-amplitude (TA), frequency-amplitude (FA), and time-frequency amplitude (TF) analysis methods to study the characteristics of audible sound signals generated during the laser cleaning of paint under different laser energy levels (Fig. 6.2). In the experiment, the audible sound signals were received by a mobile phone placed 8 cm to the left of the sample, and the acoustic signals were collected using the voice memo function of the phone (Fig. 6.3). The results indicated that the TA spectrum showed poor correlation of the audible sound signals. In the FA spectrum, the frequency corresponding to the maximum amplitude was 0.04 kHz when the cleaning threshold was reached or exceeded. The critical number of laser pulses required for paint removal without damage could be obtained from the number of peaks in the TA spectrum. Chen et al. [18] found that the characteristics of the sound signals during laser paint removal are closely related to the state of laser paint stripping. The amplitude and local standard deviation (LSD) of the acoustic signals gradually decrease and tend to stabilize as the paint is removed, which can more timely reflect the state of laser paint removal compared to the TA spectrum (Fig. 6.4e). They also pointed out that the amplitude of the sound waves is related to the smoothness of the cleaned sample surface. In the experiment, the spectral fluctuations between the first pulse and the ninth and tenth pulses were small (Fig. 6.4). This is because the first pulse of the laser irradiated a smooth interface of the paint layer, and by the ninth and tenth pulses, the paint layer had been cleaned, revealing a smoother substrate surface. Additionally, the experiment compared the laser-induced breakdown spectroscopy (LIBS) method for monitoring and found that the monitoring based on sound signals has a wider dynamic range. Dimitroulaki et al. [19] proposed a real-time monitoring technology based on photoacoustic (PA) signals to accurately track the process of ultraviolet laser removal of aged varnish coatings on cultural relics. By using a non-contact air-coupled ultrasonic transducer, they captured the acoustic signals generated by the interaction of the laser with the material, thereby enabling real-time monitoring of the laser stripping process. The study successfully determined the critical laser pulse for complete varnish removal by analyzing the time delay and amplitude variation of the PA signals (Fig. 6.5). This technology has shown good adaptability and accuracy in experiments with varnish films on different substrate materials, laying the foundation for future applications on more complex substrate structures.

During the pulsed laser cleaning process, the propagation of acoustic signals is affected by the distance between the sensor and the sound source. If the distance is too great, the signal amplitude may be reduced due to the influence of the surrounding environment and the propagation distance. Conversely, if the distance is too close, particles produced during the cleaning process can interfere with the signal. To ensure the accuracy of the vibration signals during the laser cleaning process, Chen et al. [20] constructed an acoustic monitoring experimental platform and

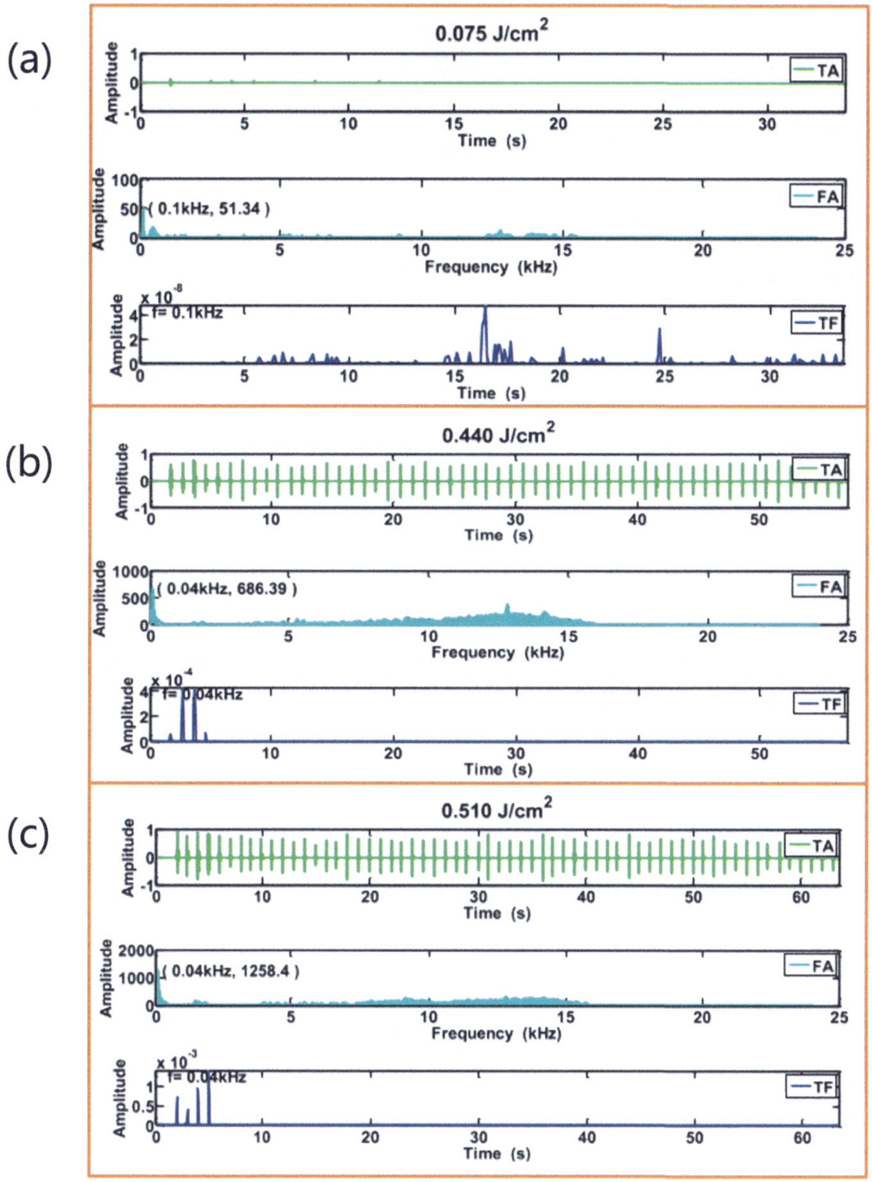


**Fig. 6.2** TA, FA and TF spectra at different powers. (**a**) Power below the wash threshold; (**b**) Power of the wash threshold; (**c**) Power above the wash threshold [17]

plotted the relationship curve between the detection distance and the signal amplitude (Fig. 6.6). They analyzed the acoustic signals generated during the laser paint removal process and extracted key characteristic parameters such as the positive pressure peak value and the duration of the pulse waveform. The results indicate

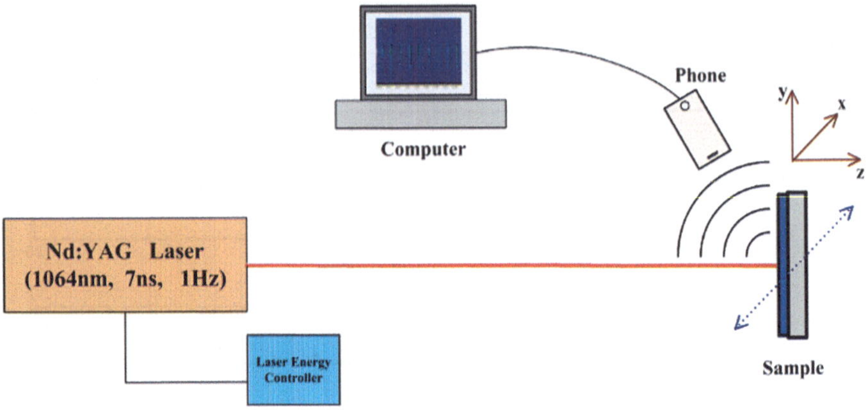


**Fig. 6.3** Schematic diagram of the acoustic monitoring experimental device during pulsed laser paint removal [17]

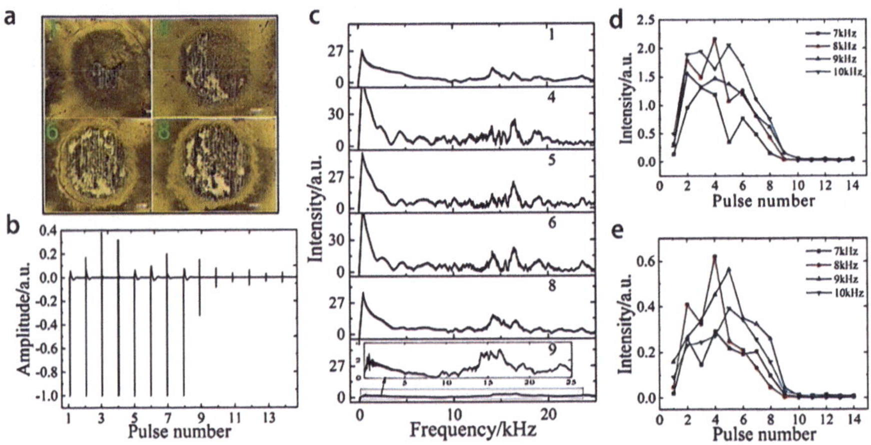


**Fig. 6.4** Surface morphology and acoustic signals under different pulse conditions. (**a**) Surface morphology of paint layer after the 1st, 4th, 6th and 8th laser pulse (45 mJ); (**b**) Time-amplitude acoustic signal changes with the number of pulses; (**c**) Frequency-amplitude waveform changes with the number of pulses (The number of pulses is from top to bottom is: 1, 4–6, 8, 9); (**d**) Signal intensity changes with the number of pulses at different frequencies; (**e**) The LSD changes with the number of pulses at different frequencies [18]

that the energy and frequency of the laser have a significant impact on the depaint sound signals. As the laser energy density increases, the intensity of the sound signals also strengthens. Acoustic monitoring technology can effectively extract characteristic information during the cleaning process (Fig. 6.7). With the established discriminant function, real-time monitoring of the laser paint removal process can be achieved, and the experimental parameters are shown in Table 6.1. Facing the issue that most microphones and other acoustic signal collection devices cannot

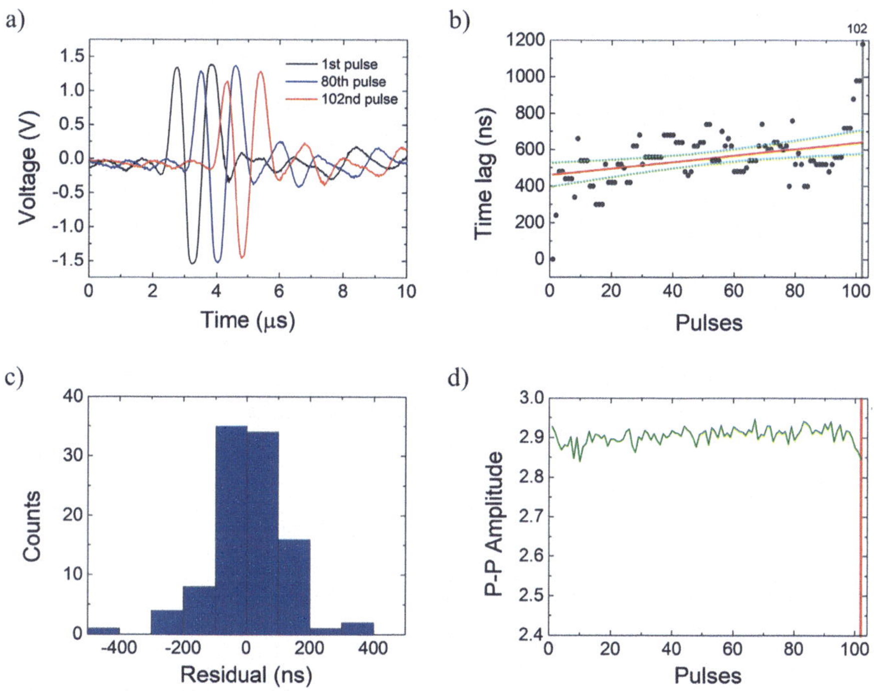


**Fig. 6.5** The film on quartz was irradiated with F = 0.8 J/cm$^2$ laser flux. (**a**) Recorded PA signals corresponding to the 1st, 80th and 102nd laser ablation pulses; (**b**) The time-lag product of each PA waveform generated by the subsequent laser pulse with the maximum cross-correlation of the first pulse PA signal; (**c**) Residuals of the time-delayed data points; (**d**) Peak-to-peak amplitude response [19]

capture ultrasonic frequency sound wave signals, Tserevelakis et al. [21] used a piezoelectric ultrasonic transducer that was in direct contact with the cleaning sample to convert the vibrations generated during the laser cleaning of black graffiti on marble into electrical signals. Although this method has a wide enough frequency range to capture the sound wave signals, the detected electrical signals are usually very weak and need to be amplified by a low-noise amplifier and recorded by a high-speed oscilloscope, as shown in Fig. 6.8b. The Hilbert transform is applied to each pulse time-domain waveform to obtain the normalized average amplitude of the acoustic signal, as shown in Fig. 6.8c. This monitoring method was further validated for its high level of accuracy in predicting the cleaning level by comparing it with traditional imaging analysis techniques such as scanning electron microscopy and confocal microscopy. Additionally, the optical absorption characteristics of the studied samples were analyzed using a spectrophotometer, further confirming the reliability of this method. Papanikolaou et al. [22] optimized the cleaning effect and avoided the discoloration issue on the stone surface by using a combination of infrared and ultraviolet laser cleaning. They proposed a photoacoustic-optical monitoring system that can track high-frequency acoustic waves generated during the laser

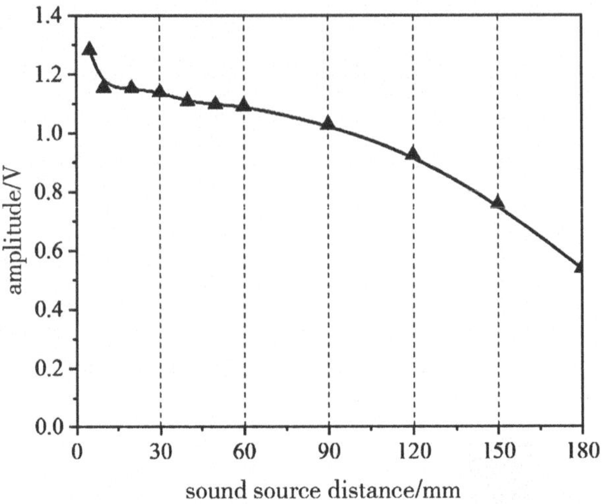


**Fig. 6.6** Acoustic amplitude and sound source distance curve [20]

ablation process in real-time and, in conjunction with high-resolution optical images (Fig. 6.9), accurately monitor the removal process of the crust on the stone surface and assess the cleaning progress and the extent of damage to the stone. The results showed that the temperature increase caused by the infrared (IR) laser was approximately twice that of the ultraviolet (UV) laser. Additionally, it was found in the experiment that there is a good linear relationship between the intensity of the photoacoustic (PA) signals and the temperature increase on the material surface caused by laser irradiation. Xie et al. [23] utilized Acoustic Emission (AE) technology to monitor the state of pulsed laser rust removal. By analyzing the time-domain and frequency-domain characteristics of the AE signals, they revealed the dynamic mechanism of the interaction between the laser and the rust layer. The experiment also integrated with a high-speed camera to further validate the reliability of the AE technology. The results showed that spectral analysis of the Root Mean Square (RMS) signal, which is used to describe the vibration characteristics after filtering through Fast Fourier Transform (FFT), decreases with the accumulation of rust removal effects and can effectively characterize the effectiveness of rust layer removal (Fig. 6.10).

Table 6.2 summarizes some laser cleaning acoustic monitoring experiments and their conclusions, showing that significant progress has been made in the field of acoustic signal monitoring for laser cleaning. However, there are still some challenges and limitations that exist. For instance, the interpretation of acoustic signals under the influence of high-energy-density lasers may become complex, and the propagation of acoustic signals may be interfered with by ambient noise. Additionally, the characteristics of the acoustic signals may vary for different types of materials and contaminants, which necessitates the development of more precise and adaptable monitoring algorithms and the adjustment of laser cleaning

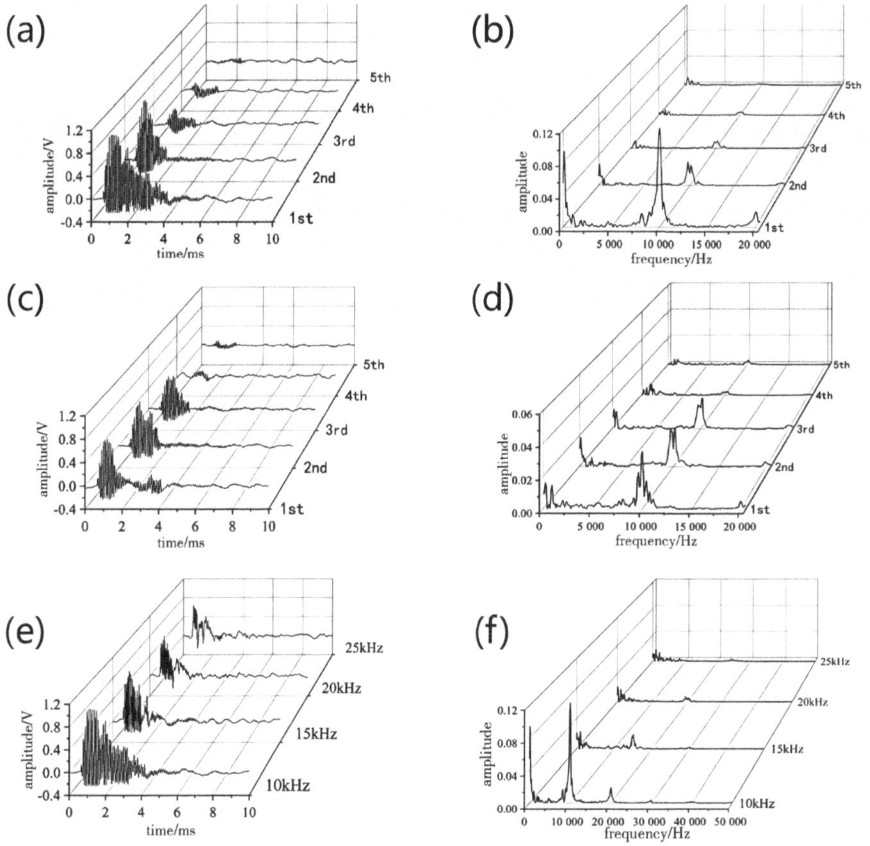


**Fig. 6.7** Characteristics of acoustic signals for laser cleaning with different power and frequency. (**a**) Acoustic signal waveform of 80 W laser paint removal; (**b**) Spectrum of acoustic signal for 80 W laser paint removal; (**c**) Acoustic signal waveform of 60 W laser paint removal; (**d**) Spectrum of acoustic signal for 60 W laser paint removal; (**e**) Acoustic signal waveform of different frequency laser paint removal; (**f**) Spectrum of acoustic signal for different frequency laser paint removal [20]

**Table 6.1** Experimental parameters of laser paint removal [20]

| Experiment | Average power (W) | Frequency (kHz) | Single cleaning laser energy (mJ) | Cleaning times |
|---|---|---|---|---|
| 1 | 100 | 10 | 500 | 5 |
| | 80 | 10 | 400 | 5 |
| | 60 | 10 | 300 | 5 |
| 2 | 100 | 15 | 333.3 | 1 |
| | 100 | 20 | 250 | 1 |
| | 100 | 25 | 200 | 1 |

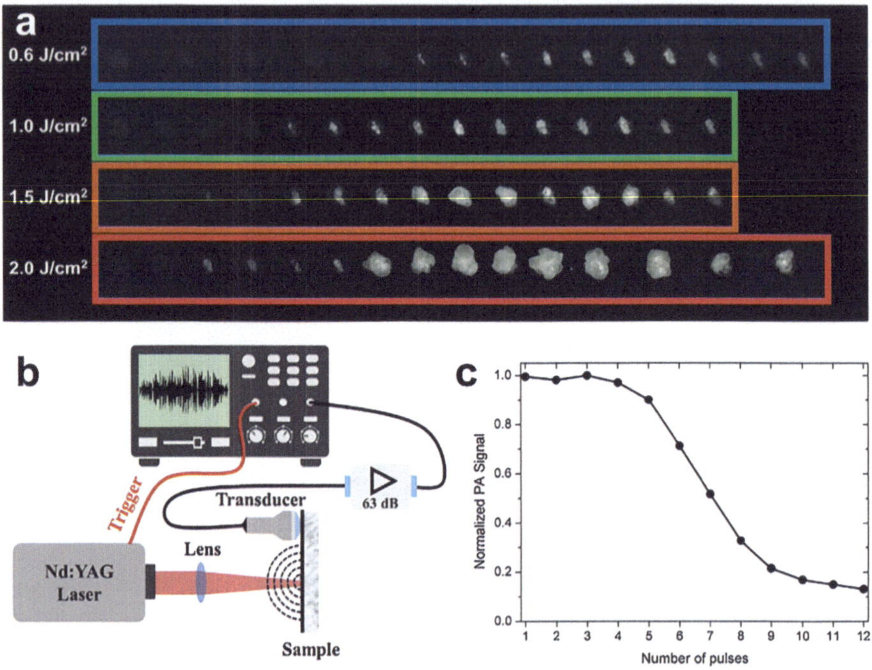


**Fig. 6.8** Online Photoacoustic Monitoring of Laser Cleaning of Black Graffiti on Marble. (**a**) Laser-assisted removal of black graffiti under different energy fluxes, with the number of applied pulses gradually increasing (1–15 pulses from left to right); (**b**) Experimental setup during the photoacoustic measurement process; (**c**) Typical evolution of normalized photoacoustic signals as a function of the number of pulses applied at 1.5 J/cm² [21]

parameters by researchers. Future studies will focus on enhancing the sensitivity and accuracy of acoustic signal monitoring techniques, as well as the development of more intelligent monitoring systems. By integrating advanced signal processing algorithms, machine learning, and artificial intelligence technologies, the predictive and control capabilities of acoustic signal monitoring can be further improved. Moreover, the development of non-contact sensors and automated monitoring systems will make the application of acoustic signal monitoring technology in laser cleaning more efficient and reliable.

## 6.3 Spectral Monitoring Technology

The high noise generated during high-power laser cleaning processes makes it difficult for some traditional monitoring techniques to accurately capture signals in real time [27]. Laser-Induced Breakdown Spectroscopy (LIBS) is an intelligent online monitoring method commonly used in the field of laser cleaning, which can

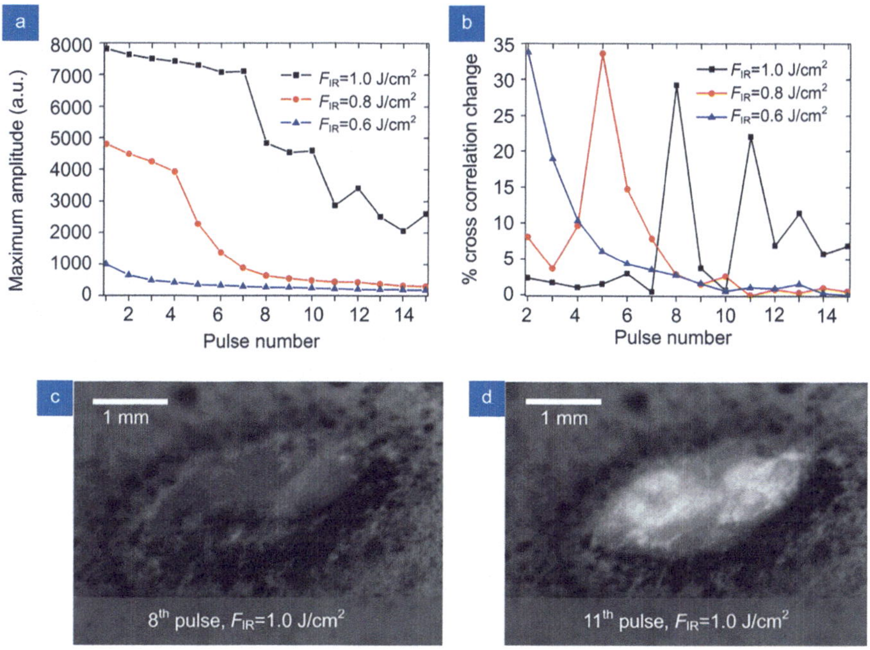


**Fig. 6.9** Application of Acoustic-Optical system in laser cleaning process. (**a**) Maximum amplitude values change with pulse number under different energy densities; (**b**) Percentage change in maximum amplitude; (**c**) Optical image corresponding to the 8th laser pulse; (**d**) Optical image corresponding to the 11th laser pulse [22]

be rapid, non-destructive or microdamaged analysis, enabling element detection of various forms, including solids, liquids and gases, suitable for complex samples, such as coatings, films and composite materials. Its principle is through the high energy pulse laser induced breakdown on the sample surface to form plasma. After the pulse ends, when the electrons in the plasma transition from a higher to a lower energy level, they emit spectral lines at specific wavelengths. By analyzing the light emitted by the plasma with discrete spectral characteristics using a spectrometer, and based on the changes in the LIBS characteristic peaks of different paint layer elements on the sample surface, the cleaning process and effect can be assessed, as shown in Fig. 6.11.

During the process of removing multi-layer coatings from aircraft skin, Yang et al. [28] monitored the removal of paint from the surface of aluminum alloy under different pulse actions using LIBS. They found that the appearance and disappearance of the characteristic peak of Ca I (422.7 nm) could serve as a basis for determining the removal of the topcoat and primer, respectively, while the appearance of the Al I (394.5 and 396.2 nm) characteristic peak indicated that the laser had begun to act on the substrate (Fig. 6.12). This work demonstrates the potential of LIBS in real-time monitoring of the laser cleaning effect and ensuring that the substrate is not damaged, providing an efficient quality control method for aviation

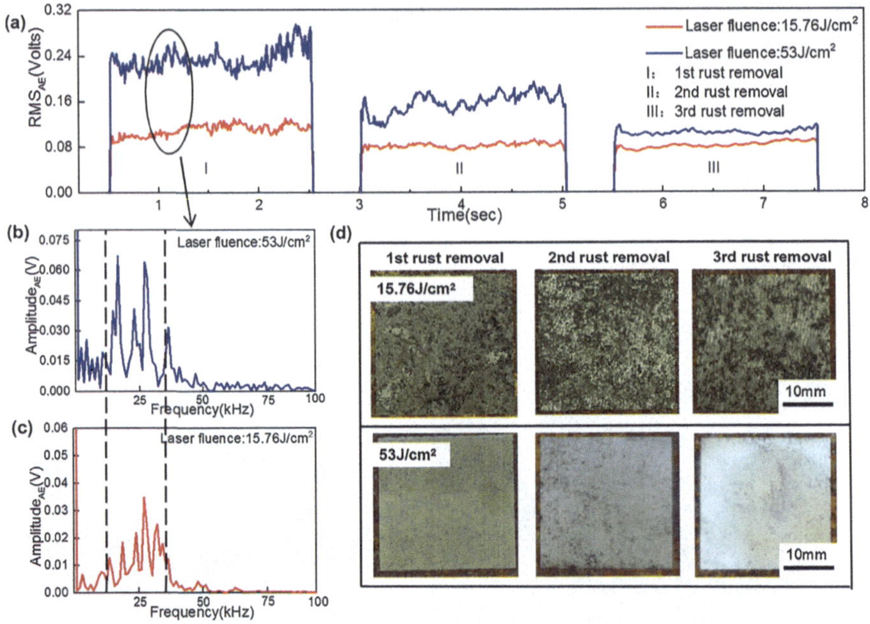


**Fig. 6.10** Analysis of RMS signal characteristics and effects in laser rust removal process. (**a**) Comparison of RMS signal characteristics during laser rust removal; Comparison within the 15 ~ 35 kHz frequency band: (**b**) Laser flux of 53 J/cm$^2$, (**c**) Laser flux of 15.76 J/cm$^2$, (**d**) Changes in the residual rust layer during different laser rust removal processes at 15.76 and 53 J/cm$^2$ [23]

maintenance. When collecting LIBS (Laser-Induced Breakdown Spectroscopy) spectra during laser cleaning processes in an atmospheric environment, it is inevitable that characteristic spectral lines of air will be introduced. In another study, Yang et al. [29] directly induced breakdown in the air and synchronized the exposure time with that of the sample to obtain an air LIBS spectrum. The experiment showed that the air spectral lines were predominantly distributed in the 740–880 nm wavelength range. Subsequently, by referring to the atomic spectral database of the National Institute of Standards and Technology (NIST), they were able to eliminate the impact of the air spectral lines on the analysis of the results. Similarly, Tong et al. [30] considering the constant proportion of oxygen and nitrogen elements in the air, conducted a statistical analysis on the intensity values of the O and N atomic spectral lines measured ten times. They found that the overall distribution pattern and the relative standard deviation remained consistent, indicating that the spectral lines of N and O atoms could be disregarded to effectively evaluate the cleaning process. Voznesenskaya et al. [31] used LIBS technology to monitor the laser cleaning process of aluminum alloys in real-time. The research results indicated that by recording the spectral lines of aluminum (Al) and chromium (Cr), the laser cleaning process could be controlled to ensure that the cleaning was stopped once the anodic oxidation layer was reached.

**Table 6.2** Laser cleaning sound monitoring experiment

| Matrix | Cleaning target | Detection distance | Frequency range | Conclusion | Reference |
|---|---|---|---|---|---|
| Aluminum plate | Paint layer | 4 cm | 20–20 kHz | Undamaged frequency band: 7–10 kHz Damage frequency band: 10–20 kHz | [18] |
| Q235 steel | Rust layer | Direct contact | 125–750 kHz | Undamaged frequency band: 15–35 kHz Undamaged frequency band: 65–85 kHz and 140–160 kHz | [23] |
| Quartz metal | Varnish | A few centimeters | MHz | The amplitude of the paint layer on the quartz substrate decreased by about 15%; on the metal substrate, it decreased by about 19% | [19] |
| Marble slab | Crust simulated layer | Direct contact | 5 MHz | The temperature increase caused by the IR laser is approximately twice that of the UV laser | [22] |
| p-Type bare silicon wafer | Silicon dioxide particles | Non-contact | 10 Hz–15 kHz | The intensity of the acoustic emission is directly proportional to the strength of the shock wave | [14] |
| Aluminum steel copper plates | Paint layer | 8 cm | 20–20 kHz | In the FA spectrum, the paint begins to be removed when the frequency corresponding to the maximum amplitude is 0.04 kHz | [17] |
| Marble slab | Acrylic black paint layer | Direct contact | 5 MHz | The acoustic-optic signal amplitude is highly negatively correlated with the cleaning effect | [21] |
| Aluminum Silicon Copper | Contaminants | Non-contact | 20–20 kHz | When the laser intensity is lower than the material ablation threshold, the acoustic wave amplitude decreases to zero | [15] |

(continued)

**Table 6.2** (continued)

| Matrix | Cleaning target | Detection distance | Frequency range | Conclusion | Reference |
|---|---|---|---|---|---|
| Iron | Paint layer | Non-contact | 8 kHz | Determine whether the paint layer is cleaned according to the change of the spectral peak | [24] |
| Iron | Rust layer | 10 cm | 10–15 kHz | When the power density is higher than the cleaning threshold, no or weak acoustic signal indicates the removal of rust scale | [25] |
| 45# Steel | Black alkyd paint | 5 cm | 20–20,000 Hz | Bayesian discriminant function monitoring is established by classifying variables of acoustic signal spectrum characteristic parameters | [26] |

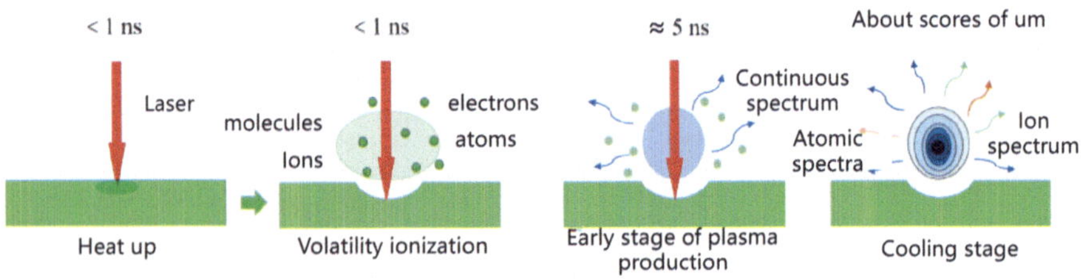


**Fig. 6.11** A schematic diagram of the LIBS technique [70]

Li et al. [32] utilized nanosecond laser for cleaning the oxide layer on the surface of hot-rolled stainless steel and employed LIBS technology for real-time monitoring. By varying the laser power from 20% to 90% of the maximum power, they studied the spectral changes during the cleaning process. Additionally, they characterized the surface before and after cleaning using a Laser Scanning Confocal Microscope (LSCM), X-ray Diffractometer (XRD), and Optical Microscope (OM), as depicted in Fig. 6.13. The results indicate that the Relative Intensity Ratio (RIR) of the emission lines for Fe (520.9 nm) and Cr (589.2 nm) can effectively determine the cleaning status of the oxide layer. When the oxide layer is not completely removed, the RIR remains essentially unchanged; however, once the oxide layer is effectively removed, the RIR will increase significantly. Choi et al. [33] developed two types of two-dimensional mapping methods based on LIBS spectra for monitoring the high-power laser removal of thick paint layers from stainless steel surfaces. The first method involves the use of element distribution maps with specific elemental peak intensities, and the second method employs distribution maps constructed using Pearson's correlation coefficient. By comparing these two types of

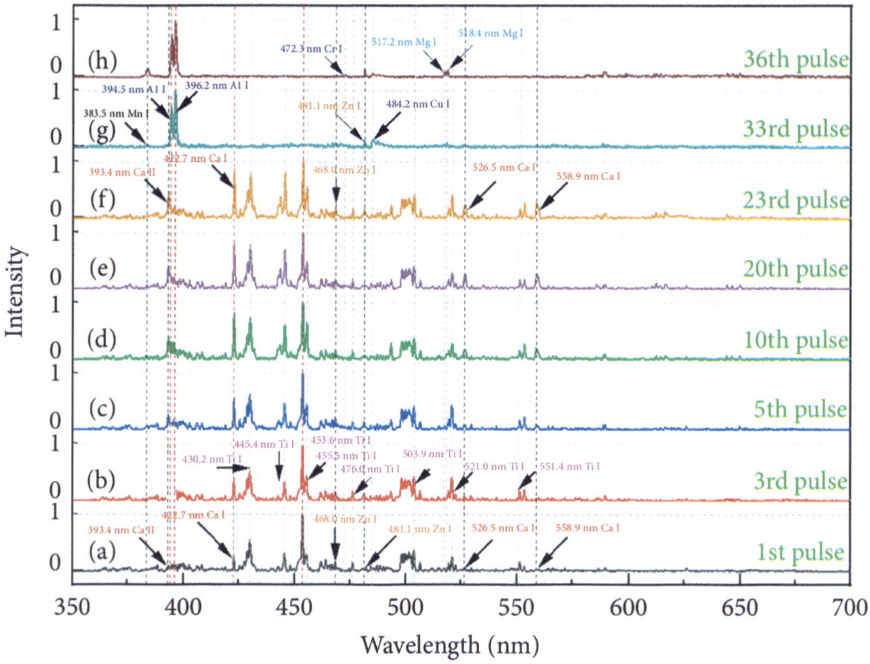


**Fig. 6.12** Laser-induced breakdown spectroscopy signal after different pulse numbers acting vertically on the topcoat surface: (a) 1, (b) 10, (c) 20, (d) 30, (e) 39, (f) 40, (g) 45, (h) 62 [28]

mappings with the results of an Electron Probe Microanalyzer (EPMA) analysis (Fig. 6.14), researchers found that the mapping method based on the correlation coefficient can more accurately reflect the cleaning effect, especially in distinguishing between areas where paint residue remains and areas that have been cleaned.

Wang et al. [34] applied LIBS for real-time monitoring of the quality of laser cleaning carbon fiber reinforced plastic (CFRP). By contrasting and analyzing the differences in elemental composition between the carbon fiber and the epoxy resin layer, as well as the LIBS spectra and other a priori information, they identified that the Na (I) characteristic peak (at a wavelength of 588.819 nm) in the LIBS spectrum of the carbon fiber could serve as a feature indicator for judging the quality of cleaning, as shown in Fig. 6.15a, b, which confirmed the feasibility of the monitoring technique. The experiment also explored a positive correlation between the average power of the laser and the LIBS characteristic spectral intensity at a wavelength of 588.819 nm. Using a piece-wise linear fitting method, they determined the optimal threshold value for the average power of laser cleaning to be 8.2 W. Choi et al. [27] found that it was difficult to distinguish samples with similar chemical compositions in the LIBS spectra, during the laser cleaning monitoring of eight different paint 304L stainless steel samples. Therefore, they developed a method based on a deep learning Convolutional Neural Network (CNN) to enhance the accuracy and speed of LIBS spectral analysis. By combining LIBS with CNN, the CNN model

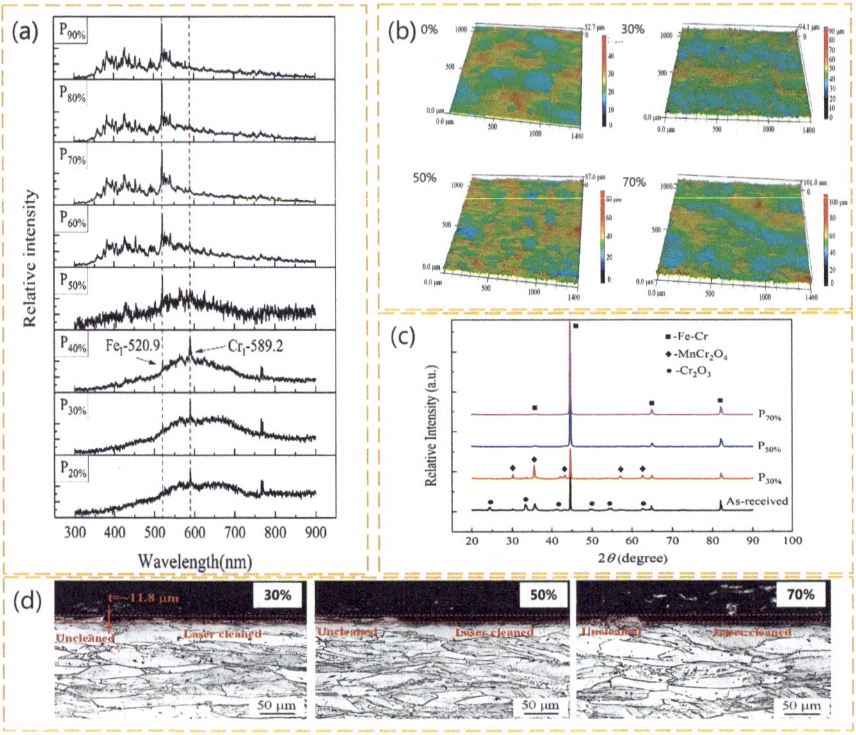


**Fig. 6.13** Spectral changes and surface characteristics during laser cleaning of oxide layer. (**a**) Optical spectra obtained during the laser cleaning of the oxide layer at varied laser powers; (**b**) LSCM images showing the 3D morphology at typical powers; (**c**) XRD patterns of the surface before and after laser cleaning at typical powers; (**d**) OM images of the cross-section of hot-rolled stainless steel after laser cleaning at typical powers [32]

for classifying laser-cleaned samples achieved an accuracy rate of 94.55%, and the LIBS spectral analysis time was reduced to only 0.09 s.

Although LIBS technology has the advantages of small sample damage and wide application range, its spectrum is vulnerable to interference, and vulnerable to the laser energy threshold of density, When the laser energy density is too low, it cannot generate any detectable LIBS signals. Only when the laser energy density reaches a level capable of inducing plasma can the LIBS spectrum be obtained, thus presenting certain limitations in the field of laser cleaning [35]. In response to this issue, when Senesi et al. [36] were cleaning black crusts from limestones using a laser, they combined two pulse lasers, each with an energy of 45 mJ, in a collinear arrangement. Employing a double-pulse (DP) method with a 500 ns delay between the pulses, they irradiated the sample surface sequentially. By comparing this to a single pulse (SP) of 90 mJ, they discovered that the DP yielded a higher intensity of LIBS emission lines (Fig. 6.16). This increase in intensity is due to the fact that the second pulse can reheat the plasma that was initially induced by the first pulse. Similarly,

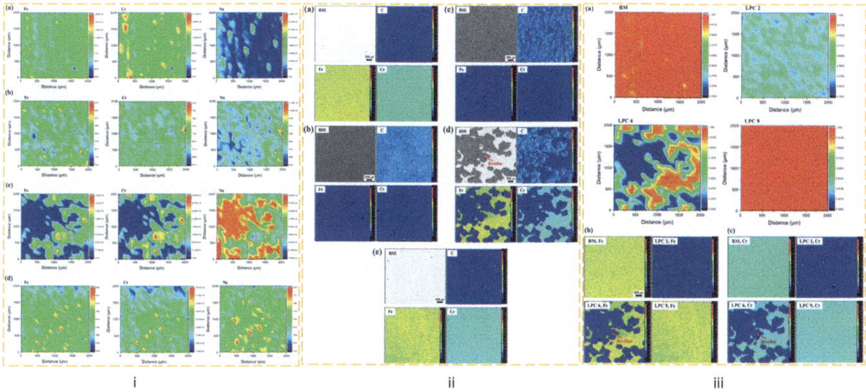


**Fig. 6.14** (**i**) LIBS spectroscopic monitoring and EMPA analysis during laser removal of paint layer (a) LIBS-based 2D correlation coefficient maps using the Pearson's correlation coefficient; (b) Fe elemental distribution maps; (c) Cr elemental distribution maps obtained from the EMPA analysis for the base metal, LPC 2, LPC 6, and LPC 9. (**ii**) BSE images and the corresponding elemental distribution maps obtained from the EMPA analysis (a) base meta; (b) paint; (c) LPC 2; (d) LPC 6; (e) LPC 9. (**iii**) (a) LIBS-based 2D correlation coefficient maps using the Pearson's correlation coefficient; (b) Fe elemental distribution maps; (c) Cr elemental distribution maps obtained from the EMPA analysis for the base metal, LPC 2, LPC 6, and LPC 9 [33]

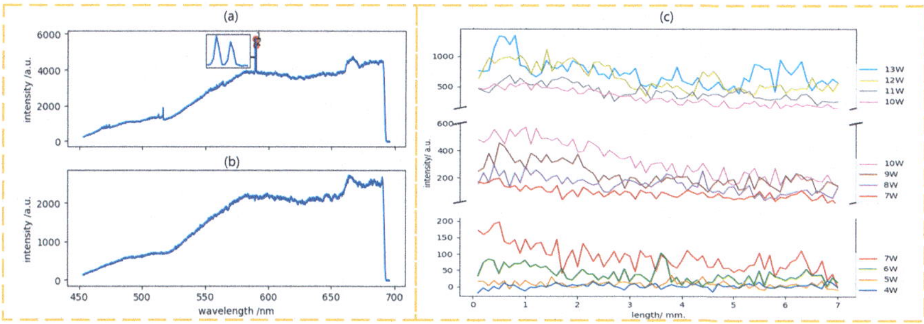


**Fig. 6.15** Analysis of LIBS spectral characteristics and strength variation during cleaning of carbon fiber and epoxy resin layer. (**a**) Reference spectra of the carbon-fiber substrate; (**b**) Reference spectra of the epoxy resin layer; (**c**) LIBS intensity curve in the cleaning area under different average laser power [34]

Skruibis et al. [37] also confirmed this point through experimentation. This approach ingeniously addressed the requirement of LIBS technology for a certain threshold of laser energy.

Gobernado-Mitre et al. [38] enhanced the detection of certain weak signals in LIBS spectra by employing Laser-Induced Fluorescence (LIF) technology, thereby improving the sensitivity of the monitoring. Marimuthu et al. [39] compared the

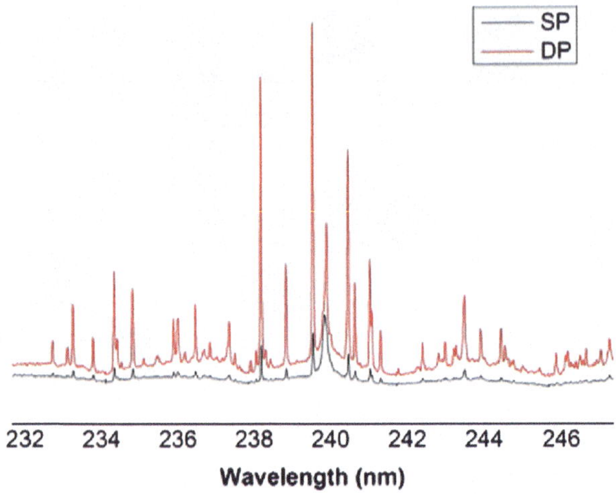


**Fig. 6.16** The SP and DP LIBS emission spectra of the black crusts 232–247 nm band [36]

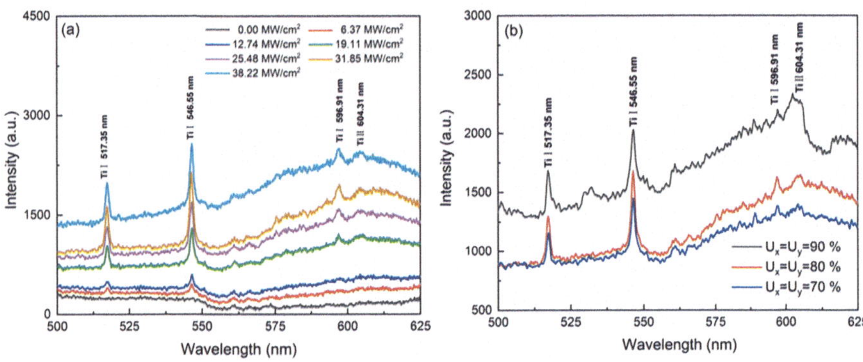


**Fig. 6.17** Spectroscopic analysis of the oxide film on the TA15 titanium alloy. (**a**) Different maximum power densities with an 80% spot overlap ratio; (**b**) different spot overlap ratios with a 31.85 $MW/cm^2$ maximum power density [40]

effectiveness of two monitoring systems through experimentation: Probe Beam Reflection (PBR) and Plume-Emission Spectroscopy (PES). The results indicated that the PBR system could effectively distinguish between cleaned and uncleaned surfaces by changes in the intensity of the reflected light, while the PES system had weaker signals within the range of laser cleaning operations and might require a spectrometer with higher gain and a larger beam size to enhance its performance. Li et al. [40] conducted real-time monitoring of the process of cleaning titanium alloy oxide films with nanosecond pulsed laser by utilizing Optical Emission Spectroscopy (OES) technology. By analyzing the changes in the spectral lines of the titanium element in the OES spectrum, such as the peak intensities of Ti I and Ti II (Fig. 6.17), they were able to promptly optimize cleaning parameters such as power density and spot overlap rate. This allowed them to determine when the oxide film was effectively removed and when cracks and other damages began to appear on the surface.

Spectral monitoring technology plays a vital role in the process of laser cleaning. By analyzing the spectral signals emitted by the plasma generated from the interaction between the laser and the material, this technology can provide detailed information about the material's composition and surface condition. This information is crucial for understanding and controlling the laser cleaning process, as it can reveal the chemical changes and physical states of the material during cleaning. In the application of laser cleaning, spectral technology enables the monitoring of processes such as the removal of paint, rust, and oxide layers, and it also demonstrates its unique value in the field of cultural heritage restoration [41].

However, despite the significant advantages of spectral monitoring technology, it also faces some challenges in practical applications. Firstly, when the laser's energy density is low, it may not effectively excite the plasma, resulting in weak spectral signals that are difficult to analyze accurately. Therefore, careful selection of laser parameters is necessary to ensure sufficient plasma excitation and clear, readable spectral signals. Secondly, spectral monitoring technology typically requires a certain degree of difference in physical or chemical properties between the sample's coating and substrate to facilitate differentiation and identification. If the coating and substrate are too similar in composition, it may cause spectral signal overlap, increasing the complexity of the analysis. Additionally, the cost of spectral detection equipment is relatively high, and it requires a certain level of technical expertise from operators, which to some extent limits its widespread application in the industrial field. To overcome these limitations, researchers are exploring new spectral detection technologies for monitoring, such as Raman spectroscopy and X-ray photoelectron spectroscopy, which can provide more material information and have less stringent requirements for laser energy.

## 6.4 Visual-Image Monitoring Technology

The most intuitive and rapid method in the monitoring process is typically direct observation using devices such as cameras, optical microscopes, and electron microscopes. Affected by environmental conditions or the limitations of macro-detection, it is usually necessary to rely on a visual monitoring system to identify, denoise, gray and other image processing technologies, divide the image into multiple regions and extract useful features such as texture and edge from them, so as to assist researchers in image classification or target recognition according to the extracted features.

Liu et al. [42] developed a laser-based intelligent rust removal system based on machine vision, which uses an industrial camera to capture images of the metal surface and obtain images of rusted areas. The system then employs image processing algorithms to automatically identify rust areas in the images and intelligently selects the appropriate laser cleaning parameters based on the degree of rusting, thereby enhancing the efficiency and quality of rust removal while also reducing health risks to the operators. Lee et al. [16] proposed a surface monitoring technology based on color modulation, which can monitor and diagnose the surface condition in real-time during the laser cleaning process. This technology utilizes the precise configuration of a broadband detector to measure the spectral feature changes of multicolored light, which can be represented by three parameters: Dominant Wavelength (DW), Energy Level (EL), and Excitation Purity (EP). In the experiment, the tungsten wire halogen lamp was focused through an optical element onto the surface to be cleaned. The reflected light was then collected by an optical fiber bundle and detected by three photodetectors after passing through three different glass filters (Fig. 6.18). The spectral responses of the three detectors correspond to the perceptual intensities of Hue, Lightness, and Saturation, respectively. The results show that there is a significant difference in the Hue value between dirty and clean surfaces, while the difference between clean and damaged surfaces is smaller. The Lightness value can clearly distinguish between clean and damaged surfaces because the underlying crystalline marble structure is revealed after cleaning, making the damaged surface much brighter than the clean surface. Shi et al. [43] employed a high-speed camera and a special light source to capture clear images of the surface, and then used a series of image processing techniques, such as illumination removal algorithms, image segmentation, and edge detection, to evaluate the cleaning effect. The experimental results demonstrated that this technology can significantly enhance the efficiency and accuracy of cleaning quality inspection, and it greatly increases the level of automation compared to manual inspection methods.

Based on the differences in physical properties between the substrate and the paint layer in the images, the collected images of the paint layer and particles are first converted into dark-field images, while the substrate is then converted into bright-field images. Subsequently, these color images are transformed into grayscale images to create a grayscale matrix, where each pixel corresponds to a specific grayscale value. By calculating the average grayscale value of all pixels in the grayscale matrix, we can determine the overall grayscale value of the image. Using the

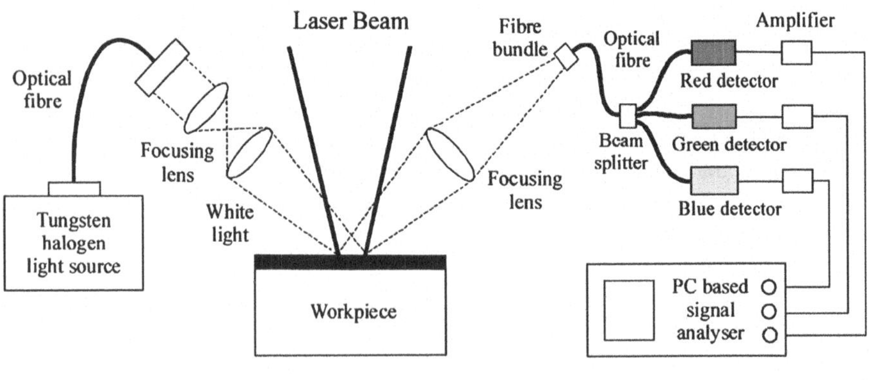


**Fig. 6.18** Schematic of the color monitoring system for surface monitoring and diagnosis during the laser cleaning process [16]

magnitude of this grayscale value, the cleanliness of the sample surface can be assessed. Duan et al. [44] performed grayscale processing on the surface of carbon steel before and after laser cleaning, then calculated the mean value of the grayscale image to quantify the quality of the cleaning effect, as shown in Fig. 6.19. Furthermore, they determined the optimal rust removal process parameters using two statistical methods: range analysis and variance analysis. Li et al. [45] collected images of the surface of rusted workpieces, calculated their Gray Level Co-occurrence Matrix (GLCM) and texture features of the convex-concave areas, then randomly generated laser parameters and used a Particle Swarm Optimization (PSO) supported Vector Machine (SVM) model to predict the cleaning performance. The experimental results showed that the cleaning qualification rate reached 92.5%. The process of extracting image grayscale values is cumbersome and complex. Therefore, Hu et al. [46] proposed a technology for visual monitoring based on deep learning algorithms to automatically evaluate the quality of cleaning and predict the condition of the surface after cleaning. This technology can identify changes in flame during the cleaning process, accurately extract key information from images, and make precise predictions. It demonstrates that the algorithm model, combined with visual image monitoring, can effectively enhance the automation and accuracy of the laser cleaning process. It allows for real-time monitoring and evaluation of

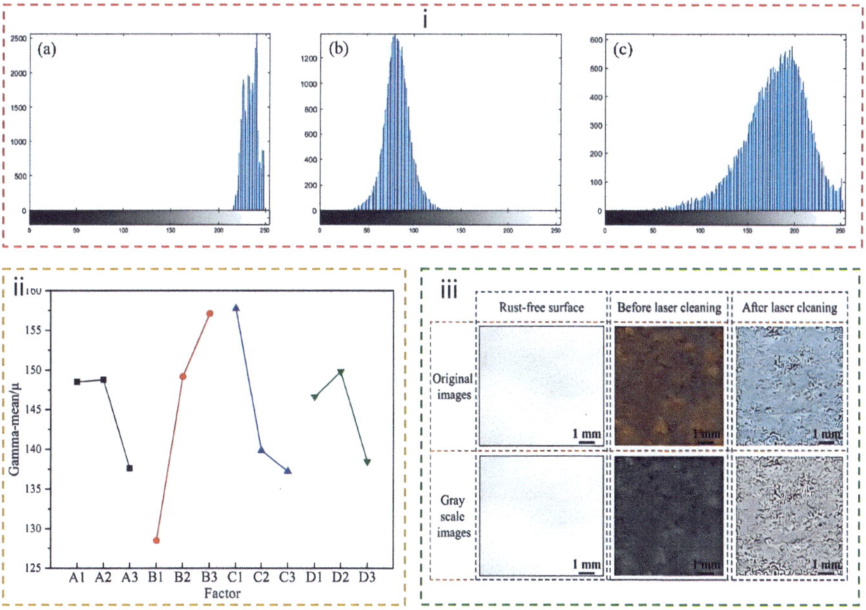


**Fig. 6.19** (**i**) Carbon steel histograms of different states: (a) Unrusty carbon steel grayscale histogram; (b) Rusty carbon steel grayscale histogram; (c) Gray scale histogram after laser rust removal. (**ii**) Grayscale mean distribution. (**iii**) Original and grayscale images before rusting, after rusting, and following laser rust removal [44]

the laser cleaning effect without contact with the material, thereby improving the efficiency and quality of the cleaning. Meng et al. [47] proposed an image processing technology combining K-means clustering algorithm in CIELAB color space and Canny Edge Detection Algorithm, which was used to monitor and analyze the characteristics and removal mechanism of paint layer in the process of laser cleaning aircraft skin. Through this method, researchers can accurately distinguish and identify different paint layers.

During the laser cleaning process, when the energy density exceeds the ablation threshold, the substrate surface temperature rapidly increases. After the base material melts or vaporizes, it generates significant steam pressure, forming a hemispherical shock pressure-wave [48] that breaks and strips the surface material. At the same time, some atoms or molecules absorb a large amount of energy, become ionized, and form plasma. These particles are ejected in a plume-like manner under the combined action of the plasma shock wave and pressure-wave, as shown in Fig. 6.20 [35]. The plume includes various particle flows such as solids, liquids,

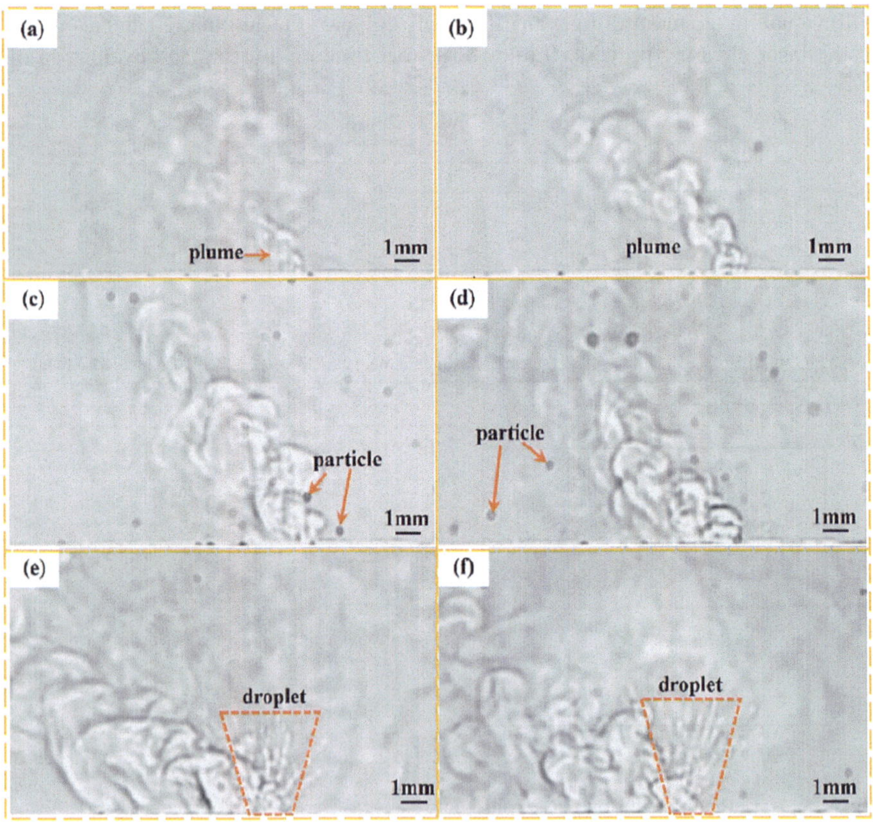


**Fig. 6.20** High-speed shadow imaging of the paint removal process by laser with different energy densities: (**a**)–(**f**) the laser energy density increases successively [35]

gases, and plasma [49]. By analyzing the composition and energy distribution within the plume, the impact of the laser treatment process on the material can be understood, and processing parameters can be further optimized to achieve better results.

High-speed shadow imaging is an advanced visual technology system, typically composed of components such as a high-speed camera, a light source, and data processing software. It is mainly used to capture and analyze the dynamic behavior of rapidly moving objects. By using a high-speed camera to capture continuous images of the paint layer disintegration to plume formation during the laser cleaning process, researchers can understand the dynamic behavior of surface particles during the cleaning process, thereby deducing the cleaning mechanisms and progress. Yu et al. [35] conducted a study on the dynamic behavior of laser paint removal using high-speed shadow imaging technology. The laser cleaning process and the high-speed shadow imaging system are depicted in Fig. 6.21. The working principle is that during the laser cleaning process, a particle flow consisting of gas, solid, and liquid at high temperatures is produced, causing a deflection of the laser beam's light source. This results in the formation of a shadow that reflects the changes in the object's position and shape. Utilizing a high-speed camera to capture these images, and through the application of a special algorithm, the positional information of the object at various times is integrated. Consequently, this process reveals the object's motion trajectory and dynamic properties. During the laser cleaning process, a substantial amount of stray light is generated that can interfere with the imaging of high-speed cameras. To mitigate this, researchers placed a specific narrow-band filter in front of the high-speed camera, effectively blocking the impact of the stray light. Through TG-IR (Thermometric-Fourier Transform Infrared Spectroscopy) analysis, it was demonstrated that the plume produced during the laser paint

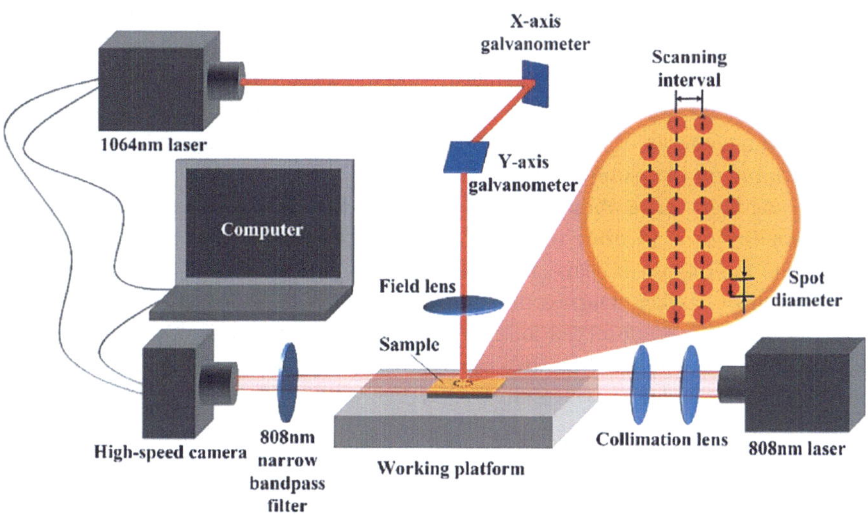


**Fig. 6.21** Laser cleaning and a high-speed shadow imaging system [35]

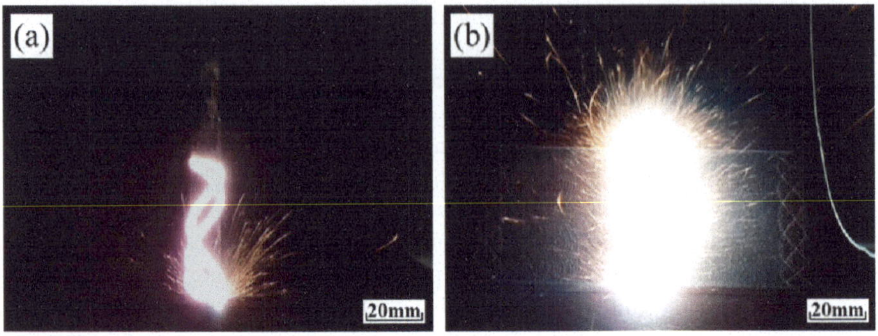


**Fig. 6.22** Laser-induced plume/plasma images under the (**a**) paint residual and (**b**) removal conditions during laser cleaning with wave clean beam pattern [51]

stripping process is primarily composed of a large quantity of carbon dioxide, along with minor amounts of water, molten droplets, and trace organic molecules. These findings are consistent with the conclusions of the research conducted by Zou et al. [50] and Gu et al. [59]. Kim et al. [51] also observed and compared the plume/plasma behavior before and after cleaning using a high-speed camera. They found that the removal of the coating exposed the metal surface directly to the laser beam, increasing the energy absorption and heat input on the metal surface. This, in turn, promoted the evaporation of the metal surface and the generation of plasma, resulting in a brighter plume/plasma phenomenon. This phenomenon can serve as an important indicator for determining whether the coating has been completely removed, as shown in Fig. 6.22. Gregorčič et al. [48] utilized high-speed shadowgraphy to observe the rapid photodynamics phenomena that occur when metal surfaces are laser-ablated. By tracking the changes in the blast wave radius over time and estimating its mechanical energy using the Jones model, they were able to calculate the optodynamic (OD) energy conversion efficiency, which is the ratio of the blast wave energy to the energy of the excitation pulse. The experimental results indicated that the optodynamic energy conversion efficiency increases with the increase in the energy of the excitation pulse. Lei et al. [52] observed the effects of laser cleaning on the paint layer on the surface of aluminum alloy at different time scales using high-speed photography, as shown in Fig. 6.23. During the millisecond laser cleaning process, they observed the production of a large amount of black smoke and melting of the paint layer at the edge of the laser spot, with the cleaned area appearing darker. During the nanosecond pulsed laser cleaning process, a stronger plasma was observed around the laser spot, and larger fragments of the paint layer were stripped from the substrate surface, revealing a brighter metallic luster in the cleaned area. The analysis and comparison of the physical phenomena in the two laser cleaning processes indicate that laser cleaning with a longer time scale is dominated by thermal ablation and vaporization, while that with a shorter time scale is primarily characterized by the thermoelastic vibration effect.

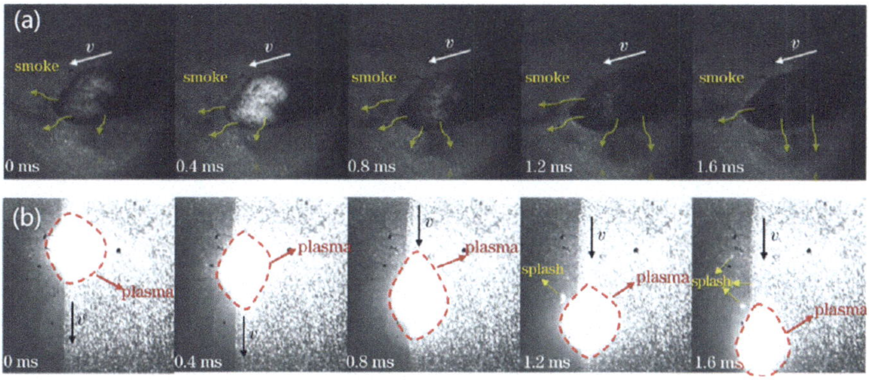


**Fig. 6.23** The laser cleaning coating process captured by the high-speed camera. (**a**) The millisecond-pulse laser cleaning process, P = 500 W; (**b**) The nanosecond-pulse laser cleaning process, P = 200 W [52]

During laser processing, Ho et al. [53] estimated the depth of drilling hole by accumulating the pixels of plasma area in each frame image. The relationship between pixel cumulative value in plasma region and drilling depth is established. This method can be considered for the process monitoring of traditional pulsed laser cleaning. However, it should be noted that the plasma and reflected light generated during laser cleaning may be very intense. Therefore, it may be necessary to use special filters or adjust the exposure settings of the camera to prevent overexposure and ensure that the image acquisition system can adapt to the high brightness environment. And laser cleaning usually requires fast response, so the monitoring system needs to have efficient image processing ability to realize real-time or near-real-time monitoring. Martí-López et al. [54] studied the shock waves and cavitation bubbles generated underwater during laser shock peening (LSP), and analyzed their dynamic behaviors by optical polarization visualization technology. Takata et al. [55] observed the dynamic characteristics of cavitation bubbles with high-speed camera and revealed the mechanism of underwater LSP shock wave. At present, there is no perfect online monitoring technology for liquid-assisted laser cleaning method. This visualization technology by analyzing the dynamic characteristics of underwater shock waves and cavitation bubbles provides a promising starting point for online monitoring of liquid-assisted laser cleaning, but further research and development are still needed to adapt to specific cleaning applications and environments.

Image monitoring technology can intuitively capture the visual image of the cleaning area through a high-resolution camera, which provides an intuitive and quantitative means for the evaluation of cleaning effect. Then the captured images are analyzed by image processing algorithm, which can effectively identify different states in the cleaning process, including areas with insufficient cleaning, proper cleaning and excessive cleaning. This technology can reveal the possible problems in the cleaning process, such as uneven cleaning effect or potential substrate damage, so that the operator can adjust the cleaning parameters in time and optimize the cleaning process. In addition, image monitoring technology also has strong anti-interference ability. In the complex industrial environment, it can resist the interference of external factors such as dust, vibration and other light sources and provide stable and reliable monitoring results. By combining advanced image processing technology, such as machine learning and deep learning algorithm, image monitoring technology can further improve its recognition accuracy and realize more intelligent cleaning process control.

## 6.5 Thermal Monitoring Technology

In the process of laser cleaning, the interaction between laser and substance will release a large number of thermal signals. With the help of infrared thermal imaging system, we can capture the thermal radiation signal and physical characteristics of the samples in the process of laser cleaning, and present them in the form of temperature change, so that we can intuitively monitor and analyze the thermal effect of the cleaning process. Heatmaps generated by the thermoimaging system can reveal regions of hot spots or uneven heating that may indicate the presence of contaminants or uneven laser energy absorption. By monitoring these thermal features, researchers can adjust laser parameters such as power, pulse frequency, and scan speed to optimize the cleaning process and ensure uniform removal of the material while avoiding damage to the underlying substrate.

Bandara et al. [56] employed Thermographic measurements to assess the thermal performance of graphene-coated filters under laser irradiation, as well as the temperature changes during the process of removing biological fouling. In the experiment, when a laser intensity of 21.6 $W/cm^2$ was used to irradiate a filter coated with 500 μg of graphene, the temperature on the surface of the filter increased by approximately 328 ± 9 °C within about 6 s, as shown in Fig. 6.24. This significant temperature rise indicates that the high thermal conductivity characteristic of

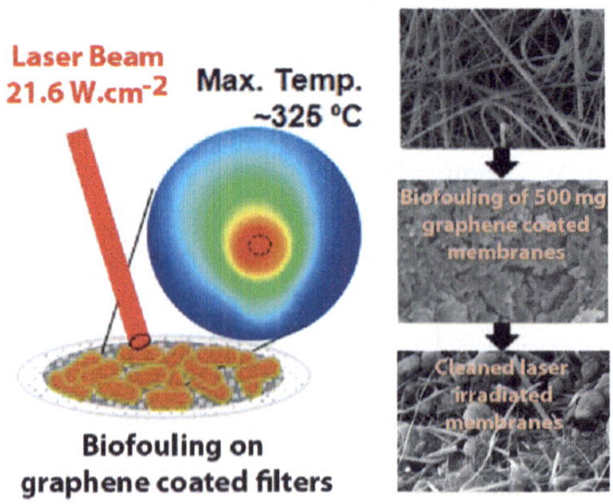


**Fig. 6.24** Membrane cleaning process with laser [56]

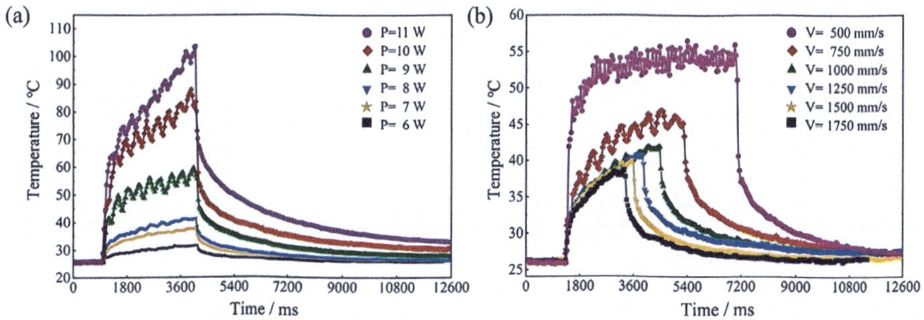


**Fig. 6.25** Temperature of sample surface during laser cleaning at different laser power (**a**) and different scanning speed (**b**) [57]

graphene can effectively convert laser energy into thermal energy, thereby enabling rapid heating and removal of biological foulin. The thermal signal generated in the laser cleaning process can be well captured and presented in the form of temperature change, and the physical characteristics of the sample can be directly monitored and analyzed.

For heat-sensitive materials, such as composite materials, thermal monitoring technology is helpful to prevent the degradation or damage of material properties caused by laser cleaning. Through real-time temperature monitoring, Gu et al. [57, 58] showed the temperature change during the peeling process of coating on the surface of honeycomb sandwich panel under different laser power and scanning speed. The research results reveal that under the condition of low power and high scanning speed, the ultraviolet picosecond laser can effectively reduce the heat affected zone and keep the surface temperature of the material at a low level (Fig. 6.25). Thereby avoiding the thermal damage to the sensitive composite substrate and realizing the fine management of the thermal effect in the cleaning process.

Gu et al. [59] observed and recorded the temperature response and the dynamic behavior of plume formation and disappearance in the process of laser cleaning the surface paint layer of carbon fiber reinforced resin matrix composite (CFRP) aircraft skin by infrared thermal imaging technology (Fig. 6.26), and judged the severity of laser cleaning according to the color and volume changes of plume. Aujeszky et al. [60] utilized thermal modeling technology based on laser thermography to accurately describe the diffusion process of heat in materials and the close relationship between this process and the thermal properties of materials. Suzuki et al. [61] used an infrared thermal imager to monitor in real-time the temperature changes on the surface of marble during the laser cleaning process. They precisely recorded the rate of temperature increase and the highest temperature reached under different laser operating modes. This also helped to determine the time required for the surface temperature to return to its initial state after laser treatment, as well as to observe the diffusion of heat around the treated area (Fig. 6.27), so as to evaluate the thermal effect caused by the interaction between laser and paint and marble

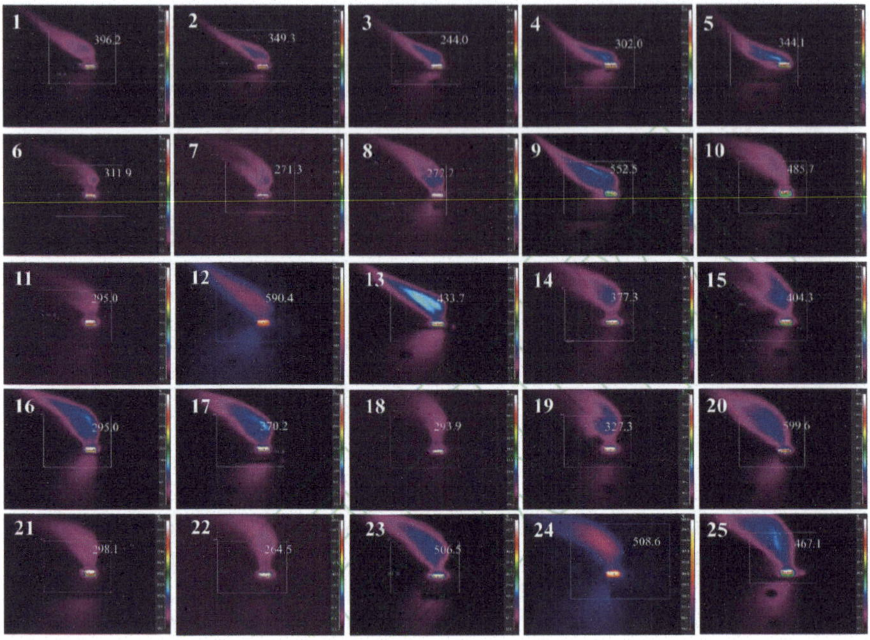

**Fig. 6.26** Infrared thermal imaging of the sample at peak temperature [59]

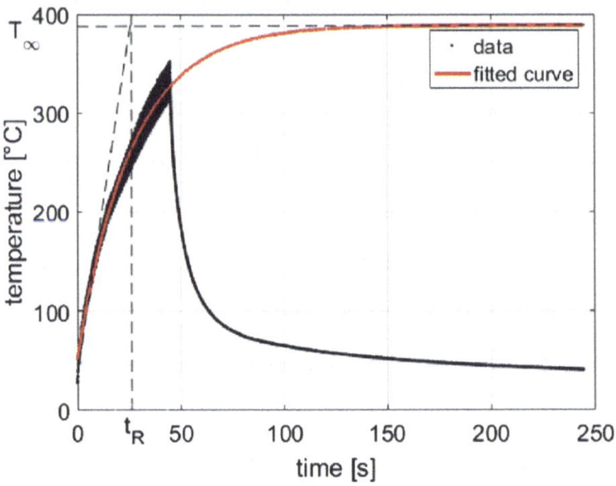


**Fig. 6.27** Red line is a typical trend of the cold/hot behavior; point is the data fitting. $T_\infty$ represents the maximum temperature value of the surface temperature; $t_R$ represents the increase in the ambient temperature caused by the surface temperature [61]

substrate, and understand the possible risk of thermal damage to materials caused by laser cleaning.

Different from laser cleaning, a series of laser processing technologies, such as laser heating, ablation and ablation, aim to remove or evaporate a part of materials through high energy of laser, involving the surface and internal structure of materials to a certain depth. However, in terms of thermal monitoring, there will also be different inspirations. For example, Sihn et al. [62] used an infrared camera and a thermocouple embedded in the material to capture the temperature changes on the surface and inside of the composite during the temperature evolution of fiber reinforced polymer matrix composites heated by laser. Zhang et al. [63] used infrared camera and thermocouple to monitor the front and back surface temperatures of CFRP laminates. The results show that the thermal conductivity of CFRP laminates is low, which leads to significant temperature difference between the front and rear surfaces. This enlightens that the laser cleaning process can be monitored by thermal infrared imaging combined with embedded thermocouple to obtain more accurate temperature information.

Thermal monitoring technology has many advantages in the field of laser cleaning. It not only provides an accurate temperature measurement method, but also deepens our understanding of the thermal response of materials during cleaning by capturing and analyzing the thermal signal generated by the interaction between laser and materials in real time. In addition, thermal monitoring technology can also help to evaluate and control the risk of thermal damage that may occur during laser cleaning. It is very important to detect and prevent overheating in time by monitoring the temperature change in the cleaning area, which is very important for protecting fragile or high-value materials. For example, in the cleaning process of heat-sensitive composite materials or electronic components, thermal monitoring can ensure that unnecessary coatings or pollutants are removed without damaging their structure and performance. However, thermal monitoring technology also has some limitations, such as the instability of data due to the fluctuation of ambient temperature; The difference of material surface reflectivity may affect the accurate measurement of temperature by infrared thermal imager; The high cost of high-quality thermal monitoring equipment may limit its popularity in some application scenarios; Although thermal monitoring can be carried out in real time, it may be difficult to capture transient temperature changes in the process of rapid change. Generally speaking, thermal monitoring technology has obvious advantages in providing accurate temperature measurement and risk control, but it also needs to pay attention to its challenges in environmental stability, cost and real-time.

## 6.6 Composite Monitoring Technology

In the continuous progress of laser cleaning technology, a single monitoring means has gradually failed to meet the requirements of high precision and high efficiency cleaning. In order to achieve a more comprehensive and deeper understanding of the cleaning process, the composite monitoring technology came into being. By combining sound wave, spectrum, image and thermal monitoring methods, it provides a multi-dimensional and all-round monitoring scheme for laser cleaning.

LIBS is a technology to analyze the elements on the surface of materials by laser-induced plasma emission spectroscopy. Although it has excellent spectral sensitivity in elemental analysis, it has some problems such as low time efficiency, limited spatial resolution and high requirements for the stability and reliability of the measurement system. Pozo-Antonio et al. [64–66] used a hyperspectral imaging technology (HSI) which combined spectral technology and image technology. The applicability of hyperspectral imaging technology in evaluating the effectiveness of laser cleaning on granite's crustose lichen, graffiti, and encrustation has been discussed. The study found that hyperspectral imaging technology can non-invasively measure reflectance changes on the cleaned surface, identifying the cleaning effectiveness of each procedure. The greatest advantage of this technology is its ability to quickly acquire and reconstruct pixel spectra. It can not only provide the spatial resolution information of the image, but also provide the detailed spectral information of each pixel in multiple spectral bands. Therefore, combining LIBS with HSI technology can effectively provide high spatial resolution and time-efficient elemental distribution analysis, thereby compensating for the spatial and temporal limitations of LIBS technology. This combined approach not only accelerates imaging speed but also enhances data processing capabilities through principal component analysis and Spearman correlation analysis. As a result, the LIBS-HSI imaging system becomes an efficient and reliable tool suitable for rapid, high-resolution visualization of elemental analysis on material surfaces [67].

Although LIBS is also non-invasive, it is a micro-destructive technique that may have minor physical effects on certain material surfaces. Cucci et al. [68] integrated fiber optic reflectance spectroscopy (FORS) with HSI. When more detailed spectral analysis is needed for specific points or areas within HSI images, the fiber optic probe of FORS can be utilized to precisely measure specific points, providing more detailed information. By simultaneously using the Spectral Angle Mapper (SAM) algorithm to classify HSI data, the classification map generated by SAM can be compared with the RGB image reconstructed from HSI data to visually show the distribution of treated and untreated areas (Fig. 6.28). FORS and HSI technologies do not require the removal or contact of materials from the surface of cultural relics, reducing the risk of potential damage to the artifacts. FORS and HSI technologies do not require the removal or contact of materials from the surface of artifacts, thereby reducing the risk of potential damage to the cultural relics.

Verhoff et al. [69] coupled a spectrometer (Acton, SP2500i) to a CCD camera (ICCD, Princeton Instruments PI-MAX) to capture the spectra emitted by the plasma (Fig. 6.29) and analyzes the dynamics of the plasma generated when femtosecond and nanosecond lasers ablate brass. The study indicated that under the same laser intensity conditions, the plasma produced by femtosecond lasers had a higher electron density and temperature near the target surface and in the early stages of evolution. During femtosecond laser ablation, the material evapor more efficiently and the plasma is more intensive; the nanosecond laser is dominated by continuum

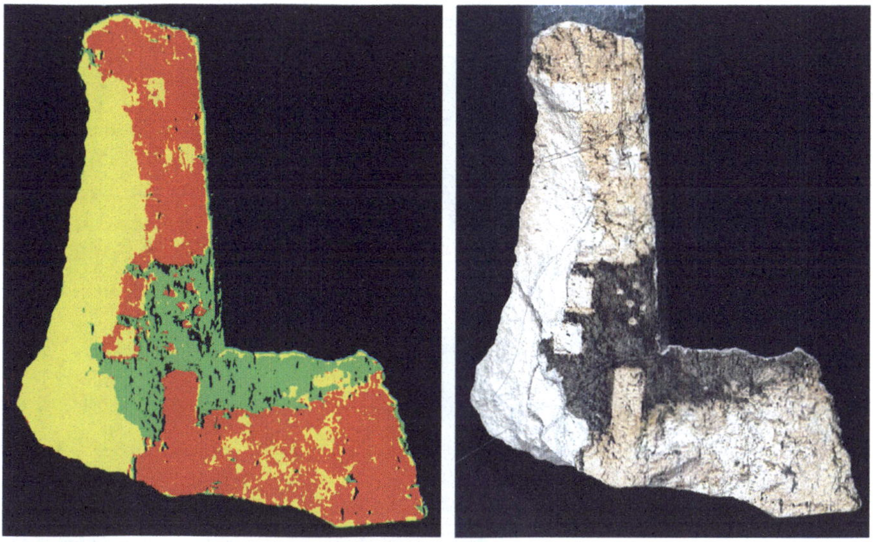


**Fig. 6.28** Spectral imaging analysis of limestone surface. (**a**) The image obtained by SAM algorithm after classifying limestone surface; (**b**) RGB image reconstructed from HSI data [68]

and ion emission, then converted to atomic emission. Lin et al. [70] combined machine vision and LIBS spectroscopy techniques for online monitoring of laser paint stripping on aircraft skin. Through high contrast imaging and image processing algorithm, the residual paint layer on the surface after paint removal can be automatically identified and monitored. At the same time, the state of matrix material can be monitored in real time by LIBS technology, and the dynamic feedback can be realized, and the laser parameters can be adjusted in real time, which improves the automation and intelligence level of paint removal. Bian et al. [71] monitored the paint layer on the surface of white marble by LIBS combined with image binarization, and helped to distinguish the paint layer from the white marble substrate according to the detailed information of the material composition, ensuring that the laser energy would not mistakenly act on the cultural relics themselves, and at the same time, converted the image into black and white for intuitive visual analysis, which enabled researchers to quickly evaluate the cleaning effect. The combination of the two can verify each other and improve the accuracy of the monitoring results. Zha et al. [72] also combined LIBS with image processing techniques in their study. They utilized the LIPS analysis method to investigate the variation trend of Pearson correlation coefficient of small area rust layers with the number of cleaning cycles.

Additionally, they determined the optimal cleaning cycles for different thicknesses of rust layers. By optimizing the laser spot overlap rate through image processing techniques, they examined the change pattern of cleanliness under different spot overlap rates, and concluded that 50% was the optimal spot overlap rate. During

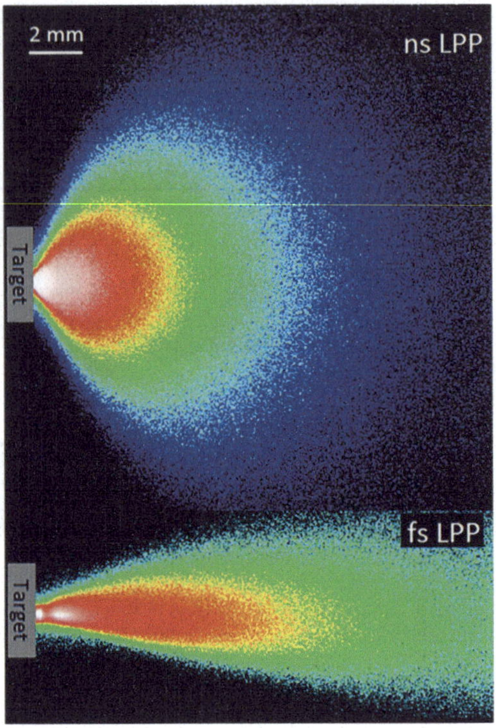


**Fig. 6.29** Time and spectrally integrated ICCD images of ns and fs plasmas recorded under similar laser fluence conditions. (The delay and integration times used for obtaining these images are 0 ns and 2 μs) [69]

**Table 6.3** Laser cleaning power parameters

| | Power (W) | | | | |
|---|---|---|---|---|---|
| Materials | C1 | C2 | C3 | C4 | C5 |
| CFRP | 30 W | 35 W | 40 W | 43 W | 45 W |

the laser paint removal process, when the laser energy is low, it is unable to excite the generation of induced plasma, resulting in a weak LIBS spectrum, which indicates that the cleaning process is predominantly vibrational; whereas at higher laser energies, plasma radiation interferes with the transmission of acoustic waves, causing the acoustic signals to be chaotic. Therefore, Li et al. [73] combined LIBS technology with acoustic signal monitoring to verify each other, avoiding the limitation of plasma caused by the weak laser energy and acoustic signal by external interference and affect the accuracy of monitoring, and the laser paint peeling mechanism can be explored through these two ways. Song et al. [8] studied the online monitoring method of infrared nanosecond laser cleaning technology in the process of removing the surface coating of carbon fiber reinforced polymer (CFRP) under different powers (laser power parameters are shown in Table 6.3). Temperature monitoring, acoustic signal analysis and high-speed photography are combined to track and analyze the physical phenomena in the cleaning process, such as thermal effect,

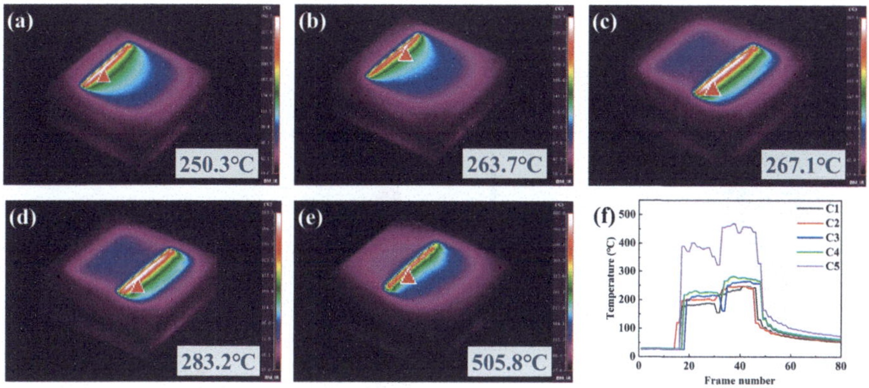


**Fig. 6.30** The maximum temperature of the sample during laser cleaning of CFRP: (**a**) C1; (**b**) C2; (**c**) C3; (**d**) C4; (**e**) C5; (**f**) Real-time temperature of each sample

vibration effect and plasma shock effect. The optimum temperature range of material surface to avoid excessive cleaning is determined (Fig. 6.30), and the integrity of cleaning is evaluated by using the characteristic frequency and amplitude intensity of acoustic signal (Fig. 6.31). In addition, high-speed photographic monitoring reveals the size change of plasma plume, which further verifies the different stages of cleaning (Fig. 6.32).

Although thermal imaging cameras can effectively reflect the thermal effects of laser cleaning and monitor temperature changes during the cleaning process, their relatively low frame rate means that they often cannot accurately track and record the rapid temperature changes that occur during the very brief action time of pulsed lasers. To overcome these limitations, Wang et al. [74] adopted the monitoring method of high-speed camera assisted thermal imaging analyzer. Utilizing high-speed photography technology to captured dynamic images during the laser cleaning process, every stage of rust removal from Q235 carbon steel is recorded at an extremely high frame rate, including phenomena such as plasma generation and particle splashing, as shown in Fig. 6.33. High-speed photography and thermal imagers form a complementary, On the one hand, it provides an intuitive visual record of the cleaning process. On the other hand, it supplements the temperature change data through the thermal imager. The combination of the two can analyze the mechanism of laser rust removal more comprehensively, including the physical process of the interaction between laser and rust, the influence of thermal effect on the substrate, and the possibility of secondary oxidation risk.

The application of composite monitoring technology in laser cleaning realizes the multi-angle evaluation of cleaning process by integrating acoustic wave, spectrum, image and thermal monitoring data. The advantage of this method is that it can provide more comprehensive feedback, help identify the insufficient and excessive cleaning areas, and monitor the risk of thermal damage to ensure the

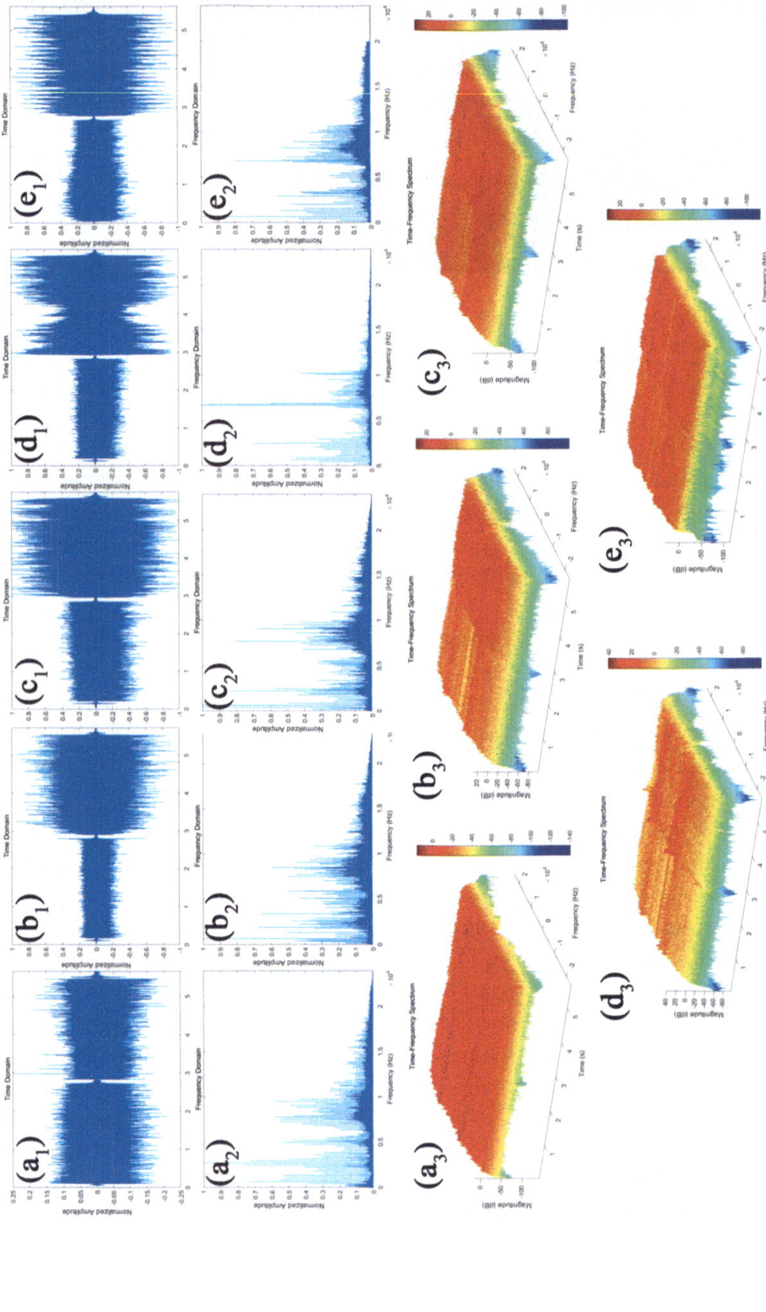


**Fig. 6.31** Sound signal monitoring during laser cleaning of CFRP: (**a**$_1$–**e**$_1$) TA spectrum; (**a**$_2$–**e**$_2$) FA spectrum; (**a**$_3$–**e**$_3$) TF spectrum

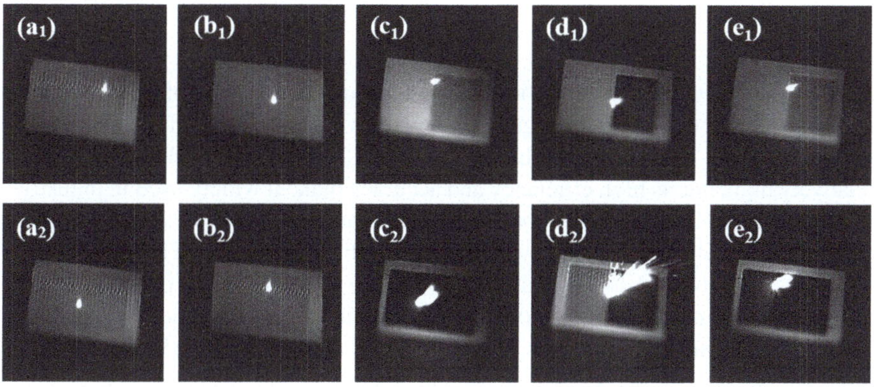


**Fig. 6.32** Dynamic behavior of laser cleaning CFRP: (**a**$_1$)–(**a**$_2$) C1; (**b**$_1$)–(**b**$_2$) C2; (**c**$_1$)–(**c**$_2$) C3; (**d**$_1$)–(**d**$_2$) C4; (**e**$_1$)–(**e**$_2$) C5

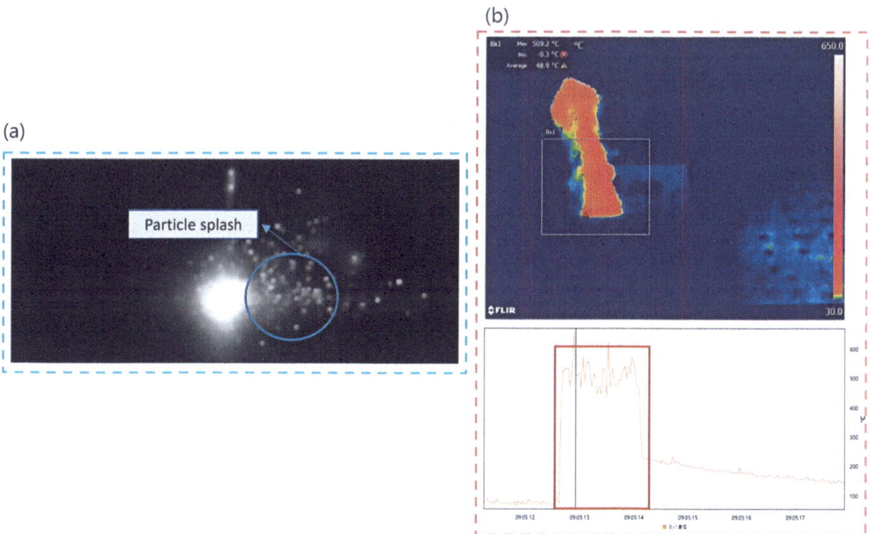


**Fig. 6.33** Pulsed laser rust removal. (**a**) High-speed photography of particle splash; (**b**) Thermal imaging for temperature monitoring [74]

safety of materials. In addition, through algorithm optimization, the composite monitoring technology can realize the dynamic adjustment of cleaning parameters and improve the consistency and reliability of cleaning. With the progress of technology, it is expected that the accuracy and intelligence of monitoring will be further improved in the future, and the application scope of laser cleaning technology will be expanded.

## 6.7 Conclusion

At present, the widely used technologies in online monitoring of laser cleaning include acoustic wave monitoring, LIBS spectrum monitoring and high-speed photographic visual monitoring. Relatively speaking, the cost of LIBS equipment is relatively high, but its high efficiency and accuracy make it widely used in industrial applications, while the microphone used for acoustic monitoring and high-speed camera for visual monitoring are widely used in laboratory research because of their low cost. Therefore, it is of great significance to develop a new economical and efficient online monitoring technology for promoting the development and application of laser cleaning technology. Laser cleaning interface monitoring is crucial to improve the laser cleaning efficiency and quality of the interface, the acoustic waves, spectrum, image and thermal signal monitoring can accurately reflect the surface cleanliness and integrity, the real-time analysis of these signal feedback can provide important basis for cleaning process parameters adjustment, so as to improve the continuity of laser cleaning, intelligent, high efficiency and high quality. Laser cleaning interface monitoring has become a hot research direction in the field of laser cleaning, which plays an important role in promoting the industrialization process of laser cleaning technology. But the following aspects deserve attention:

1. Laser cleaning interface monitoring often only as auxiliary observation, its indispensable has not caused enough attention, with the naked eye observation as the basic means of laser cleaning surface quality evaluation is not accurate, also directly affect the continuity of laser cleaning operations, has become a laser cleaning from "extensive" to "fine" the biggest obstacle. Therefore, the laser cleaning interface monitoring module should be widely added in the research and development of laser cleaning equipment, to build an intelligent response system integrating dynamic monitoring feedback and laser cleaning operation, so as to further promote the industrialization process of laser cleaning technology.
2. Dynamic monitoring data characteristics and surface cleanliness (integrity) mapping rule has not yet fully mastered, not established enough to support the dynamic monitoring results response extraction of comparison database, lead to dynamic monitoring data cannot directly generate quantitative cleaning effect evaluation conclusion, but for different matrix, different coating, different state of cleaning surface cleanliness (integrity) state database building need a lot of experimental accumulation. The addition of artificial intelligence for laser cleaning monitoring technology has brought new opportunities, the application of machine learning and deep learning algorithm, extract useful information from complex data, establish cleaning effect and the complex relationship between the monitoring signal, through the establishment of surface cleaning state standard than database, not only can improve the level of intelligent monitoring, also can achieve a higher degree of automation control.

3. Single monitoring technology has its shortcomings, such as acoustic monitoring may be disturbed by background noise, and the resolution of fine microwave vibration is low; the cost of spectral monitoring equipment is high and the laser energy can not be too low; Image recognition monitoring has a certain dependence on the light source condition and no discrimination ability when the color difference of substrate and coating is small; The response speed of thermal radiation monitoring is not enough to keep up with the frequency of laser pulses, and it is difficult to obtain the sample surface temperature under each pulse in real time. In this case, the method of multi-monitoring technology fusion can effectively avoid the defects of a single monitoring technology. Therefore, the laser cleaning researchers and practitioners should focus on the development of composite monitoring methods coupled by different monitoring technologies, so as to realize the accurate monitoring of the cleaning effect, which will be the development trend of the dynamic monitoring technology of the laser cleaning interface.

## References

1. Zhu G, Xu Z, Jin Y, et al. Mechanism and application of laser cleaning: A review [J]. Optics and Lasers in Engineering, 2022, 157: 107130.
2. Hou L, Yin F, Wang S, et al. A review of thermal effects and substrate damage control in laser cleaning [J]. Optics & Laser Technology, 2024, 174: 110613.
3. Hou S, Luo J, Zhang Q. Research on the real-time monitoring technology of laser cleaning [J]. Optical technology, 2007 (S1): 61–62.
4. Zhou Z, Sun W, Wu J, et al. The fundamental mechanisms of laser cleaning technology and its typical applications in industry [J]. Processes, 2023, 11(5): 1445.
5. Li Y, Wu Z, Chu D, et al. Research progress of laser cleaning and monitoring technology (invited) [J]. Infrared and laser engineering, 2023, 52(2): 20220784-1–20220784-19.
6. Teule R, Scholten H, Van den Brink O F, et al. Controlled UV laser cleaning of painted artworks: a systematic effect study on egg tempera paint samples [J]. Journal of Cultural Heritage, 2003, 4: 209–215.
7. Rezk R A, Abdel Ghany N A, Mostafa A M. Laser-assisted method for cleaning and analysis of archaeological metallic coins [J]. Coatings, 2022, 12(10): 1548.
8. Song Y, Wang S, Pan Y, et al. Comparative study of infrared nanosecond laser surface paint removal for carbon fiber reinforced polymer and glass fiber reinforced polymer [J]. Journal of Cleaner Production, 2024, 451: 142061.
9. Tong Y, ShangGuan J, Ren X, et al. Surface quality of laser paint removal of marine steel: a comparative study using a Gaussian beam and a flat-top beam [J]. Applied Optics, 2022, 61(9): 2237–2246.
10. Sun L X, Wang W J, Qi L F, et al. Online monitoring of laser cleaning effect of carbon fiber composite materials based on laser-induced breakdown spectroscopy technology [J]. Chin. J. Laser, 2020, 47: 299–308.
11. Liu Y, Liu W, Zhang D, et al. Experimental investigations into cleaning mechanism of ship shell plant surface involved in dry laser cleaning by controlling laser power [J]. Applied Physics A, 2020, 126: 1–17.
12. Hu C, He G, Chen J, et al. Research on cleaning mechanism of anti-erosion coating based on thermal and force effects of laser shock [J]. Coatings, 2020, 10(7): 683.

13. Zhou Y. Acoustic monitoring of laser paint removal effect based on wavelet packet transformation and K mean clustering [D]. Xiamen Institute of Technology, 2023. https://doi.org/10.27866/d.cnki.gxlxy.2022.000238.
14. Kim T, Lee J M, Cho S H, et al. Acoustic emission monitoring during laser shock cleaning of silicon wafers [J]. Optics and lasers in engineering, 2005, 43(9): 1010–1020.
15. Lu Y F, Hong M H, Chua S J, et al. Audible acoustic wave emission in excimer laser interaction with materials [J]. Journal of applied physics, 1996, 79(5): 2186–2191.
16. Lee J M, Watkins K G. In-process monitoring techniques for laser cleaning [J]. Optics and lasers in engineering, 2000, 34(4–6): 429–442.
17. Zou W, Song F, Luo Y. Characteristics of audible acoustic signal in the process of laser cleaning of paint on metal surface [J]. Optics & Laser Technology, 2021, 144: 107388.
18. Chen Y, Deng G L, Zhou Q, et al. Acoustic signal monitoring in laser paint cleaning [J]. Laser Physics, 2020, 30(6): 066001.
19. Dimitroulaki E, Tserevelakis G J, Melessanaki K, et al. Photoacoustic real-time monitoring of UV laser ablation of aged varnish coatings on heritage objects [J]. Journal of Cultural Heritage, 2023, 63: 230–239.
20. Chen Y, Huang H, Ye S. Research on the acoustic monitoring technology of the laser paint removal process [J]. Application of laser, 2020, 40(6): 1153–1159.
21. Tserevelakis G J, Pozo-Antonio J S, Siozos P, et al. On-line photoacoustic monitoring of laser cleaning on stone: Evaluation of cleaning effectiveness and detection of potential damage to the substrate [J]. Journal of Cultural heritage, 2019, 35: 108–115.
22. Papanikolaou A, Tserevelakis G J, Melessanaki K, et al. Development of a hybrid photoacoustic and optical monitoring system for the study of laser ablation processes upon the removal of encrustation from stonework [J]. Opto-Electronic Advances, 2020, 3(2): 190037-1–190037-11.
23. Xie X, Huang Q, Long J, et al. A new monitoring method for metal rust removal states in pulsed laser derusting via acoustic emission techniques [J]. Journal of Materials Processing Technology, 2020, 275: 116321.
24. Zou W, Xiong J. Acoustic wave monitoring and analysis of laser dry paint removal [J]. Journal of Gannan Normal University, 2010, 31(6): 43–45.
25. Xu J, Sun Z, Zhou W, et al. Research on the real-time monitoring technology of the laser rust removal process [J]. Photon Journal, 2002, 31(9): 1090–1092.
26. Chen Y, Huang H, Ye D, et al. Study on the acoustic monitoring method of laser paint removal based on Bayesian discrimination [J]. Laser Technology, 2022, 46(2).
27. Choi S, Park C. Convolution Neural Network with Laser-Induced Breakdown Spectroscopy as a Monitoring Tool for Laser Cleaning Process [J]. Sensors, 2022, 23(1): 83.
28. Yang W, Qian Z, Cao Y, et al. LIBS monitoring and analysis of laser-based layered controlled paint removal from aircraft skin [J]. Journal of Spectroscopy, 2021, 2021: 1–12.
29. Yang W, Qian Z, Cao Y, et al. Study on the controllability of aircraft skin laser paint removal based on LIBS spectrum and composition analysis [J]. Spectroscopy and spectral analysis, 2021, 41(10): 3233–3239.
30. Tong Y, Zhang Y, Yao H, et al. Plasma spectroscopy of the laser cleaning process in air [J]. Spectroscopy and spectral analysis, 2011,31(09):2542–2545.
31. Voznesenskaya A A, Gusev D S, Kochuev D A, et al. Laser-induced breakdown spectroscopy monitoring of aluminum alloys laser cleaning process [C]. Journal of Physics: Conference Series. IOP Publishing, 2022, 2316(1): 012006.
32. Li X, Guan Y. Real-time monitoring of laser cleaning for hot-rolled stainless steel by laser-induced breakdown spectroscopy [J]. Metals, 2021, 11(5): 790.
33. Choi S, Choi J, Park C. Multivariate analysis-based laser-induced breakdown spectroscopy for monitoring of laser paint cleaning [J]. Applied Physics B, 2023, 129(1): 8.
34. Wang W, Sun L, Lu Y, et al. Laser induced breakdown spectroscopy online monitoring of laser cleaning quality on carbon fiber reinforced plastic [J]. Optics & Laser Technology, 2022, 145: 107481.

35. Yu H, Zhang Z, Wang H, et al. Investigation on dynamic behavior of laser paint removal based on high-speed shadow imaging technology [J]. Optics & Laser Technology, 2023, 163: 109441.
36. Senesi G S, Carrara I, Nicolodelli G, et al. Laser cleaning and laser-induced breakdown spectroscopy applied in removing and characterizing black crusts from limestones of Castello Svevo, Bari, Italy: a case study [J]. Microchemical Journal, 2016, 124: 296–305.
37. Skruibis J, Balachninaite O, Butkus S, et al. Multiple-pulse Laser-induced breakdown spectroscopy for monitoring the femtosecond laser micromachining process of glass [J]. Optics & Laser Technology, 2019, 111: 295–302.
38. Gobernado-Mitre I, Prieto A C, Zafiropulos V, et al. On-line monitoring of laser cleaning of limestone by laser-induced breakdown spectroscopy and laser-induced fluorescence [J]. Applied spectroscopy, 1997, 51(8): 1125–1129.
39. Marimuthu S, Pathak S, Radhakrishnan J, et al. In-process monitoring of laser surface modification [J]. Coatings, 2021, 11(8): 886.
40. Li Z, Xu J, Zhang D, et al. Nanosecond pulsed laser cleaning of titanium alloy oxide films: Modeling and experiments [J]. Journal of Manufacturing Processes, 2022, 82: 665–677.
41. Fortes F, Cabalín L, Laserna J. The potential of laser-induced breakdown spectrometry for real time monitoring the laser cleaning of archaeometallurgical objects [J]. Spectrochimica Acta Part B: Atomic Spectroscopy, 2008, 63(10):1191–1197.
42. Liu J. Research on laser intelligent rust removal System Based on machine vision [D]. Suzhou University, 2020.
43. Shi T, Zhou L, Wang C, et al. Real-time detection system of aluminum alloy laser cleaning based on machine vision [J]. Chinese Journal of Lasers, 2019, 46(4): 0402007.
44. Duan J, Shi R, Wei Z, et al. Research on carbon steel surface based on image processing [J/OL].China laser, 1–18 [2024-04-16]. http://kns.cnki.net/kcms/detail/31.1339.TN.20240314.0840.037.html.
45. Li J, Liu H, Shi L, et al. Imaging feature analysis-based intelligent laser cleaning using metal color difference and dynamic weight dispatch corrosion texture [C]. Photonics. MDPI, 2020, 7(4): 130.
46. Hu Q, Wei X, Liang X, et al. In-process vision monitoring methods for aircraft coating laser cleaning based on deep learning [J]. Optics and Lasers in Engineering, 2023, 160: 107291.
47. Meng Y, Zhang L, He C, et al. Skin characteristics and mechanism of laser cleaning aircraft based on image processing [J/OL]. Laser technology, 1–17 [2024-04-16]. http://kns.cnki.net/kcms/detail/51.1125.tn.20231018.1024.002.html.
48. Gregorčič P, Zadravec J, Možina J, et al. Optodynamic energy conversion efficiency during laser ablation on metal surfaces measured by shadow photography [J]. Applied Physics A, 2014, 117: 353–357.
49. Bogaerts A, Chen Z, Gijbels R, et al. Laser ablation for analytical sampling: what can we learn from modeling? [J]. Spectrochimica Acta Part B: Atomic Spectroscopy, 2003, 58(11): 1867–1893.
50. Zou C, Zhao S, Yu H, et al. Plume ejection and product in laser cleaning of ferrite paint [J]. Optics & Laser Technology, 2023, 167: 109733.
51. Kim J E, Lee J M, Hyun J H, et al. A study on the laser removal of epoxy coatings on SS400 surface by beam scanning patterns [J]. Coatings, 2021, 11(12): 1510.
52. Lei Z, Sun H, Tian Z, et al. Effect of laser with different time scale on the cleaning quality of paint layer on aluminum alloy surface [J]. Chinese Journal of Lasers, 2021, 48(6): 0602103.
53. Ho C C, He J J, Liao T Y. On-line estimation of laser-drilled hole depth using a machine vision method [J]. Sensors, 2012, 12(8): 10148–10162.
54. Martí-López L, Ocaña R, Porro J A, et al. Optical observation of shock waves and cavitation bubbles in high intensity laser-induced shock processes [J]. Applied Optics, 2009, 48(19): 3671–3680.
55. Takata T, Enoki M, Chivavibul P, et al. Acoustic emission monitoring of laser shock peening by detection of underwater acoustic wave [J]. Materials transactions, 2016, 57(5): 674–680.

56. Bandara P C, Ibañez de Santi Ferrara F, Nguyen H, et al. Investigation of thermal properties of graphene-coated membranes by laser irradiation to remove biofoulants [J]. Environmental science & technology, 2018, 53(2): 903–911.
57. Gu J, Su X, Jin Y, et al. Precise paint stripping based on photochemical effect by UV laser on honeycomb sandwich panel [J]. Journal of Materials Research and Technology, 2023, 25: 3768–3779.
58. Gu J, Su X, Jin Y, et al. Towards low-temperature laser paint stripping by photochemical mechanism on CFRP substrates [J]. Journal of Manufacturing Processes, 2023, 85: 272–280.
59. Gu J, Li W, Su X, et al. Research on the laser cleaning technology of carbon fiber composite aircraft skin surface paint layer [J/OL]. China laser, 1–19 [2024-04-16]. http://kns.cnki.net/kcms/detail/31.1339.tn.20230817.1737.014.html.
60. Aujeszky T, Korres G, Eid M. Measurement-based thermal modeling using laser thermography [J]. IEEE Transactions on Instrumentation and Measurement, 2018, 67(6): 1359–1369.
61. Suzuki A, Riminesi C, Ricci M, et al. Testing of a new Yb: YAG fiber laser system for the removal of graphic vandalism from marble [J]. Heritage Science, 2023, 11(1): 132.
62. Sihn S, Pitz J, Gerzeski R H, et al. Experimentally-validated computational model for temperature evolution within laser heated fiber-reinforced polymer matrix composites [J]. Composite Structures, 2019, 207: 966–973.
63. Zhang Y, Pan J, Zhou S, et al. Experimental Investigation on Ablation Behaviors of CFRP Laminates in an Atmospheric Environment Irradiated by Continuous Wave Laser [J]. Polymers, 2022, 14(23): 5082.
64. Pozo-Antonio J S, Fiorucci M P, Rivas T, et al. Suitability of hyperspectral imaging technique to evaluate the effectiveness of the cleaning of a crustose lichen developed on granite [J]. Applied Physics A, 2016, 122: 1–9.
65. Pozo-Antonio J S, Fiorucci M P, Ramil A, et al. Hyperspectral imaging as a non destructive technique to control the laser cleaning of graffiti on granite [J]. Journal of nondestructive evaluation, 2016, 35: 1–8.
66. Pozo-Antonio J S, Fiorucci M P, Ramil A, et al. Evaluation of the effectiveness of laser crust removal on granites by means of hyperspectral imaging techniques [J]. Applied Surface Science, 2015, 347: 832–838.
67. Zhao S, Zhao Y, Hou Z, et al. Rapid and high-resolution visualization elements analysis of material surface based on laser-induced breakdown spectroscopy and hyperspectral imaging [J]. Applied Surface Science, 2023, 629: 157415.
68. Cucci C, De Pascale O, Senesi G S. Assessing laser cleaning of a limestone monument by fiber optics reflectance spectroscopy (FORS) and visible and near-Infrared (VNIR) hyperspectral imaging (HSI) [J]. Minerals, 2020, 10(12): 1052.
69. Verhoff B, Harilal S S, Freeman J R, et al. Dynamics of femto-and nanosecond laser ablation plumes investigated using optical emission spectroscopy [J]. Journal of Applied Physics, 2012, 112(9).
70. Lin D. Online monitoring and feedback study of aircraft skin laser paint removal based on machine vision and LIBS technology [D]. The Civil Aviation Flight Academy of China, 2023. https://doi.org/10.27722/d.cnki.gzgmh.2023.000171.
71. Bian T, Bai Y, Yu L, et al. Laser cleaning of paint layers on white marble surface based on cooperative use of laser-induced breakdown spectroscopy and image binarization [J]. Journal of Cultural Heritage, 2023, 62: 124–133.
72. Cha W, Yu L, Li B, et al. Monitoring of the laser cleaning process on metal surfaces based on image processing and LIPS analysis [J]. ACTA PHOTONICA SINICA, 2021, 50(7): 249–259.
73. Li S, Zhang L, He J, et al. Real-time monitoring of laser cleaning aircraft skin based on acousto-optic composite method (invited) [J]. Infrared and laser engineering, 2023, 52(2): 20220821-1–20220821-10.
74. Wang K. Study on pulse laser derusting technology of Q235 carbon steel [D]. Huazhong University of Science and Technology, 2017.

# Chapter 7
# Thermomechanical Coupling Modeling and Simulation in Laser Cleaning Surface-Interface

## 7.1 Introduction

As an advanced surface cleaning and pretreatment technology, laser cleaning utilizes the energy released by laser irradiation to remove coatings or contaminants from material surfaces. This method is renowned for its high precision, excellent cleaning efficiency, and environmentally friendly characteristics [1, 2]. Compared to traditional cleaning techniques, laser cleaning demonstrates significant advantages, particularly in industrial sectors with stringent cleanliness requirements. It is applicable not only for the removal of aged paint layers [3], but also for mold maintenance [4], artifact restoration [5], and the cleaning of miniature electronic components [6]. In fields where strict standards for cleaning precision, substrate integrity, and safety are required, the application of laser cleaning technology is particularly important. The principle of laser cleaning involves the interaction between the laser and the material surface to remove contaminants or coatings [7]. The principle of laser cleaning primarily involves the synergistic effects of thermal ablation [8], mechanical vibration [9], plasma shock [10] and photochemical reactions [11] to achieve efficient and non-destructive removal of surface coatings (Fig. 7.1).

To further explore the interaction between lasers and materials during the laser cleaning process, many researchers analyze the physicochemical phenomena involved through numerical simulations to elucidate the cleaning mechanisms. Guo et al. [12] developed a two-dimensional laser cleaning simulation model based on the thermal ablation effect, linking the laser ablation thermal power density to the normal velocity of the ablated contaminant layer boundary. This model achieved a dynamic energy conservation relationship between the latent heat consumption of the contaminant layer and the energy provided by laser ablation. Hu et al. [13] investigated the thermal effects of metal materials under pulsed laser irradiation with varying duty cycles through numerical simulations. Currently, most simulations of laser cleaning utilize the approach of defining laser irradiation as a heat flux boundary condition, which does not capture the real-time changes in surface


S. Wang, *Laser Cleaning*, https://doi.org/10.1007/978-981-95-2935-3_7

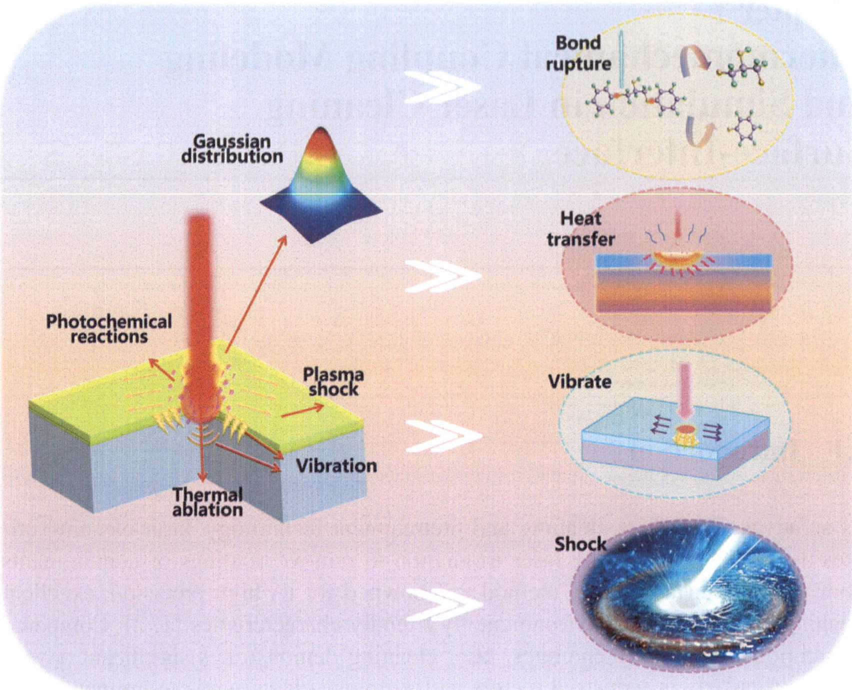


**Fig. 7.1** Schematic diagram of laser cleaning

morphology that affect the absorption and reflection of the laser beam during the cleaning process. Li et al. [14] conducted a simulation analysis of the temperature field during the laser cleaning of titanium alloy oxide films, exploring the effects of factors such as laser spot shape, energy loading method, and pulse width on the variations in the temperature field throughout the cleaning process. Guo et al. [15] expanded their simulation of aluminum alloy oxide film cleaning under different laser power levels by incorporating an analysis of the transient distribution of the convective field based on the temperature field. They investigated the influence mechanism of molten pool flow behavior on surface morphology. The study found that as the laser power increased, the depth of the depressions formed on the aluminum alloy surface gradually increased. While the former analysis of the interaction between laser and material remained at a macroscopic level, the latter did not deeply explore the impact of microscopic changes in depression morphology on laser cleaning. Ding et al. [16] conducted an investigation through numerical simulations into the application of laser cleaning technology for the removal of micron- and nano-sized particles from transparent substrates, identifying the threshold conditions for particle removal. Mei et al. [6] explored the process of removing silicon dioxide ($SiO_2$) particles from the surface of silicon wafers using nanosecond pulsed laser cleaning via numerical simulations. They provided a detailed analysis of the temperature rise and thermal expansion forces on the wafer surface following laser

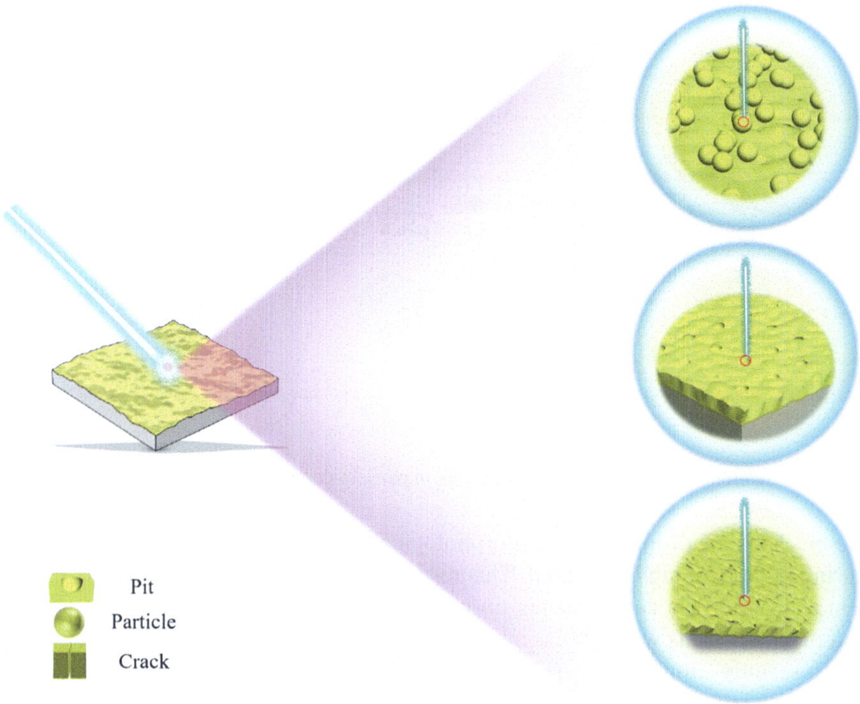


**Fig. 7.2** Schematic diagram of microscopic morphology of heterogeneous coatings

energy absorption, quantified the cleaning forces, and considered both the particle removal threshold and the wafer damage threshold. Research on laser cleaning of particles has mainly focused on exploring cleaning thresholds, while the microscopic mechanisms of particle effects on cleaning efficacy have not been thoroughly discussed, limiting the general applicability of these studies.

In general, the coating on the surface to be cleaned often exhibits non-homogeneous features such as particles, pits (with pores manifesting as pits on the surface), cracks, and dopants (Figs. 7.2 and 7.3). These heterogeneous surface morphologies can interfere with laser reflection and absorption in various ways, leading to phenomena such as localized overheating and stress concentration, which may result in over-cleaning and crack formation, ultimately causing material damage [17]. At present, research on the interaction between laser and material during the laser cleaning process, whether through numerical simulations or experiments, is primarily conducted at the macroscopic level [18, 19]. However, studies on the impact of defect in heterogeneous coatings at the microscopic level on laser cleaning effectiveness remain insufficient. Simulating and analyzing the interaction between complex morphological changes and lasers can provide crucial theoretical support for the cleaning process and subsequent treatments. Therefore, investigating

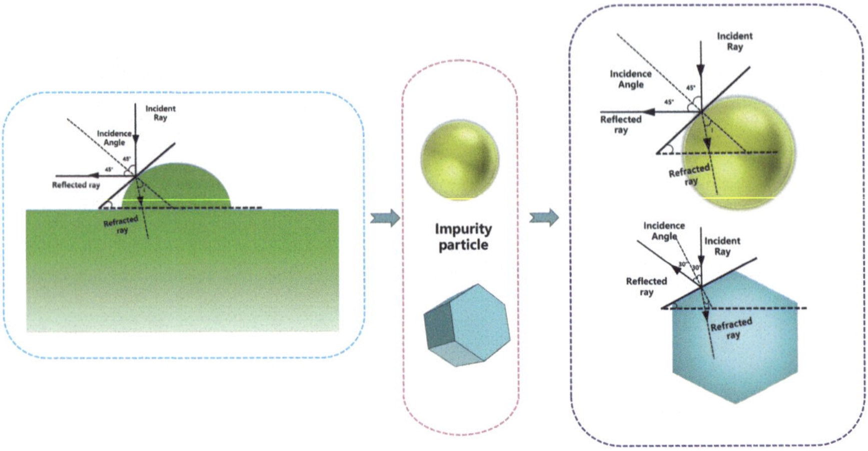


**Fig. 7.3** Theoretical analysis of ray interactions in particle-doped coatings

the mechanisms of microscopic morphological changes in heterogeneous coatings during laser cleaning is essential. This research also offers valuable insights for the removal of surface contaminants on coatings, particles on precision instruments, and other materials with similar microstructural characteristics.

This study employs the geometric optics module to explore the interaction between the propagation path of light and the microscopic morphology of the material surface. The thermal module is used to investigate the temperature distribution changes in the irradiated area after laser exposure on the coating surface. Additionally, the mechanical module simulates the changes in stress distribution following ablation during the laser cleaning process. This approach aims to enhance the understanding and analysis of the laser cleaning mechanisms for heterogeneous coating surfaces, providing a theoretical foundation for laser cleaning experiments.

## 7.2 Laser Cleaning Simulation Process

### *7.2.1 Model Establishment and Material Properties*

#### 7.2.1.1 Geometric Modeling

In this study, three physical models with different microscopic morphological features were established, including particles, pits, and cracks. The models represent a paint coating layer with micron-scale microscopic features. The models and mesh divisions are shown in Fig. 7.4, and the dimensional parameters are listed in Table 7.1.

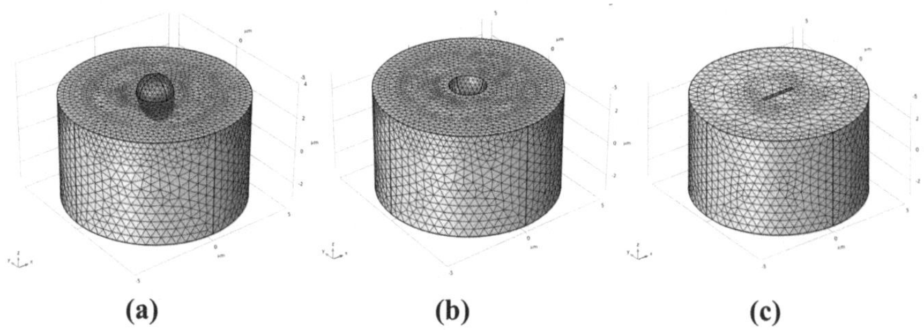

Fig. 7.4 Model and mesh division: (a) particle, (b) pit, (c) crack

Table 7.1 Dimensional parameters of microscopic morphology

| Morphological features | Dimensions (μm) |
|---|---|
| Substrate | Radius 5, height 6 |
| Semi-circular doped particles ($TiO_2$) | Radius 1 |
| Semi-circular particle | Radius 1 |
| Half-circular pit | Radius 1 |
| Triangular prism crack | Base is an isosceles triangle with base length 2 and height 1.5, thickness 2 |

#### 7.2.1.2 Mesh Division

During the finite element simulation of the laser cleaning process, the material surface experiences ablation and vaporization effects, accompanied by significant surface deformation. Therefore, it is essential to refine the mesh in the surface region to capture this dynamic process. Particularly when dealing with the deformation of small particles, pits, and cracks on the material surface, it is necessary to locally densify the mesh around the particles, pits, and cracks, as well as their neighboring areas, to accurately simulate the evolution of the microscopic morphology and its impact on the surrounding region. This adaptive mesh refinement strategy enables better resolution of the dynamic changes in the surface microstructure during laser cleaning, ensuring high accuracy and stability of the simulation.

#### 7.2.1.3 Material Properties

In practical industrial applications, there is a wide variety of non-homogeneous coatings, among which primer surfaces often exhibit relatively rough textures and contain doped particles, displaying pronounced non-homogeneous characteristics. To ensure that the simulation accurately reflects these surface properties, we utilized the thermophysical parameters of epoxy ester primer and the commonly doped $TiO_2$ particles in paint, as provided in Table 7.2, for the simulation.

**Table 7.2** Main physical properties of the material

| Physical properties | Epoxy ester primer | Doped particles ($TiO_2$) |
|---|---|---|
| Constant pressure heat capacity [J/(kg K)] | 1884 | 900 |
| Thermal expansion coefficient [1/K] | 30e−6 | 8.5e−6 |
| Density [kg/m$^3$] | 1200 | 4200 |
| Thermal conductivity [W/(m K)] | 0.5 | 9 |
| Young's modulus [Pa] | 2e9 | 3e11 |
| Poisson's ratio | 0.3 | 0.3 |
| Latent heat of vaporization [kJ/kg] | 2000 | 7500 (>5000) |
| Phase change temperature [K] | 657.55 | 3245 |
| Refractive index, real part | 1.4 | 2.9 |
| Refractive index, imaginary part | 0.01 (<0.01) | 0.01 (<0.01) |

## *7.2.2 Governing Equations*

### 7.2.2.1 Geometric Optics Governing Equations

Ray Tracing

When laser irradiates the surface of inhomogeneous material, the light will be reflected and absorbed. The irregular shape of the material surface will change the original propagation direction of light, and in the process of laser cleaning paint layer, the paint layer on the material surface will be ablated to produce ablation pits, which makes the light propagation path more complicated.

The equations for the rate of change of the light position vector and the wave vector are as follows:

$$\frac{dq}{dt} = \frac{\partial \omega}{\partial k} \tag{7.1}$$

$$\frac{dk}{dt} = -\frac{\partial \omega}{\partial q} \tag{7.2}$$

In the equations, q is the position vector, k is the wave vector, representing the direction of light propagation and wavelength information, w is the frequency of the light wave.

Equation (7.1) describes the propagation direction and speed of light, while Eq. (7.2) describes the changes in the wave vector during light propagation in a medium due to the medium's heterogeneity or reflection and refraction at interfaces. Together, these two equations form the fundamental equations of Hamiltonian optics. Specifically, treating light as particles and solving these equations allows for the determination of the propagation path and direction of light.

Reflection and Absorption Between Laser and Material

When a laser beam strikes the surface of a material, the light interacts with the material, leading to the distribution of energy among reflection, transmission, and absorption. This process follows the principle of energy conservation, which states that the total energy incident on the material interface is equal to the sum of the energy reflected, transmitted, and absorbed by the material:

$$\mathrm{E}_0 = \mathrm{E}_\mathrm{R} + \mathrm{E}_\mathrm{A} + \mathrm{E}_\mathrm{T} \tag{7.3}$$

In this context, assuming that propagation losses in the air medium are negligible, $\mathrm{E}_0$ represents the total energy released by the laser, $\mathrm{E}_\mathrm{R}$ is the energy reflected by the material, $\mathrm{E}_\mathrm{A}$ is the energy absorbed by the material, and $\mathrm{E}_\mathrm{T}$ is the energy transmitted through the material.

The essence of laser cleaning lies in the conversion of the absorbed light energy into thermal energy by the coating on the material surface under laser irradiation, resulting in the removal of the coating through thermal ablation, vibration, and impact. In simple terms, during this interaction, the light energy does not appear or disappear, it merely changes its form of existence. In the widespread application of laser cleaning technology, the substrate materials and their attached coatings are often opaque. Therefore, during laser cleaning, the transmission of incident laser light through the material can be neglected. This assumption simplifies Eq. (7.3):

$$E_0 = E_R + E_A \tag{7.4}$$

Convert according to Eq. (7.4):

$$1 = \frac{\mathrm{E}_\mathrm{R}}{\mathrm{E}_0} + \frac{\mathrm{E}_\mathrm{A}}{\mathrm{E}_0} = \mathrm{R} + \mathrm{A} \tag{7.5}$$

where R is the reflectance of the material, and A is the absorbance of the material.

According to Bouguer's Law, which describes the absorption characteristics of light as it propagates through a medium, the attenuation of light is directly proportional to the path length through the medium and the absorption coefficient of the medium. The expression is given by:

$$I = I_0 e^{-ax} \tag{7.6}$$

where, I is the intensity of light after passing through the medium, $I_0$ is the initial intensity of the incident light, $a$ is the absorption coefficient of the medium, and $x$ is the distance the light travels within the medium. This indicates that the degree of attenuation of light intensity during the propagation of the laser within the material is determined by the material's absorption coefficient $a$.

The complex refractive index n of the material is given by:

$$\mathrm{n} = \beta + \mathrm{ik} \tag{7.7}$$

where β is the real part of the complex refractive index, representing the refractive index of the medium through which light propagates, and k is the extinction coefficient.

According to Fresnel's equations, when the laser is incident on the material, the reflectance R is given by:

$$R = \left( \frac{n\cos\theta - n_1 \cos\theta_1}{n\cos\theta + n_1 \cos\theta_1} \right)^2 \tag{7.8}$$

where $n_1$ is the refractive index of the first medium through which light propagates, $\theta_1$ is the angle of refraction, and θ is the angle of incidence.

When the laser is incident vertically on a planar material from air, the angle between the incident light and the normal to the material surface is 0°. By definition, the angle of incidence is the angle between the incident light and the normal, thus the angle of incidence is 0°. Since the angle of incidence is 0°, according to Snell's law, when light passes perpendicularly from a medium with a lower refractive index (such as air, which has an approximate refractive index of 1) to a medium with a higher refractive index, the angle of refraction will also be 0°, as no refraction occurs. This means the light will return along the same path to the original medium. It is important to note that this situation assumes the material surface is a perfect plane and that the refractive index of the material is isotropic. From Eq. (7.7), it follows that:

$$R = \left| \frac{n-1}{n+1} \right|^2 = \frac{(\beta-1)+k^2}{(\beta+1)+k^2} \tag{7.9}$$

By combining Eqs. (7.5) and (7.9), it follows that:

$$A = 1 - R = \frac{4\beta}{(\beta+1)^2 + k^2} \tag{7.10}$$

The absorption rate A of the material can also be expressed by the following equation:

$$A = \frac{4\pi k}{\lambda} \tag{7.11}$$

where λ represents the laser wavelength.

It can be seen from the above that the material's absorption rate of the laser is related to the laser wavelength, the material's intrinsic properties, and the surface morphology of the sample being cleaned.

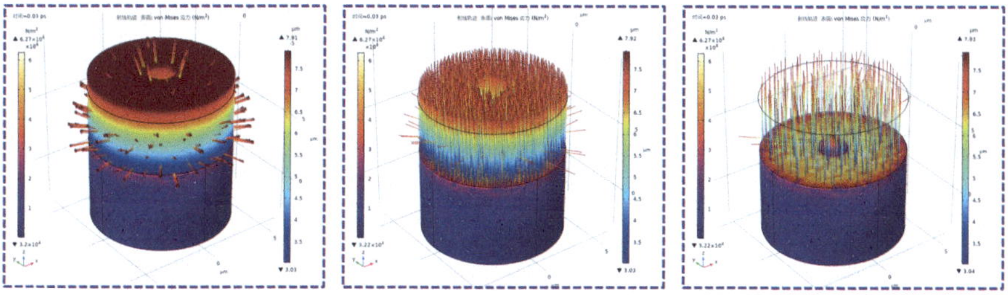


**Fig. 7.5** Ray emission model: (**a**) 20,000 rays; (**b**) 2000 rays; (**c**) 200 rays

Laser Emission and Power Deposition

The laser beam is a highly concentrated stream of photons, both temporally and spatially, with a very small divergence angle and good focusing capability. Using an optical focusing system, the laser beam can be converged into an extremely small range at the micrometer level. However, in the case of very small particles, the laser beam can still reflect and refract within the micrometer range. Therefore, we utilize the boundary release module to converge 20,000 rays within a 5 μm range (as shown in Fig. 7.5) and distribute them in a Gaussian manner to investigate the propagation and absorption of light within the laser beam at the microscopic level.

The Gaussian distribution is a typical model for describing the energy distribution of lasers. As shown in Fig. 7.1, the expression for the distribution function of laser power density is:

$$P_{(x,y)} = \frac{2P_{src}}{\pi r_{laser}^2} \exp\left(\frac{-\left(x^2 + y^2\right)}{r_{laser}^2}\right) \tag{7.12}$$

In the equation, $P_{src}$ represents the total source power of the laser, and $r_{laser}$ denotes the beam radius.

$$Q = \frac{1}{A_i} \Sigma Q_j \tag{7.13}$$

In this equation, Q represents the total deposited power per unit area $A_i$, where $Q_j$ is the power contribution from each individual ray. This accurately reflects the energy deposited by the light on the wall surface, and the total deposited power at the boundary is obtained through the accumulation and averaging of these powers.

#### 7.2.2.2 Solid Heat Transfer Equation

In the process of laser cleaning, the local temperature rise induced by laser irradiation creates a temperature gradient that causes heat flow, leading to the diffusion of temperature within the material. The heat conduction equation plays a crucial role in this process. It describes the temperature distribution and evolution of the material under laser radiation and serves as the foundation for accurately simulating and predicting temperature changes.

The heat conduction equation is:

$$\rho c_p \frac{\partial T}{\partial t} + \rho c_p u \cdot \nabla T + \nabla \cdot q = Q \tag{7.14}$$

In the equation, $\rho$ represents the material density, $c_p$ is the specific heat capacity of the material, T is the temperature, t is time, k is the thermal conductivity of the material, q is the heat flux vector, Q denotes the energy deposition of the light, and $\rho c_p u \cdot \nabla T$ represents the convection term, which describes heat transfer due to fluid motion, where u is the fluid velocity vector.

$$q = -k\nabla T \tag{7.15}$$

The Fourier's law of heat conduction describes the relationship between heat flux density and temperature gradient.

Laser heating can cause the surface material to vaporize or undergo ablation, leading to the formation of vapor or gas flow near the surface. These flow processes remove heat through convection. The convective heat flux can be expressed as:

$$q_0 = h\left(T_{surf} - T_{ref}\right) \tag{7.16}$$

In the equation, h represents the convective heat transfer coefficient, $T_{surf}$ is the temperature of the material surface, and $T_{ref}$ is the ambient temperature.

During the laser heating and ablation process, the material surface gradually decreases due to vaporization, leading to changes in morphology. This change can be described using the Moving Boundary Method. The ablation evolution rate $v_{abl}$ of the material surface morphology can be defined according to the mass balance equation:

$$v_{abl} = \frac{q_0}{\rho \cdot L_{vapor}} \tag{7.17}$$

In this equation, $L_{vapor}$ represents the latent heat of vaporization of the material.

#### 7.2.2.3 Solid Mechanics Equation

The uneven temperature distribution inside the material caused by localized heating from laser irradiation leads to thermal stress. By using the stress equation, it is possible to simulate the material's stress response to temperature changes, preventing excessive deformation or damage. During the laser cleaning and ablation process, the material surface is gradually removed, resulting in residual stress and deformation. This helps in simulating the post-ablation changes in surface morphology and stress concentration regions. The governing equation for solid mechanics motion is expressed as:

$$\rho\frac{\partial^2 u}{\partial t^2}=\nabla\cdot S \tag{7.18}$$

In the equation, ρ represents the material density, u is the displacement vector, t is time, and $\nabla\cdot S$ denotes the divergence of the stress tensor, which reflects the internal stress effects on the material.

#### 7.2.2.4 Multiphysics Thermal-Mechanical Coupling Equations

In practical engineering problems, many materials are subjected not only to external loads but also to temperature changes, leading to alterations in material morphology and subsequently affecting their mechanical properties. Therefore, it is necessary to couple thermodynamics with solid mechanics to express the interactions between thermal effects and mechanical effects.

The expression for the thermal expansion equation is:

$$\epsilon_{th}=\alpha(T)\left(T-T_{ref}\right) \tag{7.19}$$

$\epsilon_{th}$ is the thermal strain, $\alpha(T)$ is the coefficient of thermal expansion at temperature $T$, and $T$ is the current temperature.

In multiphysics coupling, these equations are used to describe the interaction between thermal and mechanical fields. When materials are heated, thermal expansion occurs, leading to thermal strain, which affects the mechanical properties of the material, such as stress and strain distribution. Meanwhile, the thermal diffusion behavior of the material influences the overall temperature distribution of the system. These interactions are simultaneously calculated through the coupling equations, resulting in more accurate simulation results.

The conventional formula for thermal diffusivity only takes into account the thermal conductivity, density, and specific heat capacity of the material, without sufficiently considering the deformation effects during temperature changes. In the thermal-mechanical coupling equations, the thermal diffusivity characteristics are described through the strain and its relationship with the rate of temperature change, with the expression for thermal diffusivity given by:

$$\alpha_t=\left(1+\epsilon_{th}\right)^{-1}\frac{\partial\epsilon_{th}}{\partial T} \tag{7.20}$$

The coupling of the heat transfer field with the thermal expansion field implies that temperature changes will influence the thermal expansion of the material, while the thermal expansion will, in turn, affect the temperature distribution.

### *7.2.3 Calculation Method*

When simulating laser cleaning of heterogeneous coatings at the microscopic level using finite element software, one of the challenges encountered is that changes in the surface morphology of the material can affect the propagation trajectory of the laser rays. To accurately reflect this interaction, this study employs an iterative calculation method. Initially, the coating's initial morphology and material properties are defined. A model of the interaction between the laser and the material is established to simulate the propagation and cleaning process of the laser on the coating, taking into account factors such as laser power density, wavelength, and incident angle.

At each time step, the morphology of the coating surface is updated based on the material removal rate and surface change patterns during the laser cleaning process. Using the updated surface morphology data, the trajectory of the laser rays is recalculated. Due to changes in the surface morphology, the propagation path and intensity distribution of the laser will also change, necessitating corresponding adjustments to the ray trajectories to reflect these variations. Through this process, we can comprehensively capture the dynamic evolution of the coating surface morphology during laser cleaning and the complex relationship between the laser and the material. This iterative calculation method not only improves the accuracy of the simulation but also facilitates a deeper understanding of the evolution of the microscopic morphology of the coating during laser cleaning, thereby optimizing laser cleaning strategies in practical applications.

## 7.3 Results and Discussion

To investigate the surface morphology and doping effects of non-homogeneous coatings on the propagation and absorption of laser light, a low-power laser (0.5 W) is used to irradiate the characteristic morphology of the coating surface, with rays in the beam exhibiting a Gaussian distribution. An iterative time step of 50 ns is set to simulate the coupled effects of microscopic morphology evolution and light propagation during the laser cleaning process. The results of the simulations on the distributions of temperature and stress fields are analyzed and discussed, revealing the key mechanisms of multiphysics coupling arising from changes in the microscopic morphology of the heterogeneous coating under laser action.

## 7.3.1 *The Influence of Particles in Laser Cleaning*

### 7.3.1.1 Ray Tracing

In the finite element simulation of ray tracing results for laser cleaning of the coating, 20 rays were randomly selected as representative objects. The analysis focused on the changes in particle surface morphology under laser irradiation at 100, 300, and 600 ns, as well as the reflections of the rays, as shown in Fig. 7.6.

As shown in Fig. 7.6a, during the initial heating stage at 100 ns of laser irradiation, the coating and surface particles have not yet deformed, and the propagation paths of the rays are mostly consistent with the theoretical model (Fig. 7.3). When the incident angle of the rays with the particles exceeds 45°, some rays are reflected to the surrounding area of the particles, while rays with an incident angle less than

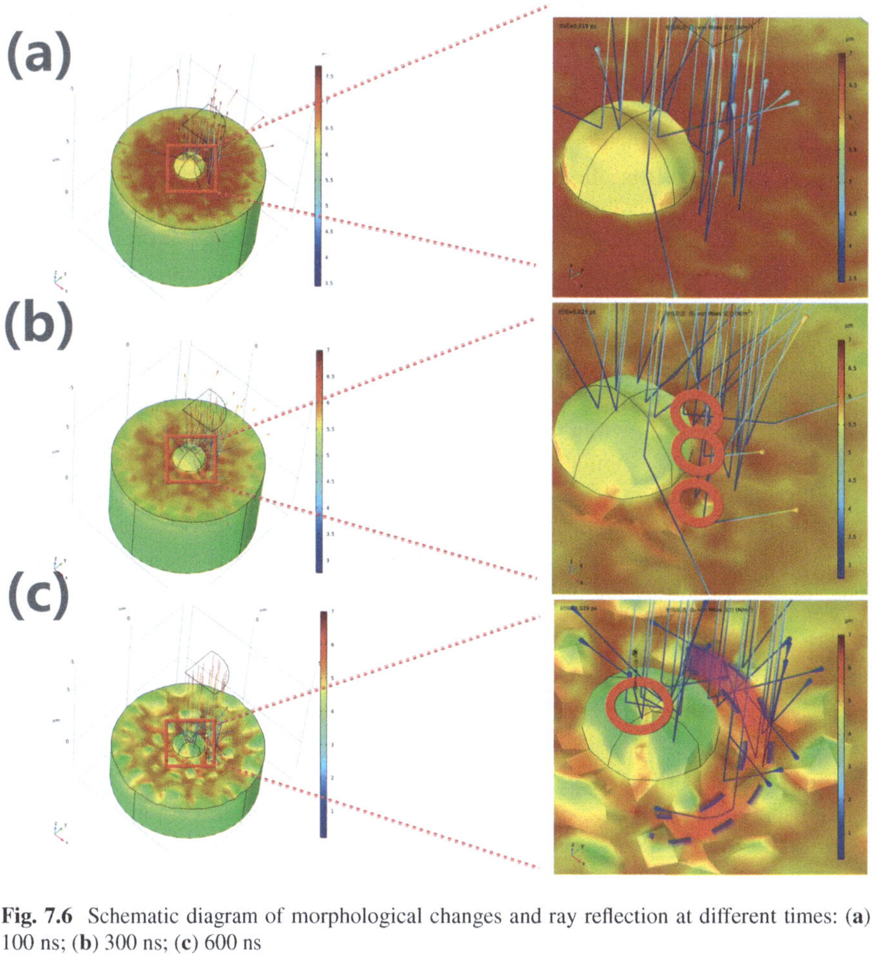


**Fig. 7.6** Schematic diagram of morphological changes and ray reflection at different times: (**a**) 100 ns; (**b**) 300 ns; (**c**) 600 ns

45° tend to scatter in all directions. After 300 ns of laser irradiation, as shown in Fig. 7.6b, the propagation directions of the rays become more chaotic. It is noteworthy that due to the structural characteristics of the particles, the surrounding area is subjected to an increased number of reflected rays, resulting in a higher temperature in this annular region compared to other areas, and the ablation depth is also more pronounced. Based on the three typical ray trajectories marked with red circles in Fig. 7.6b, it can be inferred that as the ablation depth in this region increases, multiple ray reflections will occur inside the pit, leading to further expansion and deepening of the annular region. After 600 ns of laser irradiation, as shown in Fig. 7.6c, the annular region further expands compared to the situation at 300 ns, confirming the previously predicted ablation pattern. Additionally, since the center of the particle is located in the region with the highest laser energy density, it reaches the vaporization temperature faster than other areas, resulting in a noticeable ablation pit. As indicated by the red circle in Fig. 7.6c, multiple reflections and absorptions occur inside the pit, further exacerbating the ablation depth.

#### 7.3.1.2 Temperature Field

As shown in Fig. 7.7, the temperature distribution and conductive heat flux characteristics are derived from the instantaneous results of laser irradiation at 100, 200, 300, and 600 ns. As the duration of laser irradiation increases, the surface temperature of the coating gradually rises, leading to changes in surface morphology. As shown in Fig. 7.7a, it can be observed that the heat in the edge region of the particles is transferred from the periphery to the center (the direction of the arrows indicates the direction of heat transfer). Since the center of the particles is located at the position of highest laser energy concentration, this phenomenon causes the temperature at the particle center to rise more quickly, enabling it to reach the material's ablation vaporization temperature more rapidly than other regions, thereby inducing the phenomenon of particle invagination (Fig. 7.7d).

Isothermal surface analysis was conducted for a specific temperature point in the 10–50 °C range at peak temperatures of 100, 200, 300, and 600 ns (as shown in Fig. 7.8). The results reveal that the temperature distribution is characterized by a dense high-temperature region at the particle center and its surrounding area, exhibiting an overall Gaussian distribution. This distribution feature is consistent with the Gaussian profile of laser energy density.

Through the analysis of the cross-section in Fig. 7.9, the changes in particle morphology and the distribution characteristics of the temperature field during the laser cleaning process are clearly demonstrated. The center of the particle and the surrounding annular area where reflection is concentrated show the most significant ablation depth. Combined with the results of the top arc temperature curve (as shown in Fig. 7.10), it can be observed that during the 0–200 ns time period of laser irradiation, the overall temperature is in a rising state. This is due to the energy being transferred to the material surface during laser irradiation, converting light energy into thermal energy and causing a rapid increase in local temperature.

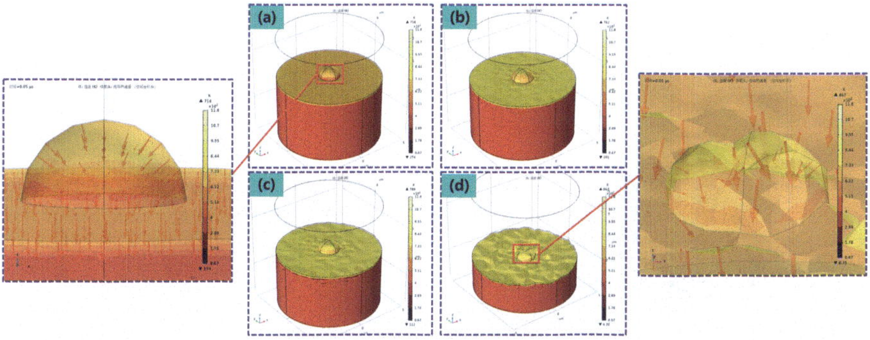


**Fig. 7.7** Schematic of temperature field distribution and conductive heat flux: (**a**) 100 ns; (**b**) 200 ns; (**c**) 300 ns; (**d**) 600 ns

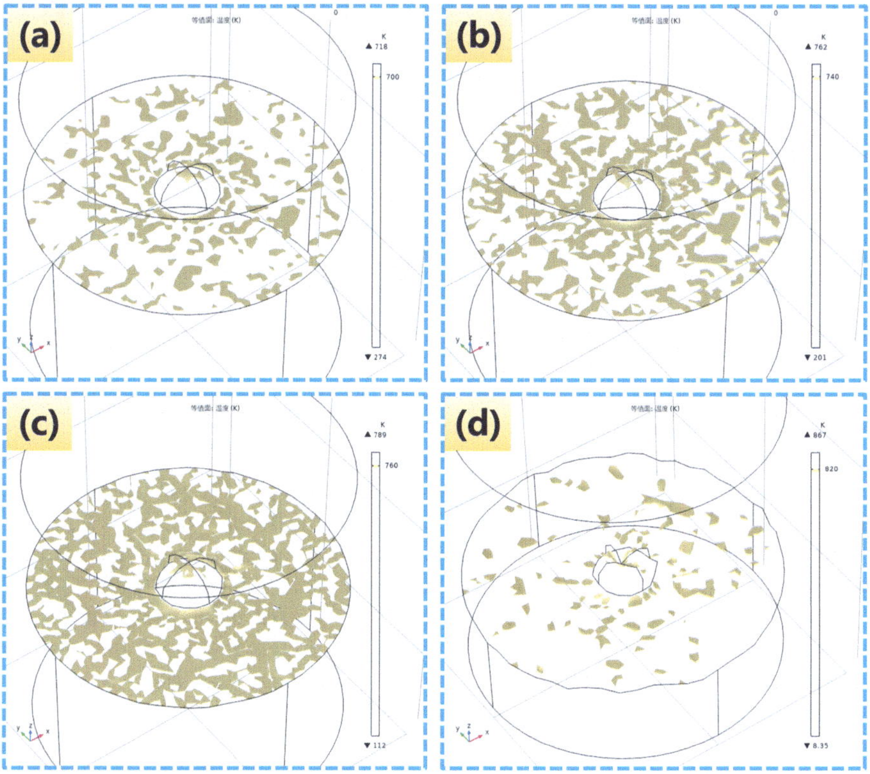


**Fig. 7.8** Schematic diagram of isothermal surfaces of the particles: (**a**) 100 ns; (**b**) 200 ns; (**c**) 300 ns; (**d**) 600 ns

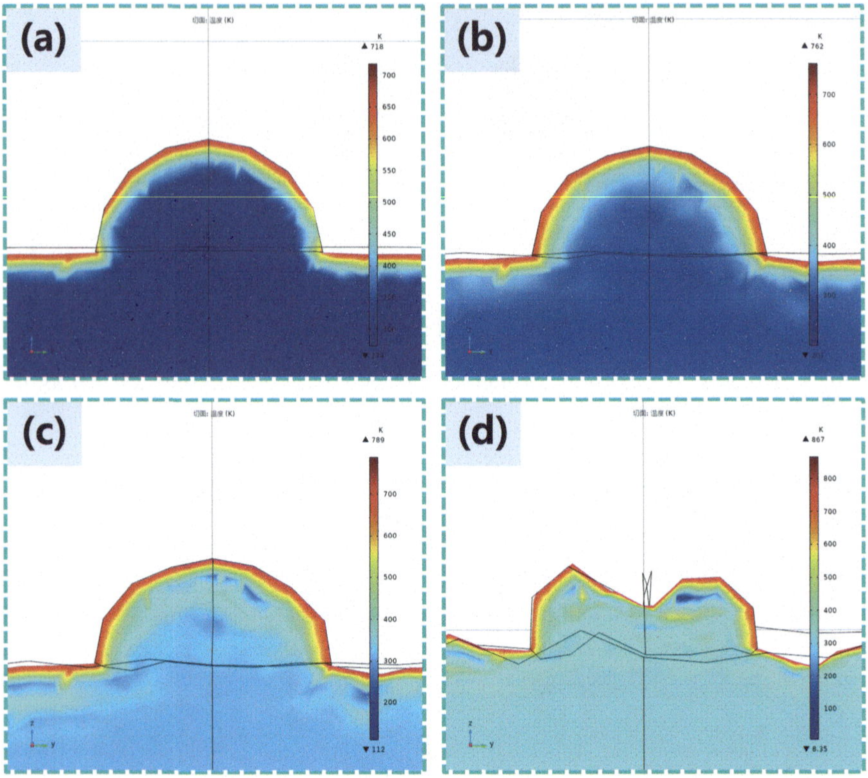


**Fig. 7.9** Schematic diagram of yz cross-sectional temperature distribution: (**a**) 100 ns; (**b**) 200 ns; (**c**) 300 ns; (**d**) 600 ns

Subsequently, heat diffuses into the material through conduction, forming a distinct temperature gradient. After 200 ns, the temperature change gradually becomes more moderate, primarily because when the surface temperature of the material reaches the vaporization temperature due to laser irradiation, the material undergoes a phase change from solid to gas. This phase change process requires the absorption of a large amount of vaporization latent heat, resulting in a significant slowdown in the rate of temperature increase. When the sum of the vaporization latent heat absorbed by the material and the heat conducted into the material reaches equilibrium with the energy absorbed at the surface, the temperature change stabilizes, and the surface temperature no longer increases significantly.

#### 7.3.1.3 Stress Field

As shown in Fig. 7.11a, the direction of the arrows indicates the displacement direction of the material due to thermal expansion. The size of the arrows represents the magnitude of the velocity at that point, indicating the displacement distance per unit

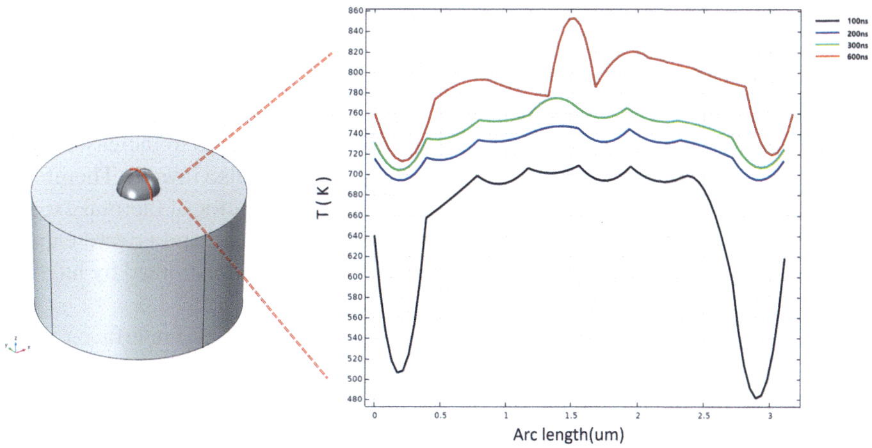


**Fig. 7.10** Schematic diagram of top arc temperature curve

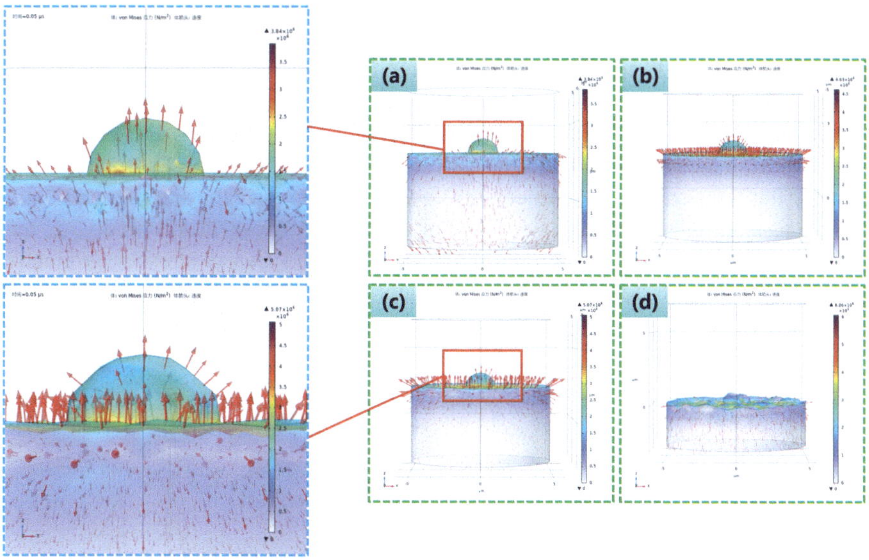


**Fig. 7.11** Schematic diagram of thermal expansion velocity: (**a**) 100 ns; (**b**) 200 ns; (**c**) 300 ns; (**d**) 600 ns

time. This shows that at the initial stage of laser irradiation, specifically at 100 ns, the surface temperature rises rapidly, and the material surface experiences thermal expansion first. At this time, the surface expands faster than the interior, leading the velocity arrows to point outward from the interior toward the surface, indicating that the surface is expanding outward. As the laser irradiation continues, as shown in Fig. 7.11c, the internal temperature of the material gradually increases, and the temperature gradient between the surface and the interior narrows. During this process, the stress generated by thermal expansion at the surface gradually transfers into the material's interior, making the expansion behavior more complex. Due to the uneven expansion of the surface and the interior, even though the surface temperature remains high, the degree of internal expansion continues to increase, causing some areas of the surface to expand not only outward but also inward. Therefore, under the influence of this accumulated stress, the velocity vector of thermal expansion points toward the surface, showing a trend of both outward and inward expansion, reflecting the complex stress state in the surface region constrained by internal expansion.

As shown in Fig. 7.12, at different time points, the stress in the interface region between the particles and the substrate surface (the circumferential part at the bottom of the particles) exhibits a wavy distribution as a function of arc length. During the initial 100 ns, the stress is relatively low because, although the surface temperature rises rapidly at the onset of laser irradiation, significant internal temperature gradients have not yet formed due to the time required for thermal diffusion and heat transfer. In this very short period, mechanical deformation lags behind temperature changes. Thus, even though the surface temperature increases, the temperature in deeper layers remains low, resulting in overall thermal expansion being relatively small and keeping the stress at a low level. As time progresses, the stress increases and stabilizes. This is due to heat gradually conducting into the material, creating a significant temperature gradient that leads to non-uniform thermal expansion, resulting in substantial thermal stress within the material. Throughout this process, the temperature difference between the surface and the interior gradually stabilizes, leading to dynamic equilibrium of internal stress. Additionally, as the material undergoes vaporization and ablation, localized excess stress is gradually released. The plastic deformation of the material and rebound pressure also help

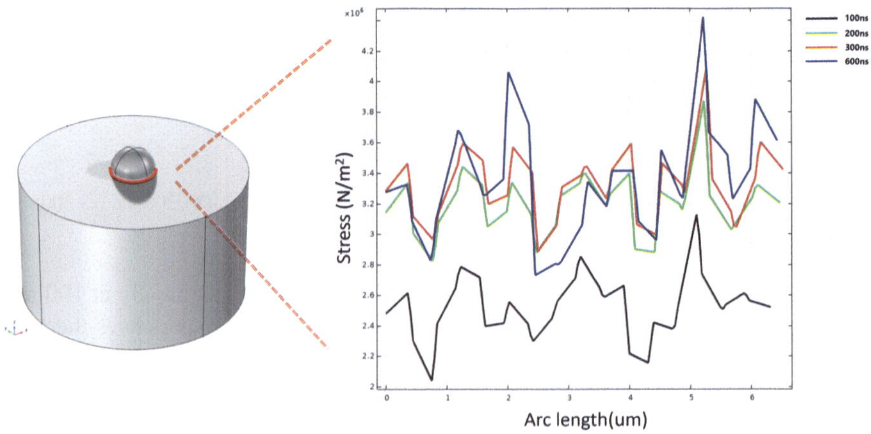


**Fig. 7.12** Schematic diagram of stress around particles changing with arc length at different times

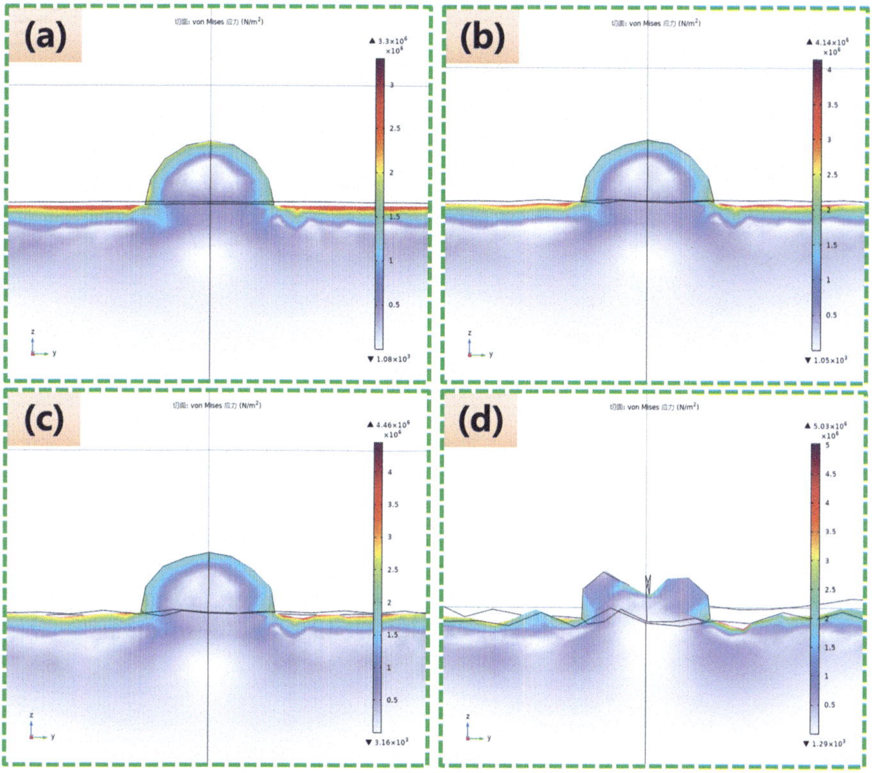


**Fig. 7.13** Schematic diagram of stress cross-section: (**a**) 100 ns; (**b**) 200 ns; (**c**) 300 ns; (**d**) 600 ns

alleviate stress concentrations, resulting in a relatively stable stress state after reaching a peak.

Thermal expansion induced by laser irradiation generates localized stress waves, which may interact with the material's structure and boundaries as they propagate internally. Due to boundary effects, complex stress concentrations can form at the interface between the particles and the substrate during heat conduction and stress transfer. The stress near the boundary exhibits periodic fluctuations with distance, ultimately resulting in a wavy stress distribution.

By comparing Figs. 7.13 and 7.14, it can be observed that regions with deeper ablation exhibit higher stress. The concentrated laser energy acting on the surface of the paint causes a rapid increase in local temperature, leading to thermal expansion of the material within and around the particles. This temperature gradient results in the generation of internal stress. During the ablation process, the paint material may undergo phase changes (such as melting or vaporization), resulting in changes in volume and further causing stress concentration. The stress distribution is influenced by various factors, including thermal conduction paths, material nonlinear responses, and geometric shapes. Since the surface of the particle is curved, the

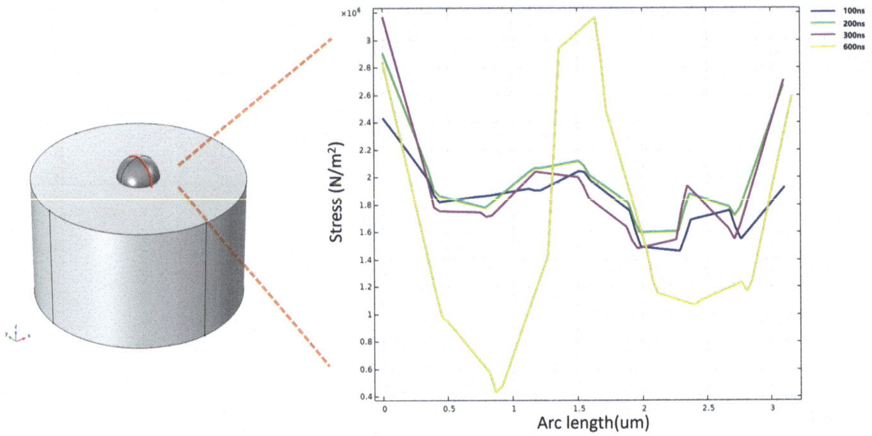


**Fig. 7.14** Schematic diagram of top arc stress curve at different times

internal stress varies with changes in curvature. Differences in curvature at different arc lengths lead to non-uniform stress distribution, with areas of larger curvature being prone to stress concentration, while regions with smaller curvature experience relatively lower stress, resulting in an overall ridge-like stress distribution.

#### 7.3.1.4 $TiO_2$ Particle Doping

As shown in Fig. 7.15a, the stress within the $TiO_2$ doped particles is significantly higher than that in the paint layer, which is due to the higher Young's modulus of the particles, resulting in increased stress under thermal or mechanical loading. Figure 7.15c indicates the presence of stress concentration at the interface between the doped particles and the paint layer, primarily caused by differences in thermal properties such as thermal conductivity, coefficient of thermal expansion, and specific heat capacity between the two materials. The higher thermal conductivity of the $TiO_2$ particles enables them to more rapidly transfer absorbed heat to cooler regions (as shown in Fig. 7.15e), thus maintaining a relatively lower temperature (as illustrated in Fig. 7.15b). Additionally, the specific heat capacity of the $TiO_2$ particles is relatively low, while the organic components of the paint typically have a higher specific heat capacity. This means that under the same energy input, the paint absorbs more heat, resulting in a faster temperature increase. Coupled with the difference in thermal expansion coefficients between the two materials, this leads to a mismatch in thermal expansion at the particle-paint interface, causing higher stress around the particle and lower stress at the center. During the laser cleaning process, the removal of $TiO_2$ particles mainly relies on factors such as the thermal expansion mismatch, mechanical stress, shock waves, and localized laser-induced vibrations. Although $TiO_2$ itself does not vaporize or decompose during the cleaning process, the stress and shock waves generated by the paint material surrounding the particles

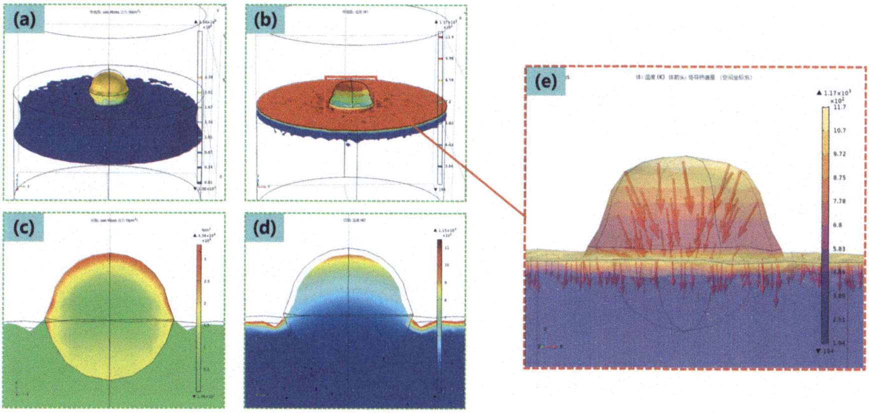

Fig. 7.15 Temperature and stress distribution at 200 ns: (a) Stress isosurface, (b) Temperature isosurface, (c) Stress cross-section on the yz-plane, (d) Temperature cross-section on the yz-plane, (e) Schematic of conductive heat flux

due to laser heating are sufficient to dislodge the particles from the paint surface, thereby achieving the cleaning effect.

## 7.3.2 *The Influence of Pits in Laser Cleaning*

### 7.3.2.1 Ray Tracing

In the demonstration of the ray tracing paths around the pit, 20–30 rays were selected as characteristic objects to observe the changes in pit morphology and ray propagation paths at 100 and 300 ns (Fig. 7.16). The ray tracing results at 100 ns are shown in Fig. 7.16a. Unlike the center point of the particle, which absorbs light energy only once during the initial irradiation, the geometric shape of the pit causes the laser energy to be highly concentrated not only at the center but also to reflect a significant amount of light towards the bottom of the pit. This results in the energy absorbed at the center of the pit being far greater than that in the side regions. As the irradiation time increases, the pit gradually deepens, and the number of reflections of rays in the vicinity of the pit bottom significantly increases (as shown in Fig. 7.16d), causing the ablation area to extend not only downward but also to expand outward from within the pit.

### 7.3.2.2 Temperature Field

At the microscopic level, local thermal expansion can induce micro-stress concentrations within the material. Similar to particles, the curvature and geometry of pits can amplify this effect, potentially leading to the formation of micro-cracks or local

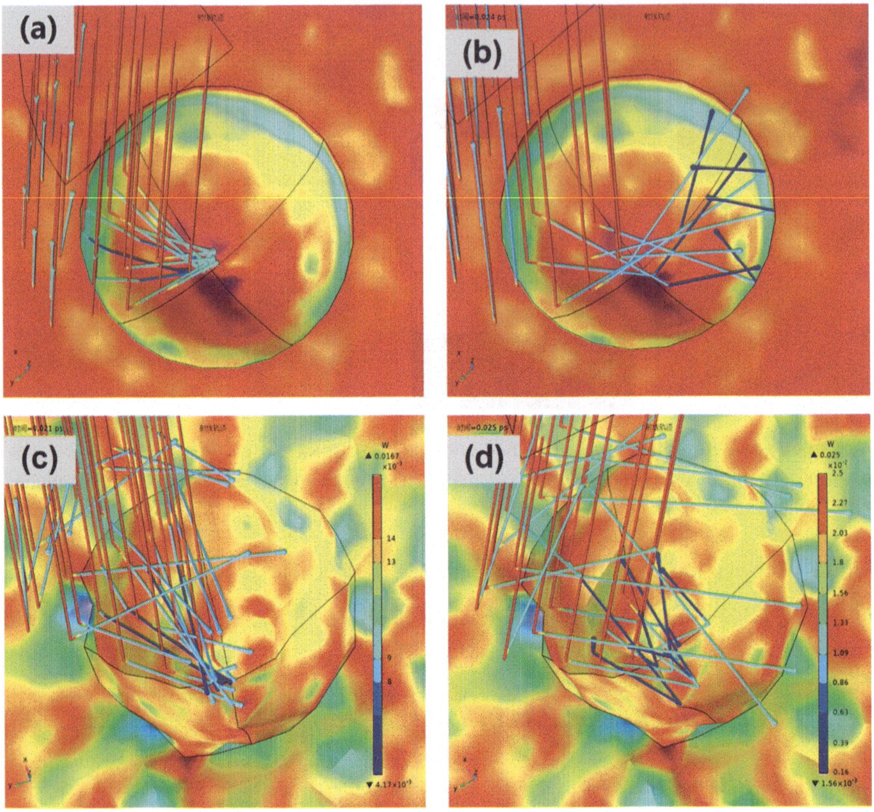


**Fig. 7.16** Demonstration of ray tracing paths around the pit: (**a**, **b**) at 100 ns; (**c**, **d**) at 300 ns

melting on the material surface. The accumulation of localized heat causes a sharp rise in the temperature gradient within the pit. As shown in Fig. 7.17, the thermal flux conduction rate within the pit is relatively uniform at 100 ns, whereas at 400 ns, the thermal flux conduction rate at the center bottom of the pit significantly increases, which may lead to uneven cleaning effects. Furthermore, excessive heat accumulation at the center bottom could damage the substrate.

Figure 7.18 illustrates the isothermal surfaces at four different time points, with the isothermal values set below 40 °C of the maximum temperature at each moment. As time progresses, the isothermal surfaces gradually shrink, indicating that the areas with temperatures exceeding a specific value are decreasing, reflecting a reduction in the high-temperature region. This phenomenon suggests that heat may become more concentrated or dissipated locally. With the narrowing range of isothermal surfaces, temperature variations become steeper, and the transition between high and low-temperature areas becomes more pronounced, indicating an increase in the temperature gradient at the local point (center of the pit). This increased temperature gradient may lead to significant thermal stress concentrations at the bottom

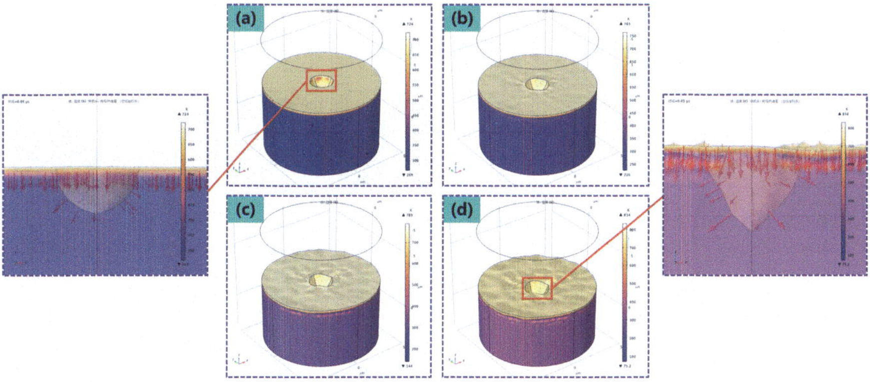


**Fig. 7.17** Schematic diagram of temperature field distribution and conductive heat flux: (**a**) 100 ns; (**b**) 200 ns; (**c**) 300 ns; (**d**) 400 ns

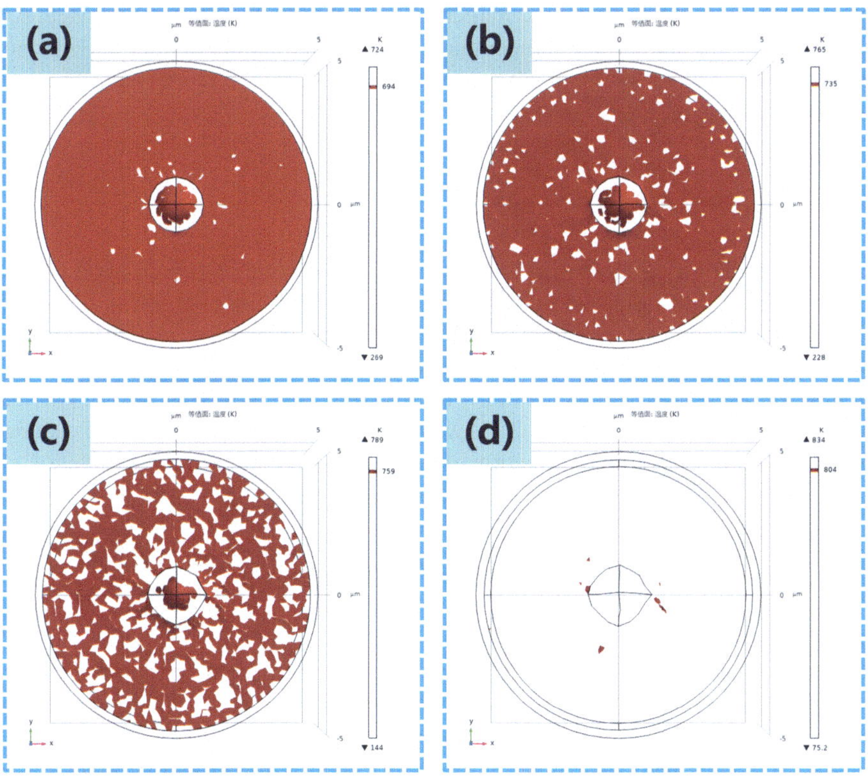


**Fig. 7.18** Schematic diagram of isothermal surfaces of the pit: (**a**) 100 ns; (**b**) 200 ns; (**c**) 300 ns; (**d**) 400 ns

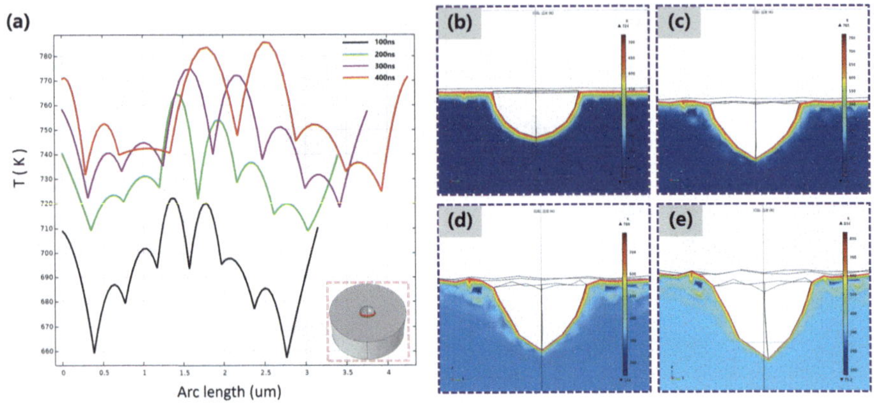


**Fig. 7.19** (**a**) Bottom arc temperature curve of yz; yz cross-sectional temperature schematic: (**b**) 100 ns; (**c**) 200 ns; (**d**) 300 ns; (**e**) 400 ns

of the pit, potentially causing thermal cracking or localized structural damage. As seen in Fig. 7.19a, the overall temperature in the pit area stabilizes after 200 ns, and its formation mechanism is largely consistent with the temperature changes in the particle region.

#### 7.3.2.3 Stress Field

Figure 7.20 shows the isosurfaces of stress in the yz plane at 100 and 300 ns. As time progresses, the initial stress concentration, which was only present at the center of the bottom of the pit, gradually expands to a larger area at the bottom of the pit. When combined with the ray tracing results, it can be analyzed that when the ablation depth reaches a certain extent, a stress impact will form between the interior and the surface (as shown in Fig. 7.21), which accelerates the peeling of the coating near the pit and leads to the formation of larger and more irregular pits.

### *7.3.3 The Influence of Crack in Laser Cleaning*

#### 7.3.3.1 Ray Tracing

By comparing the xz plane ray trajectories in Fig. 7.22a, b, it is observed that at 100 ns, light reflects multiple times within the crack, but its propagation range is limited to the inclined portions of the crack. In contrast, at 300 ns, the morphology of the crack changes; the originally vertical portions become irregular, resulting in a chaotic light propagation path that begins to concentrate towards the bottom of the crack. The elongated crack morphology gradually expands into a wider pit morphology (as shown in Fig. 7.22c).

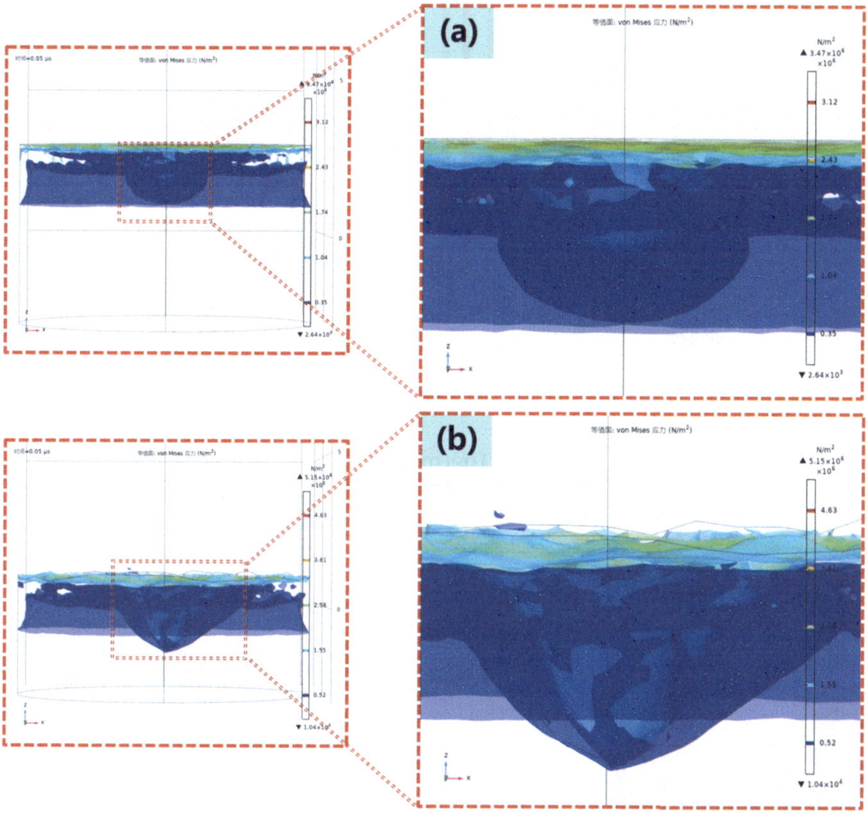


**Fig. 7.20** yz cross-sectional stress isosurfaces: (**a**) 100 ns; (**b**) 300 ns

#### 7.3.3.2 Temperature Field

Figure 7.23 illustrates the morphological evolution of the crack coating under laser irradiation at different time intervals. As time progresses, a noticeable depression around the crack can be observed, as shown in Fig. 7.23d, e. The inverted triangular depression around the crack is attributed to the higher overall temperature in the near-surface region, which reaches the vaporization temperature more quickly compared to the deeper parts of the crack.

As shown in Fig. 7.24, the temperature distribution at the crack and its variation with laser irradiation time can be visually observed. The highest temperature occurs at the interface between the crack and the plane, while the overall temperature inside the crack is lower than that of the plane. At 100 ns, due to geometric factors, edges 1 and 2 of the crack are perpendicular to the incident light, receiving less radiation. By 200 ns, the overall temperature of the crack has reached the phase transition temperature, and the morphological changes lead to an increased amount of received light. By 600 ns, the temperature in certain regions (at the position of 3 μm arc length) approaches that of the plane.

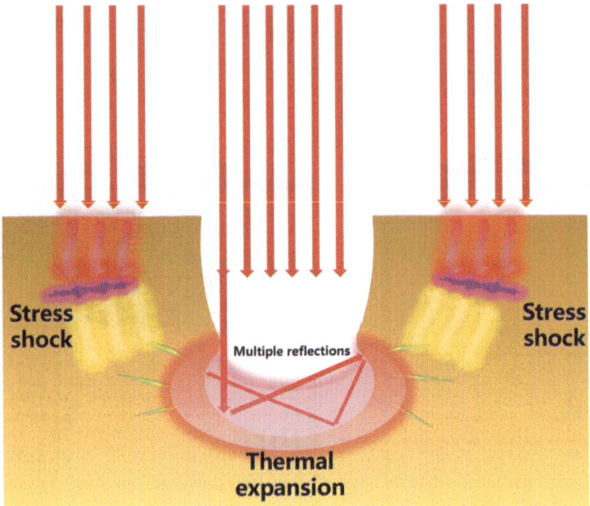


**Fig. 7.21** Schematic diagram of internal and external stress shock

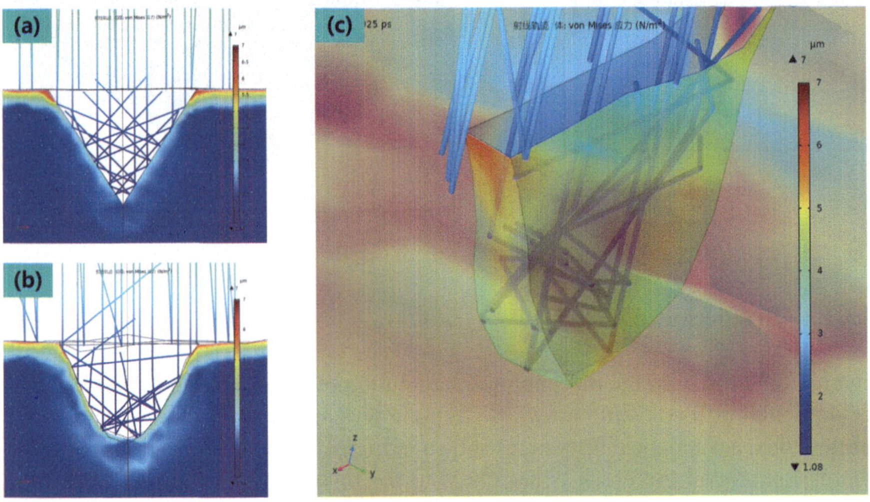


**Fig. 7.22** Ray tracing diagram of cracks: (**a**) 100 ns, xz plane; (**b**) 300 ns, xz plane; (**c**) 300 ns, magnified ray tracing diagram

#### 7.3.3.3 Stress Field

Through Fig. 7.25, it can be observed that whether under laser irradiation for 100 ns or 300 ns, the stress is highest at the junction between the crack and the plane, which is also the position with the highest temperature. On the premise of little change in the overall morphology of cracks, the stress distribution is notch-shaped. From the

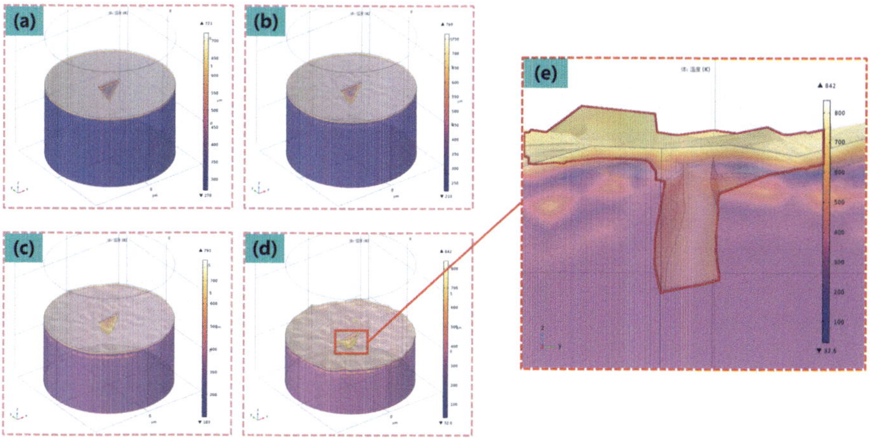


**Fig. 7.23** Temperature and morphology of cracks at different times: (**a**) 100 ns; (**b**) 200 ns; (**c**) 300 ns; (**d**) 600 ns

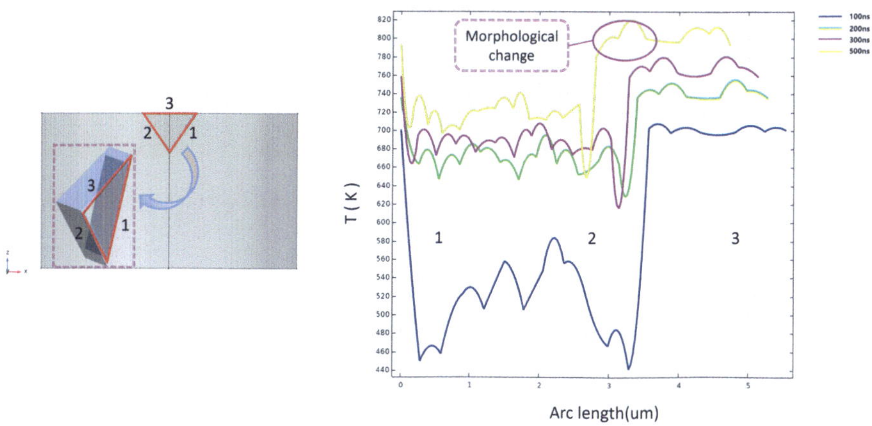


**Fig. 7.24** Temperature curve of the zx plane of the crack

stress isosurface map in the xz plane, it is evident that at 100 ns of laser irradiation, the internal stress within the crack is relatively low, and the stress at the bottom of the crack is significantly lower than in other areas (Fig. 7.26a). This shows a considerable difference compared to the stress distribution in the pit at 100 ns of laser irradiation. At 300 ns of laser irradiation, the stress distribution around the crack is similar to that of the pit morphology, with concentrated stress surrounding the defect; however, in the case of the crack, the stress is primarily concentrated in the upper-middle region of the defect (Fig. 7.26b), while in the pit, it is concentrated in the lower-middle region.

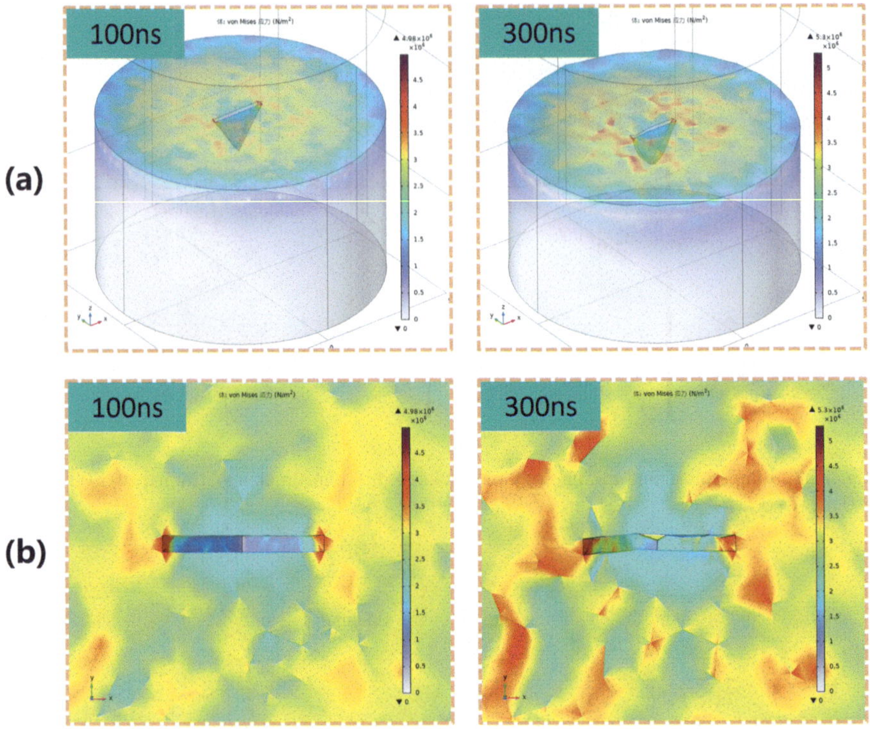


Fig. 7.25 Stress distribution of the crack: (**a**) Stress distribution at different times; (**b**) Distribution of xy stress isosurfaces at different times

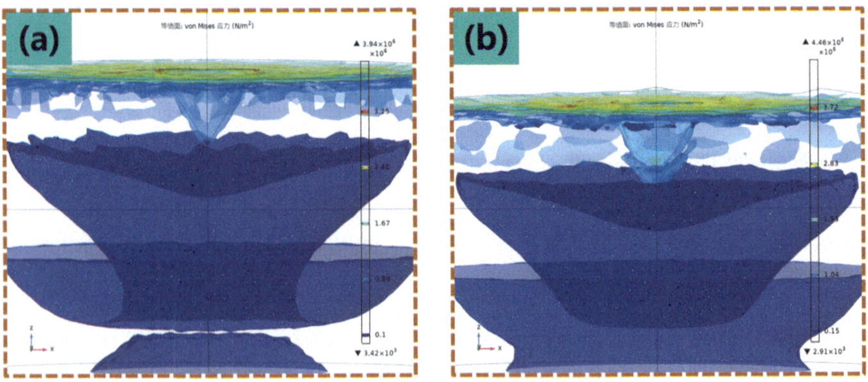


Fig. 7.26 xz plane stress isosurface of the crack: (**a**) 100 ns; (**b**) 300 ns

## 7.4 Conclusion

This study analyzes the key physical phenomena in the process of laser cleaning heterogeneous coatings through numerical simulations, revealing the impact of defects and impurities (such as particles, pits, and cracks) on the cleaning effectiveness. The results indicate that the microtopography and compositional makeup of the coating surface play a critical role in the absorption and reflection of laser energy, as well as in the local temperature and stress distribution. Consequently, the following conclusions can be drawn:

1. The local temperature changes induced by laser irradiation can vary significantly due to the geometric shapes and compositional makeup of microscopic defects in the coating (or contaminant layer). The particle and pit structure will lead to energy concentration in a specific area, resulting in local overheating, which will accelerate the ablation and cleaning process. However, this localized high temperature also increases the risk of excessive cleaning and surface damage, which means that the process parameters of laser cleaning must be reasonably adjusted according to the properties and structural characteristics of the coating.
2. The distribution of thermal stress is influenced by the surface morphology, with areas of pits and cracks being prone to stress concentration, which may lead to the formation and propagation of microcracks in the material. The propagation of these cracks can act as a driving force for coating peeling. The morphology and composition of doped particles in the coating will lead to stress concentration, which is a complex stress transfer process on the surface and inside, further aggravating the local expansion and deformation of the material and causing the coating to peel off.
3. The changes in the microtopography of the coating surface not only affect the laser absorption efficiency but also have a significant impact on the subsequent laser transmission path. The presence of pits and cracks causes multiple reflections of the light, further increasing localized energy accumulation, which may lead to uneven cleaning effects and potential substrate damage. Therefore, it is essential to timely adjust the cleaning parameters during the process to ensure uniform distribution of laser energy and reduce the risk of substrate damage.

In summary, the presence of microdefects and dopants on the surface of heterogeneous coatings significantly impacts the efficiency and safety of laser cleaning, making them critical factors in elucidating the mechanisms of laser cleaning and regulating the cleaning process. A deeper understanding of the characteristics of heterogeneous coatings can enhance the accuracy of explaining laser cleaning mechanisms and provide effective guidance for the scientific regulation of laser cleaning processes. This understanding is important for reducing the risk of material damage during cleaning and improving the controllability and precision of the overall process. Furthermore, although this study primarily uses primer as the main model, the foundational research can also lay the groundwork for the theoretical removal of other contaminants and heterogeneous coatings such as oxide films.

## References


1. Zhu G, Xu Z, Jin Y, et al. Mechanism and application of laser cleaning: A review [J]. Optics and Lasers in Engineering, 2022, 157: 107130.
2. Song Y, Wang S, Pan Y, et al. Comparative study of infrared nanosecond laser surface paint removal for carbon fiber reinforced polymer and glass fiber reinforced polymer [J]. Journal of Cleaner Production, 2024, 451: 142061.
3. Li J, Yang J, Liu J, et al. Research on Laser Cleaning Technology for Aircraft Skin Surface Paint Layer [J]. Materials, 2024, 17(10): 2414.
4. Ye Y, Jia B, Chen J, et al. Laser cleaning of the contaminations on the surface of tire mould [J]. International Journal of Modern Physics B, 2017, 31(16–19): 1744100.
5. Rezk R A, Abdel Ghany N A, Mostafa A M. Laser-assisted method for cleaning and analysis of archaeological metallic coins [J]. Coatings, 2022, 12(10): 1548.
6. Mei L, Lin L, Yan D, et al. Theoretical and experimental research on nanosecond laser cleaning of $SiO_2$ particles on silicon wafer surface [J]. The International Journal of Advanced Manufacturing Technology, 2024, 130(3): 1693–1708.
7. Li Z, Wang S, Zheng W, et al. A review of dynamic monitoring technology and application research of laser cleaning interface [J]. Measurement, 2024: 115311.
8. Chen T, Wang W, Tao T, et al. Multi-scale micro-nano structures prepared by laser cleaning assisted laser ablation for broadband ultralow reflectivity silicon surfaces in ambient air [J]. Applied Surface Science, 2020, 509: 145182.
9. Zhu G, Zhang D, Zhang L, et al. Environmentally friendly photothermal reconstruction of cement/asphalt road: Can road markings also be cleaned in such unexpected ways? [J]. Journal of Cleaner Production, 2024, 434: 140347.
10. Zhang G, Hua X, Huang Y, et al. Investigation on mechanism of oxide removal and plasma behavior during laser cleaning on aluminum alloy [J]. Applied Surface Science, 2020, 506: 144666.
11. Gu J, Su X, Jin Y, et al. Towards low-temperature laser paint stripping by photochemical mechanism on CFRP substrates [J]. Journal of Manufacturing Processes, 2023, 85: 272–280.
12. Guo J, Cai H, Han J, et al. Research on Laser Cleaning Simulation Model Based on Thermal Ablation Effect (Invited) [J]. Infrared and Laser Engineering, 2023, 52(2): 20220779-1–20220779-9.
13. Hu P, Chen F. Simulation Analysis of Thermal Effects on Metal Materials Under Repeated Frequency Pulsed Laser Irradiation [J]. Chinese Journal of Lasers, 2016, 43(10): 1001006-1.
14. Li Z, Xu J, Zhang D. Finite Element Simulation of Temperature Field During Laser Cleaning of Oxide Film on TA15 Titanium Alloy [J]. Scientia Sinica Technologica, 2022, 52(2): 318–332.
15. Guo L, Li Y, Geng S, et al. Numerical and experimental analysis for morphology evolution of 6061 aluminum alloy during nanosecond pulsed laser cleaning [J]. Surface and Coatings Technology, 2022, 432: 128056.
16. Ding K, Zhou K, Feng G, et al. Mechanism and conditions for laser cleaning of micro and nanoparticles on the surface of transparent substrate [J]. Vacuum, 2022, 200: 110987.
17. Hou L, Yin F, Wang S, et al. A review of thermal effects and substrate damage control in laser cleaning [J]. Optics & Laser Technology, 2024, 174: 110613.
18. Wang J, Yang Y, Qi J, et al. Thermodynamic simulation, surface morphology and bending property of carbon fiber reinforced polymer composite material subjected to laser cleaning [J]. Optics & Laser Technology, 2022, 152: 108099.
19. Wang T, Lu S, Liu H, et al. Numerical Simulation and Experimental Study of Nanosecond Pulsed Laser Cleaning of Aluminum Alloy Surface Paint Layer [J]. Journal of Russian Laser Research, 2024, 45(1): 106–116.

# Chapter 8
# Substrate Surface Morphology Control After Laser Paint Removal on Aluminum Alloy

## 8.1 Introduction

Aluminum alloy, as a lightweight metallic material, is widely used in various fields. To prevent unnecessary damage and corrosion, a protective paint layer is usually applied to its surface. However, during routine inspection and maintenance, the surface paint layer needs to be removed [1, 2]. Traditional cleaning methods are commonly used in industrial applications but have limitations. For example, mechanical abrasion cleaning relies on various specialized tools to remove the contaminant layer through physical contact and friction [3]. However, this direct physical interaction and the high intensity of mechanical impact can easily damage the workpiece surface, failing to meet the strict requirements of high cleanliness and precision [4], and the method is inefficient and poses poor working conditions. The high-pressure water jet cleaning method consumes a large amount of water at the initial stage, and the polluted water is difficult to recycle [5]; in addition, it carries the risk of secondary rusting on the metal surface. Chemical cleaning, although widely used, can easily cause environmental pollution and pose health risks to operators [6]. Ultrasonic cleaning is limited by the size of the cleaning tank, which restricts the size of workpieces that can be cleaned. Therefore, there is an urgent need for a novel cleaning method. With the increasing reliability of lasers and significantly reduced production and maintenance costs, laser-based removal of surface contaminants has been widely applied [7–10].

In laser cleaning of aluminum alloys, it is essential to consider the characteristics of both the contaminants and the aluminum alloy substrate [11], because different laser process parameters lead to different cleaning results [12, 13]. Zhang et al. [14] found that using a nanosecond pulsed laser with a wavelength of 1064 nm to remove red and blue paint layers, the blue paint layer had a higher absorption coefficient under the same laser energy. As a result, most of the laser energy was absorbed by the blue paint, leading to a much higher temperature than the red paint layer. When



S. Wang, *Laser Cleaning*, https://doi.org/10.1007/978-981-95-2935-3_8

the temperature reached the boiling point of the blue paint, it evaporated and was removed. For the red paint, however, the laser energy penetrated into the substrate during the cleaning process, increasing the substrate temperature. Since the paint did not evaporate, thermal stress caused by the elevated substrate temperature led to paint delamination. Tian et al. [15] used a nanosecond pulsed fiber laser to clean marine biological fouling on aluminum alloy surfaces. A laser energy density between 2.76 and 5.52 $J/cm^2$ achieved excellent surface quality, and at 5.52 $J/cm^2$, melting and microstructuring occurred on the aluminum alloy surface, imparting superhydrophobic characteristics. Jing et al. [16] employed a 100 ns laser to clean the oxide layer on 7075 aluminum alloy. Due to the significantly higher melting point of the oxide layer compared to the alloy, high energy was required for effective cleaning. When the laser power was less than 32 W, a large amount of oxide layer remained. At 48 W, the oxide layer was significantly reduced, but surface discoloration began to appear. At 80 W, the oxide removal was optimal and the surface morphology was best. However, when the laser power exceeded 96 W, cracks appeared on the alloy surface. Wang et al. [17] studied the effect of different laser energy densities on the removal of the oxide film from 7075 aluminum alloy. They found that when the energy density ranged between 1.43 and 1.82 $J/cm^2$, most of the original oxide film was effectively removed without noticeable substrate melting. As the laser energy density increased from 1.82 to 2.21 $J/cm^2$, newly formed magnesium oxide and aluminum oxide layers on the surface blocked direct contact between the salt spray and the aluminum alloy substrate, thereby enhancing post-cleaning corrosion resistance. However, excessive energy density removed the newly formed oxide films, and the resulting non-uniformly melted surface promoted salt solution accumulation, reducing the corrosion resistance after cleaning.

Previous research has shown that during laser cleaning, due to the low melting point of aluminum alloys, non-uniform contaminant distribution, high thermal conductivity, and difficulty in obtaining optimal process parameters, complete removal of contaminants via laser inevitably affects the aluminum alloy surface [18, 19]. The changes that occur on the metal surface after laser cleaning are a focus of current research, yet they have not been fully studied. In this work, both high-energy-density single-pass cleaning and low-energy-density dual-pass cleaning methods were applied to paint layers on aluminum alloy surfaces, aiming to optimize laser processing parameters by adjusting laser energy density and spot overlap ratio. The surface morphology, cleanliness, and performance of the substrate after laser cleaning were analyzed.

## 8.2 Experimental Materials and Methods

### *8.2.1 Laser Cleaning Equipment*

The laser cleaning system utilized a nanosecond pulsed laser with a wavelength of 1064 nm. A schematic diagram of the laser cleaning device is shown in Fig. 8.1. The system mainly consists of a laser source, control system, optical fiber, scanning

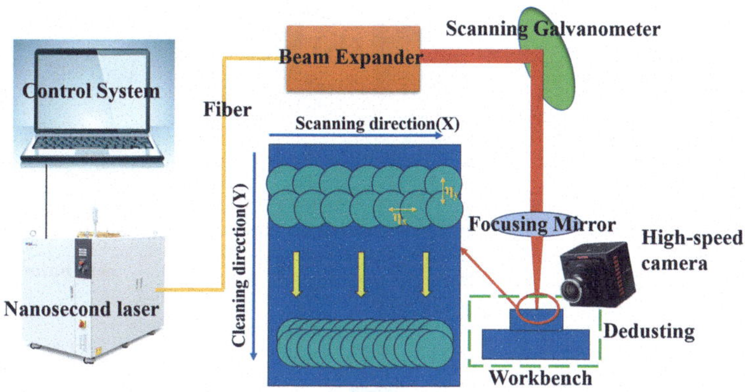


**Fig. 8.1** Schematic diagram of laser cleaning equipment

galvanometer, focusing lens, working platform, and dust removal unit. The laser beam spot conforms to a Gaussian distribution, and the scanning path follows an S-shape. A high-speed camera was used to observe the dynamic behavior of the laser cleaning process.

### 8.2.2 *Sample Preparation*

In this study, commercially available 6061 aluminum alloy sheets with a thickness of 1 mm were selected as the base material, and processed into specimens of 15 × 15 × 1 mm in size. To ensure surface cleanliness before any subsequent treatment, the specimens were subjected to ultrasonic cleaning to remove surface contaminants. After ultrasonic cleaning, the surface of the specimens was coated with a matured zinc yellow acrylic polyurethane paint, with the coating thickness controlled within the range of 70 ± 5 μm. The coating consisted of components such as polyacrylic resin, polyisocyanate resin, pigments, additives, and organic solvents. To ensure complete curing of the paint before the laser cleaning experiments, the specimens were left to stand in a well-ventilated environment for 1 week. Figure 8.2 shows the aluminum alloy specimens prepared for cleaning. Figure 8.2a shows the original surface morphology of the aluminum alloy prior to painting, revealing inherent striping caused by the rolling process. Figure 8.2b presents a cross-sectional image of the painted specimen, clearly showing a three-layer structure: the upper layer is thermoplastic embedding resin, the middle layer is the chromium yellow acrylic polyurethane paint, and the bottom layer is the aluminum alloy substrate. Figure 8.2c, d characterize the elemental composition and distribution of the substrate and paint layer by energy dispersive X-ray spectroscopy (EDS).

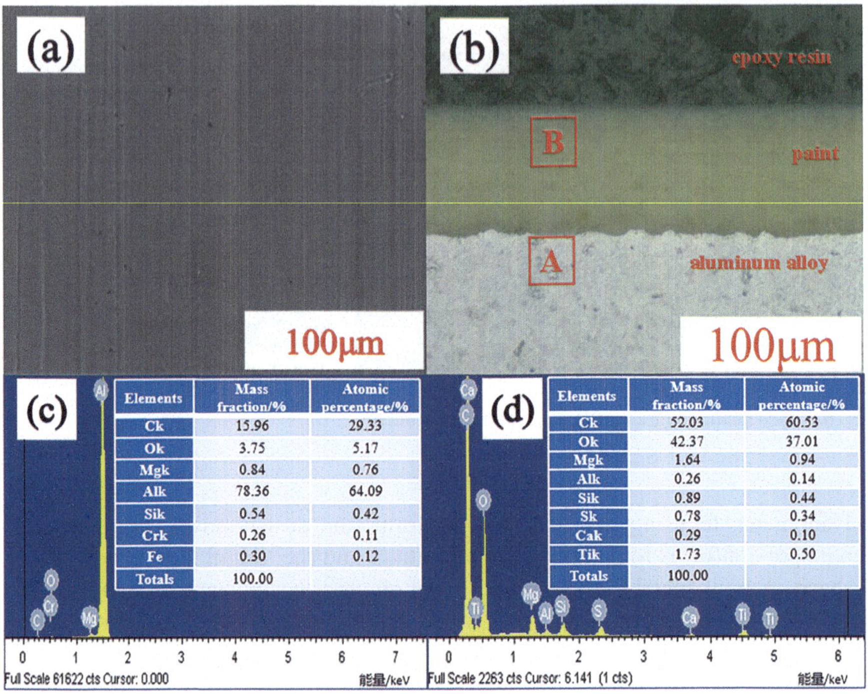


**Fig. 8.2** The materials to be cleaned: (**a**) the original surface morphology of the aluminum alloy, (**b**) the cross-sectional morphology of the aluminum alloy after painting, (**c**) EDS analysis of the substrate's elemental composition, (**d**) EDS analysis of the paint layer's elemental composition

### *8.2.3 Parameter Selection*

To evaluate the effects of different laser processing parameters on paint removal efficiency and substrate surface quality, systematic comparative experiments were conducted. The experimental design included a comparison between a single-pass high-power cleaning method and a two-pass low-power cleaning method, with variations in laser power, scanning speed, and line spacing. The specific settings are listed in Table 8.1. Considering inherent performance differences among different laser systems, energy density and spot overlap ratio were adopted as standardized indicators to ensure comparability of experimental results. The energy density was calculated using the following equation [20]:

$$\mathrm{F} = 4P_v / \left( f \pi \mathrm{D}^2 \right) \tag{8.1}$$

where Pv represents the average power, f denotes the pulse frequency, and D is the diameter of the laser spot. In this study, a laser cleaning method was adopted in which the overlap ratio of the laser spot in both the scanning and cleaning directions was adjusted simultaneously to optimize cleaning performance. The formulas for

**Table 8.1** Parameters for nanosecond pulsed laser cleaning of aluminum alloy

| Laser parameters | Numerical value |
|---|---|
| Laser power (P)/W | Single-pass cleaning: 60, 66, 72, 78, 84<br>Double-pass cleaning: 30, 36, 42, 48, 54 |
| Scanning speed (V)/(mm/s) | 1000, 1500, 2000, 2500 |
| Line spacing (L)/mm | 0.01, 0.015, 0.02, 0.025 |
| Diameter of the laser spot (D)/μm | 87 |
| Wavelength (λ)/nm | 1064 |
| Focal length (d)/mm | 254 |
| Pulse width (τ)/ns | 200 |
| Frequency (f)/kHz | 100 |

calculating the overlap ratios in the scanning and cleaning directions are as follows [21]:

$$\eta_x = \left(1 - V / \mathrm{Df}\right) \times 100\% \tag{8.2}$$

$$\eta_y = \left(1 - L / \mathrm{D}\right) \times 100\% \tag{8.3}$$

Here, V denotes the scanning speed, and L represents the line spacing. During the laser cleaning process, two modes were employed to investigate the influence of spot overlap ratio on the surface morphology and performance of aluminum alloy: (1) fixing the spot overlap ratio at 71%, and (2) fixing the energy density. In the fixed energy density mode, the energy density for single-pass cleaning was set to 11.10 J/cm$^2$, while that for dual-pass cleaning was set to 6.06 J/cm$^2$.

### 8.2.4 Characterization Methods

To evaluate the effectiveness of laser cleaning, various methods were used for characterizing cleanliness. A high-resolution field emission scanning electron microscope (SEM) and an energy-dispersive X-ray spectrometer (EDS) were employed to analyze the morphological features and surface elemental composition of the samples. A laser confocal microscope was used to measure the three-dimensional topography and surface roughness of the cleaned samples. Fourier-transform infrared spectroscopy (FT-IR), with a spectral range of 400–4000 cm$^{-1}$, was applied to analyze changes in chemical bonding. X-ray photoelectron spectroscopy (XPS) was used to detect variations in surface chemical states before and after laser cleaning. Focused ion beam (FIB) technology, combined with transmission electron microscopy (TEM), was employed to observe changes in crystal structure at the micro/nanoscale. For performance testing, a CHI660E electrochemical workstation (manufactured by Corrtest) was used to conduct electrochemical corrosion and impedance tests in 3.5% sodium chloride (NaCl) solution. The tests were carried out using a standard three-electrode system, where the working electrode was the laser-cleaned specimen with an exposed area of 100 $mm^2$ after encapsulation, the auxiliary electrode was a platinum electrode, and the reference electrode was an Ag/AgCl electrode in saturated potassium chloride solution. The samples were immersed in the solution for 1 h before measurement. The scan rate for the polarization curve was set to 5 mV/s, and electrochemical impedance spectroscopy (EIS) measurements were conducted under potentiostatic conditions within a frequency range of 100 kHz to 0.01 Hz. An SRV-4 reciprocating friction and wear tester was used to evaluate the wear resistance of the samples. The test parameters were set as follows: reciprocating stroke of 1 mm, reciprocating frequency of 10 Hz, and test duration of 30 min. A GCr15 bearing steel ball with a diameter of 10 mm and a hardness of HV750 was selected as the counterface material. A normal load of 10 N was applied during the test.

## 8.3 Results and Discussion

### 8.3.1 *Influence of Laser Cleaning on Surface Microstructure*

Figure 8.3 shows the changes in the surface microstructure of aluminum alloy after single-pass laser cleaning under different laser energy densities. In Fig. 8.3a, at a laser energy density of 10.09 J/$cm^2$, a large amount of paint remained on the aluminum surface. Due to partial peeling of the paint, many pits appeared on the coating surface, but the aluminum substrate was not yet exposed. When the energy density

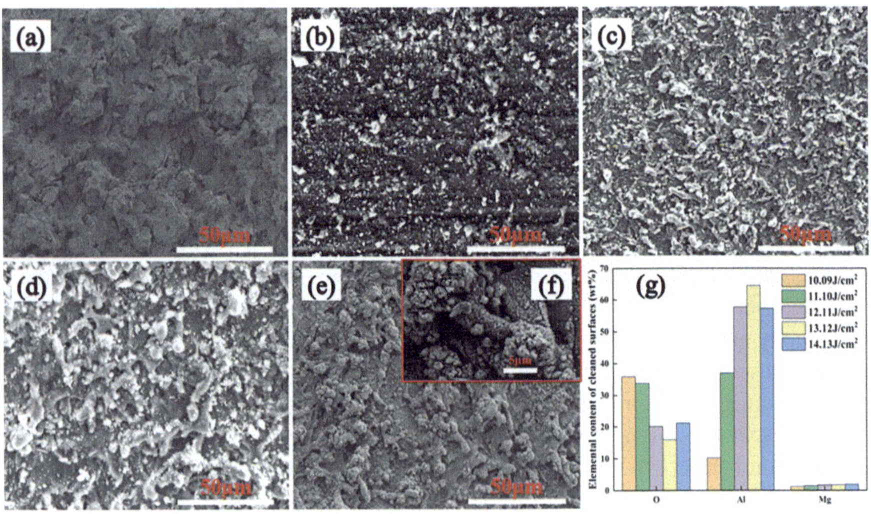


**Fig. 8.3** The surface morphology after a single-pass laser cleaning session at different energy densities: (**a**) 10.09 J/$cm^2$, (**b**) 11.10 J/$cm^2$, (**c**) 12.11 J/$cm^2$, (**d**) 13.12 J/$cm^2$, and (**e**), (**f**) 14.13 J/$cm^2$, with (**g**) the surface elemental composition

increased to 11.10 J/cm$^2$ (see Fig. 8.3b), the striping patterns on the aluminum substrate began to appear, indicating that most of the paint had been removed; however, many small paint particles remained on the surface. Figure 8.3c illustrates that under an energy density of 12.11 J/cm$^2$, a large number of small paint particles were removed, and the aluminum surface exhibited signs of remelting and oxidation. The action of the laser caused molten metal droplets to spatter from the ablation zone, forming stripe-like structures at the edges of the ablation pits. This is attributed to the extremely short pulse duration of nanosecond lasers, which prevents the droplets from coalescing before solidification. As energy density further increased, Fig. 8.3c–e show that the number of spattered metal droplets increased and accumulated, promoting the development of the striped structures and eventually filling the ablation pits. In Fig. 8.3f, the remelted metal surface appears rough and is covered with a relatively soft oxide layer. Elemental concentration analysis in Fig. 8.3g reveals that with increasing laser energy density, the concentration of oxygen (O) first decreases and then increases, while that of aluminum (Al) initially increases and then decreases. The concentration of magnesium (Mg), on the other hand, continuously increases. This phenomenon occurs because during laser treatment, the O-rich paint layer is gradually removed, initially reducing the surface oxygen content. As the paint is removed and the aluminum substrate becomes exposed, it absorbs more laser energy, resulting in a remelting and oxidation process, which increases the oxygen content while reducing the relative aluminum content. Moreover, the solubility of Mg increases at higher temperatures, but rapidly decreases during fast cooling, leading to its precipitation as compounds, which manifests as an increase in Mg concentration. Thus, the observed trends in surface elemental composition under specific laser energy densities further verify the above analysis [22].

Figure 8.4 presents the three-dimensional surface morphology of aluminum alloy after single-pass laser cleaning. At an energy density of 10.09 J/cm$^2$, a large amount of residual paint remained on the surface, resulting in a highly rough appearance. When the energy density increased to 11.10 J/cm$^2$, the paint was almost completely removed, leaving only small raised paint particles. At this point, the surface roughness of the aluminum alloy reached its minimum, appearing relatively smooth overall. As the laser energy density further increased (see Fig. 8.4c–e), remelting began to occur on the aluminum surface. During this process, the root mean square height Sq (the root mean square of the height at each point in the defined area, equivalent to standard deviation) gradually increased. The maximum height Sz (sum of the maximum peak height and maximum valley depth) initially increased and then decreased. This variation is primarily due to increased laser energy absorption by the aluminum alloy, which caused surface remelting and the formation of multiple peak-like protrusions, thereby increasing surface roughness. However, with continued energy density increase, the elevated temperature allowed molten metal to flow and fill the ablation pits, ultimately resulting in a decrease in the maximum surface feature height.

Figure 8.5 displays the surface morphology and corresponding three-dimensional profiles of aluminum alloy after single-pass laser cleaning at different spot overlap

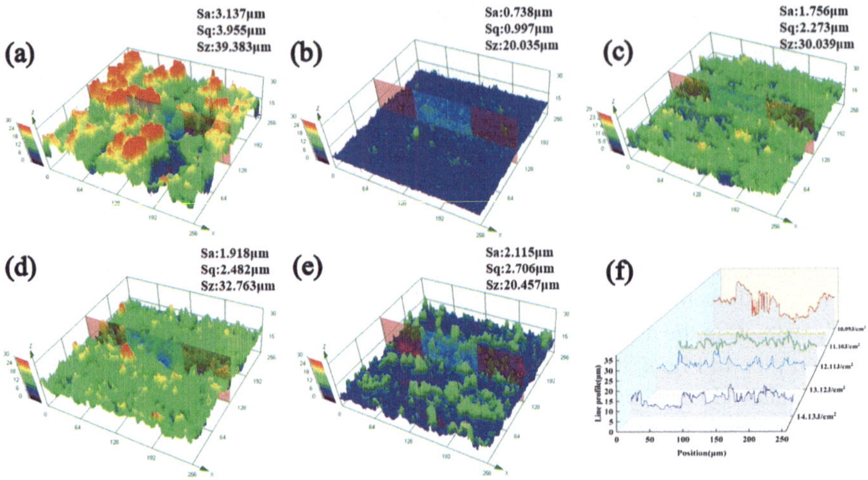


**Fig. 8.4** The three-dimensional surface morphology after a single laser cleaning session at varying energy densities: (**a**) 10.09 J/cm$^2$, (**b**) 11.10 J/cm$^2$, (**c**) 12.11 J/cm$^2$, (**d**) 13.12 J/cm$^2$, (**e**) 14.13 J/cm$^2$, (**f**) The cross-sectional morphology

ratios. As the spot overlap ratio increased, more laser energy was applied to the aluminum substrate surface, enhancing thermal accumulation effects and promoting deeper heat penetration into the material. This process not only caused surface melting but also drove the remelting phenomenon deeper into the alloy, forming deeper melt pools. The formation of melt pools led to the coalescence of larger molten metal droplets into rod-like structures covering the surface (see Fig. 8.5b). When the spot overlap ratio reached 83%, the concentrated action of multiple laser beams on the same area triggered spattering and accumulation of molten metal droplets, forming larger block-like structures and significantly increasing surface roughness. With further increase in overlap ratio, these blocky particles aggregated into larger-scale structures, ultimately resulting in the surface being covered by a cotton-like metal oxide layer. Three-dimensional morphology maps clearly showed that both the root mean square height and maximum height of the surface increased with increasing spot overlap ratio, with no observed decrease. Specifically, at an overlap ratio of 89%, the maximum surface height reached 89.063 μm, and the root mean square height was 7.493 μm (see Fig. 8.5i). Elemental concentration analysis (see Fig. 8.5j) indicated that as the spot overlap ratio increased, the concentrations of carbon (C) and aluminum (Al) gradually decreased, while those of oxygen (O) and magnesium (Mg) increased. This was mainly due to the cumulative effect of multiple laser beams intensifying the deep remelting process on the aluminum alloy surface, which further promoted the reaction between molten metal and atmospheric oxygen, leading to the formation of metal oxides.

To reduce the interaction between high-energy laser and the aluminum alloy surface, a two-pass cleaning method under low energy density was adopted. Figure 8.6 shows the evolution of aluminum alloy surface microstructure after single and

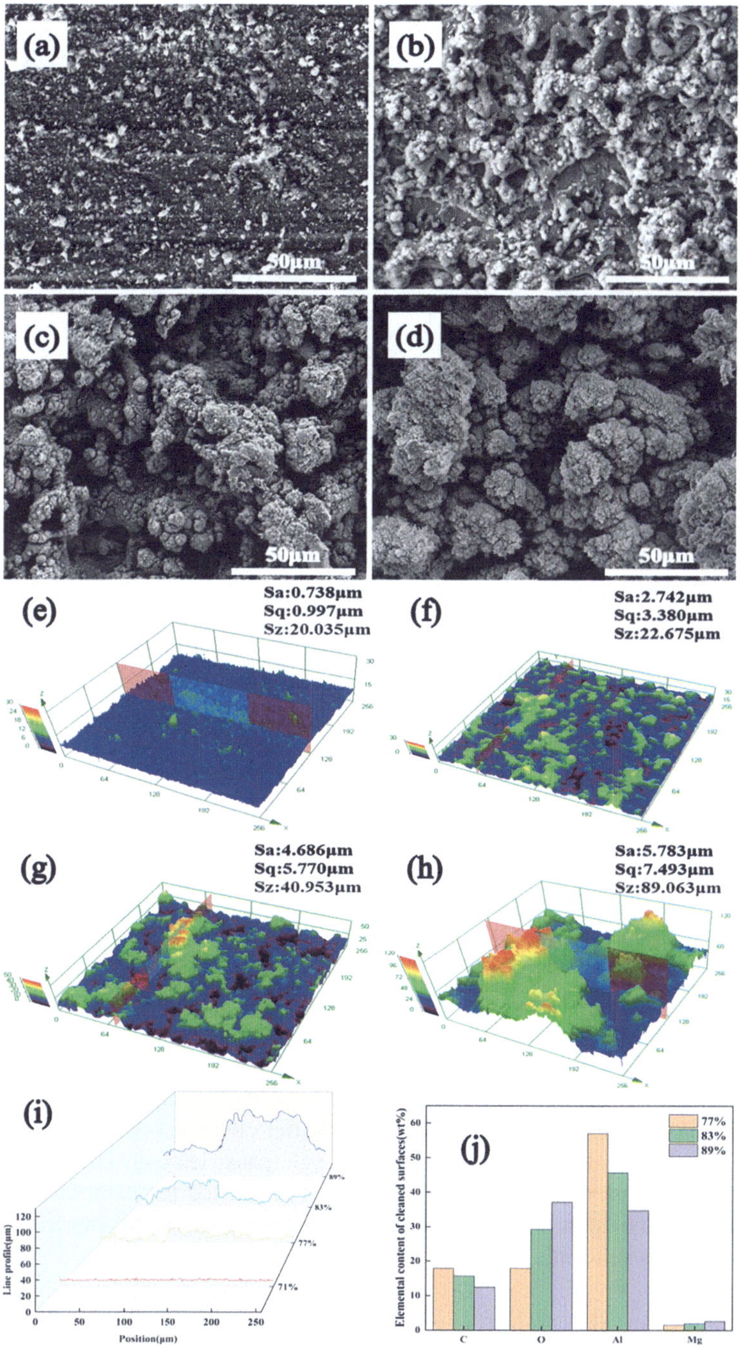


**Fig. 8.5** The surface morphology after a single laser cleaning at different spot overlap rates: (**a**) 71%, (**b**) 77%, (**c**) 83%, (**d**) 89%, (**e**)–(**h**) corresponding three-dimensional morphologies, (**i**) cross-sectional view, (**j**) surface elemental composition

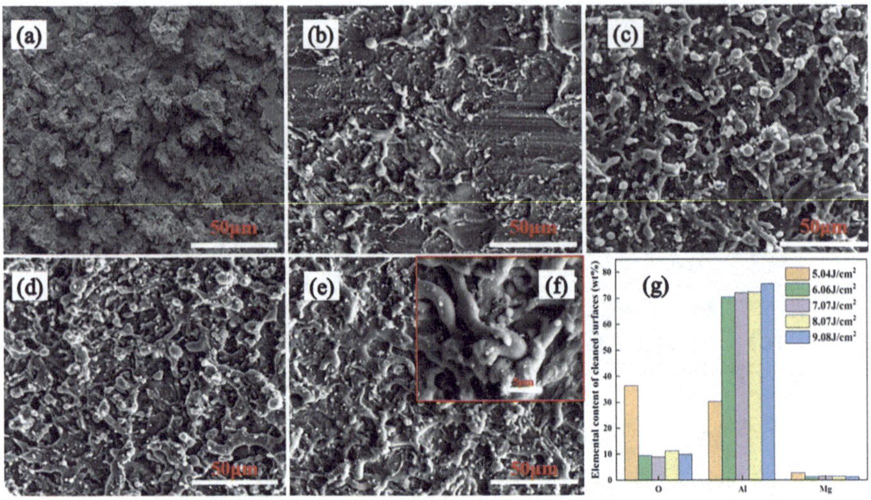


**Fig. 8.6** The surface morphology after double-pass laser cleaning at different energy densities: (**a**) 5.04 J/cm$^2$, (**b**) 6.06 J/cm$^2$, (**c**) 7.07 J/cm$^2$, (**d**) 8.07 J/cm$^2$, and (**e**), (**f**) 9.08 J/cm$^2$, (**g**) surface elemental composition

two-pass laser cleaning at various energy densities. At a laser energy density of 5.04 J/cm$^2$, most of the paint remained on the surface, indicating that the energy density was insufficient for effective paint removal. As the energy density increased to 6.06 J/cm$^2$, the paint layer was completely removed, leaving no residual particles on the substrate surface. The original strip-like surface texture of the aluminum alloy reappeared, accompanied by slight signs of remelting. This result demonstrated that, under appropriate energy density, two-pass cleaning could effectively remove the paint without significant surface damage. With further increase in energy density, surface melting became increasingly prominent. The images showed a transition from slight to obvious melting, indicating that energy density significantly influenced the extent of surface melting. At higher energy densities, the molten metal surface appeared smooth, without the formation of cotton-like metal oxides. This indicated that oxidation could be avoided within an optimized energy density range through laser parameter control. Elemental concentration analysis revealed that paint removal caused a rapid decrease in O and Mg concentrations, followed by stabilization. Conversely, the Al concentration after cleaning rose sharply, reaching a peak 12% higher than that observed after single-pass cleaning. This change indicated that two-pass cleaning not only effectively removed the paint layer but also reduced surface oxidation, thereby better preserving the original composition of the aluminum alloy substrate.

As shown in Fig. 8.7, the 3D morphology after low-energy-density secondary laser cleaning reveals that the root mean square height and the maximum height first decreased and then increased. At an energy density of 5.04 J/cm$^2$, the maximum height after two-pass cleaning was lower than that observed after single-pass cleaning at 10.09 J/cm$^2$. This is because single-pass cleaning at higher energy density

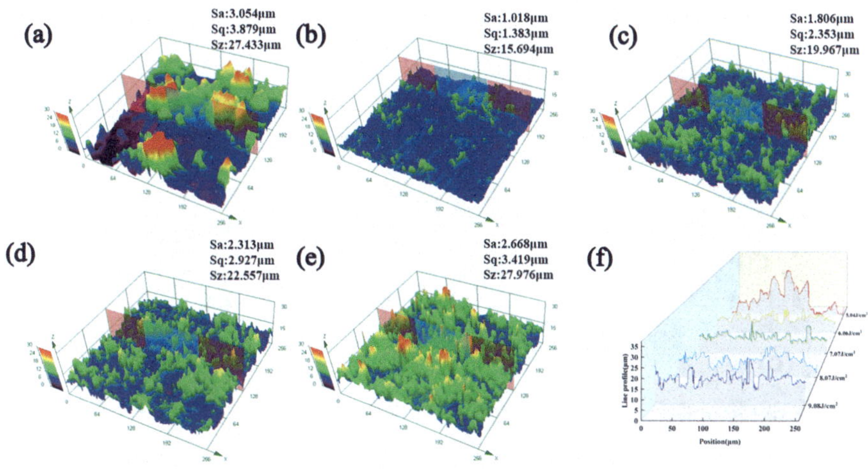


**Fig. 8.7** The three-dimensional morphology after double-pass laser cleaning at varying energy densities: (**a**) 5.04 J/cm$^2$, (**b**) 6.06 J/cm$^2$, (**c**) 7.07 J/cm$^2$, (**d**) 8.07 J/cm$^2$, (**e**) 9.08 J/cm$^2$, (**f**) Cross-sectional morphology

caused deeper ablation of the paint layer, resulting in greater surface roughness. In contrast, secondary cleaning removed the paint more uniformly, reducing residual paint fluctuations. At 6.06 J/cm$^2$, the paint layer was completely removed, and the aluminum substrate showed slight remelting with sparse small protrusions. At this point, the surface roughness was the lowest, with Sq = 1.383 μm and Sz = 15.694 μm. As the energy density increased, both Sq and Sz increased. However, due to insufficient molten metal to fill the ablation pits under low energy density in secondary cleaning, surface roughness exceeded that of single-pass cleaning. Three-dimensional morphology after secondary laser cleaning at different energy densities: (a) 5.04 J/cm$^2$, (b) 6.06 J/cm$^2$, (c) 7.07 J/cm$^2$, (d) 8.07 J/cm$^2$, (e) 9.08 J/cm$^2$, (f) cross-sectional morphology (Fig. 8.7). Figure 8.8 illustrates the surface morphology and 3D profile of aluminum alloy after secondary cleaning under different spot overlap ratios. As overlap ratio increased, thermal accumulation intensified, resulting in evident surface melting and resolidification. The prolonged laser-substrate interaction provided sufficient temperature and duration for molten metal to coalesce into ridge-like structures. Notably, no significant oxide formation was observed in Fig. 8.8d, h. Compared to single-pass cleaning, the substrate after secondary cleaning appeared more metallic and glossy. During this process, fine metal particles tended to aggregate but did not fully merge into larger blocks or form loose oxides. According to 3D morphology analysis, both Sq and Sz increased with overlap ratio and showed no decline throughout the range. At 89% overlap, the maximum height was 42.798 μm, half of the value from single-pass cleaning. As shown in Fig. 8.8j, C and O concentrations first decreased then increased, while Al concentration initially increased then declined, and Mg remained stable. These results indicate that compared to single-pass cleaning, secondary cleaning significantly reduced C and O content, while enhancing Al presence. This demonstrates that secondary cleaning

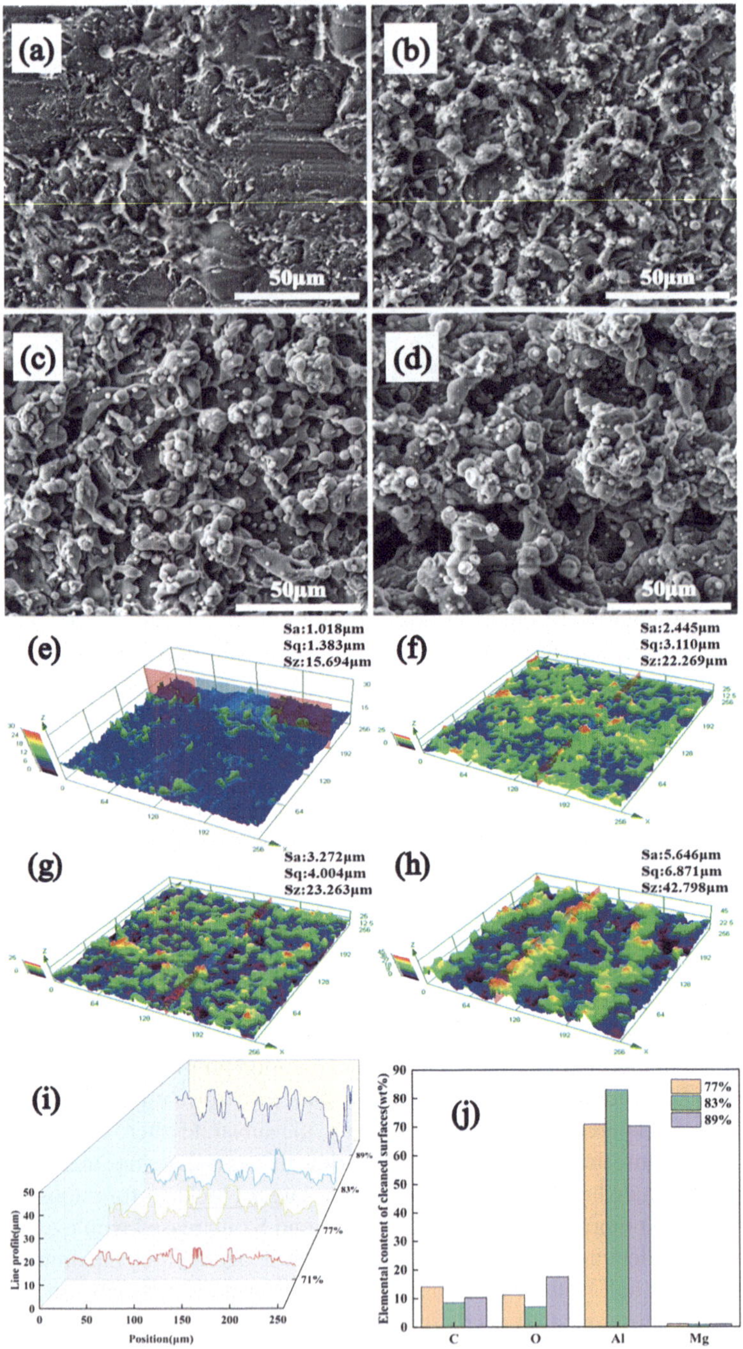


**Fig. 8.8** The surface morphology after laser double-pass cleaning at different spot overlap rates: (**a**) 71%, (**b**) 77%, (**c**) 83%, (**d**) 89%, (**e**) ~ (**h**) Corresponding three-dimensional morphologies, (**i**) Cross-sectional view, (**j**) Surface elemental composition

more thoroughly removes surface contaminants and reduces oxidation, better preserving the aluminum alloy's original composition and properties.

Figure 8.8 illustrates the surface morphology and 3D profile of aluminum alloy after secondary cleaning under different spot overlap ratios. As overlap ratio increased, thermal accumulation intensified, resulting in evident surface melting and resolidification. The prolonged laser-substrate interaction provided sufficient temperature and duration for molten metal to coalesce into ridge-like structures. Notably, no significant oxide formation was observed in Fig. 8.8d, h. Compared to single-pass cleaning, the substrate after secondary cleaning appeared more metallic and glossy. During this process, fine metal particles tended to aggregate but did not fully merge into larger blocks or form loose oxides. According to 3D morphology analysis, both Sq and Sz increased with overlap ratio and showed no decline throughout the range. At 89% overlap, the maximum height was 42.798 μm, half of the value from single-pass cleaning. As shown in Fig. 8.8j, C and O concentrations first decreased then increased, while Al concentration initially increased then declined, and Mg remained stable. These results indicate that compared to single-pass cleaning, secondary cleaning significantly reduced C and O content, while enhancing Al presence. This demonstrates that secondary cleaning more thoroughly removes surface contaminants and reduces oxidation, better preserving the aluminum alloy's original composition and properties.

The above experiments examined the influence of laser energy density and spot overlap ratio on surface microstructure and elemental composition, comparing single-pass and secondary laser cleaning. Results showed that at low energy densities, only partial paint removal occurred, with the substrate still covered. As energy increased, the paint was more effectively removed, but remelting and oxidation phenomena emerged, with metal droplets forming stripe-like structures. High energy density intensified this effect, leading to rough surfaces with soft oxide layers. 3D morphology analysis showed that optimal energy densities yielded the lowest roughness, whereas excessive energy caused Sz to increase then decline. Increased overlap ratio enhanced thermal accumulation, promoting deeper melt pools and larger structures, increasing surface roughness. Secondary cleaning at low energy densities was superior in completely removing the paint, reducing oxidation, and preserving the original material. By adjusting laser parameters, cleaning effectiveness can be optimized while minimizing substrate damage.

To further understand these changes, surface elemental concentrations under different conditions were analyzed. With rising energy density, O content showed a decrease-then-increase trend, Al increased then decreased, and Mg continuously increased, reflecting paint removal, substrate exposure, and subsequent oxidation. Furthermore, secondary cleaning notably reduced C and O content and enhanced Al presence, indicating improved decontamination and oxidation suppression, better preserving substrate properties.

### *8.3.2 Characterization After Surface Cleaning*

To evaluate substrate damage under different cleaning methods, Fig. 8.9 presents cross-sectional profiles after laser cleaning. Given the pulse duration of 200 ns, thermal diffusion was limited, and the affected zone was relatively shallow. As shown in Fig. 8.9(a1), (b1), (c1), and (d1), proper parameter control effectively removed the paint without damaging the underlying substrate, demonstrating that laser paint removal can be safely implemented when parameters are well-controlled. In single-pass cleaning, increased energy density led to the formation and deepening of ablation pits. Similarly, higher overlap ratios in single-pass cleaning prolonged laser-material interaction, aggravating damage and causing pits and grooves. Notably, at 89% overlap, the substrate appeared fragmented due to intensified thermal and mechanical effects, with a damage depth of ~109.3 μm. In contrast, after secondary cleaning, regardless of energy or overlap, surfaces remained relatively smooth, with no clear ablation pits. Especially at 89% overlap, the damage depth was only ~55 μm, half of that from single-pass cleaning, indicating secondary cleaning reduces damage while ensuring effective decontamination.

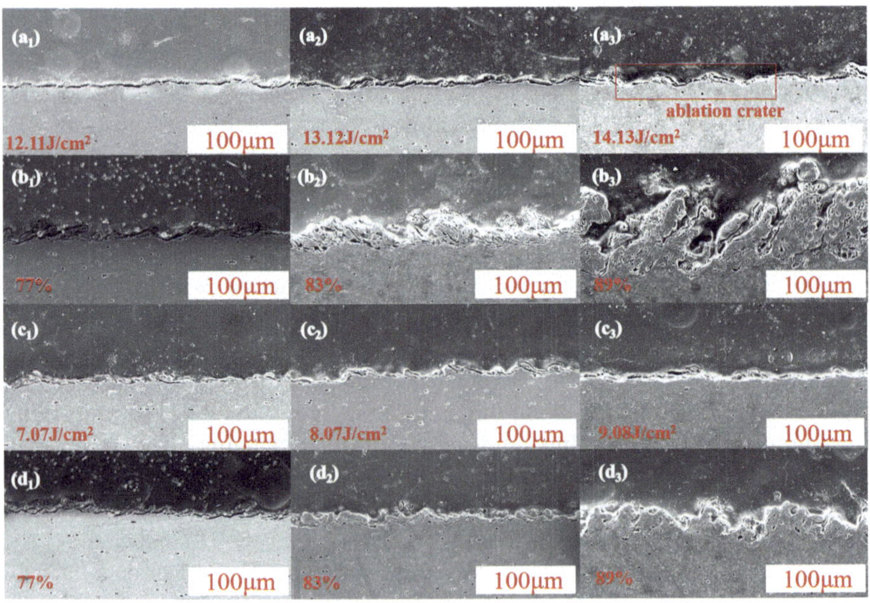


**Fig. 8.9** Cross-sectional images of aluminum alloy after laser over-cleaning (**a**) single-pass cleaning at various energy densities, (**b**) single-pass cleaning at different spot overlap rates, (**c**) double-pass cleaning at various energy densities, and (**d**) double-pass cleaning at different spot overlap rates

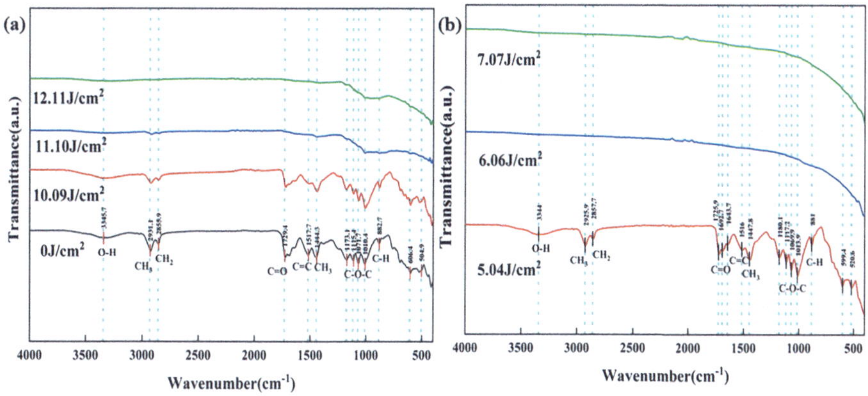


**Fig. 8.10** The FT-IR spectra following laser cleaning at varying energy densities: (**a**) after a single-pass laser cleaning session, and (**b**) after double-pass laser cleaning sessions

#### 8.3.2.1 Evaluation of Cleaning Efficiency

Figure 8.10 presents FT-IR spectra after single-pass and secondary laser cleaning at different energy densities. The spectrum at 0 J/cm$^2$ corresponds to the uncleaned painted surface, showing vibrational features of organic functional groups: 3345 cm$^{-1}$ for O–H stretching, 2931 and 2855 cm$^{-1}$ for saturated C–H stretching, 1729 cm$^{-1}$ for C=O stretching, 1517 cm$^{-1}$ for symmetric C=C stretching, 1444 cm$^{-1}$ for in-plane C–H bending, 1115 cm$^{-1}$ for C–O–C stretching, and 881 cm$^{-1}$ for out-of-plane C–H bending. In Fig. 8.11a, during single-pass cleaning, characteristic peaks significantly weakened as energy increased, indicating effective paint removal. However, at 12.11 J/cm$^2$, weak peaks from organic compounds persisted, suggesting residual contamination. By contrast, Fig. 8.10b shows that after secondary cleaning at 7.07 J/cm$^2$, no organic peaks were observed, indicating more thorough paint removal. This aligns with SEM results.

To further examine surface elemental changes, four representative samples were selected for XPS analysis: uncleaned painted surface, single-pass at 12.11 J/cm$^2$, secondary cleaning at 7.07 J/cm$^2$, and secondary cleaning at 89% overlap and 6.06 J/cm$^2$. Figure 8.11 shows the full XPS spectra. After laser cleaning, peaks for Na1s, Zn2p, C1s, and Cl2p diminished or disappeared. Under 12.11 J/cm$^2$ single-pass cleaning, new peaks for Mg1s, Al2p, and Ti2p emerged, along with intensified O1s, indicating oxidation of metal elements due to rapid heating. In contrast, under 7.07 J/cm$^2$ secondary cleaning, Mg1s and Ti2p did not appear, suggesting milder oxidation. However, under 6.06 J/cm$^2$ and 83% overlap, Mg1s and Ti2p reappeared, indicating that increased overlap prolonged laser-material interaction and induced oxidation through thermal accumulation.

Figures 8.12 and 8.13 present detailed XPS spectra for single-pass and secondary cleaning, respectively. C peaks were corrected and fitted using Avantage software. Al2p shows closely spaced spin-orbit components, typically appearing as

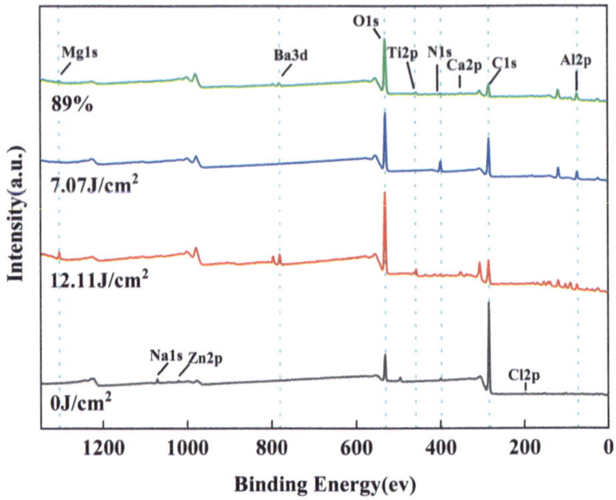


**Fig. 8.11** XPS full spectra of laser-cleaned and uncleaned samples

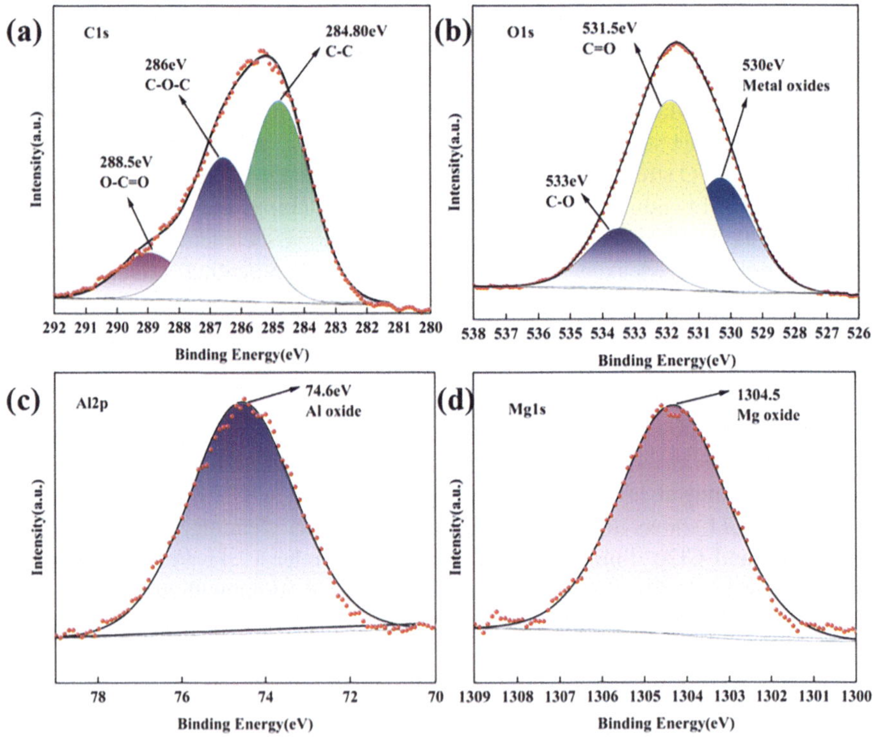


**Fig. 8.12** Detailed XPS spectra of single-pass laser cleaning at an energy density of 12.11 J/cm$^2$: (**a**) C1s, (**b**) O1s, (**c**) Al2p, and (**d**) Mg1s regions

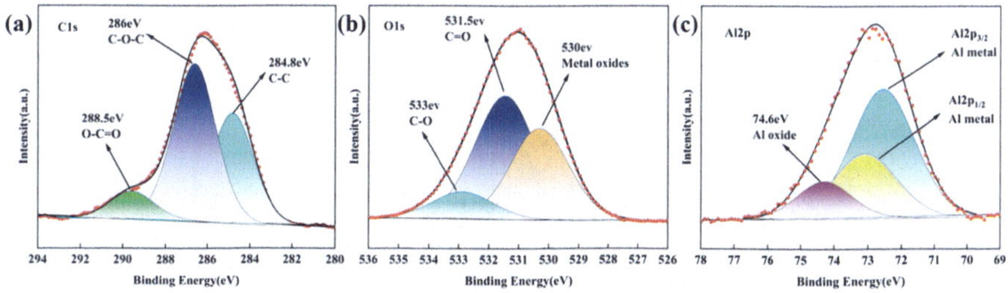


**Fig. 8.13** Detailed XPS spectra of double-pass laser cleaning at an energy density of 7.07 J/cm$^2$: (**a**) C1s, (**b**) O1s, and (**c**) Al2p regions

asymmetric peaks rather than distinct splitting under standard conditions. For aluminum compounds like $Al_2O_3$, the splitting is negligible. In Fig. 8.12, high-energy single-pass cleaning led to Al and Mg oxidation, evidenced by strengthened O1s and emergence of Mg1s and Ti2p peaks, confirming rapid oxidation. In contrast, Fig. 8.13 shows that in secondary cleaning, metallic Al dominated the Al2p peak, indicating limited oxide formation and lower oxidation under reduced energy density.

In the subsequent experiments, a laser energy density of 14.13 J/cm$^2$ was applied to clean the aluminum alloy, and its surface composition and crystal structure were analyzed in detail. As shown in Fig. 8.14a, b, a thin cross-sectional sample was prepared using focused ion beam (FIB) technique, followed by transmission electron microscopy (TEM) analysis. Figure 8.14c illustrates the formation of a new structural layer on the surface of the aluminum alloy after laser cleaning. As shown in Fig. 8.14d, this layer consists of two distinct sublayers: the top layer is a newly formed oxide layer generated by remelting and redeposition of the aluminum alloy; the bottom layer is the aluminum alloy substrate. To further analyze the changes in crystal structure after laser cleaning and to precisely characterize the interface between the oxide layer and the aluminum substrate, high-resolution electron microscopy was used to observe atomic arrangements, and the corresponding selected area electron diffraction (SAED) patterns were provided, as shown in Fig. 8.14e, f. The SAED pattern of the oxide layer region exhibits diffuse ring features, indicating that the newly formed layer is predominantly amorphous. In contrast, across the boundary shown in Fig. 8.14d, the region of the aluminum alloy substrate displays a series of clear and discrete bright spots, reflecting the lattice structure and symmetry of aluminum. This indicates that after laser cleaning, the crystal structure of the aluminum alloy surface underwent a significant transformation, forming an amorphous nanoscale oxide layer. In addition, Fig. 8.14g shows a large number of dislocations generated in the aluminum alloy after laser cleaning. The formation of these dislocations may be attributed to the thermal effect of the laser, which caused slight surface melting followed by rapid solidification. At the same time, shock waves generated by the laser-induced plasma could propagate into the material, further promoting the formation of these dislocations [23].

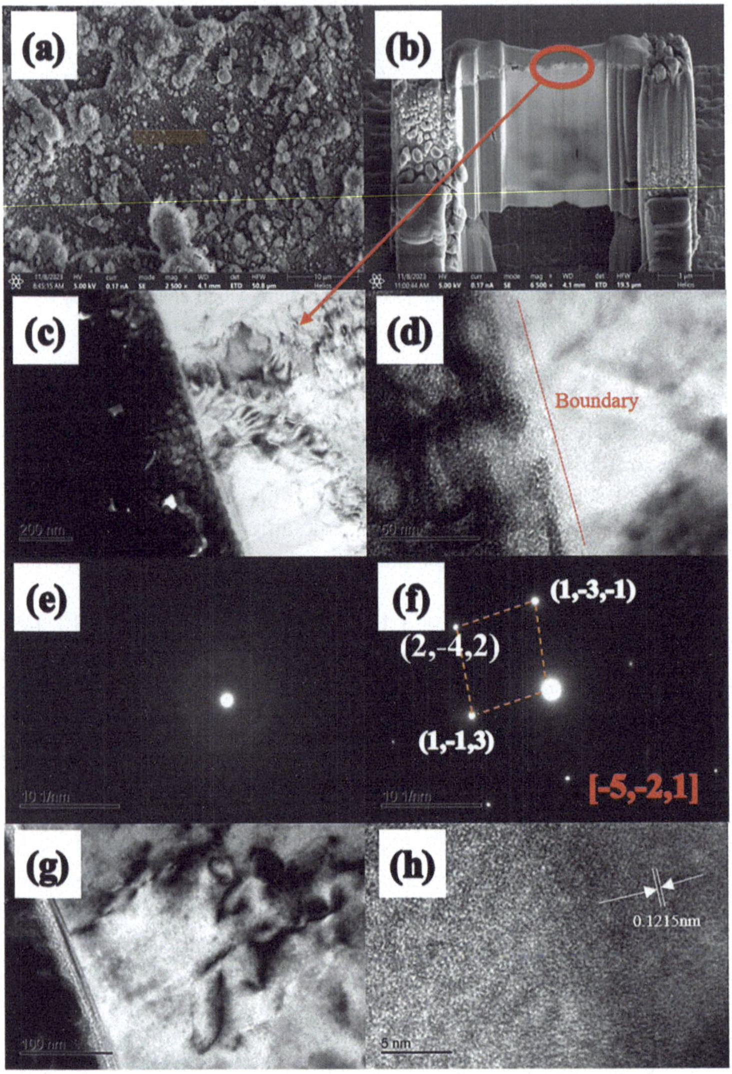


**Fig. 8.14** FIB and TEM analyses: (**a**–**b**) the sampling locations prepared by FIB, (**c**–**d**) magnified images at the cleaning interface, (**e**–**f**) the SAED patterns for the oxide layer and the aluminum alloy, respectively, (**g**) Aluminum alloy bright field image, (**h**) Atomic arrangement of aluminum alloy cross-section

#### 8.3.2.2 Performance Evaluation After Surface Treatment

As shown in the polarization curves in Fig. 8.15, the laser-cleaned samples exhibited significantly improved electrochemical corrosion performance after immersion in 3.5% NaCl solution for 1 h. The samples treated with single-pass laser cleaning at high energy density showed not only a lower corrosion potential (Ecorr),

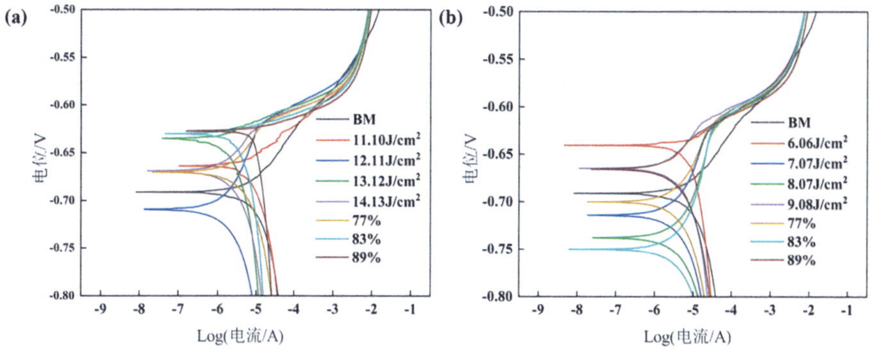


**Fig. 8.15** Polarization curves for laser cleaning: (**a**) single-pass cleaning, (**b**) double-pass cleaning

**Table 8.2** Fitted electrochemical parameters for single-pass laser-cleaned samples and BM

| Experimental sample | $E_{corr}$ (V) | $I_{corr}$ (A/cm$^2$) | ba (mV) | bc (mV) |
|---|---|---|---|---|
| BM | −0.69122 | $5.9445 \times 10^{-5}$ | 64.273 | 399.97 |
| 11.10 J/cm$^2$ | −0.66327 | $1.3283 \times 10^{-5}$ | 47.401 | 404.42 |
| 12.11 J/cm$^2$ | −0.70894 | $4.1499 \times 10^{-6}$ | 160.22 | 302.23 |
| 13.12 J/cm$^2$ | −0.63529 | $3.4571 \times 10^{-6}$ | 52.136 | 316.51 |
| 14.13 J/cm$^2$ | −0.66867 | $2.2213 \times 10^{-6}$ | 51.177 | 180.83 |
| 77% | −0.66983 | $3.1681 \times 10^{-6}$ | 53.477 | 159.98 |
| 83% | −0.63009 | $3.4662 \times 10^{-6}$ | 14.956 | 246.3 |
| 89% | −0.62705 | $9.9382 \times 10^{-6}$ | 14.377 | 389.06 |

indicating enhanced thermodynamic stability and reduced corrosion tendency, but also a lower corrosion current density (Icorr) compared to the untreated base material (BM), suggesting a slower corrosion kinetics. With increasing laser energy density, a Tafel passivation region appeared in the polarization curves, indicating the formation of a protective oxide film on the aluminum alloy surface. This film effectively inhibited the penetration of corrosive media and significantly reduced the corrosion rate. For instance, at an energy density of 14.13 J/cm$^2$, the minimum corrosion current density measured was $2.2213 \times 10^{-6}$ A/cm$^2$. However, under a fixed energy density of 11.10 J/cm$^2$, an increase in spot overlap ratio resulted in a gradual narrowing of the passivation region and an increase in corrosion current density. This may be attributed to the higher laser energy absorption caused by the increased spot overlap ratio, which damaged the formed oxide layer and reduced corrosion resistance.

To further analyze the surface modification effect after laser cleaning, two equivalent circuits were used to fit the EIS data, as shown in Fig. 8.16. The original sample was fitted using the equivalent circuit shown in Fig. 8.16c, which reflects the condition where no oxide film is present on the aluminum alloy surface. The laser-cleaned samples were fitted using the equivalent circuit in Fig. 8.16d, which describes the presence of an oxide film. These equivalent circuits are applicable to

**Table 8.3** Fitted electrochemical parameters for double-pass laser-cleaned samples and BM

| Experimental sample | $E_{corr}$ (V) | $I_{corr}$ (A/cm$^2$) | ba (mV) | bc (mV) |
|---|---|---|---|---|
| BM | −0.69122 | $5.9445 \times 10^{-5}$ | 64.273 | 399.97 |
| 6.06 J/cm$^2$ | −0.64035 | $9.0445 \times 10^{-6}$ | 31.477 | 312.42 |
| 7.07 J/cm$^2$ | −0.71382 | $9.0204 \times 10^{-6}$ | 160.11 | 304.01 |
| 8.07 J/cm$^2$ | −0.73765 | $7.1175 \times 10^{-6}$ | 197.98 | 311.87 |
| 9.08 J/cm$^2$ | −0.66462 | $4.1265 \times 10^{-6}$ | 55.517 | 191.62 |
| 77% | −0.69963 | $5.459 \times 10^{-6}$ | 91.685 | 205.9 |
| 83% | −0.7498 | $5.2002 \times 10^{-6}$ | 241.88 | 335 |
| 89% | −0.66552 | $4.3435 \times 10^{-6}$ | 54.612 | 182.95 |

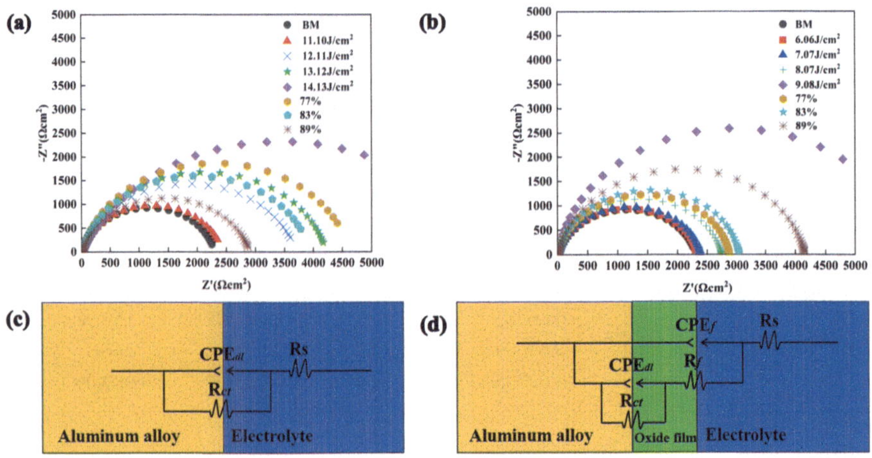


**Fig. 8.16** EIS Analysis: (**a**) single-pass laser cleaning, (**b**) double-pass laser cleaning, (**c**) BM equivalent circuit, (**d**) equivalent circuit for laser-cleaned samples

specimens with or without passivation layers and can be used to compare electrochemical properties under different surface conditions [24]. The equivalent circuit used in this study consists of solution resistance (Rs), constant phase element of the oxide film (CPEf), oxide film resistance (Rf), double-layer constant phase element (CPEdl), and charge transfer resistance (Rct). As shown in Fig. 8.16a, b, the Nyquist semicircle diameter of the laser-cleaned samples is larger than that of the original sample, indicating that laser cleaning has a positive effect on corrosion resistance. Further analysis of the corrosion resistance trend in single-pass cleaned samples reveals that the Nyquist semicircle diameter gradually increases with increasing laser energy density. This suggests that at a spot overlap ratio of 71%, the thermal effect of the laser makes the oxide film on the aluminum alloy surface denser and improves corrosion resistance. When the laser energy density is 11.10 J/cm$^2$, the Nyquist semicircle diameter gradually decreases with increasing spot overlap ratio. This is because, with a further increase in spot overlap ratio, in addition to thermal effects, mechanical effects become more significant, damaging the dense oxide film layer and reducing corrosion resistance. Figure 8.16b shows the EIS results of two-pass low-power laser-cleaned samples. It can be seen that the Nyquist semicircle diameter is concentrated around 3000 Ω cm², which is significantly smaller than the approximately 4000 Ω cm² observed for single-pass high-power cleaned samples. This is due to the relatively low laser energy density, which is insufficient to produce a fully developed oxide film on the aluminum alloy surface. As laser energy density and spot overlap ratio increase, the Nyquist semicircle diameter gradually increases, indicating improved corrosion resistance.

As shown in Fig. 8.17, the friction coefficient curves of the aluminum alloy surface after single-pass and two-pass laser cleaning under different energy densities and spot overlap ratios are presented. From Fig. 8.17a, it can be seen that at an energy density of 11.10 J/cm², due to a large number of small residual paint particles on the aluminum alloy surface, the friction coefficient is lower than that of the untreated sample [25]. At an energy density of 12.11 J/cm², as the paint layer is removed and the laser acts on the aluminum substrate, surface melting occurs under high temperature, followed by resolidification, increasing surface roughness and thus the friction coefficient. As the energy density increases further, a thicker metal oxide layer gradually forms on the aluminum alloy surface, resulting in a decreasing

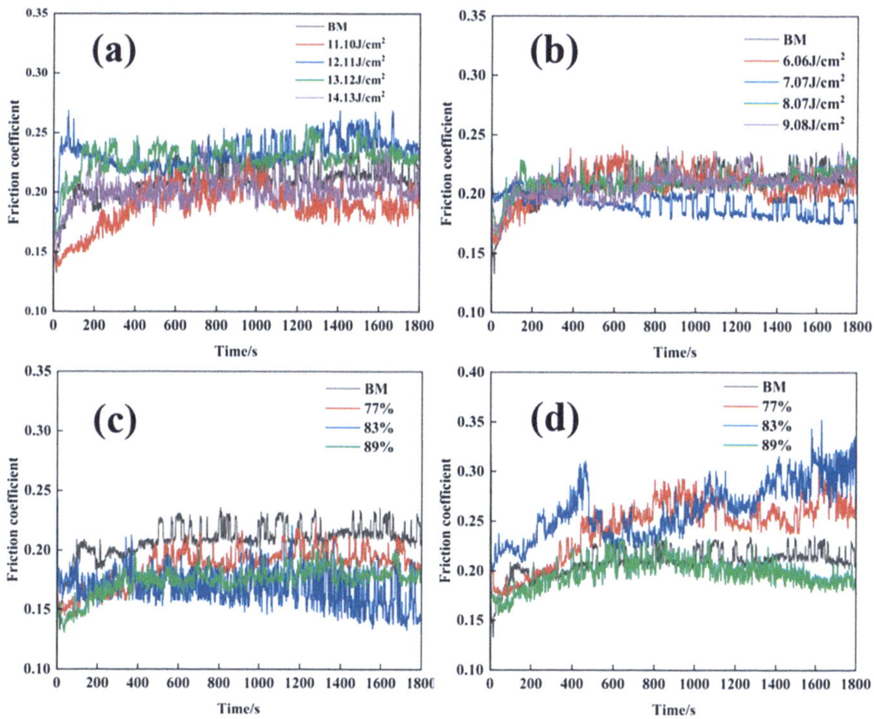


**Fig. 8.17** Coefficient of friction of aluminum alloy surfaces before and after laser cleaning, (**a**), (**c**) single-pass cleaning, (**b**), (**d**) double-pass cleaning

trend in the friction coefficient. From Fig. 8.17b, it can be seen that the friction coefficient of the two-pass low-energy-density cleaned samples is not significantly different from that of the uncleaned samples, indicating that the effect of two-pass low-energy-density laser cleaning on the sample surface is minimal. From Fig. 8.17c, in the case of single-pass cleaning, the friction coefficient of laser-cleaned samples decreases as the spot overlap ratio increases compared to the uncleaned sample. This is because, under high power and low scanning speed, the laser has a longer interaction time with the substrate and higher energy input, which facilitates the formation of a metal oxide film on the aluminum alloy surface, thereby reducing the friction coefficient. However, as the scanning speed further decreases, the oxide film on the aluminum alloy surface is damaged under the combined thermal and mechanical effects of the laser, significantly increasing the surface roughness and thus the friction coefficient again. From Fig. 8.17d, it can be seen that the friction coefficient shows a trend of first increasing and then decreasing. This may be because under the two-pass laser cleaning process, due to the combined thermal and mechanical effects, the laser significantly influences the surface morphology of the aluminum alloy. Under this overlap ratio, the relatively low laser energy density is insufficient to produce an oxide film, but the surface roughness is greater than that of the base material, leading to a gradual increase in the friction coefficient. As the spot overlap ratio increases, the laser interaction time increases, and the oxide film on the aluminum alloy gradually forms, resulting in a decreasing trend in the friction coefficient。.

## 8.4 Research Findings and Technical Highlights

In this study, high-energy-density single-pass and low-energy-density two-pass cleaning methods were explored for removing paint using nanosecond lasers. By optimizing two key laser process parameters—laser energy density and spot overlap ratio—we investigated the evolution characteristics of surface morphology on aluminum alloy substrates under single and double laser cleaning conditions. The main conclusions are summarized as follows:

1. Under high-energy-density single-pass cleaning conditions, due to coating ablation and residual laser irradiation, the surface of the aluminum alloy substrate experienced thermal accumulation, resulting in the formation of an oxide film. As the laser energy density increased, the remelting phenomenon on the substrate surface became more pronounced, leading to a reduction in surface roughness. It can be seen that the single-pass cleaning method can produce a substrate surface with corrosion-protective oxide film and allows for selective adjustment of surface roughness.
2. Under low-energy-density two-pass cleaning conditions, the surface cleanliness was significantly higher, and minimal metal oxide layer was formed on the aluminum alloy surface. The aluminum content exceeded 80%, indicating that

thermal accumulation on the substrate surface was significantly reduced during the two-pass cleaning process. Therefore, for applications where remelting of the substrate surface is not desired, low-energy-density two-pass cleaning is recommended. This cleaning method is particularly suitable for substrates requiring strict control of interface temperature, such as carbon fiber-reinforced composites.

3. During the laser cleaning process, increasing the laser energy density while simultaneously increasing the scanning speed can effectively mitigate thermal erosion damage to the substrate surface caused by laser irradiation. However, increasing the spot overlap ratio significantly affects the depth of substrate damage. These findings suggest that effective cleaning of the aluminum alloy surface without causing damage can be achieved by reasonably increasing the laser energy density and reducing the spot overlap ratio. By adjusting the two parameters—laser energy density and spot overlap ratio—an amorphous nanostructured modification layer can be formed on the surface of the aluminum alloy substrate after laser cleaning. This layer improves the corrosion resistance and reduces the friction coefficient of the aluminum alloy surface, thus enabling a transition from non-destructive cleaning to surface strengthening.

## References


1. Zhou L, Zhao H C, Zhang Q, et al. Experimental study and mechanism analysis of a pulsed laser cleaning aluminum alloy process [J]. Applied optics, 2024, 63(6): A32–A43.
2. Zhao H C, Qiao Y L, Du X, et al. Laser cleaning performance and mechanism in stripping of Polyacrylate resin paint [J]. Applied physics A, 2020, 126(5): 360.
3. Kumar A, Gumma S, Roychowdhury S, et al. Laser-assisted removal of weld heat tints from stainless steel surface [J]. Journal of laser applications, 2022, 34(1): 012003.
4. Miao R, Wang T, Yao T, et al. Experimental and numerical simulation analysis of laser paint removal of aluminum alloy [J]. Journal of laser applications, 2022, 34(1): 012002.
5. Miao X J, Zhang C L, Wu M P, et al. Application of a water jet for cleaning grease and improving the surface adhesion properties of galvanized steel wire ropes [J]. Scientific reports, 2022, 12: 9680.
6. Baij L, Hermans J, Ormsby B, et al. A review of solvent action on oil paint [J]. Heritage science, 2020, 8(1): 43.
7. Gu J Y, Su X, Jin Y, et al. Precise paint stripping based on photochemical effect by UV laser on honeycomb sandwich panel [J]. Journal of materials research and technology, 2023, 25: 3768–3779.
8. Wang Q D, Kainuma S, Zhuang S S, et al. Laser cleaning on severely corroded steel members: Engineering attempt and cleanliness assessment [J]. Journal of cleaner production, 2022, 376: 134224.
9. Zhu G D, Zhang D H, Zhang L, et al. Environmentally friendly photothermal reconstruction of cement/asphalt road: Can road markings also be cleaned in such unexpected ways? [J]. Journal of cleaner production, 2024, 434: 140347.
10. Hsu H T, Lin J. Thermal–mechanical analysis of the surface waves in laser cleaning [J]. International journal of machine tools and manufacture, 2005, 45(7/8): 979–985.
11. Roberts D E. Pulsed laser coating removal by detachment and ejection [J]. Applied physics A, 2004, 79(4): 1067–1070.

12. Zou C, Zhao S S, Yu H J, et al. Plume ejection and product in laser cleaning of ferrite paint [J]. Optics & laser technology, 2023, 167: 109733.
13. Hu C Y, He G Y, Chen J, et al. Research on cleaning mechanism of anti-erosion coating based on thermal and force effects of laser shock [J]. Coatings, 2020, 10(7): 683.
14. Zhang D H, Xu J, Li Z C, et al. Removal mechanisms of nanosecond pulsed laser cleaning of blue and red polyurethane paint [J]. Applied physics A, 2022, 128(2): 170.
15. Tian Z, Lei Z L, Chen X, et al. Nanosecond pulsed fiber laser cleaning of natural marine micro-biofoulings from the surface of aluminum alloy [J]. Journal of cleaner production, 2020, 244: 118724.
16. Shao J, Zhang R L, Dong H, et al. The energy density threshold evaluation for high-precision laser cleaning of the oxidized layer on 7075 aluminum alloy with 100ns laser [J]. The international journal of advanced manufacturing technology, 2022, 120(3): 1901–1907.
17. Wang W, Shen J, Liu W J, et al. Effect of laser energy density on surface physical characteristics and corrosion resistance of 7075 aluminum alloy in laser cleaning [J]. Optics & laser technology, 2022, 148: 107742.
18. Sun Q, Zhou J Z, Meng X K, et al. Mechanical properties and microstructure characteristics of 2024-T351 aluminum alloy specimen subjected to paint removal by laser cleaning [J]. Vacuum, 2023, 211: 111927.
19. Nie J H, Zhang H, Zhang D H, et al. Removal mechanism of laser cleaning for inorganic thermal control coatings on aluminum alloys [J]. Applied surface science, 2023, 633: 157578.
20. Zhang G X, Hua X M, Li F, et al. Effect of laser cleaning process parameters on the surface roughness of 5754-grade aluminum alloy [J]. The international journal of advanced manufacturing technology, 2019, 105(5): 2481–2490.
21. Zhu L X, Gao Q, Sun B T, et al. Nanosecond laser cleaning for enhanced zinc coating quality of HSLA steel [J]. Optics & laser technology, 2021, 143: 107311.
22. He Z R, Shen Y Z, Xiong W B, et al. Uncovering the roles of laser action modes in surface mechanical properties of 2024 aluminum alloy [J]. Applied surface science, 2023, 613: 156032.
23. Yoo H J, Baek S, Kim J H, et al. Effect of laser surface cleaning of corroded 304L stainless steel on microstructure and mechanical properties [J]. Journal of materials research and technology, 2022, 16: 373–385.
24. Zhuang S S, Kainuma S, Yang M Y, et al. Characterizing corrosion properties of carbon steel affected by high-power laser cleaning [J]. Construction and building materials, 2021, 274: 122085.
25. Zhu G D, Wang S R, Cheng W, et al. Corrosion and wear performance of aircraft skin after laser cleaning [J]. Optics & laser technology, 2020, 132: 106475.

# Chapter 9
# Control of Substrate Surface Morphology During Laser Paint Removal from EH36 Steel Surface

## 9.1 Introduction

Ships operating in marine environments are frequently exposed to harsh conditions such as humidity and salt spray, which can readily lead to corrosion of steel surfaces. Typically, a protective coating is applied to the ship's surface to prevent corrosion caused by seawater, oxygen, and other media, thereby extending the vessel's service life. Additionally, during navigation, ships are subjected to natural factors such as wind and sunlight; long-term exposure to the external environment tends to cause discoloration and oxidation of the surface. Coating the surface effectively enhances the weather resistance of the steel and mitigates these adverse effects, maintaining a good exterior appearance. Moreover, marine environments host a wide variety of organisms such as algae, shells, and barnacles, which tend to adhere to the hull surface, forming fouling layers that increase hydrodynamic resistance and reduce sailing efficiency [1, 2]. Marine coatings can reduce biofouling, keep the hull surface clean, lower resistance, and improve fuel efficiency [3]. In addition to protection, coatings enhance the vessel's visual appeal, thereby increasing commercial attractiveness and market value. Painted surfaces are smoother and easier to clean and maintain, reducing maintenance costs and labor intensity. Therefore, the primary purposes of painting EH36 marine steel are to protect the steel surface, enhance weather resistance, reduce biofouling, improve appearance, and facilitate maintenance, all of which contribute to the safe operation and extended service life of the vessel. However, before performing further treatments such as spraying, plating, or welding, the paint layer must be removed to expose a clean steel surface. Moreover, during service, the paint layer may degrade due to aging, damage, or other factors, necessitating its regular removal and reapplication to prevent corrosion. This chapter explores in detail the application of nanosecond laser cleaning technology for paint removal from the surface of EH36 marine steel, as well as the influence of laser parameters on the post-cleaning surface morphology.

 
S. Wang, *Laser Cleaning*, https://doi.org/10.1007/978-981-95-2935-3_9

In this study, the surface of each EH36 steel sample was coated with a zinc yellow acrylic polyurethane paint layer with a thickness of 70 ± 5 μm, serving as the target for laser removal. Except for the replacement of the base material from aluminum alloy to EH36 marine steel, all other experimental conditions—including sample pretreatment, laser cleaning equipment, parameter selection principles, and characterization methods—were consistent with those described in Chap. 8 and are therefore not reiterated here.

## 9.2 Influence of Different Laser Cleaning Parameters on the Surface Morphology of EH36 Steel

### *9.2.1 Macroscopic Morphology*

The results of laser paint removal from the EH36 steel surface under different process parameters are shown in Fig. 9.1. When the spot overlap rate was fixed at 71% and the laser energy density was 8.07 J/cm$^2$, a significant amount of paint residue remained on the aluminum alloy surface, with a residual coating thickness of approximately 20 μm. At a laser energy density of 9.08 J/cm$^2$, the substrate surface of the EH36 steel became visible and exhibited a metallic luster, indicating that the paint layer appeared to be macroscopically removed. As the energy density increased, the substrate surface color shifted from a bright white metallic luster to a dark brown appearance. When the laser energy density was fixed at 9.08 J/cm$^2$, variations in surface morphology were investigated by changing the spot overlap rate. At overlap rates of 59% and 65%, the lower overlap resulted in shorter laser interaction time and insufficient energy input to completely remove the paint layer

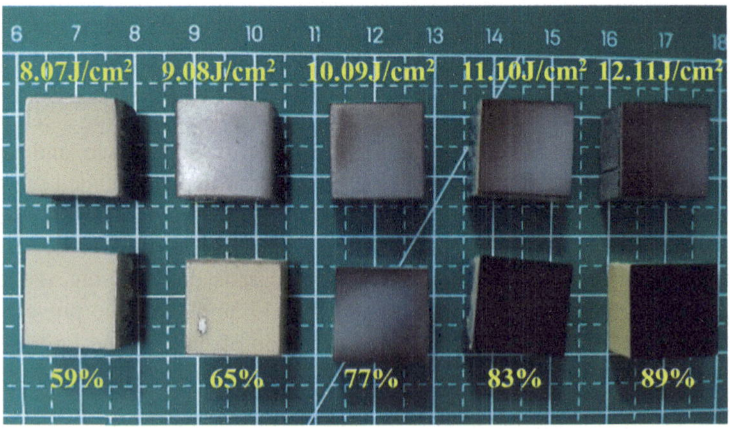


**Fig. 9.1** Macroscopic morphology of samples with different energy densities and spot overlap rates

from the EH36 steel surface. As the spot overlap rate increased, the substrate surface color changed from brown to black.

### 9.2.2 Microscopic Morphology

The surface morphology of samples cleaned at a fixed spot overlap ratio of 71% with varying laser energy densities ranging from 8.07 to 12.11 J/cm$^2$ is shown in Fig. 9.2. At a laser energy density of 8.07 J/cm$^2$, a large amount of paint remained on the EH36 steel surface. Partial removal of the paint led to a very rough coating surface, with numerous pores and cracks. No significant ablation traces were observed, and the substrate surface was not exposed. At 9.08 J/cm$^2$, the original striped structure of the steel became visible, with no paint residues observed, and slight spreading droplet-like remelting occurred. At 10.09 J/cm$^2$, the original surface morphology of the steel nearly disappeared, replaced by splattered and resolidified metallic droplets. With further increases in energy density, the remelting became more pronounced, and a larger number of splattered droplets covered the steel surface, forming a layered structure. In the magnified image of Fig. 9.2e, both the droplets and the steel substrate were densely covered with fuzzy metallic oxides. Elemental content spectra showed that at 8.07 J/cm$^2$, due to the presence of paint, high levels of carbon (C) and oxygen (O) were detected on the steel surface. At 9.08 J/cm$^2$, the Fe content suddenly increased, while C and O decreased significantly, indicating complete paint removal. As the energy density continued to increase, C and Fe levels slightly decreased, and O content increased, suggesting that the molten metal underwent reoxidation to form an oxide film under laser irradiation.

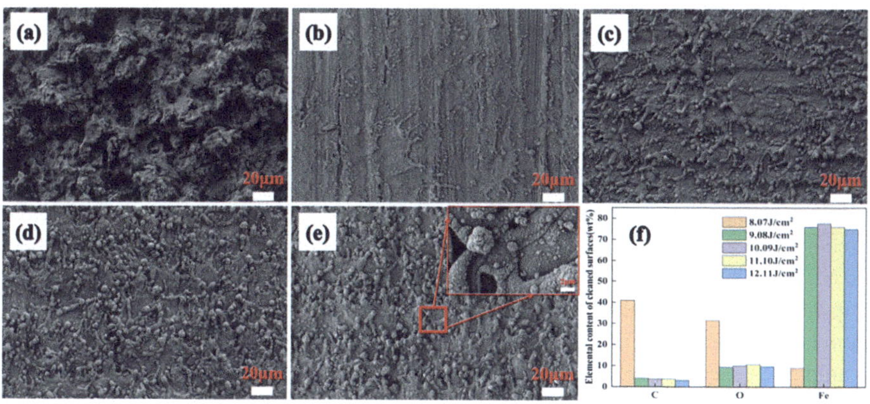


**Fig. 9.2** SEM images of cleaning at different energy densities (**a**) 8.07 J/cm$^2$ (**b**) 9.08 J/cm$^2$ (**c**) 10.09 J/cm$^2$ (**d**) 11.10 J/cm$^2$ (**e**) 12.11 J/cm$^2$ (**f**) surface element composition

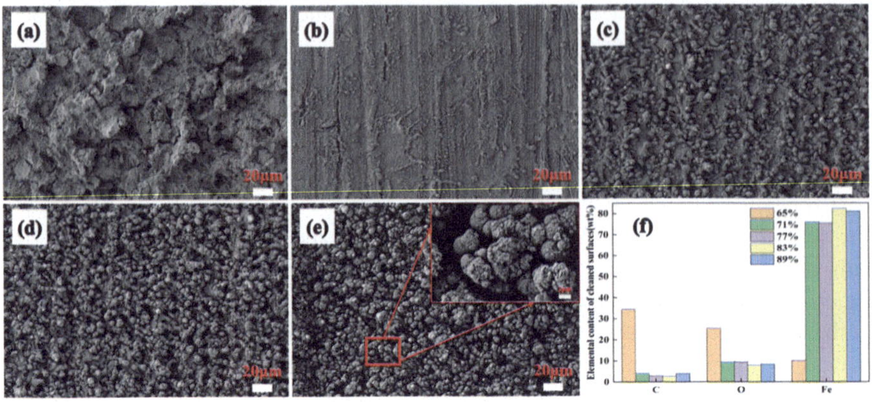


**Fig. 9.3** SEM images of cleaning with different spot overlap rates (**a**) 65% (**b**) 71% (**c**) 77% (**d**) 83% (**e**) 89% (**f**) surface element composition

At a laser energy density of 9.08 J/cm$^2$, the surface morphology of samples with varying spot overlap ratios from 65% to 89% is shown in Fig. 9.3. At 65% overlap, a large amount of paint remained on the steel surface. The paint layer appeared very rough, with many pits and cracks. The small overlap area between adjacent spots resulted in a short laser interaction time per unit area, sufficient only to remove the outer layer of the coating. As the overlap ratio increased, the paint was completely removed, and remelting began to occur on the steel surface, with a large number of splattered metallic droplets. With further increases in overlap ratio, the droplets aggregated but did not fully fill the surface, leaving numerous pits and cracks. In the magnified image of Fig. 9.3e, many metallic spheres with smooth surfaces were observed, indicating that at high overlap ratios, the oxide film did not completely cover the steel surface. Elemental spectra (f) showed that at 65% overlap, high levels of C and O were detected due to residual paint. At 71% overlap, C and O significantly decreased, while Fe increased sharply, confirming that the paint had been removed. As the overlap ratio continued to increase, the levels of C, O, and Fe first decreased and then increased, indicating the formation of an oxide film. However, at high overlap, the oxide film was not dense due to incomplete reaction between metal and oxygen.

The three-dimensional morphology of the steel surface after laser cleaning at different energy densities is shown in Fig. 9.4. Surface roughness was significantly affected by laser process parameters and initially decreased before increasing. At 8.07 J/cm$^2$, due to a large amount of residual paint, the surface displayed curved strip-like structures. The root mean square height (Sq) was 3.179 μm, and the maximum height (Sz) was 27.891 μm. At 9.08 J/cm$^2$, the removal of the paint resulted in a rapid decrease in roughness. The striped pattern of the base material was observable, and the surface roughness reached its minimum, with Sq = 1.488 μm and Sz = 14.805 μm. As energy density increased further, the striped morphology

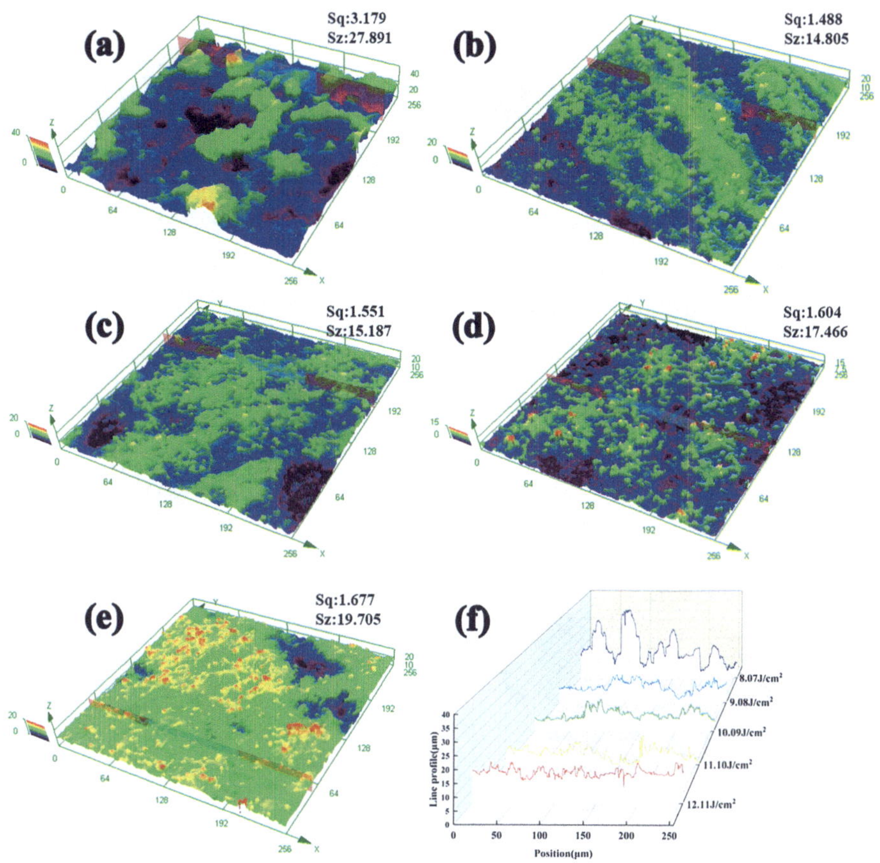


**Fig. 9.4** Three-dimensional morphology at different energy densities (**a**) 8.07 J/cm$^2$ (**b**) 9.08 J/cm$^2$ (**c**) 10.09 J/cm$^2$ (**d**) 11.10 J/cm$^2$ (**e**) 12.11 J/cm$^2$ (**f**) cross-sectional morphology

gradually disappeared, and the laser introduced microstructuring on the surface, forming small protrusions and creating ridge- and wave-like structures that increased surface roughness.

The influence of spot overlap ratio on surface morphology is further illustrated in Fig. 9.5. As the overlap ratio increased, surface roughness initially decreased and then increased. In Fig. 9.5a, at 9.08 J/cm$^2$ with 65% overlap, surface roughness was slightly higher than that at 8.07 J/cm$^2$ with 71% overlap. This was due to larger spacing between spots and higher single-spot energy, which led to deeper coating removal pits. At 71% overlap, the coating was removed, and the roughness dropped significantly. As the overlap ratio continued to increase, the laser interaction time was extended, causing more pronounced surface microstructuring with obvious protrusions. Compared with increasing energy density, increasing the overlap ratio resulted in a more rapid rise in roughness.

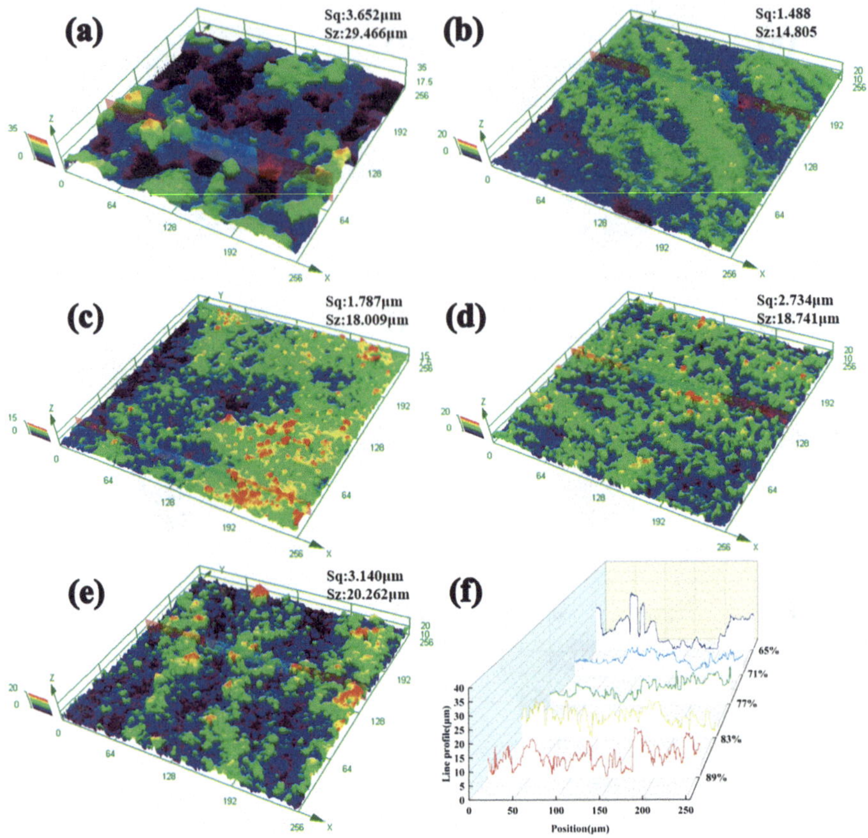


**Fig. 9.5** Three-dimensional morphology of cleaning with different spot overlap rates (**a**) 65% (**b**) 71% (**c**) 77% (**d**) 83% (**e**) 89% (**f**) cross-sectional morphology

## *9.2.3 Cleanliness Characterization*

To evaluate the cleanliness of the steel surface after laser cleaning, Fourier-transform infrared (FTIR) spectra were obtained, as shown in Fig. 9.6. The infrared spectrum of the residual paint-coated sample exhibited several prominent absorption peaks: a broad and strong peak at 3339.4 $cm^{-1}$ corresponds to the stretching vibration of O–H bonds (intermolecular hydrogen bonding), often referred to as a "bun-shaped" absorption band; the peaks at 2923.9 $cm^{-1}$ and 2864.8 $cm^{-1}$ are associated with the C–H stretching vibrations of $-CH_2$ and $-CH_3$ groups, respectively; the peak at 1729.7 $cm^{-1}$ is a characteristic absorption of C=O stretching in esters; the 1515.4 $cm^{-1}$ peak corresponds to symmetric stretching of C=C bonds; 1443.2 $cm^{-1}$ is related to in-plane bending of C–H; 1117.3 $cm^{-1}$ is attributed to C–O–C ether bond stretching; and 876.7 $cm^{-1}$ is associated with out-of-plane bending of C–H. As the laser energy density and spot overlap ratio increased, these characteristic peaks

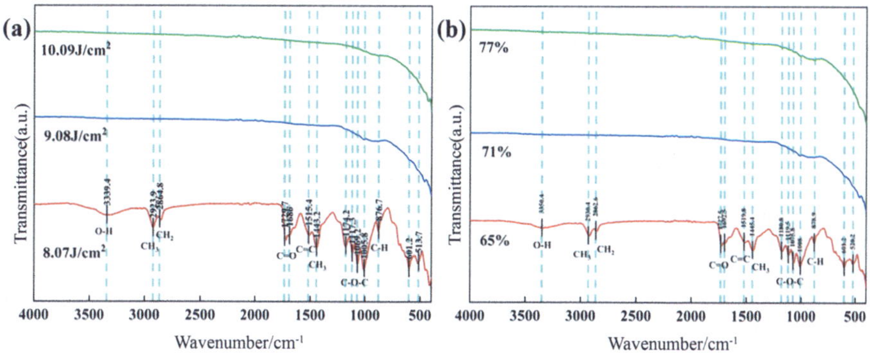


**Fig. 9.6** FTIR spectra of laser-cleaned sample surfaces (**a**) different energy densities (**b**) different spot overlap rates

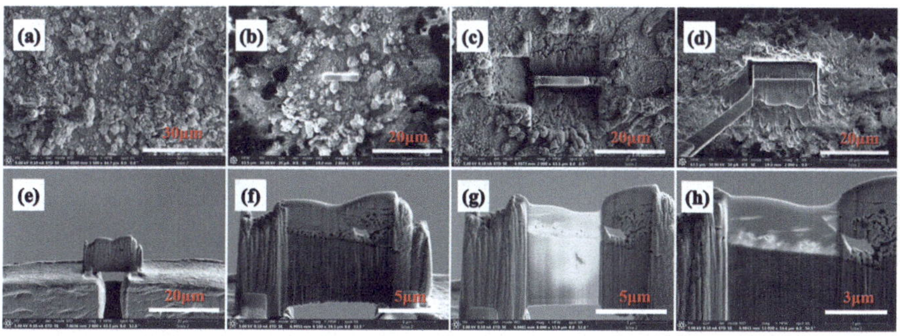


**Fig. 9.7** FIB sample preparation steps: (**a**) Site selection; (**b**) Protective layer deposition; (**c**) Coarse milling; (**d**) Fine milling; (**e**) Lamella extraction; (**f**) Final thinning; (**g**) High-magnification imaging; (**h**) Ultra-high-magnification imaging

of the coating significantly diminished, indicating that increasing these two parameters can effectively remove the paint layer from the steel surface.

## 9.3 Cross-Sectional Analysis

To investigate the cross-sectional morphology of samples after laser cleaning, focused ion beam (FIB) technology was employed to prepare thin cross-sectional slices of the sample cleaned at an energy density of 12.11 J/cm$^2$. The FIB preparation process is shown in Fig. 9.7 and includes the following steps: (1) Region selection and alignment on the sample; (2) Deposition of a protective platinum (Pt) layer above the target area to prevent damage during milling; (3) Coarse milling using a high-energy ion beam to remove bulk material around the target area for exposure and isolation; (4) Fine milling within the narrowed area to further reduce sample dimensions for TEM analysis; (5) Extraction and transfer of the sample using a FIB micromanipulator; (6) Final thinning of the lamella.

The FIB-prepared sample was then analyzed using transmission electron microscopy (TEM). Figure 9.8 shows high-resolution TEM bright-field images of the thin cross-section of the EH36 steel surface. The oxide layer was approximately 1 μm thick. In nanosecond laser surface microprocessing, the interaction time between the laser and material is short, and thermal conduction is limited, resulting in a shallow impact depth. The formation of the nanostructured layer is attributed to steel remelting and redeposition of flying nanoparticles from the plasma plume, forming a metallurgically bonded layer. The enlarged view in Fig. 9.8b of the interface between the oxide layer and steel substrate reveals a ~10 nm thick dark band at the boundary, likely caused by rapid cooling under significant undercooling, leading to

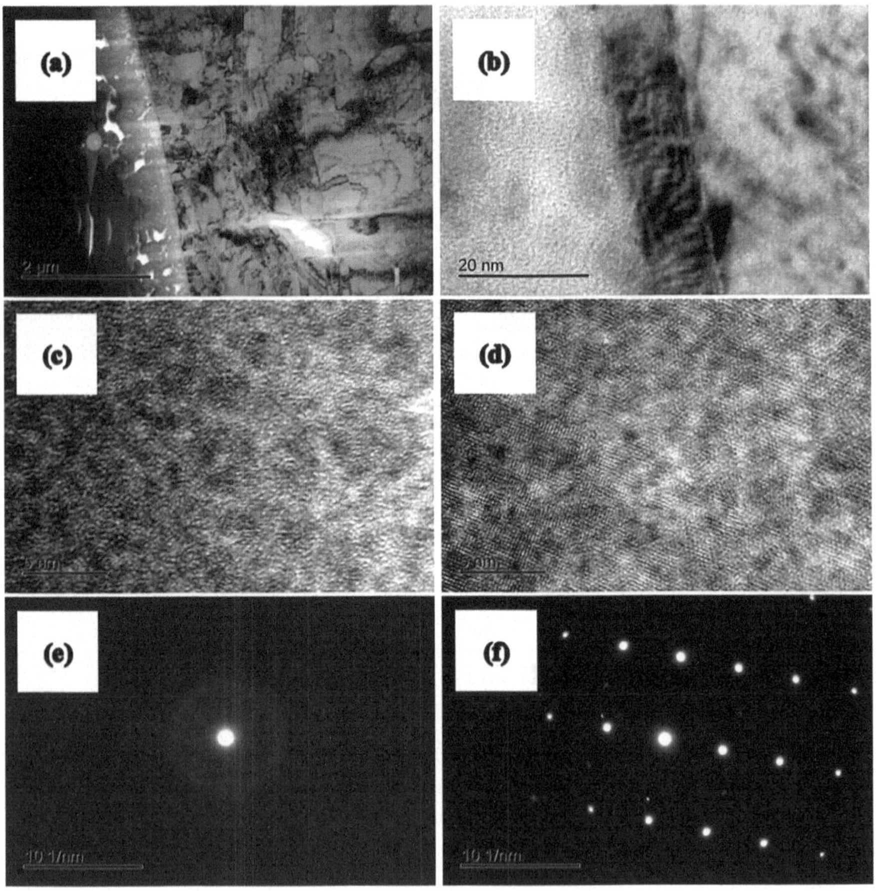


**Fig. 9.8** TEM analysis of samples after cleaning at an energy density of 12.11 J/cm$^2$. (**a**) and (**b**) show the bright-field images of the oxide layer and steel matrix interface, respectively. (**c**) and (**e**) depict the atomic arrangement and SAED patterns of the oxide layer, while (**d**) and (**f**) present the atomic arrangement and SAED patterns of the steel

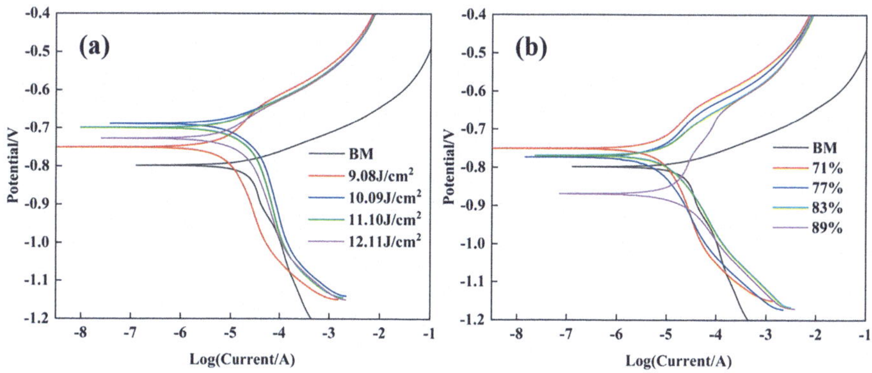


**Fig. 9.9** Polarization curves of EH36 steel: (**a**) Effect of laser energy density (9.08–12.11 J/cm$^2$); (**b**) Effect of spot overlap ratio (71%–89%).

grain refinement. During laser cleaning, multiple physical and chemical processes occurred, including surface ablation, melting, recrystallization, nanoparticle spattering, and interaction between molten steel and atmospheric oxygen, which ultimately resulted in a rough oxide layer replacing the previously smooth surface. To analyze the crystal structure of the nanostructured oxide layer, high-resolution electron microscopy was used to observe atomic arrangements, along with selected area electron diffraction (SAED) patterns shown in Fig. 9.8c–f. The results indicate that the nanostructured layer is amorphous, as evidenced by diffraction rings distinctly different from the discrete diffraction spots of the steel substrate, confirming a transformation in the steel's crystal structure during the laser cleaning process.

## 9.4 Corrosion Resistance Analysis

To evaluate the corrosion resistance of EH36 steel before and after laser cleaning, electrochemical tests were conducted on the base material and samples cleaned under different energy densities and spot overlap ratios, as shown in Fig. 9.9. From the fitted polarization curves, the corresponding corrosion potential (Ecorr) and corrosion current density (Icorr) values were obtained and listed in Table 9.1. The base material exhibited a corrosion potential of −0.81314 V and a corrosion current density of $3.4946 \times 10^{-4}$ A/cm$^2$. After laser cleaning, the corrosion potential of the samples significantly increased, with a maximum value of −0.68794 V, indicating improved thermodynamic stability under electrochemical corrosion conditions. With increasing laser energy density, the corrosion current density gradually decreased, suggesting that corrosion resistance improved due to the formation of a corrosion-resistant oxide film on the steel surface induced by laser irradiation.

As the spot overlap ratio increased, the corrosion current density first decreased and then increased. This is because greater overlap prolongs the laser–material interaction time, enabling not only effective removal of the paint layer but also

**Table 9.1** Fitted electrochemical parameters of EH36 samples after laser cleaning compared to original samples

| Experimental sample | $E_{corr}$ (V) | $I_{corr}$ (A/cm$^2$) | ba (mV) | bc (mV) |
|---|---|---|---|---|
| BM | −0.81314 | $3.4946 \times 10^{-4}$ | 80.231 | 175.7 |
| 9.08 J/cm$^2$ | −0.7496 | $1.2526 \times 10^{-5}$ | 188.6 | 399.03 |
| 10.09 J/cm$^2$ | −0.68794 | $1.1168 \times 10^{-5}$ | 82.421 | 260.57 |
| 11.10 J/cm$^2$ | −0.69866 | $9.3824 \times 10^{-6}$ | 85.26 | 254.64 |
| 12.11 J/cm$^2$ | −0.72704 | $9.2165 \times 10^{-6}$ | 91.95 | 200.02 |
| 77% | −0.77262 | $6.2112 \times 10^{-6}$ | 118.12 | 215.71 |
| 83% | −0.76802 | $8.6032 \times 10^{-6}$ | 104.83 | 177.14 |
| 89% | −0.86845 | $3.0607 \times 10^{-5}$ | 470.33 | 197.89 |

microstructuring of the steel surface. The laser's thermal effect led to the formation of an oxide layer that impeded corrosion. However, at 83% overlap, the combined thermal and mechanical effects of the laser became more intense, damaging the previously dense oxide film. The resulting increase in surface roughness significantly enlarged the contact area between the electrolyte and the steel surface, thereby increasing the corrosion current density. Based on the above analysis, the best corrosion resistance was achieved at an energy density of 9.08 J/cm$^2$ and a spot overlap ratio of 77%, where the corrosion current density reached a minimum of $6.2112 \times 10^{-6}$ A/cm$^2$.

To further clarify the corrosion resistance of steel before and after laser cleaning, electrochemical impedance spectroscopy (EIS) was used to evaluate the samples. The results were fitted using two types of equivalent circuits: one without an oxide film and one with an oxide film, as shown in Fig. 9.10. Figure 9.10a shows the Nyquist plots obtained in a 3.5 wt% NaCl solution. Generally, the diameter of the Nyquist arc represents the reaction resistance, where a larger diameter indicates higher corrosion resistance. Based on this principle, it can be determined that the sample cleaned at a laser energy density of 9.08 J/cm$^2$ and a spot overlap ratio of 77% exhibits excellent corrosion resistance. Further details can be seen in the Bode plot in Fig. 9.10b. In the high-frequency region, $|Z|$ values among samples are relatively close, but significant differences appear in the low-frequency region, where the sample with 9.08 J/cm$^2$ and 77% overlap shows the highest $|Z|$ value.

## 9.5 Wear Resistance Analysis

To evaluate the wear resistance of the samples before and after laser cleaning, fretting wear tests were conducted on the base material and samples treated at various energy densities and spot overlap ratios. The variations in friction coefficient over time are shown in Fig. 9.11. It can be observed that the friction coefficient of the laser-cleaned samples is significantly lower than that of the base material. As shown

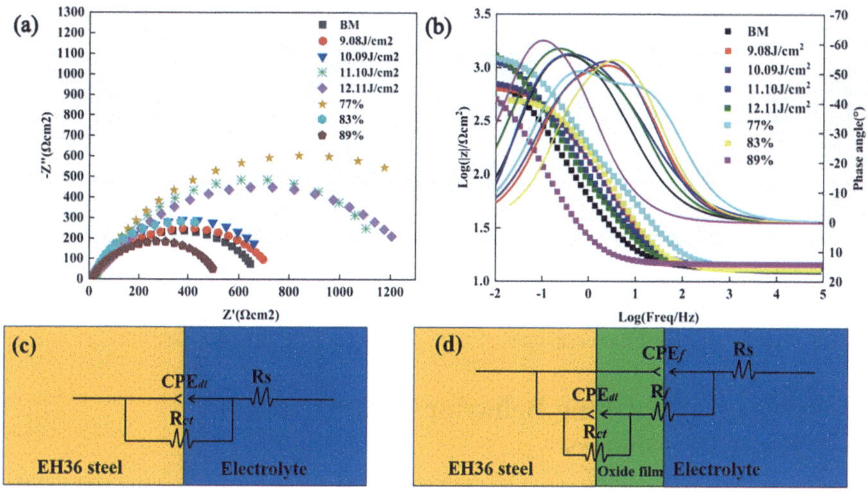


**Fig. 9.10** EIS analysis (**a**) Nyquist plot (**b**) Bode plot (**c**) equivalent circuit without oxide film (**d**) equivalent circuit with oxide film

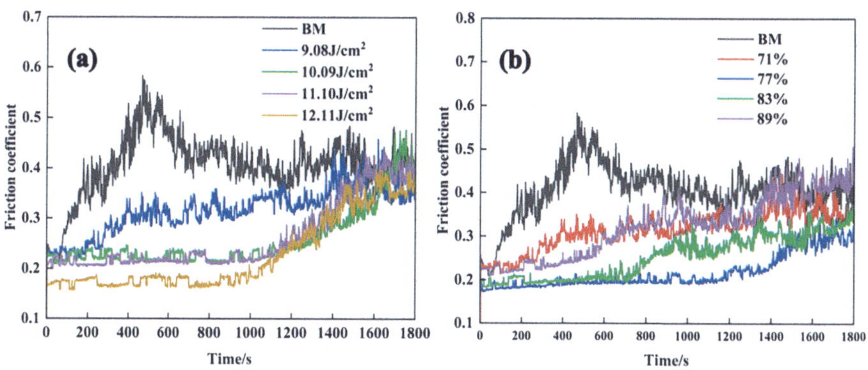


**Fig. 9.11** Friction coefficient variation over time during fretting wear tests: (**a**) Effect of laser energy density (9.08–12.11 J/cm²); (**b**) Effect of spot overlap ratio (71%–89%)

in Fig. 9.11a, the friction coefficient decreases with increasing energy density, which is attributed to the formation of a denser oxide film on the steel surface under higher energy input. At an energy density of 12.11 J/cm$^2$, the friction coefficient remains relatively stable for the first 1000 s, after which it gradually increases and eventually approaches that of the base material. This trend indicates that the oxide film helps reduce friction initially, but as it gradually wears off and eventually breaks down, direct contact between the base material and counter surface leads to an increase in friction.

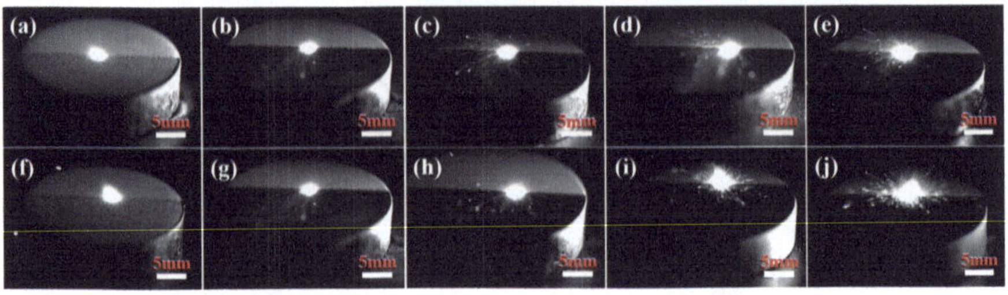


**Fig. 9.12** Dynamic cleaning behavior on steel surfaces (**a**) 8.07 J/cm$^2$ (**b**) 9.08 J/cm$^2$ (**c**) 10.09 J/cm$^2$ (**d**) 11.10 J/cm$^2$ (**e**) 12.11 J/cm$^2$ (**f**) 65% overlap rate (**g**) 71% overlap rate (**h**) 77% overlap rate (**i**) 83% overlap rate (**j**) 89% overlap rate

## 9.6 Dynamic Cleaning Behavior Under Different Laser Parameters

To investigate the behavior of the paint layer during laser cleaning of the steel surface, Fig. 9.12 illustrates the dynamic cleaning process under different laser parameters. The area above the bright spot represents the uncleaned region, while the area below represents the cleaned region. The bright spot corresponds to the plasma plume generated by laser interaction with the paint layer and steel. This plume is accompanied by particle collisions, as evidenced by visible particle spattering. The size of the plasma plume reflects the intensity of plasma impact effects. However, due to the millisecond-level shutter speed and frame rate of the camera, it is difficult to capture the real-time formation and expansion of the plasma. Thus, each captured frame can be considered to represent the approximate plasma size under identical conditions. Figure 9.12a–e show the cleaning behavior under different energy densities. At 9.08 J/cm$^2$, significant paint residue remains on the steel surface. Due to the low mass of paint particles, no apparent spattering is observed. At 10.09 J/cm$^2$, the paint is completely removed, and the contrast between the upper and lower regions of the spot becomes visible, accompanied by metal particle spattering due to plasma impact. As the energy density increases further, the interaction between laser and steel intensifies, slightly enlarging the bright spot diameter, while its shape remains nearly elliptical.

## 9.7 Key Findings and Application Potential

This chapter investigated the application of nanosecond laser cleaning technology for the removal of paint coatings from the surface of EH36 marine steel, with particular focus on the influence of laser energy density and spot overlap ratio on cleaning performance. Based on macroscopic and microscopic morphology analyses, it was found that the laser cleaning process effectively removes the paint layer, and under optimized parameters, damage to the substrate can be minimized.

FTIR analysis confirmed that increasing laser energy density and overlap ratio leads to more complete paint removal. Electrochemical testing and wear analysis further revealed that laser-cleaned surfaces exhibit improved corrosion resistance and wear performance. This is attributed to the formation of a compact oxide film on the steel surface, which reduces direct metal contact and enhances surface properties.

In addition, high-speed imaging was used to observe the dynamic laser cleaning process, revealing the formation of plasma plumes and metal particle spattering. These observations provided deeper insight into the laser cleaning mechanism.

Overall, this study demonstrates that nanosecond laser cleaning offers an effective solution for paint removal from EH36 marine steel surfaces, achieving high cleaning efficiency with minimal substrate damage, while also improving surface functionality. Future research may focus on further optimizing the laser cleaning process for broader industrial application and exploring its potential in treating other material surfaces.

## References

1. Lu Y, Ding Y, Wang M L, et al. An environmentally friendly laser cleaning method to remove oceanic microbiofoulings from AH36 steel substrate and corrosion protection [J]. Journal of Cleaner Production, 2021, 314: 127961.
2. Tian Z, Lei Z L, Chen X, et al. Nanosecond pulsed fiber laser cleaning of natural marine micro-biofoulings from the surface of aluminum alloy [J]. Journal of cleaner production, 2020, 244: 118724.
3. Choubey A, Ali S, Yadav S, et al. Investigating laser surface texturing on SS 304 for self-cleaning applications [M]//Next Generation Materials and Processing Technologies. Singapore: Springer Singapore, 2021: 221–232.

# Chapter 10
# Control of Substrate Surface Morphology After Laser Cleaning of CFRP Coating

## 10.1 Introduction

With the increasing demand for "lightweight, high speed, and high reliability" in aerospace equipment, the application of composite material skins in aircraft manufacturing has become increasingly widespread. For example, the skin of the Boeing 787 aircraft is made of full carbon fiber reinforced plastic (CFRP) structure [1–3]. However, in actual service, due to the effects of complex environmental factors such as space radiation, airflow erosion, and vibration fatigue, the aircraft skin coating is prone to aging, cracking, functional failure, or even peeling, requiring partial paint removal repair or complete repainting [4]. Therefore, efficiently removing the skin coating and ensuring the performance of the re-coating interface have become key aspects of aircraft maintenance.

Traditional chemical, mechanical, and sandblasting paint removal methods, while achieving some cleaning effect, have significant drawbacks such as high toxicity, severe pollution, low efficiency, and the risk of secondary damage to the substrate, making them difficult to meet the strict requirements for cleaning aerospace composite material skins [5, 6]. Laser cleaning technology, as a green, non-destructive, and efficient new cleaning method, can precisely remove surface contaminants through the combined effects of light pressure, impact pressure, vibration waves, and plasma bursts, while maximizing the preservation of the substrate surface integrity [7–9]. This chapter investigates the application of nanosecond and picosecond laser technology in CFRP surface coating cleaning, focusing on the impact of laser parameters on the substrate surface morphology and properties, as well as how to achieve precise morphology control. Through microscopic morphology observation and three-dimensional morphology analysis, the cleaning effects are preliminarily evaluated, and Fourier Transform Infrared Spectroscopy (FT-IR) is used to further characterize the surface cleanliness. Additionally, contact angle tests and pull-off tests are conducted to assess the wettability and adhesion properties of the cleaned material surfaces, which are directly related to the morphology and



S. Wang, *Laser Cleaning*, https://doi.org/10.1007/978-981-95-2935-3_10

**Table 10.1** Chemical composition of the coating

| Elements | C | O | Mg | Al | Si | S | Ti | Ce |
|---|---|---|---|---|---|---|---|---|
| Content (%) | 66.46 | 19.47 | 1.64 | 0.22 | 1.36 | 1.83 | 7.34 | 1.68 |

functionality of the substrate surface. Furthermore, various online monitoring techniques are implemented for the nanosecond pulse laser cleaning process to monitor and optimize the cleaning process in real-time, ensuring cleaning efficiency and substrate surface integrity.

## 10.2 Experimental Materials and Testing Methods

### *10.2.1 Experimental Materials*

The material used in this chapter is a T300 CFRP laminated composite plate, cut into rectangular thin sheets with dimensions of 15 × 15 × 2.5 mm. The coating applied is a zinc yellow acrylic polyurethane paint (TB06-9), which is commonly used for the protection of aircraft skins and composite materials. The paint was sprayed onto the CFRP surface to form a coating layer with a thickness of 80 ± 2 μm. The prepared samples were cured in air for 1 week and then sealed in sample bags for subsequent experiments.

The elemental composition of the paint layer was analyzed by Energy Dispersive Spectroscopy (EDS), and the results are shown in Table 10.1.

To investigate the decomposition rate of the paint layer at various temperatures, thermogravimetric analysis (TGA) was conducted. The TGA results are shown in Fig. 10.1. The thermogravimetric (TG) curve indicates that the decomposition temperature range of the paint is 200–660 °C, while the derivative thermogravimetric (DTG) curve shows that the maximum decomposition rate occurs at 471.6 °C.

### *10.2.2 Experimental Equipment and Parameters*

A nanosecond infrared pulsed laser cleaner (YDFLP-120-M7-L1-R, Xi'an) with a wavelength of $\lambda = 1064$ nm, adjustable pulse width of 10–350 ns, rated power of 100 W, and a spot diameter of 50 μm was employed. A picosecond infrared pulsed laser (YFL-PP-25-GT150-S-R-2-2, Xi'an) with a wavelength of $\lambda = 1060$–1080 nm, adjustable pulse width of 5–3500 ps, rated power of 25 W, and a spot diameter of 50 μm was also used.

Figure 10.2 presents a schematic diagram of the laser paint removal system. The output laser beam is first expanded into a parallel beam using a beam expander, then directed through a scanning galvanometer, focused by a lens, and finally irradiated onto the sample surface on the experimental platform. After cooling in air, the

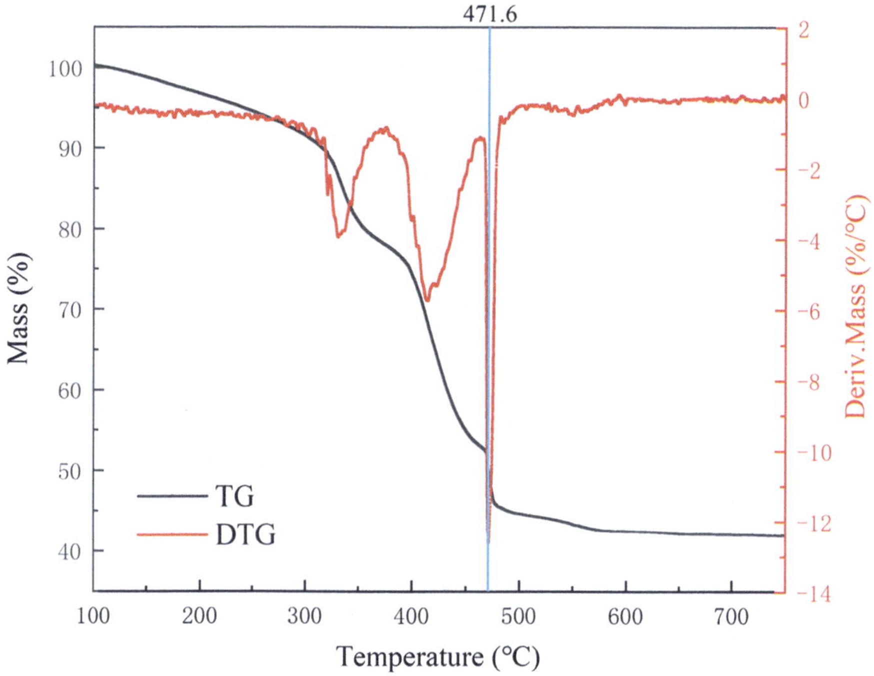


**Fig. 10.1** Thermogravimetric analysis of the coating

cleaned samples were sealed in sample bags for subsequent testing. Dust generated during the cleaning process was collected by the integrated dust extraction system.

The laser cleaning parameters are listed in Table 10.2. For nanosecond laser cleaning of CFRP, the average laser powers used were 30 W, 35 W, 40 W, 43 W, and 45 W, corresponding to sample labels C1 to C5. Since picosecond lasers have shorter pulse durations, a multi-pass cleaning strategy (3–7 passes) at rated power was adopted, and samples were labeled P1 to P5 accordingly. The uncleaned control sample is labeled AR.

### *10.2.3 Characterization Methods*

High-resolution field emission scanning electron microscopy (SEM, Nano SEM 50, American FEI) was used to observe the surface morphology after laser cleaning. A laser confocal microscope (LEXT OLS 4000) was employed to observe the 3D morphology of the cleaned samples and to measure surface roughness. For each sample, three different locations were measured and the average value was taken. Fourier Transform Infrared Spectroscopy (FT-IR, IRTracer-100, Shimadzu) was used to characterize the variation of surface functional groups before and after

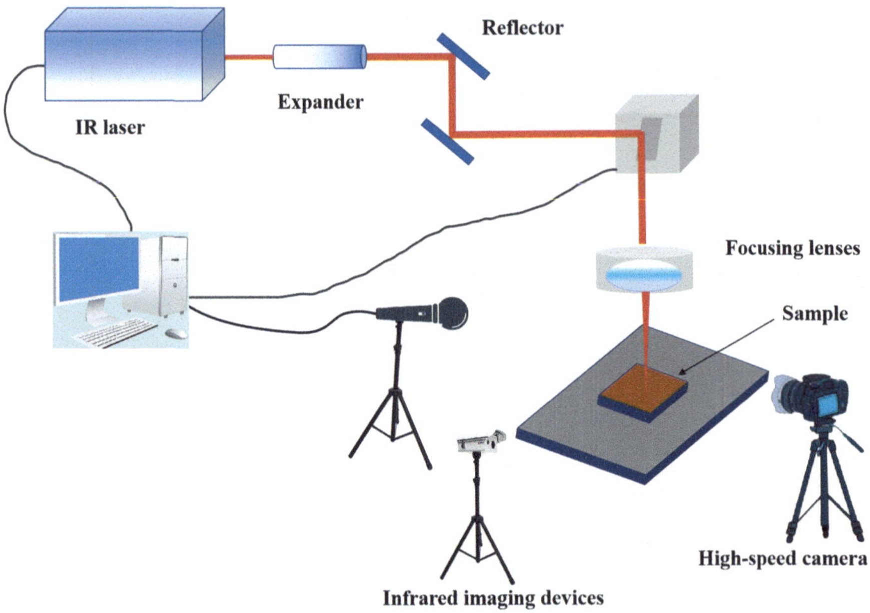


**Fig. 10.2** Schematic diagram of laser cleaning equipment

**Table 10.2** Laser cleaning parameters

| Material | Laser type | Power (W) | Repetition frequency (kHz) | Scanning speed (mm/s) | Scanning interval (mm) | Number of passes |
|---|---|---|---|---|---|---|
| CFRP | Nanosecond | 30 ~ 45 | 200 | 3000 | 0.02 | 2 |
| CFRP | Picosecond | 25 | 100 | 2500 | 0.01 | 3 ~ 7 |

cleaning, with a scanning range of 4000–400 $cm^{-1}$. A contact angle measurement instrument (FH-SDC100, China) was used to measure the contact angle of the cleaned sample surfaces. A universal tensile testing machine (C5.5104, China) was used to measure the adhesion strength of repainted samples, following the GB/T 5210-2006 standard for pull-off testing of coatings and paints. Thermogravimetric (TG) analysis curves were obtained using a thermal analyzer (STA 449 F3 Jupiter, Germany NETZSCH) under an argon atmosphere, from room temperature to 800 °C, at a heating rate of 10 °C/min. Acoustic signals generated during laser cleaning were collected using a microphone and processed with MATLAB to generate time-amplitude (TA), frequency-amplitude (FA), and time-frequency (TF) plots. An infrared thermal imager (FLIR-A600, USA) was used to monitor surface temperature variations during the cleaning process. A high-speed camera (FASTCAM Mini AX200) was used to record dynamic changes on the material surface during cleaning.

## 10.3 Characterization of CFRP Surface Cleanliness After Nanosecond Laser Cleaning

### *10.3.1 Surface Micromorphology*

Figure 10.3 shows the surface morphology of CFRP samples after laser cleaning at different average powers. As shown in the images, carbon fibers are exposed on the material surface under all selected parameters. Due to the relatively low average laser power, the surfaces of samples C1 and C2 were not completely cleaned, and a large amount of resin residue remained, as seen in Fig. 10.3a, b. The observed fiber breakage on the surface may have resulted from the plate cutting process. When P = 40 W, the amount of resin residue on the surface of sample C3 was reduced, basically meeting the cleaning requirements, as shown in Fig. 10.3c. At P = 43 W, the resin on the surface of sample C4 was completely removed, exposing relatively clean and smooth carbon fibers, as shown in Fig. 10.3d. Some large resin blocks were found on the carbon fiber surface and in the gaps, possibly due to splashed particles being deposited on the material surface by thermal or plasma shock during the cleaning process. When the power increased to 45 W, the higher laser energy caused the carbon fibers to melt and break, indicating over-cleaning, as shown in Fig. 10.3e. Over-cleaning can degrade the mechanical properties of carbon fibers, and such conditions should be avoided during actual cleaning operations.

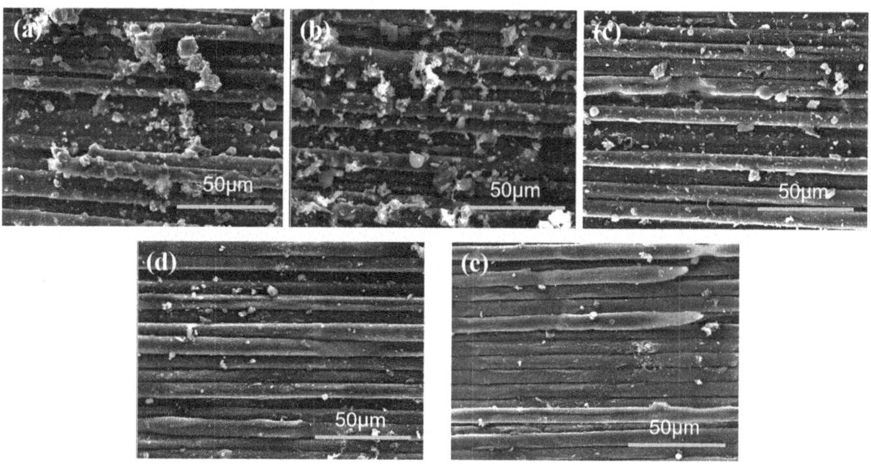


**Fig. 10.3** Microstructure of CFRP samples with different average laser power: (**a**) C1; (**b**) C2; (**c**) C3; (**d**) C4; (**e**) C5

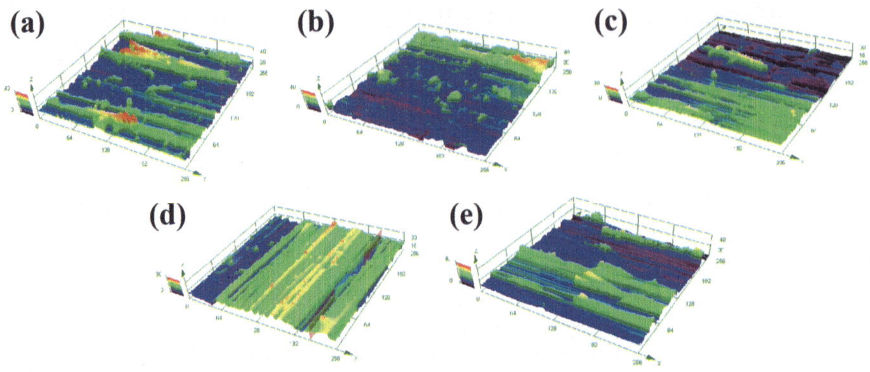


**Fig. 10.4** 3D morphology of CFRP samples under different average laser power: (**a**) C1; (**b**) C2; (**c**) C3; (**d**) C4; (**e**) C5

### *10.3.2 Three-Dimensional Morphology and Surface Roughness Analysis*

Figure 10.4 presents the 3D surface morphology of CFRP samples after cleaning under various average laser powers. As seen in Fig. 10.4a, b, significant resin blocks remained on the surfaces of C1 and C2. The coexistence of residual resin and exposed fibers led to large height differences in localized areas. As the average power increased, the large resin blocks were removed. In sample C3, only small resin particles were present, resulting in a reduced surface height difference, as shown in Fig. 10.4c. When the average power reached $P = 43$ W, the surface of C4 was completely cleaned, revealing a smooth and flat fiber surface. With further increase in average power, the surface of C5 showed fiber breakage and slight bulging due to higher energy absorption, leading to an increase in height difference again, as shown in Fig. 10.4e.

Figure 10.5 shows the surface roughness profiles of CFRP samples after cleaning with various average laser powers. It is evident that surface roughness increased significantly after cleaning due to fiber exposure. Before cleaning, the AR sample had only a paint layer on the surface, resulting in the lowest roughness of 1.61 μm. After cleaning, sample C1 showed the highest roughness of 3.588 μm, which is attributed to the low average laser power and the large quantity and size of residual resin blocks. As the power increased, the resin was mostly removed, and the surface roughness of C2 and C3 decreased. After complete cleaning, the surface roughness of sample C4 increased again, likely due to laser-induced impact on the fiber surface. In sample C5, the increased surface roughness was caused by both laser shock during cleaning and fiber breakage and bulging post-cleaning.

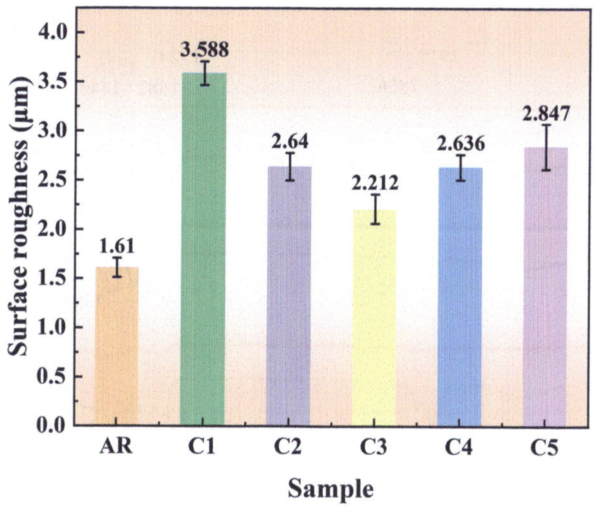


**Fig. 10.5** Surface roughness of CFRP samples after cleaning with different average laser power

### *10.3.3 FT-IR Spectroscopic Characterization*

To further determine whether the laser cleaning effectively removed the paint layer, FT-IR spectroscopy was performed on the CFRP samples, as shown in Fig. 10.6. The AR sample, which was uncleaned, served as the control group. The locations and corresponding assignments of absorption peaks are labeled in Fig. 10.6. The absorption peak near 1100 $cm^{-1}$ corresponds to the strong broad C–O–C stretching of aliphatic ethers; the 1244 $cm^{-1}$ peak corresponds to C–O stretching vibrations in polyesters. The peaks at 1380 $cm^{-1}$ and 1440 $cm^{-1}$ are attributed to $CH_3$ and $–CH_2$ bending vibrations, respectively. The 1526 $cm^{-1}$ peak corresponds to the N–H bending vibration of amide bonds. Peaks at 1682 $cm^{-1}$ and 1724 $cm^{-1}$ are associated with the C=O stretching in isocyanurate and ester groups, respectively. The 2856 $cm^{-1}$ and 2919 $cm^{-1}$ peaks represent $–CH_2$ stretching vibrations. Among these, the peaks at 1526 $cm^{-1}$, 1682 $cm^{-1}$, and 1724 $cm^{-1}$ are characteristic of the polyurethane coating. In Fig. 10.6, these three peaks are essentially absent in the cleaned samples, indicating that the paint layer on the sample surface has been effectively removed.

## 10.4 Evaluation of CFRP Surface Properties After Nanosecond Laser Cleaning

### *10.4.1 Changes in Surface Wettability*

Changes in the contact angle of the sample surface after paint removal significantly affect surface wettability, which is critical in the repainting process. Contact angle measurements were performed on the cleaned CFRP samples, with the uncleaned

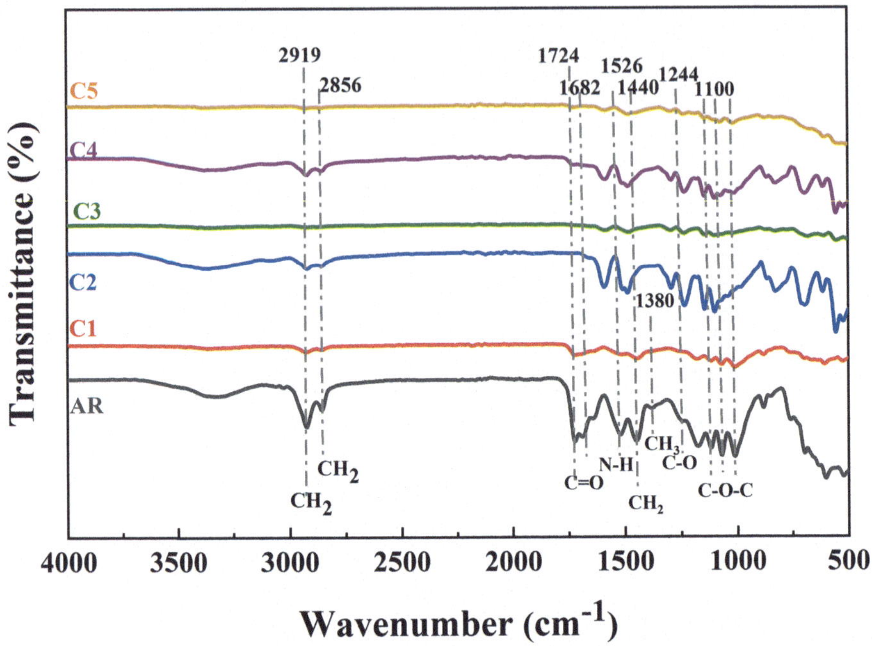


**Fig. 10.6** FT-IR spectra of different CFRP samples after laser cleaning

AR sample serving as the control. As shown in Fig. 10.7, the contact angle of the uncleaned CFRP surface was 76.9°. After laser cleaning, a decreasing trend was observed, with the fully cleaned C4 sample showing the lowest contact angle of 60.3°. The primary reason for this decrease is the effective removal of surface contaminants along with the paint layer during laser cleaning, resulting in a cleaner surface that is more easily wetted by liquids, thereby reducing the contact angle. Additionally, the laser's direct action on the material surface increases the surface energy, enhancing hydrophilicity and further lowering the contact angle. Moreover, as indicated in Sect. 10.3.2, surface roughness increased after cleaning, and this also contributed to the reduction in contact angle. Overall, the decrease in contact angle improves the surface wettability, thereby enhancing the adhesion of the repainted coating.

### *10.4.2 Adhesion Strength of Repainted Samples*

In engineering applications, structural components that have undergone paint removal need to be repainted before reuse. The repainted layer must exhibit strong adhesion to the aircraft skin to prevent peeling or detachment, thus ensuring effective surface protection and maintaining appearance. The pull-off test method was employed to evaluate the adhesion strength of the paint layer.

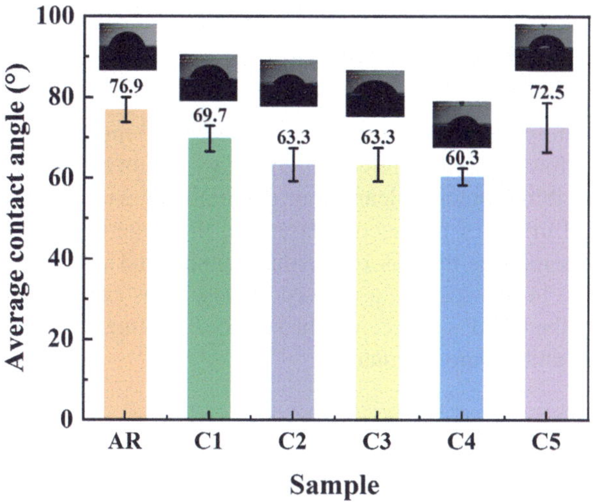


**Fig. 10.7** Surface contact angles of different samples after nanosecond laser cleaning of CFRP

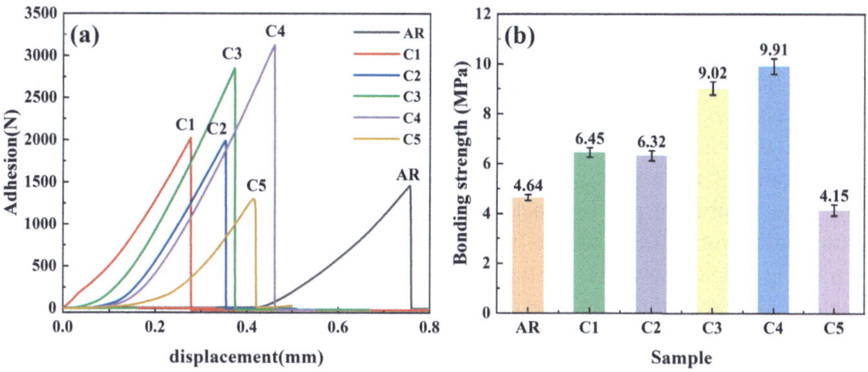


**Fig. 10.8** CFRP samples after nanosecond laser cleaning: (**a**) Pull-off test curves; (**b**) Bond strength chart

Figure 10.8 shows the tensile test curves and corresponding bond strengths for CFRP samples before and after cleaning, with the uncleaned AR sample as the control. In Fig. 10.8a, the bonding force of AR before cleaning was 1457.7 N, corresponding to a bond strength of 4.64 MPa. After laser cleaning, all samples exhibited varying degrees of improvement in bonding force. The fully cleaned sample C4 showed the highest bonding force of 3113 N, corresponding to a bond strength of 9.91 MPa, which is a 113% increase compared to the AR sample. This improvement is attributed to the increased surface roughness and lower contact angle of the fully cleaned sample, as described in Sect. 10.2.1, which enhance the adhesion of the paint layer. Notably, the excessively cleaned sample C5 exhibited the lowest bonding force and strength, due to surface damage and fiber breakage caused by over-cleaning, which reduced the bonding capability between the paint layer and the substrate surface.

## 10.5 Characterization of CFRP Surface Cleanliness After Picosecond Laser Cleaning

Compared to nanosecond lasers, picosecond lasers have much shorter pulse durations ($10^{-12}$ s), allowing them to release high energy within an extremely short time span, thereby enabling precise cleaning of materials. Due to the short pulse duration, the thermal effect of picosecond lasers on the material surface is minimal, which effectively reduces the risk of thermal damage and deformation during the cleaning process. In this section, a picosecond laser was employed to remove the paint layer from the CFRP surface, and the surface cleanliness and properties of the material after cleaning were evaluated.

### *10.5.1 Surface Micromorphology*

Considering the short pulse duration and relatively low power output of the picosecond laser, a multiple-pass cleaning strategy was adopted. Figure 10.9 presents the surface morphology of CFRP samples cleaned using picosecond laser at different numbers of passes. When the number of cleaning passes was three, heat accumulation was minimal, resulting in a large amount of residual paint on the material surface and no fiber exposure, as shown in Fig. 10.9a. As the number of cleaning passes increased to four, partial exposure of carbon fibers was observed, although a significant amount of resin particles remained on the surface, as shown in Fig. 10.9b. When the number of passes reached five, the exposed fibers appeared intact and clean, indicating that the surface was completely cleaned under this condition, as seen in Fig. 10.9c. However, further increasing the number of cleaning passes led to excessive heat accumulation, causing localized melting and breakage of the exposed fibers, as observed in Fig. 10.9d, e.

By comparing the surface morphology of CFRP samples after nanosecond laser cleaning in Fig. 10.3 and picosecond laser cleaning in Fig. 10.9, it can be concluded that picosecond lasers, owing to their low thermal impact and high precision, demonstrate superior cleaning performance. Nevertheless, due to the relatively low rated power of the picosecond laser, only a multi-pass cleaning approach can be adopted, which inevitably affects the overall cleaning efficiency.

### *10.5.2 Three-Dimensional Morphology and Surface Roughness Analysis*

Figure 10.10 presents the 3D surface morphology of CFRP samples cleaned with a picosecond laser under different numbers of cleaning passes. In Fig. 10.10a, with fewer cleaning passes, a large amount of residual paint remained on the surface,

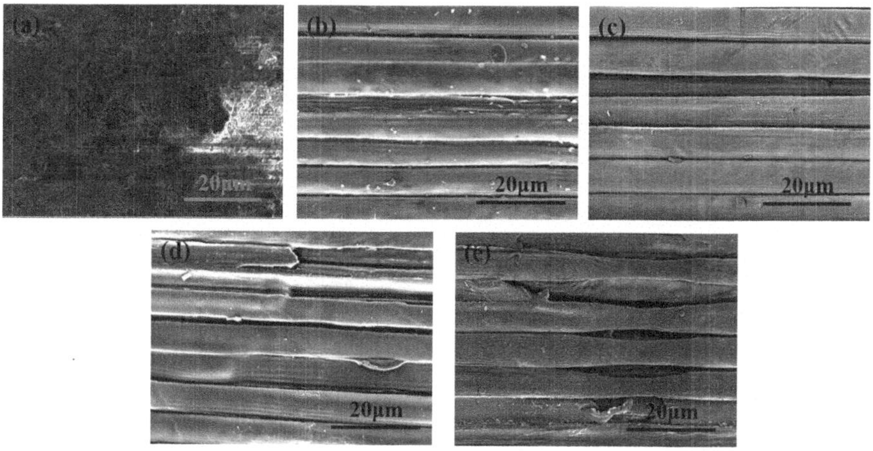


**Fig. 10.9** Microstructure of CFRP samples after picosecond laser cleaning at different cleaning times: (**a**) P1; (**b**) P2; (**c**) P3; (**d**) P4; (**e**) P5

resulting in a significant height difference across the surface. As the number of cleaning passes increased, the paint layer on the material surface was progressively removed, exposing the underlying fibers. Figure 10.10b shows that although resin particles were still present, the fiber surface was relatively flat, and the surface height variation was clearly reduced compared to the three-pass case. When the number of cleaning passes reached five, Fig. 10.10c shows a complete and clean carbon fiber surface. However, in Fig. 10.10d, e, excessive cleaning resulted in fiber damage, causing localized areas to exhibit larger height differences.

Figure 10.11 presents the surface roughness profiles of CFRP samples after picosecond laser cleaning with different numbers of passes. Before cleaning, the surface roughness was only 1.61 μm. After three passes, the surface of sample P1 retained paint residues, and the roughness increased to 1.731 μm. When the number of passes increased to four, the paint layer on the surface was completely removed, with only a few resin particles remaining, and the surface roughness decreased to 1.65 μm. After five passes, the surface was thoroughly cleaned, and the roughness slightly increased again, which is attributed to the mechanical impact of the picosecond laser. However, with further increases in the number of passes, the surface roughness increased again. Based on Figs. 10.9 and 10.10d, e, this rise in roughness is due to localized fiber damage. Furthermore, it was observed that the increase in surface roughness after picosecond laser cleaning was less pronounced than that after nanosecond laser cleaning. This is because the thermal effect of the picosecond laser is weaker, and its shorter pulse width enables more precise cleaning of the targeted area.

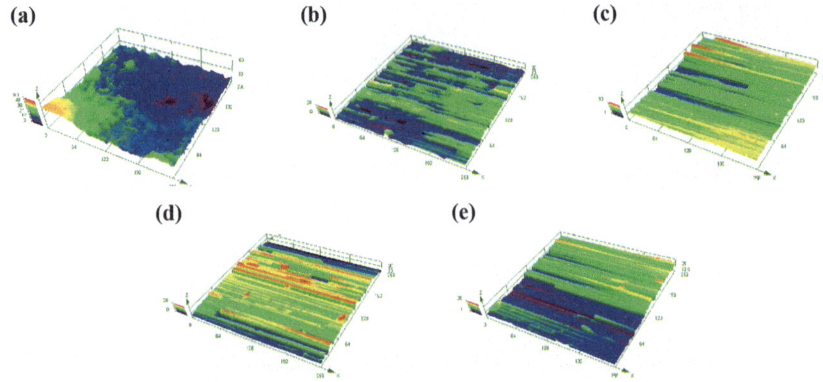


**Fig. 10.10** 3D morphology of CFRP samples after picosecond laser cleaning at different cleaning times: (**a**) P1; (**b**) P2; (**c**) P3; (**d**) P4; (**e**) P5

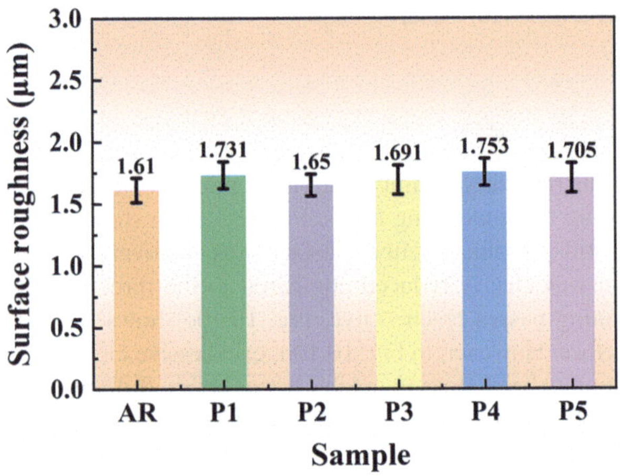


**Fig. 10.11** Surface roughness of CFRP samples after picosecond laser cleaning at different cleaning times

## *10.5.3 FT-IR Spectroscopic Characterization*

To evaluate the cleanliness of the CFRP surface after picosecond laser cleaning, FT-IR spectroscopy was conducted. The results are shown in Fig. 10.12, and the corresponding absorption peak positions are listed in Table 10.3. In the control group (AR), the absorption peaks at 1526 $cm^{-1}$, 1682 $cm^{-1}$, and 1724 $cm^{-1}$ are characteristic of polyurethane coatings. After three passes, all three peaks showed a decline in sample P1, indicating partial removal of the paint layer. With an increased number of cleaning passes, samples P2 through P5 showed complete disappearance of these three characteristic peaks, confirming the complete removal of the paint layer.

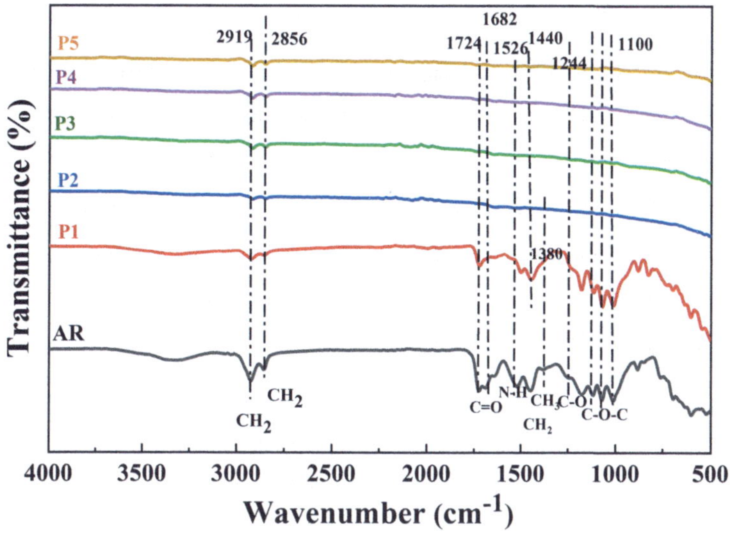


**Fig. 10.12** FT-IR spectra of CFRP samples after picosecond laser cleaning at different cleaning times

**Table 10.3** Absorption peaks at various wavenumbers

| Wavenumber ($cm^{-1}$) | Structure | Vibration type |
|---|---|---|
| 1100 | C–O–C | Stretching vibration |
| 1244 | C–O | Stretching vibration |
| 1380 | $CH_3$ | Bending vibration |
| 1440 | $-CH_2$ | Bending vibration |
| 1526 | N–H | Bending vibration |
| 1682 | C=O | Stretching vibration |
| 1724 | C=O | Stretching vibration |
| 2856 | $-CH_2$ | Symmetric $CH_2$ stretching vibration |
| 2919 | $-CH_2$ | Symmetric $CH_2$ stretching vibration |

## 10.6 Evaluation of CFRP Surface Properties After Picosecond Laser Cleaning

### *10.6.1 Changes in Surface Wettability*

Contact angle measurements were conducted on the surface of CFRP samples cleaned by picosecond laser, as shown in Fig. 10.13. Since sample P1 still retained a residual paint layer on its surface, it was excluded from testing. Only samples P2 to P5 were tested, all of which had no paint residue and exhibited fully exposed

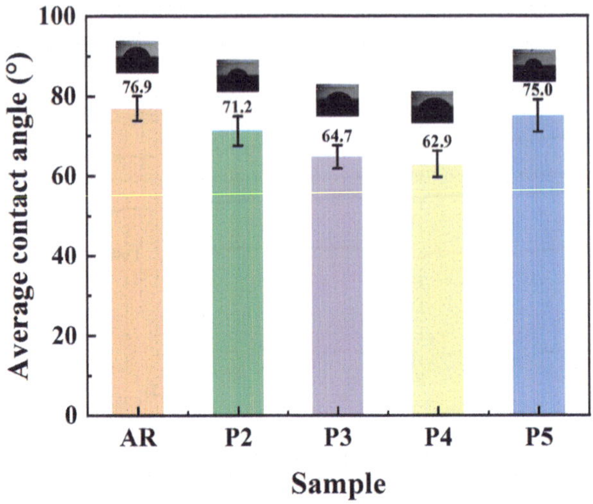


**Fig. 10.13** Surface contact angles of different samples after picosecond laser cleaning of CFRP

carbon fibers after cleaning. From the figure, it can be observed that the contact angle of the CFRP surface decreased to varying extents after cleaning. Upon complete cleaning, the surface contact angle decreased to 64.7°. Compared to nanosecond laser cleaning, the reduction in contact angle was smaller, which is attributed to the relatively slight increase in surface roughness caused by picosecond laser cleaning, resulting in a less significant improvement in wettability.

## *10.6.2 Adhesion Strength of Repainted Samples*

The CFRP samples cleaned by picosecond laser were repainted, and the adhesion strength of the surface after cleaning was evaluated using the pull-off method. The corresponding test curves and results are shown in Fig. 10.14. It can be observed that sample P3 exhibited the highest interfacial bonding strength, reaching 9.57 MPa. The analysis suggests that after complete cleaning, the surface roughness of the CFRP increased, thereby enhancing the wettability at the interface. This allowed the adhesive to achieve a more sufficient contact area with the surface, resulting in stronger adhesion. It is noteworthy that the bonding strength of the control sample AR was only 4.64 MPa, whereas after complete cleaning, the interfacial bonding strength increased by 106%. In comparison, nanosecond laser cleaning resulted in a 113% increase, indicating that the bonding strength after picosecond laser cleaning is comparable to that of nanosecond laser cleaning. However, it should be noted that in Fig. 10.14b, the bonding strengths of samples P1 and P5 were lower than that of the uncleaned sample. The analysis suggests that residual paint on sample P1 resulted in weak adhesion between the new and old coatings during repainting,

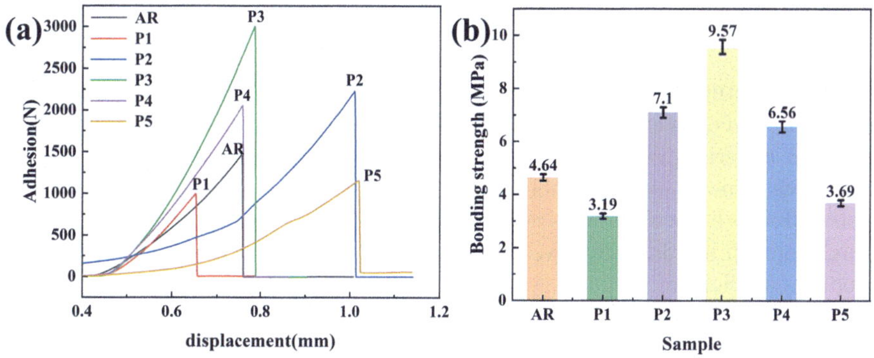


**Fig. 10.14** Different GFRP samples after picosecond laser cleaning: (**a**) Test curve by opening method (**b**) Bond strength diagram

while in sample P5, localized fiber damage caused bending deformation, and the adhesion at deformed regions was weaker than that on smooth fiber surfaces. Therefore, complete cleaning of CFRP is critical for improving the bonding strength during repainting. Higher interfacial adhesion not only increases material service life but also reduces safety hazards.

## 10.7 Online Monitoring of CFRP During Nanosecond Laser Cleaning

Online monitoring, as a technical approach, enables real-time tracking and assessment of the cleaning effect during laser cleaning. Various sensors, imaging devices, or other monitoring systems are used to collect acoustic, thermal, optical, and other signals during the process, allowing for a real-time understanding of the material's cleaning state, thereby ensuring cleaning efficiency and quality. CFRP is highly sensitive to thermal, direct impact, and other physical-chemical interactions at the surface and interface during laser cleaning, and thus places high demands on post-cleaning surface quality. Rapid surface quality assessment and real-time monitoring during laser cleaning are especially important for achieving a high-quality clean substrate. Real-time monitoring not only allows for timely determination of cleaning quality and prevents substrate damage from over-cleaning caused by uncontrollable factors, but also facilitates the investigation of different lasers' cleaning characteristics and mechanisms via monitoring of temperature, spectra, images, and acoustic signals during the process. In this study, multiple online monitoring techniques were integrated, including temperature monitoring, acoustic signal monitoring, and high-speed imaging, to track the nanosecond pulsed laser cleaning process on CFRP and GFRP in real-time, and analyze the characteristics of various cleaning states in detail.

### 10.7.1 Temperature Monitoring

As a highly efficient cleaning method, laser cleaning relies heavily on its thermal effect. However, excessive heat during cleaning may cause ablation damage to carbon fiber materials, thereby impairing their properties. To preserve the integrity of CFRP, real-time temperature monitoring can be employed to adjust process parameters and ensure that cleaning is performed within an appropriate temperature range. This approach not only improves cleaning quality but also effectively guarantees production efficiency.

Figure 10.15 shows the maximum temperature during the cleaning process for each sample and their corresponding real-time temperature curves. In Fig. 10.15a–e, the maximum temperature during cleaning appeared along the active cleaning path, while the temperature of completed paths dropped rapidly. The maximum temperatures for samples C1 to C4 ranged between 250.3 and 283.2 °C. Based on the thermal gravimetric analysis of the paint layer shown in Fig. 10.1, it can be inferred that the paint began to decompose within this temperature range, although the decomposition rate remained low. Therefore, a two-pass cleaning strategy was adopted in the experiment. First, this helped prevent excessive heat accumulation from high single-pass power, which could lead to damage as observed in sample C5. Second, the gradual thermal accumulation from two passes accelerated the decomposition of the paint layer. From Fig. 10.15f, it can be seen that the overall temperature during the second pass was approximately 50 °C higher than during the first pass due to cumulative heating. However, the over-cleaned sample C5 reached a peak temperature of 505.8 °C, which caused rapid decomposition of the paint. The excessive thermal input resulted in significant local overheating. As heat could not be dissipated in time, localized melting and fiber breakage occurred.

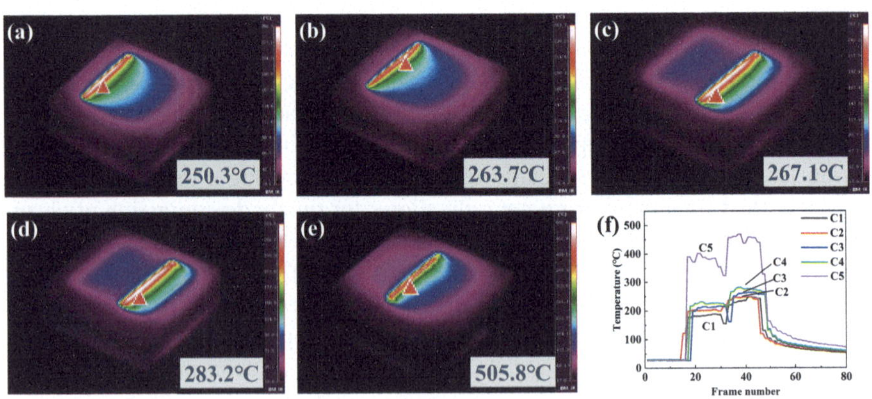


**Fig. 10.15** The maximum temperature of the sample during laser cleaning of CFRP: (**a**) C1; (**b**) C2; (**c**) C3; (**d**) C4; (**e**) C5; (**f**) Real-time temperature of each sample

### 10.7.2 Acoustic Monitoring

Figure 10.16 shows the TA, FA, and TF spectra of the acoustic signals during CFRP cleaning, with samples C1–C5 corresponding to Figures (a) through (e). In the TA spectra, the amplitude of both cleaning passes in Fig. 10.16a1 was nearly identical, indicating that the laser power was too low to clean the surface completely, leaving a large amount of resin blocks. In Fig. 10.16b1–e1, the amplitude in the first cleaning pass was significantly lower than that in the second, and all first-pass amplitudes were below 0.4. Theoretically, the first pass mainly interacted with the paint layer, and the amplitude when the laser acts on paint is expected to be below 0.4. As the laser power increased, the amplitude during the second pass gradually increased, reaching nearly 1 in sample C5. It is worth noting that in Fig. 10.16e1, the amplitude of the second cleaning of C5 was high, and the spacing between adjacent amplitudes was large. In actual cleaning, a sharp and harsh noise was heard. The analysis suggests that damage on the surface caused these areas to absorb part of the sound wave energy, resulting in large fluctuations in amplitude.

In the FA spectra, the maximum amplitude in Fig. 10.16a2 appeared at 2000 Hz, while that in Fig. 10.16d2 was at 6500 Hz. The maximum frequencies in other samples were lower. Therefore, the frequency at which the peak amplitude occurs can be used to determine whether the sample has been completely cleaned. When the highest frequency is observed at 6500 Hz, it indicates that the material is fully

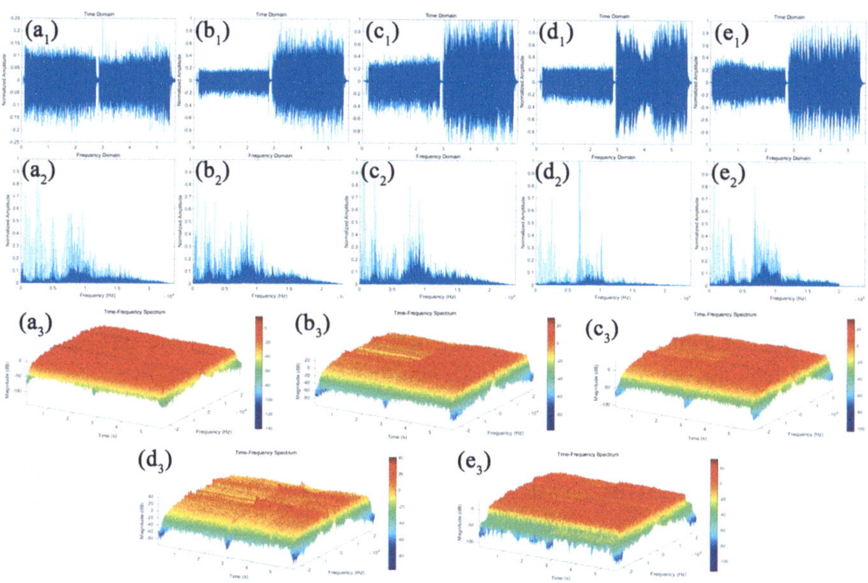


**Fig. 10.16** Sound signal monitoring during laser cleaning of CFRP: (**a1**–**e1**) TA spectrum; (**a2**–**e2**) FA spectrum; (**a3**–**e3**) TF spectrum

cleaned. Additionally, in Fig. 10.16e2, amplitudes beyond 20,000 Hz disappear, suggesting that over-cleaning may have occurred if no amplitude is detected above 20,000 Hz.

Figure 10.16a3–e3 show the 3D TF spectra. Compared with TA and FA spectra, TF spectra allow direct reading of amplitude values. In Fig. 10.16a3, the maximum amplitude of sample C1 is below 10 dB, which is consistent with the TA spectrum, confirming the overall low amplitude. For samples C2 and C3, some amplitudes reached 20 dB. Sample C4 had a maximum amplitude of 40 dB, though most of its values remained below 20 dB. The average amplitude of sample C5 during the second cleaning exceeded 20 dB. Therefore, both the peak amplitude value and its distribution in the TF spectrum can be used to assess the degree of paint removal.

### *10.7.3 High-Speed Imaging Monitoring*

To capture more detailed dynamic changes during laser cleaning, high-speed imaging was used to observe the CFRP cleaning process. Figure 10.17 presents high-speed images captured during two-pass cleaning of CFRP, with the cleaning direction from right to left. Figure 10.17a1–e1 correspond to the first cleaning pass (N1), and Fig. 10.17a2–e2 to the second pass (N2). When the laser irradiates the coating surface, it ionizes the surrounding gas medium and generates plasma. Collisions between plasma regions produce visible plasma plumes, seen as bright spots in the images. These plumes are also accompanied by shock waves. When the impact force exceeds the adhesion between the coating and substrate, paint is removed in the form of small particles. The shape and size of the plasma plume vary with cleaning stages and reflect the strength of plasma-induced shock effects.

Figure 10.18 shows the bright spot sizes during different cleaning stages. During the first pass, the laser mainly irradiated the paint layer, with most energy used for its ablation and gasification, resulting in small bright spots. During the second pass, the incompletely cleaned samples C1 and C2 still showed small bright spots (~0.7 mm), likely due to weak plasma effects at low power. As laser power increased, bright spot size also increased, reaching a maximum of 1.4 mm for sample C4, indicating increased gas ionization from vaporized paint. In Fig. 10.18e2, the spot size decreased again, as part of the laser energy was absorbed by the substrate due to over-cleaning, causing surface damage and reduced plasma generation.

## 10.8 Experimental Results and Technical Contributions

This chapter systematically studied the effects of nanosecond and picosecond laser cleaning on the paint layer removal from CFRP surfaces and examined the precise control of substrate surface morphology via laser parameter optimization. The results showed that effective removal of paint could be achieved while maintaining

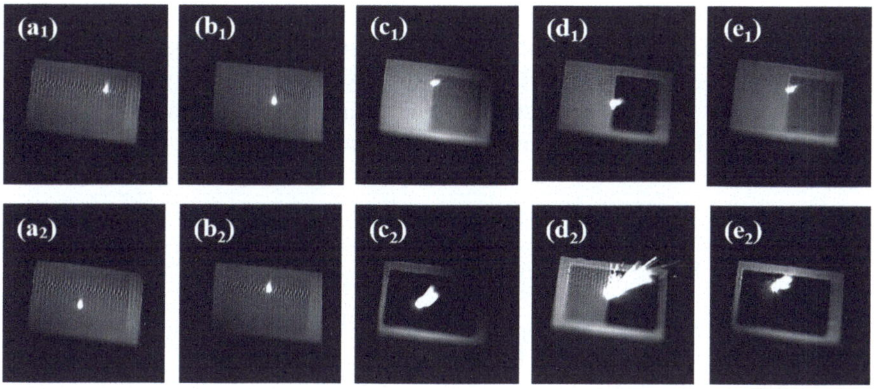


**Fig. 10.17** Dynamic behavior of laser cleaning CFRP: (**a1**)–(**a2**) C1; (**b1**)–(**b2**) C2; (**c1**)–(**c2**) C3; (**d1**)–(**d2**) C4; (**e1**)–(**e2**) C5

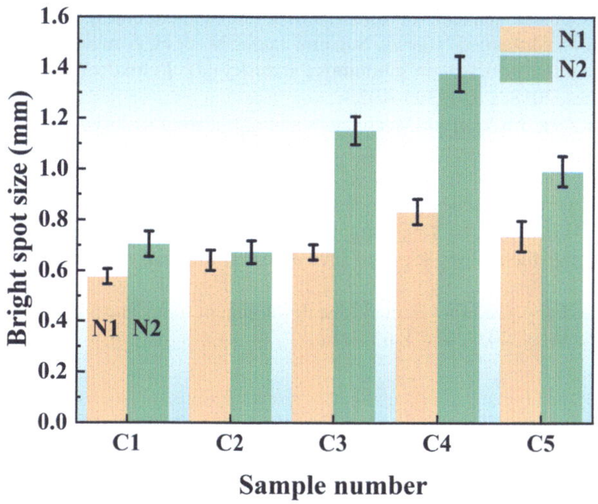


**Fig. 10.18** Spot size of nanosecond laser cleaning CFRP process

surface integrity and function by fine-tuning laser power. As laser power increased, the CFRP surface underwent a transition from incomplete to complete and then to over-cleaning, with fully cleaned samples exhibiting optimal morphology and performance.

Surface wettability and adhesion performance after cleaning were evaluated through contact angle and pull-off tests, which confirmed significant improvement in surface wettability and coating adhesion strength due to laser cleaning. Moreover, the application of online monitoring technologies provided valuable data for real-time control and process optimization, ensuring accurate surface morphology regulation.

In summary, this study demonstrated the potential of laser cleaning in CFRP coating removal and offers a novel solution for maintenance and repair of composite materials in the aerospace field. Future work may further explore laser parameter optimization to achieve more efficient and refined cleaning, thereby enabling improved control over substrate morphology and properties.

## References


1. Zhang Gaobin. Application of composite materials in civil aviation aircraft [J]. Papermaking equipment and materials, 2024, 53(1): 75–77.
2. Ji Zhengjiang, Dong Jiachen, Liang Liang, et al. Integrated layered structure design of carbon fiber prepreg absorber for aircraft skin [J]. Journal of Composite Materials: 1–10.
3. Gu Junyi, Li Wenqin, Su Xuan, et al. Research on laser cleaning process of carbon fiber composite aircraft skin surface paint layer [J]. China Laser, 2024, 51(12): 80–93.
4. Zheng Guang. Research on aircraft skin laser depainting technology [D]. Beijing: Graduate School of the Chinese Academy of Sciences (Institute of Electronics), 2005.
5. Razab M K A A, Mohamed Noor A, Suhaimi Jaafar M, et al. A review of incorporating Nd: YAG laser cleaning principal in automotive industry [J]. Journal of radiation research and applied sciences, 2018, 11(4): 393–402.
6. Zhu G D, Xu Z H, Jin Y, et al. Mechanism and application of laser cleaning: A review [J]. Optics and lasers in engineering, 2022, 157: 107130.
7. Mandolfino C, Lertora E, Genna S, et al. Effect of laser and plasma surface cleaning on mechanical properties of adhesive bonded joints [J]. Procedia cirp, 2015, 33: 458–463.
8. Jiang Yuhong, He Yuyang, Fu Yonghong, et al. Development Status of Large-scale Application of Laser Cleaning Technology (Special Invitation) [J]. Infrared and Laser Engineering, 2023, 52(2): 75–82.
9. Li Haoyu, Yang Feng, Guo Jiawei, et al. The development status and prospects of laser cleaning [J]. Laser technology, 2021, 45(5): 654–661.

# Chapter 11
# Surface Morphology Control of the Matrix After Laser Cleaning of GFRP Coatings

Glass fiber reinforced plastics (GFRP) are favored in the aerospace and aviation industries as key structural and surface coating materials due to their cost-effectiveness, good durability, heat resistance, and excellent processing capabilities [1–4]. The application of GFRP not only helps reduce the structural weight of aircraft and rockets but also plays a crucial role in improving the performance and service life of aircraft engines. However, over time, the coating on the GFRP surface may degrade or become damaged, necessitating effective cleaning and repair processes to maintain its performance. Therefore, more practical, green, non-destructive, and low-cost new cleaning technologies must be adopted to meet the specific cleaning sensitivity requirements of fiber-reinforced composite aircraft skins [5, 6]. Currently, research on laser cleaning of CFRP surface coatings has made some progress, while studies on laser cleaning of GFRP surface coatings are relatively limited, offering ample room for further exploration in this field [7].

This chapter explores the application of nanosecond pulsed laser technology in the cleaning of GFRP surface coatings, with a particular focus on the influence of laser parameters on the morphology of the substrate surface and how to achieve precise morphological control. Cleanliness after cleaning will be assessed through microscopic morphology observation, three-dimensional morphology analysis, and Fourier-transform infrared spectroscopy (FT-IR). In addition, contact angle testing and pull-off tests will be used to evaluate the wettability and bonding performance of the cleaned material surfaces. The application of online monitoring technology will provide real-time feedback during the cleaning process. Comparisons with CFRP online monitoring will be used to identify similarities and differences in the cleaning behaviors of different composite materials, ensuring cleaning efficiency and the integrity of the substrate surface.



S. Wang, *Laser Cleaning*, https://doi.org/10.1007/978-981-95-2935-3_11

## 11.1 Characterization of Surface Cleanliness of GFRP After Nanosecond Laser Cleaning

### *11.1.1 Microscopic Surface Morphology*

The materials used in this chapter are EPGC201 GFRP composite laminates, which were cut into rectangular thin-plate specimens with dimensions of 15 × 15 × 3 mm. The experimental equipment, type of paint, and coating thickness were consistent with those described in Chap. 10. The average laser powers used were 60 W, 65 W, 70 W, and 75 W, corresponding to samples labeled G1 to G4, respectively. Other process parameters included a frequency of 200 kHz, scanning speed of 3000 mm/s, scanning pitch of 0.02 mm, and two scanning passes.

Figure 11.1 shows the surface morphology of GFRP samples after laser cleaning at different average powers. When the power P = 60 W, although the paint layer on the surface is removed, a large amount of resin remains, as shown in Fig. 11.1a. When P = 65 W, relatively intact glass fibers become visible. Due to the high thermal resistance of glass fibers and their ability to transmit part of the laser beam, small amounts of resin remain on the fiber surfaces, as seen in Fig. 11.1b. When P = 70 W, the resin is completely removed, but the surface fibers become fractured,

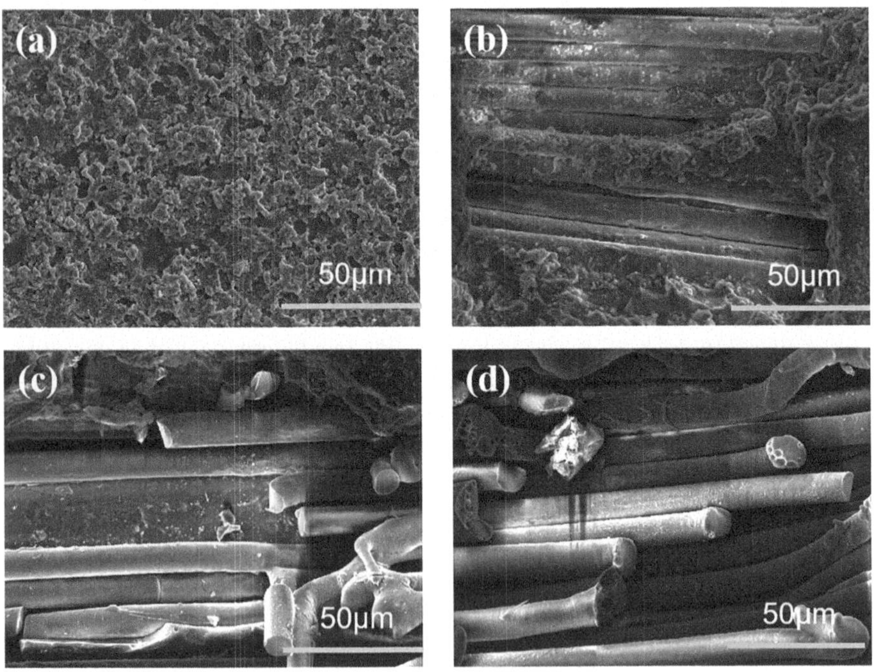


**Fig. 11.1** Microstructures of GFRP samples under different average laser powers: (**a**) G1; (**b**) G2; (**c**) G3; (**d**) G4

failing to meet cleaning requirements, as shown in Fig. 11.1c. With further increase in power, the damage to the glass fibers becomes more severe; while the surface fibers are fractured, the underlying fibers undergo bending deformation, as depicted in Fig. 11.1d.

Given the superior thermal resistance of glass fibers, the cleaning power required is higher than that for carbon fibers. Complete cleaning of CFRP exposes relatively intact carbon fibers, whereas complete cleaning of GFRP reveals only partial exposure of fibers. In cases of over-cleaning, only the surface carbon fibers in CFRP are damaged due to the significantly higher axial thermal conductivity compared to the radial direction, allowing the lower layers to remain largely unaffected. However, over-cleaning of GFRP results not only in damage to surface glass fibers but also causes bending deformation in the underlying fibers, severely compromising the material's performance.

### *11.1.2 Three-Dimensional Morphology and Surface Roughness Analysis*

Figure 11.2 presents the three-dimensional morphology of GFRP samples after cleaning under various levels of average laser power. In Fig. 11.2a, due to the relatively low average power, no fibers are exposed on the surface of sample G1, which exhibits a relatively flat topography. As the average laser power increases, exposed glass fibers begin to emerge on the surface of sample G2, as shown in Fig. 11.2b. These fibers are mostly intact, though some remain partially embedded in resin, resulting in pronounced height differences in certain areas. In Fig. 11.2c, as the resin covering the exposed fiber surfaces is further removed, the surface height variation decreases. However, due to the higher energy input, the exposed fibers in sample G3 begin to fracture and bend. In Fig. 11.2d, with a further increase in laser power, the surface fibers experience more severe breakage and bending, while the underlying fibers also suffer substantial damage, leading to a sharp increase in surface height variation.

Figure 11.3 shows the surface roughness images of GFRP samples cleaned at different average laser powers. The uncleaned sample AR, used as a control, had the lowest surface roughness of 0.42 μm. Due to the low average power, sample G1 remained entirely covered by resin without fiber exposure, resulting in a surface roughness higher than that of sample AR. When the average power increased to 65 W, some areas on the surface of sample G2 exhibited exposed intact fibers, although portions of resin remained. Consequently, the surface roughness significantly increased to 2.288 μm. As the power further increased and the resin was largely removed, the surface roughness of G3 decreased again. Sample G4 exhibited the highest surface roughness at 5.075 μm, attributed to fiber ablation and fracture on both surface and sub-surface layers caused by excessive laser energy input.

Due to the relatively high thermal resistance of GFRP, higher laser power is required for cleaning compared to CFRP. However, this also means that the thermal

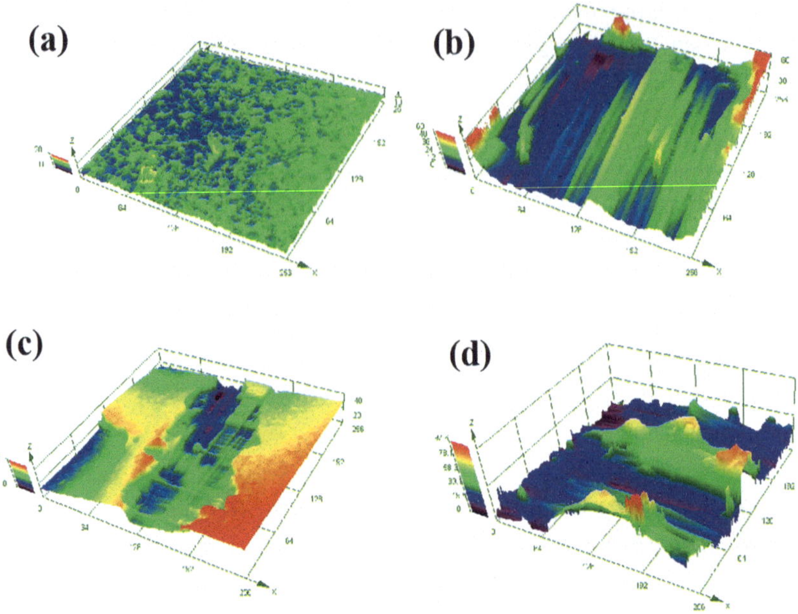


**Fig. 11.2** 3D morphologies of GFRP samples cleaned with different average laser powers: (**a**) G1; (**b**) G2; (**c**) G3; (**d**) G4

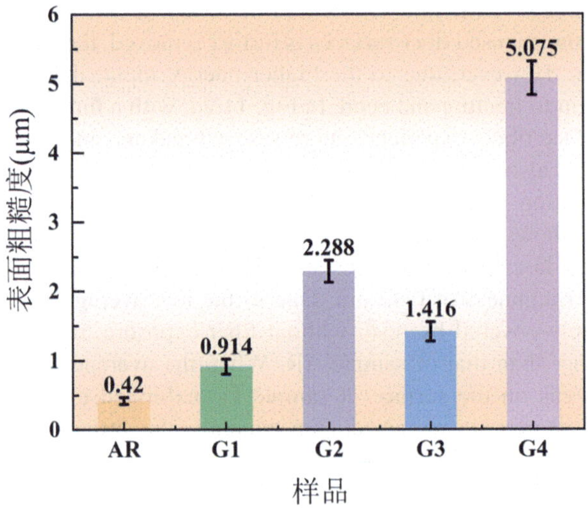


**Fig. 11.3** Surface roughness of GFRP samples after cleaning under different average laser powers

input is more difficult to control. From the observed post-cleaning 3D morphologies and roughness data, it is evident that excessive energy input, combined with the lack of high axial thermal conductivity as found in carbon fibers, causes heat to conduct in both axial and radial directions in glass fibers. This can lead to fracture of the lower-layer fibers. Therefore, it is recommended to retain a portion of the resin layer on the surface during GFRP cleaning. During recoating, these resin residues may increase the contact area with the new paint layer, thereby enhancing the bonding strength and improving cleaning effectiveness while reducing the risk of material damage.

### *11.1.3 FT-IR Characterization*

To verify the cleanliness of the paint surface after laser cleaning, FT-IR analysis was conducted on the GFRP samples. The results are shown in Fig. 11.4. The uncleaned sample AR served as the control, and the positions and corresponding names of the absorption peaks are labeled in the figure. The characteristic absorption peaks of the polyurethane coating include the deformation vibration of the amide N–H bond at 1526 cm$^{-1}$, the C=O bond of isocyanurate at 1682 cm$^{-1}$, and the ester C=O bond at 1724 cm$^{-1}$. In the FT-IR spectra of the cleaned samples shown in Fig. 11.4, these three characteristic peaks are nearly absent, indicating that the paint layer on the sample surface has been effectively removed.

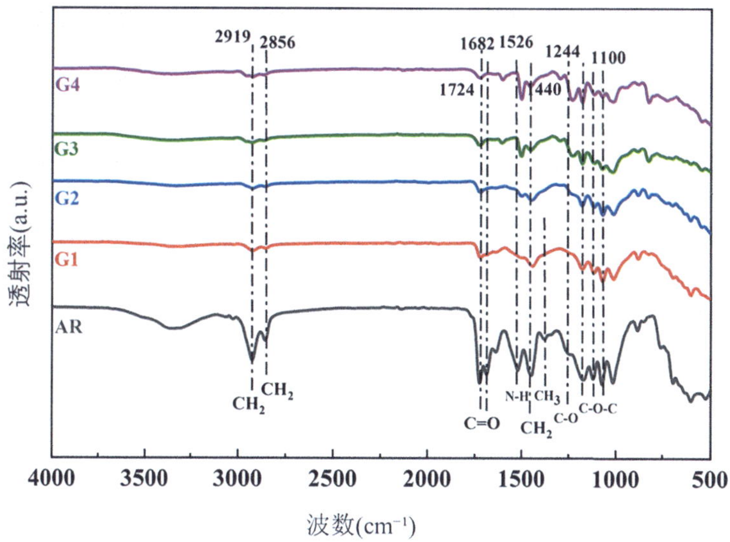


**Fig. 11.4** FT-IR spectra of different GFRP samples after laser cleaning

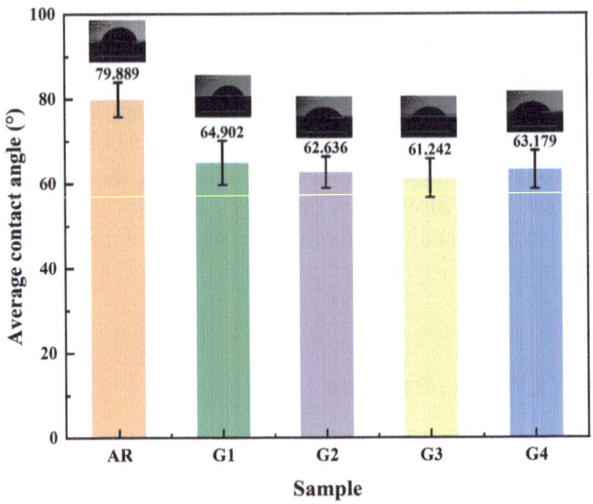


**Fig. 11.5** Surface contact angles of different GFRP samples after nanosecond laser cleaning

## 11.2 Evaluation of GFRP Surface Properties After Nanosecond Laser Cleaning

### *11.2.1 Variation in Surface Wettability*

Changes in contact angle on the sample surface after paint removal significantly affect the surface wettability of the material, which plays a critical role during the recoating process. Contact angle measurements were performed on the laser-cleaned GFRP samples, and the results are shown in Fig. 11.5. The uncleaned GFRP sample, labeled AR, served as the control group.

As shown in Fig. 11.5, the surface contact angle of the uncleaned GFRP sample was 79.9°. After laser cleaning, the contact angle exhibited a noticeable decreasing trend. Similar to CFRP, the energy applied during laser removal of the paint layer impacts the material surface, increasing its surface energy and thereby reducing the contact angle. In the case of CFRP, complete cleaning reduced the contact angle from 76.9° to 60.3°, while for GFRP, it decreased from 79.9° to 61.2°, indicating that laser cleaning has a comparable effect in enhancing the surface wettability of both materials.

### *11.2.2 Bonding Strength of Recoated Samples*

Figure 11.6 displays the pull-off test curves and corresponding bonding strength values for GFRP samples before and after laser cleaning. The sample AR, which was uncleaned, served as the control group. As shown in the figure, AR had the

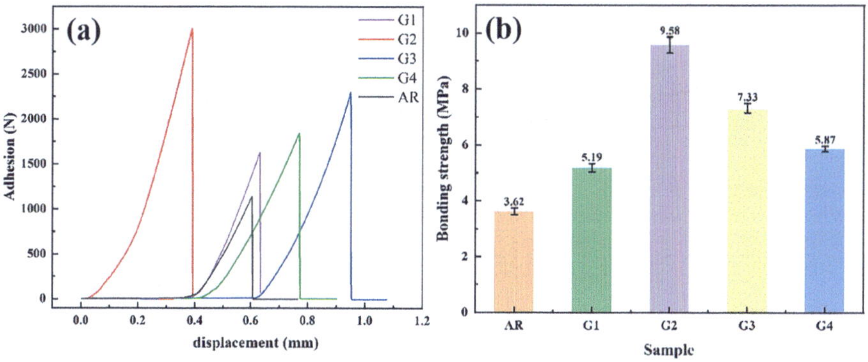


**Fig. 11.6** GFRP samples after nanosecond laser cleaning: (**a**) pull-off test curves; (**b**) bonding strength results

lowest bonding force, measured at only 1138 N, with a corresponding bonding strength of just 3.62 MPa. After laser cleaning, the bonding force of the GFRP samples showed an initial increase followed by a decrease. Among them, sample G2, cleaned to expose undamaged glass fibers, exhibited the highest bonding force of 3008 N and bonding strength of 9.58 MPa, representing a 164% improvement compared to AR. However, as the average laser power continued to increase, varying degrees of surface damage were observed, and the bonding strength began to decline. Notably, in the tests, over-cleaned sample C5 showed a lower bonding force than AR, which not only wastes resources but also introduces safety risks for subsequent recoating. Therefore, over-cleaning must be strictly avoided during the cleaning process.

Based on the bonding strength results for both CFRP and GFRP, laser cleaning significantly enhances the adhesion of recoated paint layers. The cleaned samples exhibited higher bonding strength compared to uncleaned ones, due to the surface modification induced by laser impact. As previously discussed regarding surface roughness and contact angle, the interfacial bonding properties of the cleaned surfaces were notably improved.

It is worth noting that in the GFRP cleaning process, part of the resin layer was intentionally retained—i.e., partial rather than complete cleaning—whereas CFRP was subjected to full removal. According to the pull-off test results, the bonding strength of GFRP increased by 164% compared to uncleaned samples, while CFRP increased by only 113%. Therefore, for materials with relatively stable physicochemical properties, partial cleaning is recommended. This approach not only achieves strong interfacial adhesion while reducing energy consumption, but also preserves some of the original resin layer, helping maintain material performance stability.

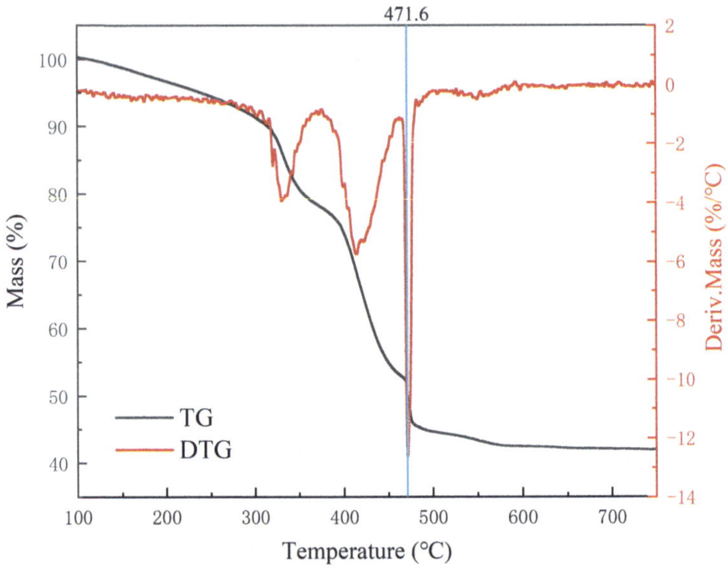


**Fig. 11.7** Thermogravimetric analysis of paint

## 11.3 Online Monitoring of Nanosecond Laser Cleaning of GFRP

### *11.3.1 Temperature Monitoring*

Figure 11.7 presents the maximum temperatures and real-time temperature curves of different GFRP samples during the cleaning process. The highest temperatures are concentrated along the laser scanning path or near the laser focal spot, while temperatures in the cleaned regions drop rapidly afterward. As shown in Fig. 11.7c, d, samples G3 and G4 were over-cleaned. According to Fig. 11.7c–e, the fiber burning temperature of GFRP exceeds 500 °C. Notably, sample G2, which showed the best cleaning performance, reached a peak temperature of 338.7 °C. According to the thermogravimetric curve in Fig. 2.2, the paint decomposition rate is high at this temperature, but without sufficient thermal accumulation to damage the fibers, indicating an optimal cleaning outcome.

Based on the temperature monitoring results in Figs. 10.13 and 11.8, it is evident that reaching a reaction temperature of 500 °C during laser cleaning of CFRP and GFRP leads to material damage and fiber fracture. Unlike CFRP, GFRP experiences more pronounced temperature rises after two instances of thermal accumulation during the cleaning process, as shown in Fig. 4.7e. With increasing laser power, the rate of temperature rise in GFRP also becomes more intense. Therefore, optimal cleaning temperatures can be obtained from Figs. 10.13 and 11.8: for CFRP, the

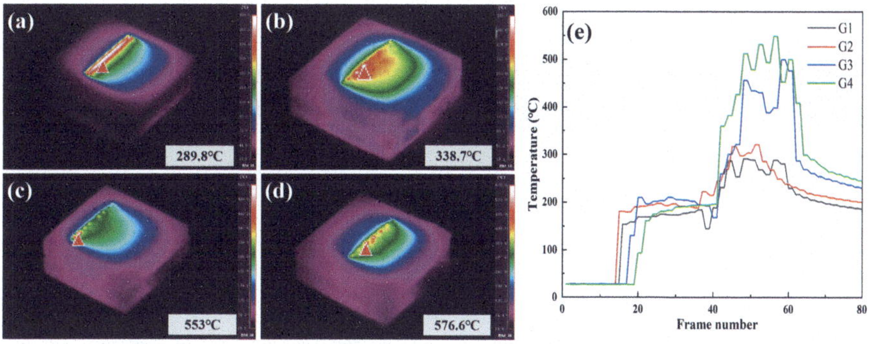


**Fig. 11.8** Maximum temperatures of CFRP samples during laser cleaning: (**a**) G1; (**b**) G2; (**c**) G3; (**d**) G4; (**e**) real-time temperatures of all samples

recommended temperature range is approximately 260–280 °C, while for GFRP it is around 330 °C.

## *11.3.2 Acoustic Monitoring*

Figure 11.9 shows the TA, FA, and TF spectra during the laser cleaning of GFRP samples. Subfigures (a)–(d) correspond to samples G1 to G4, respectively. According to the TA spectrum, the amplitude during the first cleaning cycle is greater than that during the second cycle, indicating that the sound intensity from removing the paint layer is higher than that from laser irradiation on the substrate. In Fig. 11.9a1, the second cleaning cycle of G1 involves laser irradiation of the resin layer, where the amplitude increases gradually. Once intact fibers are exposed, the amplitude stabilizes around 0.4, as seen in Fig. 11.9b1. As the power increases, the amplitude gradually decreases (Fig. 11.9c1) until large fluctuations occur in Fig. 11.9d1, accompanied by sharp, harsh sounds during the actual cleaning process. Just like CFRP, over-cleaning of GFRP results in large amplitude fluctuations, indicative of excessive ablation. However, unlike CFRP, the amplitude during GFRP paint removal is higher than that during direct laser irradiation on the GFRP substrate.

From the FA spectrum, the highest cleaning amplitude for sample G2 occurs at 3000 Hz, while samples G1, G3, and G4 exhibit peak amplitudes at 8000 Hz. Thus, the frequency of the maximum amplitude can serve as a criterion to assess the completeness of GFRP cleaning. Unlike CFRP, over-cleaned GFRP samples show residual amplitudes beyond 20,000 Hz, with slight variations in the frequency of the maximum amplitude.

The TF spectrum provides a more intuitive representation of "high decibel" regions at 3000 and 8000 Hz, seen as two distinct dark-colored, prominent zones. Consistent with the TA spectrum, Fig. 11.9a3 shows that the amplitude during the first cleaning cycle is clearly higher than during the second. In Fig. 11.9b3, the

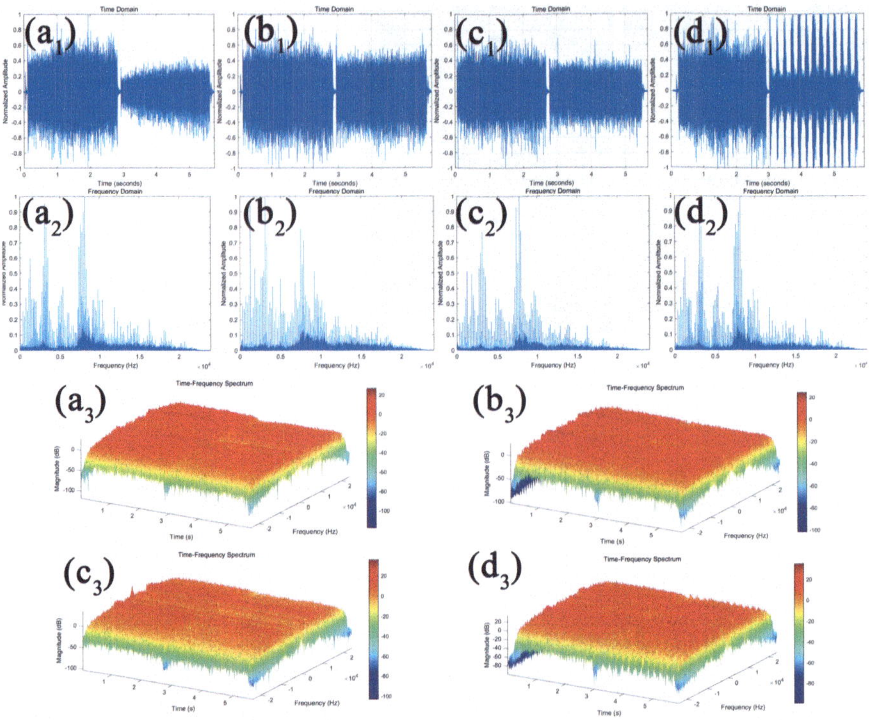


**Fig. 11.9** Acoustic signal monitoring during laser cleaning of GFRP: (**a1**–**d1**) TA spectra; (**a2**–**d2**) FA spectra; (**a3**–**d3**) TF spectra

amplitudes of both cleaning cycles are nearly identical. The highest amplitude values for the over-cleaned G3 and G4 samples are close to 30 dB, while G2 reaches just above 20 dB. Therefore, the TF spectrum can also be used to evaluate the cleaning level of the samples by analyzing the frequency and magnitude of the highest amplitude values.

For both CFRP and GFRP, relying on a single spectrum is insufficient to comprehensively determine the cleaning stages. Thus, it is necessary to integrate the unique features of all three spectra—such as amplitude variation, frequency of peak amplitude, and overall amplitude strength—for a more accurate assessment. During over-cleaning, direct laser interaction with the composite matrix produces sharp and piercing sounds, which serve as a direct auditory indication of excessive cleaning.

### *11.3.3 High-Speed Imaging Monitoring*

Figure 11.10 presents the high-speed imaging of the GFRP laser cleaning process. Images (a1) to (d1) correspond to the first cleaning cycle, while images (a2) to (d2) correspond to the second cleaning cycle. Figure 4.10 shows the bright spot

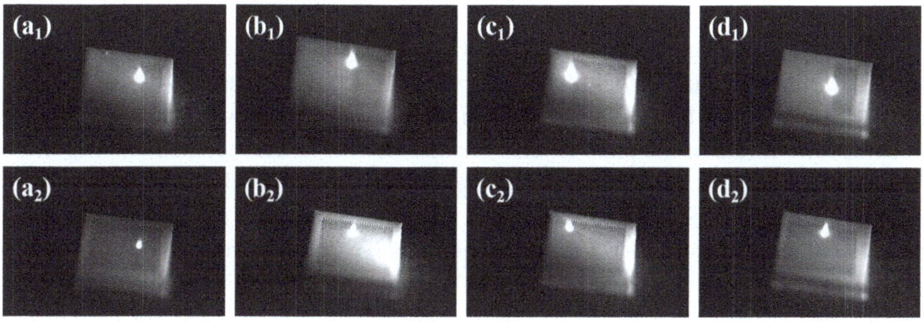


**Fig. 11.10** Dynamic behavior monitoring of nanosecond laser cleaning on GFRP: ($\mathbf{a_1}$)–($\mathbf{a_2}$) G1; ($\mathbf{b_1}$)–($\mathbf{b_2}$) G2; ($\mathbf{c_1}$)–($\mathbf{c_2}$) G3; ($\mathbf{d_1}$)–($\mathbf{d_2}$) G4

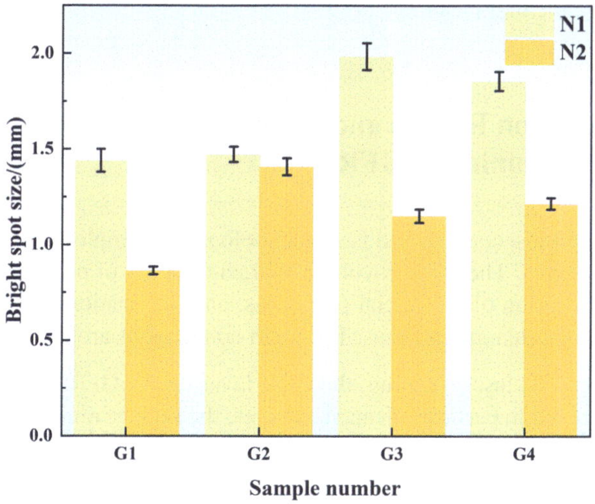


**Fig. 11.11** Spot size during the nanosecond laser cleaning process of GFRP

dimensions at different cleaning stages for each sample, where the first cleaning is denoted as N1 and the second as N2. Based on the image data in Fig. 11.10 and the quantitative results in Fig. 11.11, it can be observed that the plasma plume generated during the first cleaning is larger than that during the second cleaning. This phenomenon is mainly due to the fact that, during the first cleaning, the laser primarily irradiates the paint layer. The evaporation of part of the paint increases the gas concentration above the material surface, which, after ionization by the laser beam, leads to a higher plasma density and a larger bright spot. During the second cleaning, however, the amount of evaporated paint is reduced, the gaseous medium on the surface diminishes, and part of the paint layer has already been removed. Moreover, the glass fibers begin to transmit part of the laser energy, resulting in a reduced paint removal rate.

It is worth noting that for samples G3 and G4, the bright spot size during the first cleaning is larger than that of G2. However, after the second cleaning, the spot size for G3 and G4 is smaller than that of G2. This outcome is attributed to the higher energy input received by G3 and G4 during the first cleaning, which caused a faster decomposition rate of the paint layer compared to G2. In the second cleaning, most of the laser energy is absorbed by the fibers, resulting in fiber damage, whereas G2 was still in the process of removing paint from deeper layers, thereby producing a larger spot size than G3 and G4. A comprehensive comparison of the high-speed imaging results between CFRP and GFRP reveals that for CFRP, the spot size during the first cleaning is smaller than during the second; conversely, for GFRP, the opposite is observed. This discrepancy is mainly due to the timing of paint removal: in CFRP, the paint layer is primarily removed during the temperature rise of the second thermal accumulation, while in GFRP, the surface paint layer is largely removed during the first cleaning.

## 11.4 Application Results and Significance of Nanosecond Laser Cleaning of GFRP

In this chapter, nanosecond pulsed laser technology was employed to remove paint from GFRP surfaces. The study involved characterization of post-cleaning surface cleanliness, evaluation of interfacial properties, and application of online monitoring throughout the cleaning process. The main conclusions are as follows:

1. After nanosecond laser cleaning, the paint layer on the GFRP surface was completely removed. In partially cleaned samples, the surface remained covered with resin; in fully cleaned samples, only partial fiber exposure was observed. In over-cleaned samples, fiber breakage occurred, and further increases in power led to bending deformation of the sub-surface fibers.
2. After nanosecond laser cleaning, the surface roughness of GFRP increased from 0.42 to 2.288 μm, and the contact angle decreased, indicating enhanced interfacial wettability. Pull-off tests showed that the bonding force of fully cleaned samples increased by 164% compared to uncleaned ones. However, samples with bent sub-surface glass fibers exhibited reduced bonding strength, which is attributed to fiber damage caused by over-cleaning, impairing adhesion with the new paint layer.
3. The cleaning process was monitored in real time and compared with CFRP to identify similarities and differences. Temperature monitoring showed that, similar to CFRP, the temperature of over-cleaned samples was significantly higher than that of normally cleaned ones. Additionally, the temperature increase during the second cleaning cycle in GFRP was more pronounced than in CFRP. Acoustic monitoring revealed that the amplitude fluctuation in over-cleaned GFRP was greater; unlike CFRP, the amplitude during the first cleaning of GFRP was larger than that of the second. High-speed imaging showed that the plasma plume size

initially increased and then decreased with increasing power. Unlike CFRP, the plasma plume of GFRP was larger during the first cleaning cycle than during the second.

In summary, this study demonstrates the potential of nanosecond pulsed laser technology for cleaning GFRP surface coatings and offers a new approach for the maintenance and repair of composite materials in the aerospace sector. Future research may further optimize laser parameters to achieve more efficient and refined cleaning results, thereby enhancing the control over surface morphology and overall substrate performance.

## References

1. Bhat A, Budholiya S, Aravind Raj S, et al. Review on nanocomposites based on aerospace applications [J]. Nanotechnology reviews, 2021, 10(1): 237–253.
2. Zaaba N F, Ismail H, Saeed A M. A review: Metal filled thermoplastic composites [J]. Polymer-plastics technology and materials, 2021, 60(10): 1033–1050.
3. Murugesan S, Scheibel T. Copolymer/clay nanocomposites for biomedical applications [J]. Advanced functional materials, 2020, 30(17): 1908101.
4. Rong X D, Li Y, Chen X F, et al. Achieving high mechanical properties and corrosion resistance of Al-Zn-Mg matrix composites via regulating intragranular reinforcements [J]. Journal of materials science & technology, 2023, 153: 1–7.
5. Zou W F, Xie Y M, Xiao X, et al. Application of thermal stress model to paint removal by Q-switched Nd: YAG laser [J]. Chinese physics B, 2014, 23(7): 074205.
6. Ortiz P, Antúnez V, Ortiz R, et al. Comparative study of pulsed laser cleaning applied to weathered marble surfaces [J]. Applied surface science, 2013, 283: 193–201.
7. 周礼君. 树脂基复合材料表面涂层激光清洗应用技术研究 [D]. 南昌: 南昌航空大学, 2017.

# Chapter 12
# Controllable Texturing and Performance of Magnesium Alloy Substrate Surface After Laser Cleaning

Magnesium alloys exhibit excellent properties, including lightweight characteristics, machinability, corrosion resistance, damping capacity, dimensional stability, and impact resistance, making them widely used in aerospace, transportation, electronics, and biomedical industries. However, high-quality and efficient cleaning and pretreatment during manufacturing, maintenance, and remanufacturing remain challenging [1]. Traditional cleaning methods, including mechanical friction, chemical corrosion, high-pressure water jetting, and sandblasting, are associated with drawbacks such as environmental pollution, high time and material consumption, and substrate damage [2, 3]. These limitations hinder continuous surface cleaning and pretreatment operations, resulting in low work efficiency [4]. Therefore, there is an urgent need to explore greener, more efficient, and magnesium alloy-compatible methods for surface cleaning and pretreatment. The emergence of laser cleaning technology offers a novel approach to addressing the challenges of magnesium alloy surface cleaning and pretreatment.

The concept of laser cleaning was first introduced by Beadair et al. in 1969 [5]. Its principle involves irradiating the surface of the material to be cleaned with a high-energy-density laser beam, causing contaminants on the substrate surface to be removed through mechanisms such as vibration, ablation, plasma shock, and photochemical reactions, thereby achieving surface cleaning. At present, laser cleaning technology has been widely applied in areas such as paint removal [6], rust removal [7], oil removal [8], removal of micron- and submicron-scale particles [9], removal of marine microorganisms [10], and cultural relic restoration [11], achieving excellent cleaning results.

Researchers have found that laser cleaning technology is highly suitable for the surface cleaning of metallic components, such as ferrous alloys [12, 13], aluminum alloys [14, 15], and titanium alloys [16]. It not only effectively removes surface contaminants (or coatings) but also achieves surface modification effects. For example, Yoo et al. [17] observed that under laser parameters of 850 W power, 1064 nm wavelength, and a scanning speed of 4 m/s, paint removal treatment of 304L



S. Wang, *Laser Cleaning*, https://doi.org/10.1007/978-981-95-2935-3_12

stainless steel resulted in the formation of dense dislocation structures and subgrains on the substrate surface, significantly increasing its microhardness. Tian et al. [18] demonstrated that during the laser cleaning of marine biological contaminants from aluminum alloy surfaces, surface pollutants were effectively removed. Simultaneously, a nanostructured layer composed of amorphous and nanocrystalline phases was formed on the aluminum alloy substrate, showing great potential in anti-adhesion, antifouling, and corrosion resistance. Li et al. [16] observed during laser cleaning of titanium alloy oxide films that at a lower laser energy density of 1.99 $J/cm^2$, surface roughness decreased, and the oxide film slightly melted and re-oxidized. As the energy density increased, the oxide film was completely removed, and periodic "ridge" structures were formed on the substrate surface. These microstructures exhibit potential for superhydrophobicity, anti-adhesion, antifouling, and corrosion resistance.

However, current research mainly focuses on the removal of surface contaminants using either continuous lasers or pulsed lasers [19]. While continuous lasers exhibit significant ablation effects and high cleaning efficiency, they are prone to damaging the substrate. In contrast, pulsed laser cleaning minimizes substrate damage but suffers from low cleaning efficiency [20]. Therefore, balancing the respective advantages and disadvantages of continuous and pulsed lasers to enhance the efficiency and quality of paint layer removal from metal surfaces remains a critical challenge in laser cleaning research. In recent years, the introduction of hybrid laser cleaning technology has offered a novel approach to addressing this issue [21]. Hybrid laser cleaning combines continuous and pulsed laser beams to emit a hybrid laser with superior overall performance. Unlike traditional single-laser cleaning methods, this technology combines the advantages of both continuous and pulsed lasers: the thermal ablation effect of the continuous laser rapidly evaporates the coating, while the pulsed laser efficiently removes the coating through thermo-mechanical vibration and plasma shock effects. This complementary integration of both lasers [22, 23] significantly enhances cleaning efficiency and effectively improves surface quality.

Although single-laser cleaning technology has been extensively studied, systematic exploration of continuous-pulse hybrid laser cleaning technology remains relatively insufficient. Therefore, further research in this area is of significant importance. Meanwhile, femtosecond laser technology has garnered significant attention due to its extensive application in the preparation of microstructures on substrate surfaces [9, 24, 25]. Combining hybrid laser cleaning technology with the removal of surface contaminants and the preparation of functionalized microstructures offers the potential to not only achieve efficient contaminant removal but also to form functionalized textured layers, thereby significantly enhancing substrate surface performance. This integration could open new possibilities for the widespread application of this technology across various critical fields.

This study combines continuous and pulsed laser beams to systematically investigate the synergistic effects and efficient cleaning mechanisms of hybrid lasers in paint removal from AZ31B magnesium alloy surfaces, while analyzing their impact on substrate surface integrity and functionalization. The mechanism of hybrid laser cleaning for removing paint layers from AZ31B magnesium alloy surfaces is revealed for the first time. Additionally, five distinct microstructures were successfully fabricated, achieving functionalized modification of the substrate surface. These findings provide an innovative approach to the efficient removal of paint layers and surface functionalization of magnesium alloys, while also offering new perspectives and practical evidence for the comprehensive development and industrial application of this technology.

## 12.1 Materials and Methods

### *12.1.1 Experimental Materials*

In this study, commercially available AZ31B magnesium alloy plates were used as the experimental substrate material, primarily composed of Mg and Al elements, with a striped surface texture. Before the experiments, the magnesium alloy was cut into block samples with dimensions of 20 mm × 20 mm × 3 mm. The surface coating used H06-2 epoxy ester primer, commonly applied in aircraft painting. The substrates underwent grinding, spraying, drying, and thickness measurement as part of the pretreatment process, with the coating thickness controlled at (50 ± 5) μm. Figure 12.1 shows the flowchart for magnesium alloy sample preparation. Figure 12.2 provides a comparison of the magnesium alloy samples before and after preparation.

### *12.1.2 Hybrid Laser Cleaning Equipment*

In this experiment, a continuous-pulse hybrid laser was used to clean the paint layer on the surface of magnesium alloys. Figure 12.3 illustrates the schematic diagram of the hybrid laser cleaning equipment. The equipment aligns and shapes two laser beams separately using fast-axis collimation and beam shaping, with an independent control system enabling precise parameter adjustment. In the coupling cavity, optical devices efficiently superimpose the beams in both spatial and temporal domains. The resulting hybrid laser beam combines the high-energy-density characteristics of pulsed lasers with the stability of continuous lasers, achieving a composite beam output.

In the preliminary experiments, an orthogonal testing method was employed to evaluate the effects of continuous laser power, pulsed laser power, scanning speed, and scanning interval on the paint removal performance of magnesium alloy surfaces. The results indicated that insufficient laser energy, caused by either low laser power or excessive scanning speed and interval, failed to effectively evaporate the surface paint layer. Conversely, excessive laser power or overly small scanning

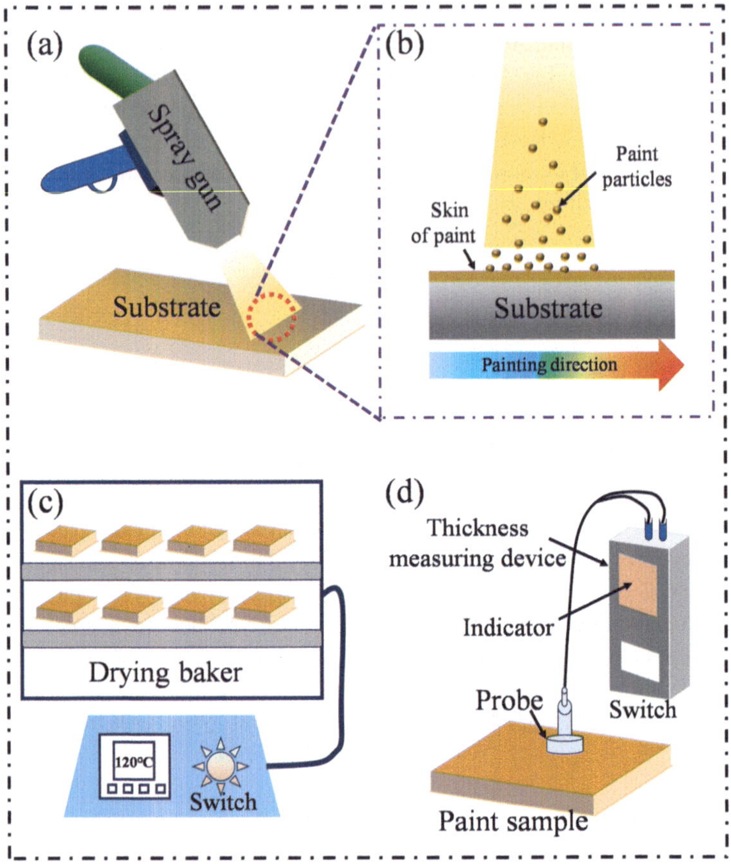


**Fig. 12.1** Flowchart of magnesium alloy sample preparation. (**a**) Paint spraying; (**b**) Magnified view of the sprayed surface; (**c**) Sample drying; (**d**) Coating thickness measurement

speed and interval may lead to substrate surface damage. Based on the preliminary results, the optimal parameters for hybrid laser cleaning and paint removal were determined, as shown in Table 12.1.

Additionally, the laser spot diameter was determined based on the calculation formula for spot diameter, as shown in Eq. (12.1). The laser spot diameter for the hybrid laser equipment was calculated to be 45 μm, where $f$ represents the focal length of the field lens, λ is the wavelength of the incident laser, and D is the diameter of the incident beam.

$$W = \frac{2f\lambda}{D} = \frac{2 \times 126 \times 10^3 \times 1064 \times 10^{-3}}{6 \times 10^3} \approx 45\,\mu\text{m} \tag{12.1}$$

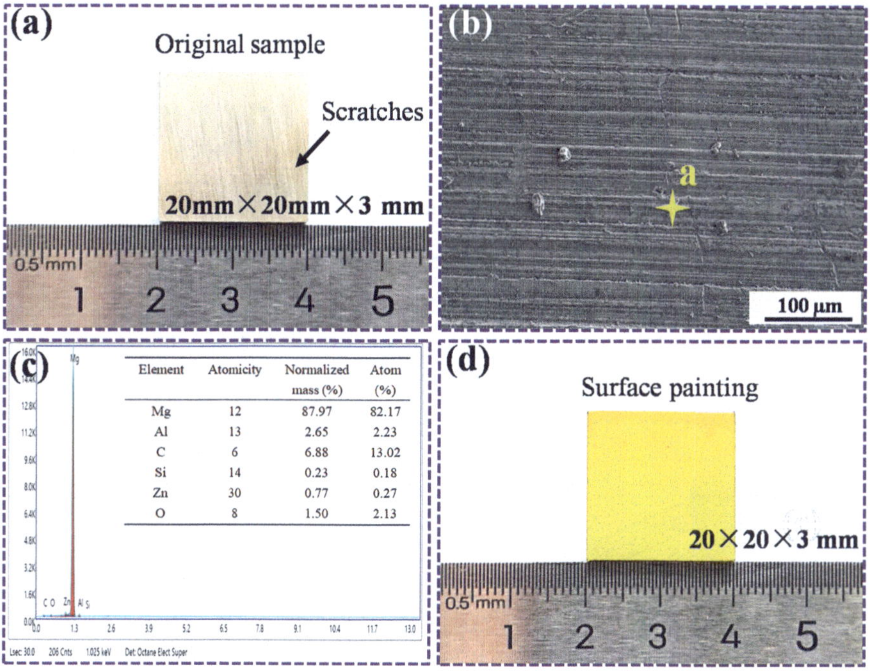


**Fig. 12.2** Comparison of magnesium alloy samples before and after preparation. (**a**) Original sample; (**b**) Surface morphology of the original sample (**c**) Energy spectrum of point a; (**d**) Sample after coating

### *12.1.3 Path Design for Hybrid Laser Cleaning*

In this study, five different laser scanning paths—linear, circular, square, loop, and bow-shaped—were designed to investigate the effects of surface texturing on the substrate after removing the paint layer from the magnesium alloy surface. The five laser scanning paths are illustrated in Fig. 12.4. The scanning intervals for the linear and square paths were set to 0.12 mm, while the circular and bow-shaped paths had intervals of 0.06 mm. Based on the bow-shaped path, the loop microstructure was fabricated using a single nanosecond laser technique. For clarity in discussion, the samples corresponding to the five laser scanning paths were labeled C1 to C5, while the control group sample was labeled AR.

### *12.1.4 Structural Characterization and Performance Testing of Samples*

For structural characterization, the macroscopic surface morphology of all samples was captured using a stereomicroscope (Leica, Germany), while the microscopic morphology was observed using a field emission scanning electron microscope

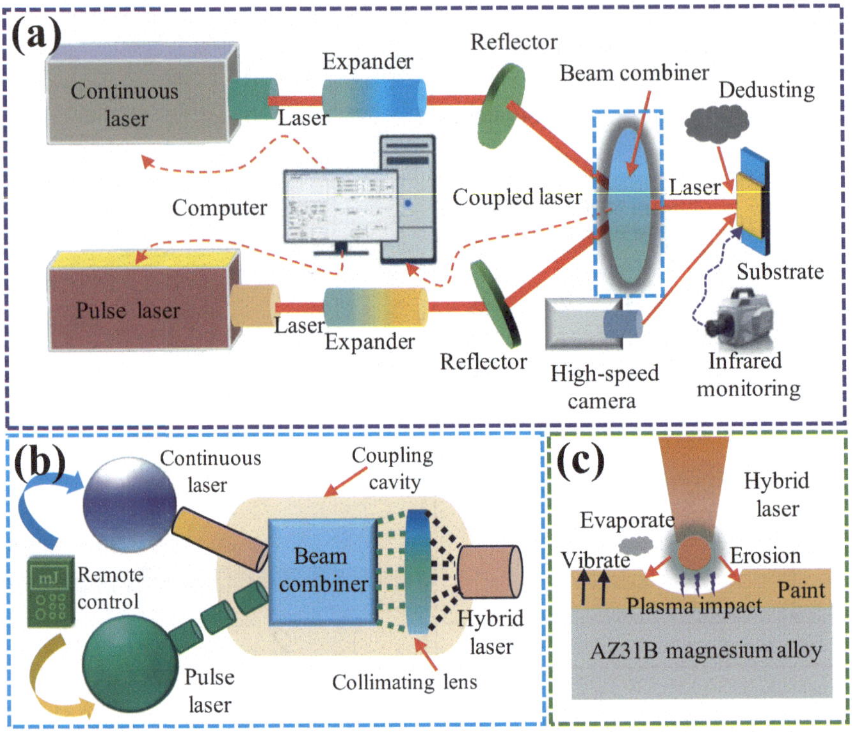


**Fig. 12.3** Schematic diagram of hybrid laser cleaning equipment. (**a**) Cleaning process using the hybrid laser coupler; (**b**) Schematic of the laser coupler principle; (**c**) Illustration of the hybrid laser cleaning process

**Table 12.1** Basic experimental parameters for hybrid laser cleaning and microstructure preparation

| Parameter | Value | Parameter | Value |
|---|---|---|---|
| Continuous laser power $P$ (W) | 90 | Pulsed laser power $P$ (W) | 45 |
| Continuous laser wavelength $\lambda$ (nm) | 1080 | Pulsed laser wavelength $\lambda$ (nm) | 1060 |
| Defocus amount $f$ (mm) | 126 | Pulse width $\tau$ (ns) | 200 |
| Spot diameter $D$ (μm) | 45 | Pulse frequency $f$ (kHz) | 8 |
| Scanning speed $N$ (mm/S) | 2700 | Scan count $N$ (c) | 2 |

(FE-SEM, Melin Compact, Zeiss, Germany). A laser scanning confocal microscope (OLS-3000, HITACHI OLYMPUS, Japan) was used to measure the surface roughness of the samples after laser cleaning. Electron backscatter diffraction (EBSD) measurements were performed using a Gemini560 system equipped with a SymmetryS2 detector to analyze the grain size changes in the substrate after laser cleaning and coating removal. The cross-sectional morphology and crystal structure after laser cleaning were observed using a transmission electron microscope (TEM, Talos F200X, USA). Additionally, focused ion beam (FIB) techniques were employed to prepare TEM samples for a more precise observation of the amorphous nanocrystalline regions in the magnesium alloy cross-section.

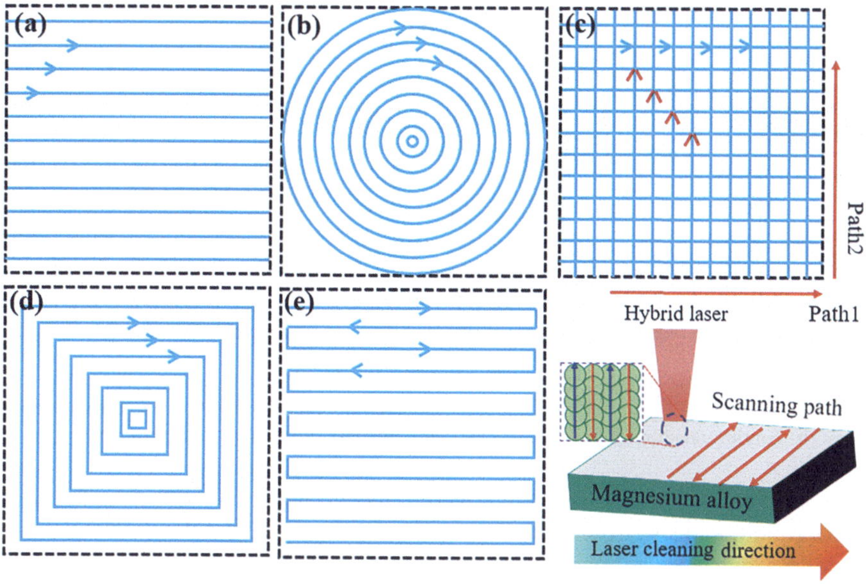


**Fig. 12.4** Schematic diagram of different scanning paths for hybrid cleaning and microstructure fabrication; (**a**) C1: linear; (**b**) C2: circular; (**c**) C3: square; (**d**) C4: loop; (**e**) C5: bow-shaped

For performance testing, an X-ray residual stress analyzer (μ-X360, PULSTEC, Japan) was used to measure the residual stress on the sample surface before and after laser cleaning. A contact angle measurement instrument (DSA, LAUDA Scientific, Germany) was used to measure the hydrophilicity of the substrate surface after cleaning. A pull-off tester (Posi Test AT-M, USA) was employed to measure the adhesion strength of the recoated paint. A tribometer (TriboLab CMP, USA) was used to conduct wear resistance testing on the samples before and after laser cleaning. The test was carried out under a load of 5 N, at a temperature of 30 °C, with a frequency of 2.5 Hz and a stroke length of 800 μm, using a reciprocating friction setup. An electrochemical workstation (Ivium, Switzerland) was used to evaluate the corrosion resistance of the samples before and after laser cleaning. The test was conducted with a scanning range of −2.2 to −0.8 V and a scanning interval of 0.01 V/s.

## 12.2 Results

### *12.2.1 Surface Texture Morphology Analysis After Hybrid Laser Cleaning*

Figure 12.5a shows the macroscopic morphology of the original sample and the five microstructured samples after hybrid laser cleaning. It can be observed that the paint layer on the cleaned sample surfaces has been largely removed, revealing a

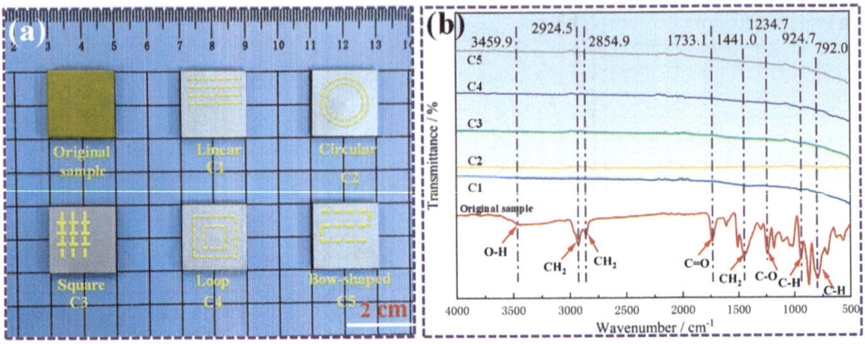

**Fig. 12.5** Analysis of surface cleanliness of magnesium alloy after mixed laser cleaning. (**a**) Macroscopic morphology of microstructures prepared by hybrid cleaning with different scanning paths; (**b**) Fourier transform infrared (FTIR) spectral analysis of different microstructures

relatively bright surface. To further verify the cleaning effectiveness, Fourier-transform infrared spectroscopy (FTIR) was used to analyze the chemical composition of organic substances on the sample surfaces, as shown in Fig. 12.5b. Previous studies [26] have shown that the original sample's paint layer contains various organic compounds, with characteristic absorption peaks at 3459.9 $cm^{-1}$ and 1174.5 $cm^{-1}$ (hydroxyl stretching vibration) and 2924.5 $cm^{-1}$ and 2854.9 $cm^{-1}$ (saturated hydrocarbon C–H stretching vibration), all of which are typical organic absorption peaks [27]. In the five microstructured samples after cleaning, these characteristic absorption peaks largely disappeared, indicating that the paint layer and residual organic substances on the surface were effectively removed, achieving efficient surface paint cleaning.

Further observation using a scanning electron microscope (SEM), as shown in Fig. 12.6, revealed that the cleaned sample surfaces exhibited five typical textured structures formed according to the predefined scanning paths: linear, circular, square, loop, and bow-shaped. These findings demonstrate the high controllability of hybrid laser cleaning technology in surface texturing design. Additionally, they provide an important reference for the customized design of paint removal and functionalized microstructures using hybrid laser cleaning technology.

It is noteworthy that, to successfully fabricate the linear and square microstructures during paint removal, the scanning interval must be set greater than the laser spot diameter. Although the FTIR results confirmed that the organic substances on the substrate surface were effectively removed, energy dispersive spectroscopy (EDS) was conducted on representative areas of the five microstructured samples to further evaluate the cleaning efficiency and surface cleanliness. The analysis aimed to detect any residual elemental composition or inorganic substances on the cleaned surfaces. The EDS analysis is shown in Fig. 12.7.

The results indicate that the inorganic carbon content on the surfaces of the circular (Fig. 12.7), loop (Fig. 12.7), and remelted (Fig. 12.7) samples is consistent with that of the uncoated magnesium alloy sample, demonstrating that the paint

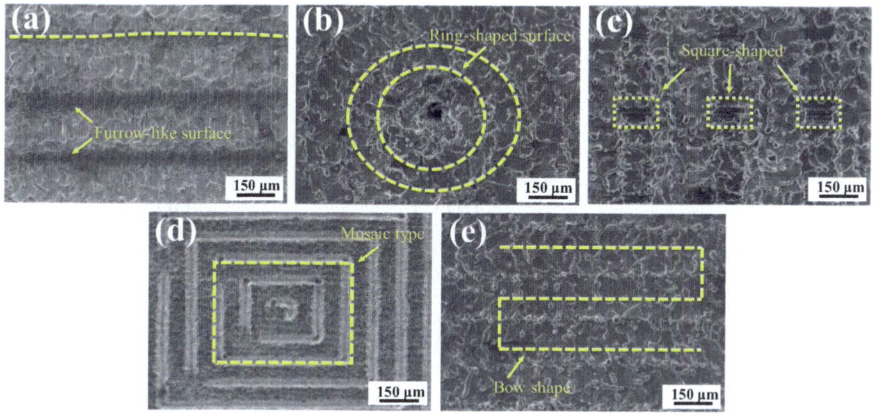


**Fig. 12.6** Microscopic morphology of microstructures prepared by hybrid cleaning with different scanning paths; (**a**) C1: linear; (**b**) C2: circular; (**c**) C3: square; (**d**) C4: loop; (**e**) C5: bow-shaped

layer was completely removed and the cleaning was highly effective. In contrast, for the linear (Fig. 12.7) and square (Fig. 12.7) samples, the larger scanning interval resulted in a slight increase in surface carbon content, as revealed by surface scanning analysis, with the carbon concentrated in localized areas. However, this residue was primarily trace amounts of carbon deposition and did not significantly affect the cleaning performance.

On the one hand, in hybrid laser cleaning, the continuous laser softens or decomposes the surrounding coating through heat conduction and thermal accumulation effects, while the nanosecond laser, with its high peak power and pulsed effects, expands the cleaning range via plasma shock waves and thermal effects. Additionally, the Gaussian energy distribution and nonlinear effects further enhance the cleaning performance [28, 29]. Moreover, the overlap of two scanning paths and the energy coverage at the edges of the laser spot effectively compensated for the gaps in the scanning interval, significantly improving the cleaning efficiency [30]. On the other hand, the small amount of carbon residue has minimal impact on the coating, welding, or bonding performance of the substrate [31]. Therefore, despite the relatively large scanning intervals of the linear and square paths, the cleaning process still demonstrated strong removal capabilities. Furthermore, it provides important insights for optimizing scanning parameters and fabricating textured surface microstructures in future studies.

Figure 12.8 illustrates the three-dimensional morphology of the substrate surface after laser cleaning. The results demonstrate that, under specific cleaning parameters, the laser effectively removed the paint layer, forming five distinct textured morphologies and significantly enhancing surface roughness. According to previous studies [32], the paint removal process is primarily driven by laser ablation and plasma shock effects. Notably, the dual-beam overlap effect of the hybrid laser significantly enhanced thermal ablation and plasma shock effects, leading to localized melting on the magnesium alloy surface. The molten magnesium atoms rapidly

**Fig. 12.7** EDS analysis after hybrid laser cleaning. (**a**) C1; (**b**) C2; (**c**) C4; (**d**) Energy spectrum at point a of C1; (**e**) Energy spectrum at point b of C2; (**f**) Energy spectrum at point c of C4; (**g**) C3; (**h**) Magnified view of a typical region of C3; (**i**, **j**) Surface scanning of the typical region of C3; (**k**) C5; (**l**) Magnified view of a typical region of C5; (**m**, **n**) Surface scanning of the typical region of C5

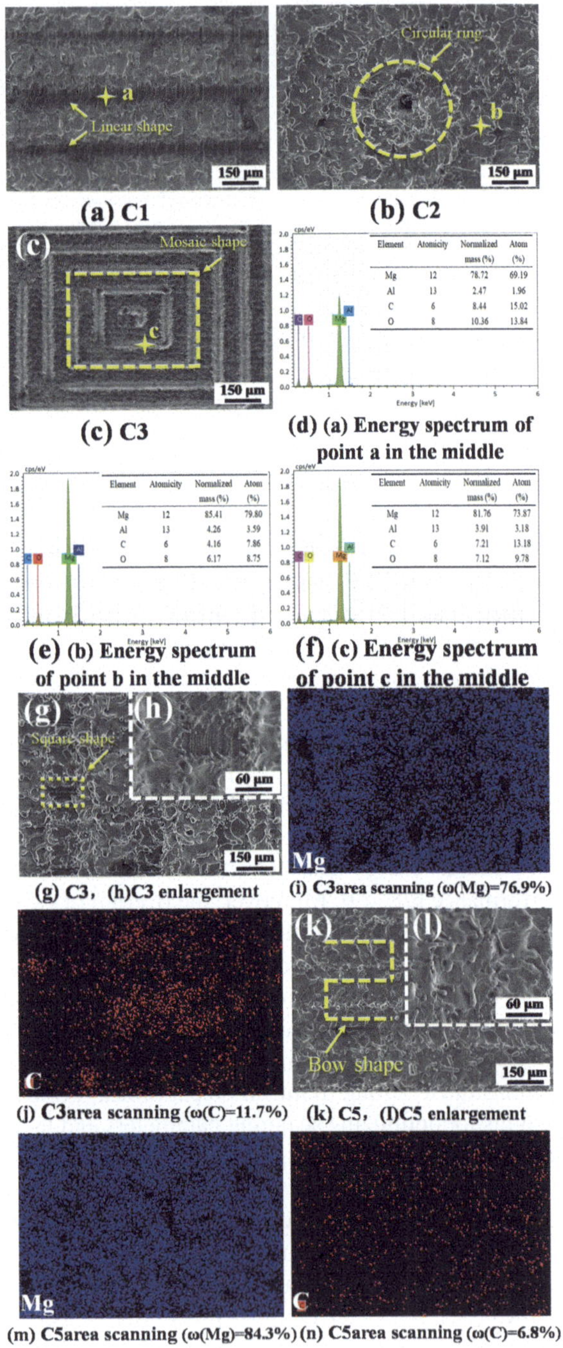

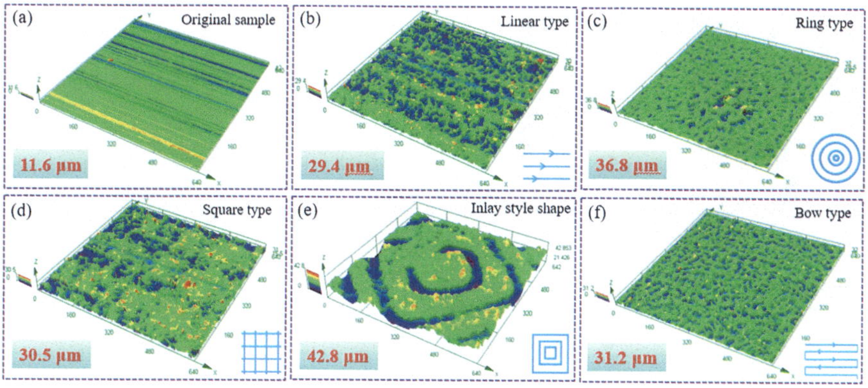


**Fig. 12.8** Three-dimensional morphology of AZ31B magnesium alloy after paint removal under different scanning paths (**a**) AR; (**b**) C1; (**c**) C2; (**d**) C3; (**e**) C4; (**f**) C5

cooled in air, forming fine droplets uniformly distributed on the substrate surface [33]. Additionally, differences in the scanning paths further accentuated the height variations between the "peaks" and "valleys" in the microstructures, significantly increasing surface roughness. The experimental results demonstrate that hybrid laser cleaning technology is highly effective in both paint layer removal and substrate roughening.

### 12.2.2 *Effects of Hybrid Laser Cleaning on Substrate Microstructure*

To better understand the microstructural morphology of the substrate surface texture after hybrid laser paint removal, electron backscatter diffraction (EBSD) analysis was performed on the cross-section of the completely cleaned substrate. Since all five scanning paths exhibited surface remelting after paint removal, this study focused on observing the grain size in the remelted region of the substrate cross-section cleaned using the bow-shaped scanning path (as shown in Fig. 12.9).

The results indicate that a remelted layer with a thickness of approximately 10–15 μm was formed on the substrate surface by hybrid laser cleaning, and significant grain refinement was observed in this remelted layer (Fig. 12.9a). Figure 12.9b presents the statistical distribution of grains in the cross-section of the substrate region. The histogram indicates that the grain size and count in this region are primarily concentrated below 6 μm. Additionally, it was calculated using software that, under the effect of laser cleaning, the average grain size in the remelted layer was approximately 1.71 μm, which was significantly smaller than the original grain size of the substrate. This phenomenon is attributed to the dual-energy output characteristics of the hybrid laser: the continuous laser provides a high-energy, sustained input, while the pulsed laser induces intermittent plasma shocks.

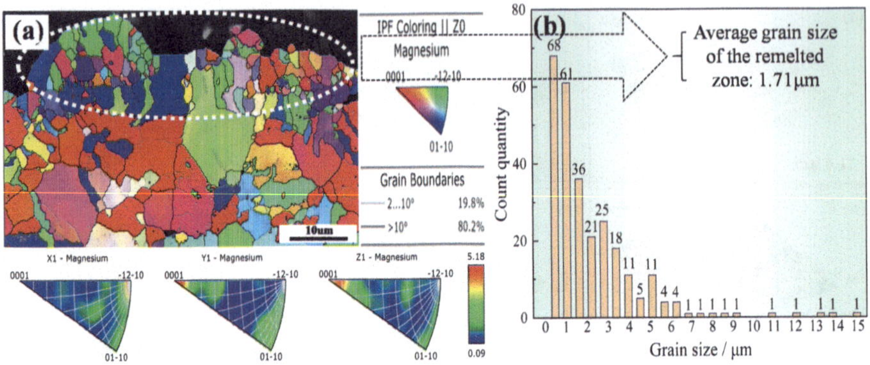


**Fig. 12.9** EBSD image and grain size distribution of the substrate cross-section after hybrid laser cleaning; (**a**) EBSD image of the cross-section; (**b**) Grain size distribution chart

On the one hand, the combined effects of high-energy continuous laser and intermittent plasma shocks induced by the pulsed laser generated a significant temperature gradient in the vertical direction of the molten substrate surface, promoting preferential grain growth along the longitudinal direction [34]. On the other hand, due to the limited ability of the magnesium alloy surface to transfer heat efficiently into the material through thermal conduction, a significant peak temperature variation occurred. This created a substantial degree of undercooling between the substrate and the remelted layer [25, 35]. This high undercooling environment significantly increased the nucleation rate of crystals in the remelted layer while effectively restricting the grain growth space [36]. It can be concluded that the synergistic effect of the high-energy continuous laser input and the intermittent plasma shocks induced by the pulsed laser is the primary reason for the formation of a high-temperature gradient and the resulting extreme grain refinement. This mechanism directly led to a significant reduction in grain size.

In the subsequent study, the cross-sectional morphology of FIB-cut samples was characterized using a scanning electron microscope (SEM), and the features of the nanostructured layer were further analyzed using a transmission electron microscope (TEM). Figure 12.10a–c shows the selected FIB cutting position in the SEM view and the low-magnification cross-sectional images before and after ion beam thinning and polishing. Figure 12.10d presents a TEM bright-field image of the magnesium alloy cross-section after hybrid laser cleaning, showing the formation of an oxide layer approximately 100 nm thick on the surface. This is attributed to the high-temperature melting and rapid cooling effects during the laser cleaning process [37]. The line scan EDS analysis and elemental distribution maps in Fig. 12.10e–g further reveal that the surface oxygen content is significantly higher than that of the internal region, with an oxide layer thickness of approximately 110 nm. This indicates that during the laser cleaning process, oxidation occurred on the substrate surface, and the high temperature facilitated the diffusion of oxygen into the surface layer of the magnesium alloy [38]. As the depth increased, the concentration of

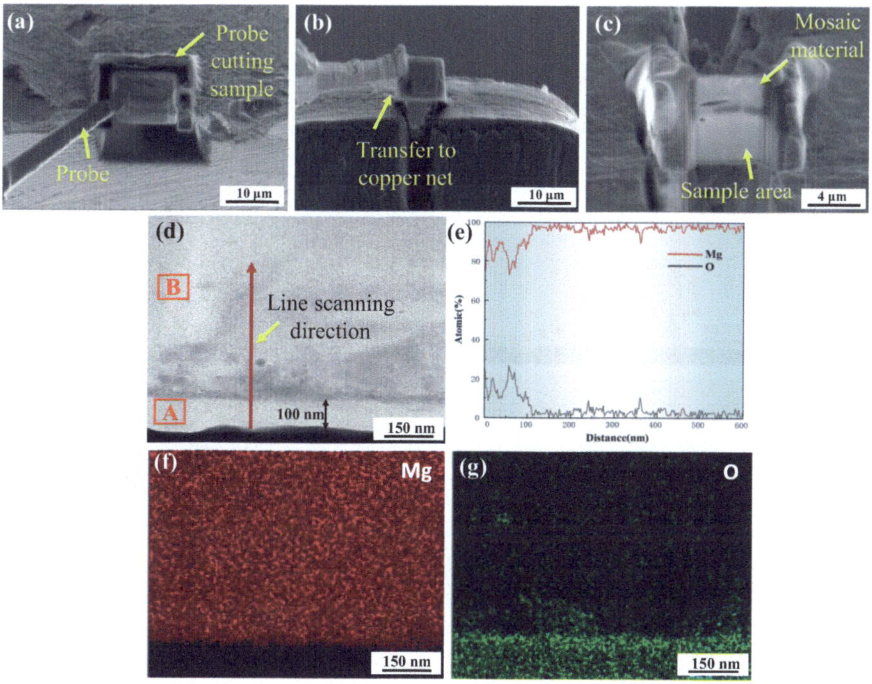


**Fig. 12.10** (**a**–**c**) FIB sample preparation position and surface morphology after ion beam polishing; (**d**) TEM image of the cross-sectional nanostructured layer; (**e**) line scan EDS analysis along the red arrow in figure (**d**); (**f**, **g**) elemental surface scanning results of the region in figure (**d**)

oxygen decreased significantly, consistent with the line analysis results. The formation of this oxide layer may have a dual effect on surface performance: on the one hand, it could enhance corrosion resistance, but on the other, it may negatively impact mechanical properties.

To further analyze the interface characteristics between the heat-affected zone (HAZ) and the magnesium alloy substrate, high-resolution TEM was used to observe atomic arrangements, and selected area electron diffraction (SAED) was performed for verification. The results are shown in Fig. 12.11. After laser cleaning, the magnesium alloy cross-section formed a nanometer-scale oxide layer with a bright band feature (Fig. 12.11a), and a distinct interface boundary was observed between the oxide layer and the substrate. The bright band layer exhibited typical amorphous structural characteristics (Fig. 12.11b), with non-periodic atomic arrangements (Fig. 12.11c). The diffuse diffraction rings further confirmed the formation mechanism of the amorphous oxide layer under rapid cooling conditions [39]. The formation of this structure is attributed to the stable thermal input from the continuous laser, which induces melting and oxidation reactions, and the transient high-power effects of the pulsed laser, which result in rapid melting, extreme cooling, and disordered atomic arrangements. The interplay between thermal stability

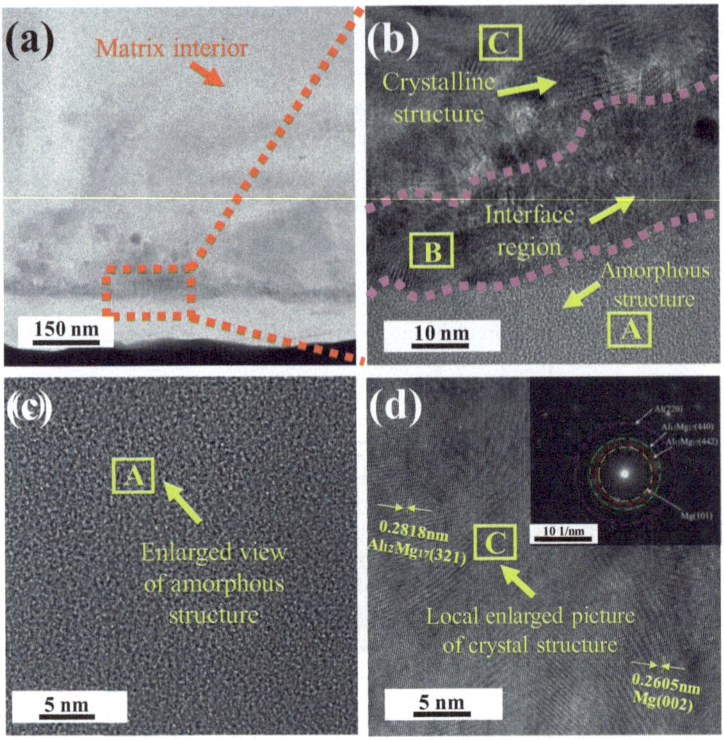


**Fig. 12.11** (**a**) Magnified image of the interface between the amorphous-nanocrystalline region of the nanostructured oxide layer and the crystalline region of the magnesium substrate; (**b**) Magnified view of the red-highlighted region in (**a**); (**c**) Magnified view of the amorphous structure in region A; (**d**) Magnified view of the crystalline structure in region C

and kinetic balance facilitated oxygen enrichment, ultimately leading to the formation of the bright-banded amorphous oxide layer [35].

In addition, the internal region of the magnesium substrate (Fig. 12.11d) is characterized by a highly ordered crystal structure and typical close-packed hexagonal (HCP) features, indicating that the effects of hybrid laser cleaning on the substrate structure are primarily confined to the heat-affected zone and its interface region [40, 41], with no significant impact on the internal substrate structure. These findings demonstrate that hybrid laser cleaning not only effectively removes surface contaminants but also facilitates the formation of an optimized surface layer structure, resulting in the formation of an amorphous nano-oxide layer approximately 100 nm thick at the substrate cross-section. However, to further enhance surface performance, it is necessary to optimize the process parameters to precisely control the extent of the heat-affected region and the properties of the oxide layer.

Figure 12.12 illustrates the microscopic morphology at the interface between the nanostructured oxide layer's amorphous-nanocrystalline region and the magnesium alloy substrate, showing the presence of numerous dislocations that aggregate to

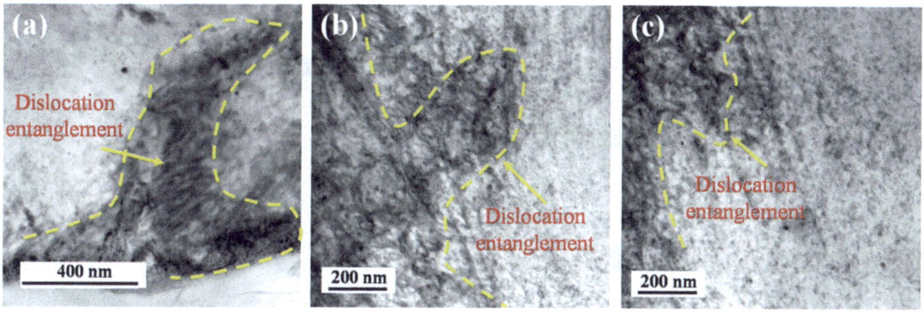


**Fig. 12.12** Dislocation morphology at the interface between the amorphous-nanocrystalline region of the nanostructured oxide layer and the magnesium substrate (**a**) area 1 (**b**) area 2 (**c**) area 3

form dislocation entanglements. Notably, the formation of dislocation entanglements is primarily attributed to two factors: first, the internal stresses induced by rapid cooling during the hybrid laser cleaning process. The rapid cooling rate causes inconsistent contraction across different regions of the material, preventing atoms from moving and rearranging uniformly, ultimately leading to the aggregation of dislocations into entanglements [42]. Second, laser irradiation causes localized heating, forming plasma, which rapidly expands and cools to generate shock waves. These shock waves induce atomic displacement within the material, further promoting the formation of dislocation entanglements [43]. These dislocation structures provide critical microstructural insights into the mechanical behavior and properties of the material.

### 12.2.3 *Effects of Hybrid Laser Cleaning on Substrate Surface Properties*

The hybrid laser technology not only efficiently removed the paint layer from the magnesium alloy surface but also formed textured microstructures on the substrate surface, significantly altering the substrate's microstructural characteristics. To further investigate the effect of surface textures formed by hybrid laser cleaning on substrate surface properties, contact angle and recoating adhesion tests were conducted (as shown in Figs. 12.13 and 12.14). It was observed that, compared to the untreated original sample, the contact angle of the laser-cleaned surface decreased significantly, indicating improved surface wettability. The original sample exhibited a contact angle of 67.916°, and when circular and loop-shaped scanning paths were applied during laser cleaning, the contact angles were reduced to 47.163° and 43.811°, respectively. This phenomenon is primarily attributed to the increased surface roughness after laser cleaning, which enlarged the contact area between the liquid and the substrate, thereby enhancing liquid adhesion and capillary effects. Furthermore, the improvement in surface wettability induced by laser cleaning has a profound impact on the recoating adhesion performance of the substrate surface.

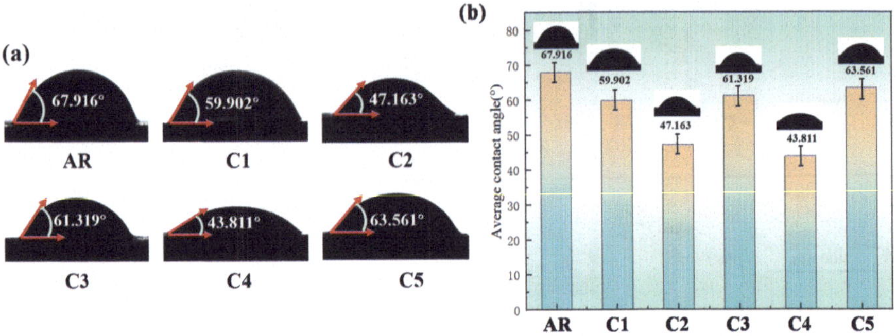


**Fig. 12.13** Contact angle test results of the original sample and hybrid laser-cleaned substrate surfaces with different textures (**a**) Sample contact angle (**b**) Contact angle histogram

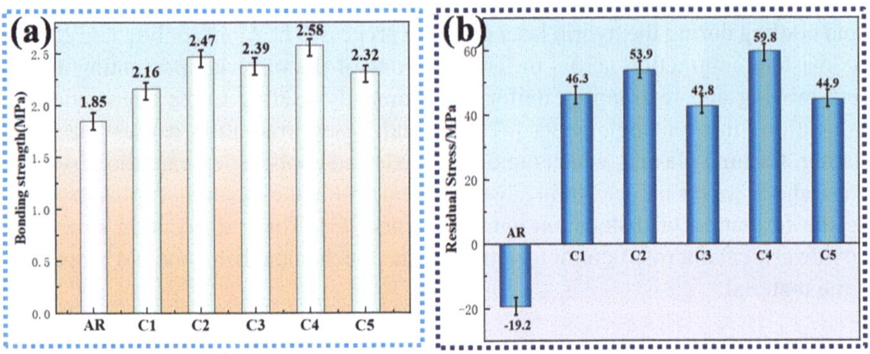


**Fig. 12.14** (**a**) Recoating adhesion strength of the original sample and the substrate after hybrid laser cleaning and paint removal; (**b**) residual stress of the original sample and the substrate after hybrid laser cleaning and paint removal

The adhesion test results (Fig. 12.14a) further confirmed this observation. The results indicated that the adhesion strength of the laser-cleaned samples was significantly improved compared to the original sample. Specifically, the adhesion strength of the loop-textured and circular-textured surfaces increased by 39.5% and 33.5%, respectively, compared to the original sample.

This improvement is primarily attributed to two factors: first, the textured layer formed after laser cleaning significantly increased surface roughness, thereby enhancing the mechanical interlocking capability between the substrate surface and the paint layer. Second, the enhanced surface wettability allowed the paint layer to spread and penetrate more easily, further increasing the interfacial adhesion strength [38].

To clarify the effects of hybrid laser cleaning on the mechanical properties of the substrate surface, the residual stress of the original and laser-cleaned samples was analyzed. Figure 12.14b shows that the laser-cleaned substrate surface exhibited tensile residual stress, while the internal region displayed compressive residual

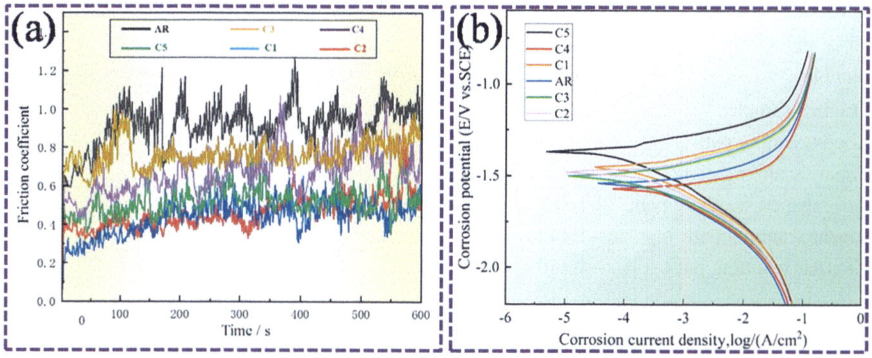


**Fig. 12.15** (**a**) Friction coefficient curves of the original sample and magnesium alloy substrate surface under five different scanning paths; (**b**) potentiodynamic polarization curves of the original sample and laser-cleaned samples under five different scanning paths

stress. According to previous studies, during the laser cleaning process, plasma shock effects generate compressive residual stress, while thermal effects result in tensile residual stress [35]. Notably, hybrid laser technology produces dual energy output characteristics: the continuous laser provides a high-energy, sustained input, while the pulsed laser induces intermittent plasma shocks. During the cleaning process, the higher power of the continuous laser results in compressive stress being insufficient to counterbalance tensile stress, ultimately leading to tensile residual stress in the substrate [44]. Furthermore, tensile stress has a positive effect in resisting deformation under external pressure [45].

Figure 12.15 illustrates the friction coefficient curves of the magnesium alloy substrate surface for the original sample and the five different scanning paths. The results indicate that, after laser cleaning, the friction coefficients of all five textured surfaces were lower than those of the original sample, demonstrating a significant improvement in wear resistance. This improvement is primarily attributed to the formation of amorphous and nanocrystalline regions on the surface [39], where grain refinement enhanced wear resistance. Additionally, the oxide layer generated by laser cleaning acts as a protective layer, filling surface defects, reducing fretting wear, and further lowering the friction coefficient [18]. The friction coefficients varied across different scanning paths, with the square and loop patterns exhibiting higher coefficients, while the bow-shaped, linear, and circular patterns showed lower coefficients. This variation was primarily attributed to surface roughness, stress concentration, and changes in surface micro-nano structures.

Additionally, the corrosion resistance test results (Fig. 12.15b and Table 12.2) show that, after laser cleaning, the corrosion potential of all scanning paths, except for the loop path, was higher than that of the original sample, indicating an improvement in corrosion resistance. The changes in corrosion resistance can be attributed to several factors. On the one hand, during the laser cleaning process, the thermo-mechanical coupling effect caused the rapid removal of the paint layer, and partial remelting and quenching of the substrate surface resulted in the formation of an

**Table 12.2** Corrosion current density and corrosion potential of the original sample and samples with different scanning paths

| Sample | Corrosion potential (mV) | Corrosion current density (V(A/cm$^2$)) |
|---|---|---|
| Original sample | −1.539 | $8.409 \times 10^{-5}$ |
| Loop scanning path | −1.569 | $1.504 \times 10^{-4}$ |
| Square scanning path | −1.498 | $2.614 \times 10^{-5}$ |
| Bow-shaped scanning path | −1.369 | $1.123 \times 10^{-5}$ |
| Linear scanning path | −1.449 | $7.341 \times 10^{-5}$ |
| Circular scanning path | −1.480 | $2.341 \times 10^{-5}$ |

amorphous-nanocrystalline oxide layer [46], which enhanced corrosion resistance. On the other hand, the increased surface roughness provided a larger contact area for the corrosive solution, while the texturing enhanced surface wettability, accelerating the diffusion of the corrosive solution and thereby reducing corrosion resistance [47].

Overall, the surface properties of magnesium alloys after laser cleaning are influenced by multiple factors, including crystalline structure, oxide layer, surface roughness, and wettability. To achieve superior surface properties, an appropriate balance among these factors must be identified.

## 12.3 Discussion

Hybrid laser cleaning, as an emerging technology, not only efficiently removes the paint layer from the surface of AZ31B magnesium alloy but also enables the fabrication of microstructures on the substrate surface and achieves surface modification. The underlying mechanisms hold significant importance for the scientific optimization of process methods. To further investigate the mechanism of high-efficiency paint removal and the surface impact-like strengthening effect of hybrid laser cleaning, Fig. 12.16 presents high-speed photographic images of the hybrid laser cleaning process on the paint layer of AZ31B magnesium alloy. The images reveal that the cleaning mechanism of hybrid laser cleaning on the surface paint layer is more complex than that of single laser cleaning.

On the one hand, the high energy density of the continuous laser acting on the paint layer causes the surface paint layer to soften and gradually separate from the substrate. Meanwhile, the sustained high temperatures may cause the substrate surface to reach the evaporation temperature of the paint layer. With further energy input, a quenching-like heat treatment effect may be induced, leading to surface remelting and modification [48]. On the other hand, the impact of the pulsed laser locally generates plasma on the magnesium alloy surface, with the plasma plume diffusing into the surrounding regions [26]. The intermittent and high-frequency impacts of the pulsed laser further accelerate the detachment of the softened or

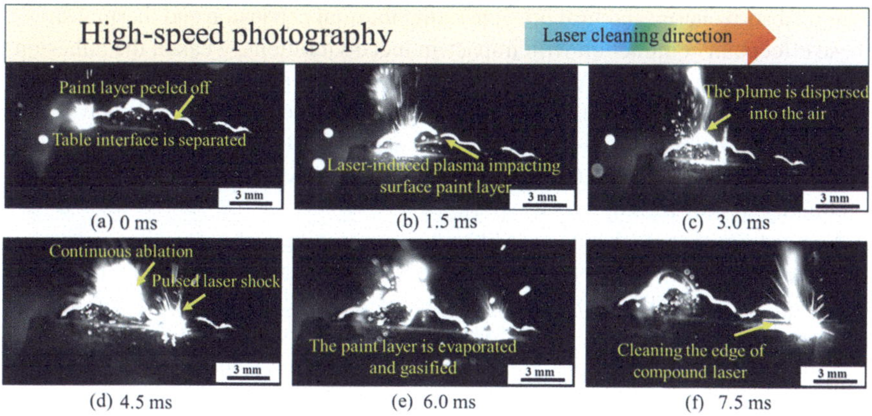


**Fig. 12.16** Dynamic monitoring of the hybrid laser cleaning process on the surface paint layer of AZ31B magnesium alloy (**a**) 0 ms (**b**) 1.5 ms (**c**) 3.0 ms (**d**) 4.5 ms (**e**) 6.0 ms (**f**) 7.5 ms

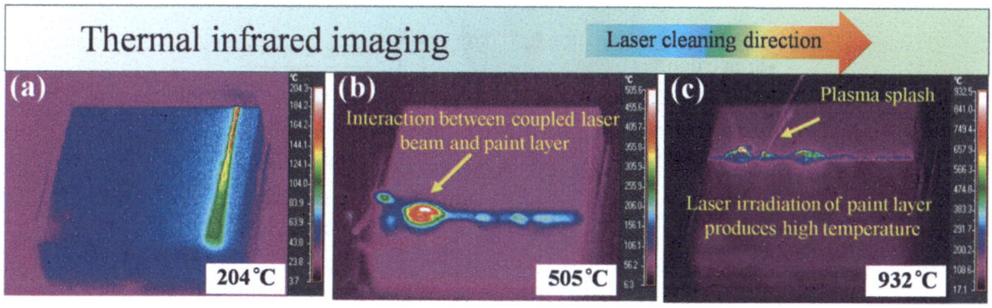


**Fig. 12.17** . Dynamic monitoring of the hybrid laser cleaning process on the surface paint layer of AZ31B magnesium alloy using thermal infrared imaging (**a**) Laser cleaning initial stage (**b**) Laser cleaning intermediate section (**c**) Laser cleaning the last section

evaporated paint layer from the substrate surface, resulting in a more efficient and higher-quality cleaning effect.

The surface interface temperature changes during the laser cleaning process were monitored in real-time using thermal infrared imaging technology, revealing the mechanisms of three key processes: heat-induced expansion, vaporization, and high-frequency impact detachment of the paint layer (Fig. 12.17). Notably, the initial longitudinal laser scan significantly increased the interface temperature (peaking at 204 °C), inducing heat expansion of the paint layer and weakening its adhesion to the substrate. The subsequent transverse scan further utilized the preheating effect, enhancing the paint layer's absorption of hybrid laser energy and facilitating vaporization and detachment under the action of high-frequency shock waves [38]. This process highlights the significant advantages of hybrid laser cleaning technology in temperature control and improving cleaning efficiency.

Therefore, hybrid laser cleaning technology exhibits multiple mechanisms in the removal of paint layers from magnesium alloys. Laser irradiation causes the paint layer to absorb energy and heat up, generating thermal expansion and thermal stress. These effects, in conjunction with impact-induced vibrations, weaken the adhesion, enabling efficient removal. As the temperature rises, the paint layer undergoes melting, vaporization, and cavitation, forming a molten pool. The shock waves and plasma generated by the pulsed laser further accelerate the detachment of the paint layer. The synergy between thermal and mechanical effects ensures high efficiency and quality in the cleaning process [49].

Another significant advantage of hybrid laser cleaning technology lies in its ability to create a textured layer on the substrate surface, enabling functional modification (Fig. 12.18). This process is similar to laser texturing and laser surface micro-etching technologies. After hybrid laser cleaning, the high-energy density laser induces melting and rapid solidification on the material surface, resulting in an

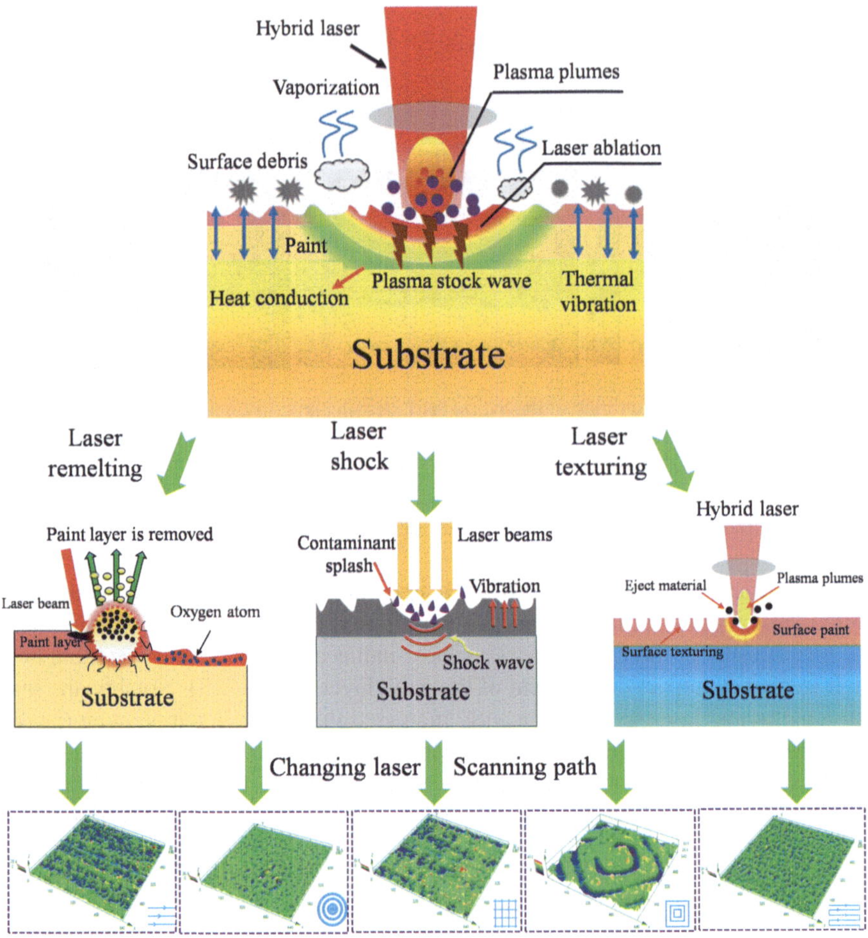


**Fig. 12.18** Schematic diagram of hybrid laser cleaning and surface texturing mechanism

extreme heat-and-rapid-cooling effect. By defining specific scanning paths, regular textured patterns can be generated on the magnesium alloy surface. Additionally, by adjusting the scanning speed and laser spot diameter, the depth and width of the texture can be precisely controlled.

The formation of this textured layer not only significantly enhances the surface properties of magnesium alloys but also improves their recoating adhesion and the bonding performance required for maintenance processes, such as welding and adhesive bonding. Notably, in the field of medical devices, this technology can also impart excellent hydrophobicity and corrosion resistance to magnesium alloy surfaces. This functional modification capability offers broad application prospects for the future innovation of laser cleaning technology, particularly in optimizing surface performance and expanding functional applications.

Additionally, during the hybrid laser cleaning process, the synergistic effects of the thermal action of the continuous laser and the impact effect of the pulsed laser enable surface remelting and rapid cooling of the magnesium alloy. Simultaneously, the exposed substrate fully interacts with oxygen and water vapor in the air at high temperatures, forming a uniform and dense oxide film (Fig. 12.19). In this process, the continuous laser increases the surface temperature, causing the paint layer to soften, melt, and vaporize, detaching from the substrate. Subsequently, the plasma shock generated by the pulsed laser further accelerates paint layer removal and induces surface atomic rearrangement. The exposed magnesium alloy surface reacts with oxygen and water vapor to form magnesium oxide (MgO) and magnesium hydroxide ($Mg(OH)_2$), which gradually deposit and cool to form a dense oxide film. This oxide film not only improves the surface's corrosion resistance but also

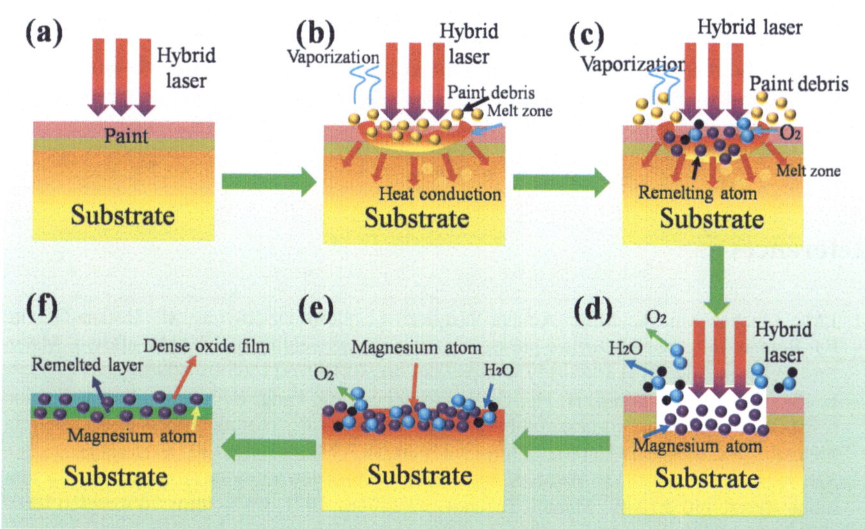


**Fig. 12.19** Schematic diagram of oxide film formation mechanism during hybrid laser cleaning of magnesium alloy surface paint layer (**a**) Laser began to remove paint.(**b**) Laser contact with paint layer (**c**) Surface remelting (**d**) Contact with air (**e**) Interatomic combination (f) Forming oxide film

enhances the mechanical strength and stability of the substrate, providing effective protection for the long-term application of magnesium alloys in complex environments.

## 12.4 Conclusion

In this study, hybrid laser cleaning technology was employed to successfully remove the paint layer from the magnesium alloy surface under specific process parameters, while also creating functional textured structures on the substrate surface.

1. The hybrid laser (combining continuous and pulsed lasers) effectively removed a paint layer approximately 50 ± 5 μm thick without damaging the substrate and simultaneously created functional textures with various patterns on the surface.
2. The formation of surface textures was accompanied by microstructural changes, including the development of a remelted layer approximately 10–15 μm thick in the substrate cross-section, along with grain refinement. Additionally, a 100 nm-thick bright band nanometer-scale oxide layer, exhibiting typical amorphous structural characteristics, was observed in the cross-section.
3. The textures formed through hybrid laser cleaning significantly reduced the surface contact angle, improving wettability, recoating adhesion, wear resistance, and corrosion resistance. Specifically, the adhesion strength of loop-textured and circular-textured surfaces increased by 39.5% and 33.5%, respectively. Additionally, the residual compressive stress on the cleaned sample surface was converted into tensile stress.
4. Real-time monitoring revealed that the synergistic effects of the thermal action of the continuous laser and the impact-induced vibrations of the pulsed laser were critical for paint layer vaporization, detachment, and plasma generation, while the direct action of the pulsed laser was the primary driving force behind texture formation.

## References


1. J.M. Sánchez-Amaya, M.R. Amaya-Vázquez, L. González-Rovira, M. Botana-Galvin, F.J. Botana, Influence of surface pre-treatments on laser welding of Ti6Al4V alloy, J. Mater. Eng. Perform. 23 (2014) 1568–1575. https://doi.org/10.1007/s11665-014-0967-y.
2. L. Zhong, Z. Li, G. Wang, H. He, G. Wei, S. Zheng, G. Feng, N. Xie, R. Zhang, Erosion resistance of valve core surface combined with WC-10Co-4Cr coating process under different pretreatments, Materials 15 (2022). https://doi.org/10.3390/ma15228140.
3. S.T. Auwal, S. Ramesh, F. Yusof, S.M. Manladan, A review on laser beam welding of titanium alloys, Int. J. Adv. Manuf. Technol. 97 (2018) 1071–1098. https://doi.org/10.1007/s00170-018-2030-x.

4. G. Baiocco, S. Genna, D. Salvi, N. Ucciardello, Laser texturing to increase the wear resistance of an electrophoretic graphene coating on copper substrates, Materials 16 (2023). https://doi.org/10.3390/ma16155359.
5. S.M. Bedair, Jr. Smith Harold P., Atomically clean surfaces by pulsed laser bombardment, J. Appl. Phys. 40 (1969) 4776–4781. https://doi.org/10.1063/1.1657288.
6. Z. Zhao, X. Liu, Z. Chen, Y. Tian, M. Chen, L. Liu, F. Song, Evaluation of laser cleaning effect for the removal of paint on aluminum alloys, Int. J. Adv. Manuf. Technol. 126 (2023) 3193–3203. https://doi.org/10.1007/s00170-023-11224-4.
7. Y. Lu, Y. Ding, G. Wang, L. Yang, M. Wang, Y. Wang, B. Guo, Ultraviolet laser cleaning and surface characterization of AH36 steel for rust removal, J. Laser Appl. 32 (2020) 032023. https://doi.org/10.2351/1.5127569.
8. M.P. Mateo, G. Nicolas, V. Piñon, A. Ramil, A. Yañez, Laser cleaning: An alternative method for removing oil-spill fuel residues, Applied Surface Science 247 (2005) 333–339. https://doi.org/10.1016/j.apsusc.2005.01.086.
9. T. Chen, W. Wang, T. Tao, A. Pan, X. Mei, Multi-scale micro-nano structures prepared by laser cleaning assisted laser ablation for broadband ultralow reflectivity silicon surfaces in ambient air, Appl. Surf. Sci. 509 (2020) 145182. https://doi.org/10.1016/j.apsusc.2019.145182.
10. Z. Tian, Z. Lei, X. Chen, Y. Chen, Evaluation of laser cleaning for defouling of marine biofilm contamination on aluminum alloys, Appl. Surf. Sci. 499 (2020) 144060. https://doi.org/10.1016/j.apsusc.2019.144060.
11. R. Xiang, J. Wang, Y. Wang, X. Chao, B. Mai, Y. Zhang, J. Cao, Study on the application of laser technology in the restoring and transformation of PbS on painted cultural relics, J. Cult. Herit. 66 (2024) 455–463. https://doi.org/10.1016/j.culher.2024.01.002.
12. Y. Lu, Y. Ding, M. Wang, L. Yang, Y. Wang, An environmentally friendly laser cleaning method to remove oceanic micro-biofoulings from AH36 steel substrate and corrosion protection, J. Cleaner Prod. 314 (2021) 127961. https://doi.org/10.1016/j.jclepro.2021.127961.
13. L.Y. Guo, S.N. Geng, J. Pang, Y.H. Hu, S. Lan, C.M. Wang, W.M. Wang, Structural transformation and property improvement of Fe78Si9B13 amorphous ribbon by pulsed laser processing, Mater. Design 160 (2018) 538–548. https://doi.org/10.1016/j.matdes.2018.09.045.
14. F.D. Zhang, H. Liu, C. Suebka, Y.X. Liu, Z. Liu, W. Guo, Y.M. Cheng, S.L. Zhang, L. Li, Corrosion behaviour of laser-cleaned AA7024 aluminium alloy, Appl. Surf. Sci. 435 (2018) 452–461. https://doi.org/10.1016/j.apsusc.2017.11.141.
15. R.A. Rajan, S. Rao Konda, C. Sajed Saraj, Y. Hang Lai, G. Verma, Z. Yu, W. Yu, D. Yan, J. Yang, Long-term seawater anti-corrosion properties of Al alloy triggered by femtosecond laser structuring with phase change, Appl. Surf. Sci. 573 (2022) 151612. https://doi.org/10.1016/j.apsusc.2021.151612.
16. Z. Li, D. Zhang, X. Su, S. Yang, J. Xu, R. Ma, D. Shan, B. Guo, Removal mechanism of surface cleaning on TA15 titanium alloy using nanosecond pulsed laser, Opt. Laser Technol. 139 (2021) 106998. https://doi.org/10.1016/j.optlastec.2021.106998.
17. H.J. Yoo, S. Baek, J.H. Kim, J. Choi, Y.-J. Kim, C. Park, Effect of laser surface cleaning of corroded 304L stainless steel on microstructure and mechanical properties, Journal of Materials Research and Technology 16 (2022) 373–385. https://doi.org/10.1016/j.jmrt.2021.11.147.
18. Z. Tian, Z. Lei, X. Chen, Y. Chen, L.-C. Zhang, J. Bi, J. Liang, Nanosecond pulsed fiber laser cleaning of natural marine micro-biofoulings from the surface of aluminum alloy, J. Cleaner Prod. 244 (2020) 118724. https://doi.org/10.1016/j.jclepro.2019.118724.
19. C. Zhou, H. Li, G. Chen, G. Wang, Z. Shan, Effect of single pulsed picosecond and 100 nanosecond laser cleaning on surface morphology and welding quality of aluminium alloy, Opt. Laser Technol. 127 (2020) 106197. https://doi.org/10.1016/j.optlastec.2020.106197.
20. J. Zhao, P. Jiang, S. Geng, Y. Wang, B. Xu, Effect of pulsed laser pretreatment induced pit-structure on the formation of intermetallic compounds in titanium-aluminum dissimilar welded joints, Opt. Laser Technol. 167 (2023) 109589. https://doi.org/10.1016/j.optlastec.2023.109589.

21. L. Hou, F. Yin, S. Wang, J. Sun, H. Yin, A review of thermal effects and substrate damage control in laser cleaning, Opt. Laser Technol. 174 (2024) 110613. https://doi.org/10.1016/j.optlastec.2024.110613.
22. Y. Wang, Z. Yu, L. Yu, J. Zhuo, Z. Chen, J. Yao, L. Wang, Study on removal mechanism of TC4 oxide film by nanosecond pulsed laser cleaning in air environment, Optics & Laser Technology 181 (2025) 111856. https://doi.org/10.1016/j.optlastec.2024.111856.
23. J. Ruan, L. Huang, H. Zhang, Z. Wang, L. Wang, B. Li, Superhydrophobic synergistic antibacterial $ZrO_2$-based surfaces with tertiary hierarchical structures fabricated by femtosecond laser texturing and surface modification, J. Alloys Compd. 1002 (2024) 175331. https://doi.org/10.1016/j.jallcom.2024.175331.
24. J. Xu, D. Yan, T. Zou, L. Li, R. Zhang, L. Hu, J. Yang, Strong modification of double-pulsed femtosecond laser to reinforce anticorrosion of al alloy surface, Opt. Laser Technol. 181 (2025) 111800. https://doi.org/10.1016/j.optlastec.2024.111800.
25. Z. He, Y. Shen, W. Xiong, J. Tao, S. Shu, X. Huang, Z. Shen, Uncovering the roles of laser action modes in surface mechanical properties of 2024 aluminum alloy, Appl. Surf. Sci. 613 (2023) 156032. https://doi.org/10.1016/j.apsusc.2022.156032.
26. Y. Song, S. Wang, Y. Pan, Z. Li, L. Yu, Comparative study of infrared nanosecond laser surface paint removal for carbon fiber reinforced polymer and glass fiber reinforced polymer, J. Cleaner Prod. 451 (2024) 142061. https://doi.org/10.1016/j.jclepro.2024.142061.
27. T. Zhang, T. Liu, S. Zhao, Z. Zhang, Y. Liu, C. Zhong, Numerical simulation and experimental study of cleaning mechanisms in multipulse laser paint removal of acrylic polyurethane composite paint layers on aluminum alloy, Optics & Laser Technology 181 (2025) 112048. https://doi.org/10.1016/j.optlastec.2024.112048.
28. P. Lorenz, M. Ehrhardt, A. Lotnyk, J. Griebel, K. Zimmer, J. Zajadacz, M. Himmerlich, E. Bez, M. Taborelli, S. Rosenow, R. Tepper, A.M. Breul, Cleaning of laser-induced periodic surface structures on copper by gentle wet chemical processing, Applied Surface Science 679 (2025) 161115. https://doi.org/10.1016/j.apsusc.2024.161115.
29. Y. Cai, X. Song, C. Chen, J. Yang, J. Gao, K. Zhang, Effect of energy distribution on laser cleaning quality of 30Cr3 ultra-high strength steel, Journal of Materials Research and Technology 30 (2024) 4868–4880. https://doi.org/10.1016/j.jmrt.2024.04.194.
30. T. Zhang, T. Liu, G. Ban, J. Zou, Z. Zhang, Y. Liu, C. Zhong, Effect of scanning speed on laser cleaning of composite paint layer on aluminum alloy, Optics & Laser Technology 171 (2024) 110470. https://doi.org/10.1016/j.optlastec.2023.110470.
31. J. Gacs, E. Sára Bogya, L. Kocsis, T. Jacob, Application of laser cleaning on AlMg4.5Mn0.4 sheets for adhesive bonding, International Journal of Adhesion and Adhesives 115 (2022) 103132. https://doi.org/10.1016/j.ijadhadh.2022.103132.
32. D. Zhang, J. Xu, Z. Li, K. Li, C. Wang, D. Shan, B. Guo, Removal mechanism of blue paint on aluminum alloy substrate during surface cleaning using nanosecond pulsed laser, Opt. Laser Technol. 149 (2022) 107882. https://doi.org/10.1016/j.optlastec.2022.107882.
33. J. Nie, H. Zhang, D. Zhang, J. Xu, J. Zhang, D. Shan, B. Guo, Removal mechanism of laser cleaning for inorganic thermal control coatings on aluminum alloys, Applied Surface Science 633 (2023) 157578. https://doi.org/10.1016/j.apsusc.2023.157578.
34. Y. Li, C. Wang, G. Mi, Influence of cleaning modes on the microstructure and performance of 5083 alloy substrate, J. Mech. Sci. Technol. 35 (2021) 3943–3949. https://doi.org/10.1007/s12206-021-0807-6.
35. J. Nie, D. Zhang, X. Su, H. Zhang, J. Zhang, J. Xu, D. Shan, B. Guo, Laser cleaning for inorganic thermal control coatings on aluminum alloys: Microstructure evolution and mechanical properties of substrate, Appl. Surf. Sci. 659 (2024) 159939. https://doi.org/10.1016/j.apsusc.2024.159939.
36. A. Zhang, Z. Guo, B. Jiang, J. Du, C. Wang, G. Huang, D. Zhang, F. Liu, S. Xiong, F. Pan, Multiphase and multiphysics modeling of dendrite growth and gas porosity evolution during solidification, Acta Materialia 214 (2021) 117005. https://doi.org/10.1016/j.actamat.2021.117005.

37. Z. Li, X. Chen, S. Yang, D. Zhang, J. Xu, R. Ma, D. Shan, B. Guo, Removal mechanism of liquid-assisted nanosecond pulsed laser cleaning TA15 titanium alloy oxide film, Journal of Materials Research and Technology 19 (2022) 4986–4997. https://doi.org/10.1016/j.jmrt.2022.06.165.
38. W. Li, Y. Jin, J. Gu, Z. Zeng, X. Su, J. Xu, B. Guo, Critical surface characteristics for coating adhesion and friction behavior of aluminum alloys after laser cleaning, J. Mater. Process. Technol. 332 (2024) 118549. https://doi.org/10.1016/j.jmatprotec.2024.118549.
39. Li Z., Xu J., Zhang D., Xu Z., Su X., Jin Y., Shan D., Chen Y., Guo B., Nanosecond pulsed laser cleaning of titanium alloy oxide films: Modeling and experiments, Journal of Manufacturing Processes 82 (2022) 665–677. https://doi.org/10.1016/j.jmapro.2022.08.033.
40. M. Shamsujjoha, S.R. Agnew, M.A. Melia, J.R. Brooks, T.J. Tyler, J.M. Fitz-Gerald, Effects of laser ablation coating removal (LACR) on a steel substrate: Part 1: surface profile, microstructure, hardness, and adhesion, Surf. Coat. Technol. 281 (2015) 193–205. https://doi.org/10.1016/j.surfcoat.2015.01.071.
41. G. Zhu, S. Wang, W. Cheng, Y. Ren, D. Wen, Corrosion and wear performance of aircraft skin after laser cleaning, Opt. Laser Technol. 132 (2020) 106475. https://doi.org/10.1016/j.optlastec.2020.106475.
42. G.-L. Shi, K. Zhang, X.-G. Li, Y.-J. Li, M.-L. Ma, J.-W. Yuan, H.-J. Zhang, Dislocation configuration evolution during extension twinning and its influence on precipitation behavior in AZ80 wrought magnesium alloy, J. Magnesium Alloys 11 (2023) 2442–2457. https://doi.org/10.1016/j.jma.2021.08.032.
43. L. Liu, J. Wang, J. Zhou, Effects of laser shock peening on mechanical behaviors and microstructural evolution of brass, Vacuum 148 (2018) 178–183. https://doi.org/10.1016/j.vacuum.2017.11.024.
44. B. Liu, G. Mi, C. Wang, Modification of TA15 alloy surface by high-pulse-frequency laser cleaning, J. Laser Appl. 32 (2020) 032019. https://doi.org/10.2351/7.0000139.
45. J. Li, J. Guo, W. Lu, L. Liu, S. Huang, M. Ji, Y. Zhao, Multiscale simulation combined with experimental investigation on residual stress and microstructure of TC6 titanium alloy treated by laser shock peening, Mater. Today Commun. 40 (2024) 109484. https://doi.org/10.1016/j.mtcomm.2024.109484.
46. Q. Sun, J. Zhou, X. Meng, P. Li, Mechanical properties and microstructure characteristics of 2024-T351 aluminum alloy specimen subjected to paint removal by laser cleaning, Vacuum 211 (2023) 111927. https://doi.org/10.1016/j.vacuum.2023.111927.
47. W. Wang, Q. Jiang, W. Liu, F. Xing, W. Wang, T. Yao, Effect of laser fluence on the cleaning quality and surface properties of TA15 titanium alloy surface paint layer, Optics & Laser Technology 181 (2025) 112032. https://doi.org/10.1016/j.optlastec.2024.112032.
48. W. Wang, L. Sun, Y. Lu, L. Qi, W. Wang, H. Qiao, Laser induced breakdown spectroscopy online monitoring of laser cleaning quality on carbon fiber reinforced plastic, Optics & Laser Technology 145 (2022) 107481. https://doi.org/10.1016/j.optlastec.2021.107481.
49. J. Gu, X. Su, Y. Jin, W. Li, Z. Zeng, D. Zhang, J. Xu, B. Guo, Towards low-temperature laser paint stripping by photochemical mechanism on CFRP substrates, J. Manuf. Processes 85 (2023) 272–280. https://doi.org/10.1016/j.jmapro.2022.11.041.

The manufacturer's authorised representative in the EU is Springer Nature Customer Service Centre GmbH, Europaplatz 3, 69115 Heidelberg, Germany. If you have any concerns regarding our products, please contact ProductSafety@springernature.com

Printed and bound by CPI Group (UK) Ltd, Croydon, CR0 4YY
21/09/2026
02220008-0017